Mathematics for Christian Living Series

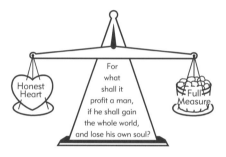

Mathematics for Christian Living Series

Applying Mathematics

Grade 8

Teacher's Manual
Part 2

Rod and Staff Publishers, Inc.
P.O. Box 3, Hwy. 172
Crockett, Kentucky 41413
Telephone: (606) 522-4348

Acknowledgements

We are indebted to God for the vision of the need for a *Mathematics for Christian Living Series* and for His enabling grace. Charitable contributions from many churches have helped to cover the expenses for research and development.

This revision was written by Brother Glenn Auker. The brethren Marvin Eicher, Jerry Kreider, and Luke Sensenig served as editors. Brother John Mark Shenk and Sisters Marian Baltozer, Amy Herr, and Christine Collins drew the illustrations. The work was evaluated by a panel of reviewers and tested by teachers in the classroom. Much effort was devoted to the production of the book. We are grateful for all who helped to make this book possible.

—*The Publishers*

Copyright, 2004

By Rod and Staff Publishers, Inc.
Crockett, Kentucky 41413

Printed in U.S.A.

ISBN 978-07399-0488-6

Catalog no. 13892.3

Materials for This Course

Books and Worksheets
Pupil's Textbook
2 Teacher's Manuals
Quizzes and Speed Tests
Chapter Tests

Tools Needed for Each Student
English ruler
Metric ruler
Protractor
Compass

To the Teacher

Basic Philosophy

Mathematics is a fundamental part of God's creation. Although a study of mathematics does not in itself lead to God and salvation, it can heighten one's awareness of the perfection in God's order. Because of that perfection, unchanging mathematical principles can be used to solve a host of everyday problems pertaining to quantity, length, size, proportion, and many other matters. The goal of this mathematics course is to develop in students the ability to solve mathematical problems so that they will be better able to serve the Lord.

May the Lord bless you as you teach this phase of the laws in His orderly world. Important as the mathematical concepts are that are presented in this course, they are superseded by the patience and wisdom that you will reveal as you present them. If you teach students to calculate well, yet influence them in a wrong direction, you will teach them to calculate for wrong purposes. But if you teach them to calculate well *and* teach godly virtues by word and example, you will help to train valuable servants for the Lord. May your teaching be fruitful to this end.

Plan of the Course

This book contains 170 lessons in twelve chapters. The last lesson of each chapter is a test which is bound in a separate, consumable booklet. This text is intended to be taught at the rate of one lesson per day. If you cannot have math class every school day, you will need to omit or combine some lessons in order to cover the material.

Pupil's Book

The lesson text presents explanations and illustrations of the new concepts. In teaching the lesson, briefly discuss the concepts in the explanation and then move promptly into solving several problems. Students learn most quickly by studying the examples, watching the teacher solve a few problems on the chalkboard, and then solving some problems themselves.

Class Practice is a set of problems to be used in teaching the lesson and providing class drill for the students. Solve one or two problems yourself, and then have the students solve the others either on paper or at the chalkboard. Check their work to make certain they are using the correct procedures.

Written Exercises are the homework problems, usually divided into Part A, Part B, and so on. The first several parts contain problems that develop the new concepts taught in the lesson. These are followed by a set of reading problems (usually six). Some of the reading problems relate to the new material, and some deal with review concepts.

Review Exercises give review of concepts taught in previous lessons. This textbook is designed to review more recent material more frequently and less recent material less frequently. Because of the review pattern and the pattern for introducing new material, it is important that the lessons are covered in consecutive order.

Challenge Exercises appear at the end of a few lessons. They are provided for the challenge of the more able and independent students, and need not be considered part of the basic assignment for all.

Pairing of problems. Most of the problems are arranged in pairs. That is, problems 1 and 2 are of the same kind, problems 3 and 4 are similar, and so on. If a lesson or set of problems is too long for the available time, the teacher can assign only the even-numbered or the odd-numbered problems with the confidence that all the main concepts will be drilled in those exercises. In general, each pupil should do at least the even- or odd-numbered problems in every part of the exercises.

Teacher's Manual

The Teacher's Manual provides a skeleton outline for each lesson. The order of the headings gives the suggested order to follow in presenting the lesson to the class.

Objectives lists the main concepts to be taught in the lesson. Watch especially for objectives that are marked with a star; they indicate concepts that are new in this math series. Plan to give special attention to these concepts as you teach.

Review gives concepts to be reviewed from previous lessons. Most often this review material corresponds to review exercises in the pupil's lesson. Use the examples given to prepare the students for the review portion of their lesson assignment. Spend more time with difficult concepts such as percentages, and less time with simpler concepts such as carrying in addition.

When reviewing, be sure to allow sufficient time for the main part of the lesson—the new material.

Introduction provides an illustration, a theme, or a train of logic with which to introduce the lesson. The Introduction is not intended to be a lesson of its own, but rather a springboard to the lesson itself.

Teaching Guide contains a set of numbered points to be taught, along with helps for presenting them to the class. Boldface is used for main lesson concepts, and regular type for further explanations and other points. After each main teaching point are several related exercises for illustration and practice.

Plan ahead so that you have sufficient time to develop this part of the lesson. Use both these exercises and the Class Practice problems in the pupil's text to reinforce the concepts being taught. In math, experience is usually the best teacher.

An Ounce of Prevention warns of wrong ideas or procedures that students need to avoid. Be sure to point these out to your students. Even though they may seem trivial, they can be real pitfalls.

Further Study presents material that is intended to broaden the teacher's base of knowledge. Although you may occasionally decide to present these facts to the class, this section is designed primarily for the teacher's benefit.

The *answer key* generally appears as colored text on the teacher's copy of the pupil pages. When more space is needed, answers are placed on the facing teacher's page.

Solutions for most reading problems are provided on facing pages for your reference.

When directions call for students to show their work, and for more advanced exercises such as square roots or algebraic equations, full solutions are given.

General Class Procedure

Following is the recommended procedure for teaching a typical lesson in this course.

1. Correct the homework from the previous lesson.
2. Review previous lessons, using the Review section of the teacher's guide.
3. Teach the material for the new lesson, using the Class Practice problems to make certain the lesson is well learned.
4. Assign the homework.
5. Administer any Quiz or Speed Test that is called for. If you have a multigrade classroom, it may fit your schedule better to give Quizzes or Speed Tests before math class while you are teaching another class.

Naturally, an experienced teacher will sometimes vary his approach to teaching the lesson. However, it is recommended that beginning teachers give careful consideration to the instructions in the Teaching Guide.

Quizzes and Speed Tests

The Quizzes and Speed Tests (found in a separate booklet) are an integral part of this course. Quizzes and Speed Tests are numbered according to the lessons with which they belong, and should be given as indicated in the Review section of the teacher's guide.

Quizzes

The purpose of the forty Quizzes is to review things that students have previously learned. These Quizzes are not timed, but are rather intended to reveal the understanding of concepts.

Speed Tests

The purpose of the eight Speed Tests is to stimulate and maintain the students' skill with basic computation. They drill basic arithmetic facts or concepts that should be so thoroughly mastered that they take little thought in the general math assignments. A good foundation in these basics is very important for accuracy.

Chapter Tests

A Chapter Test (found in a separate booklet) is to be given at the end of each chapter. Each test counts as a numbered lesson in the course.

Following are a few pointers for testing. These pointers also appear in the Teacher's Manual for the Chapter 1 Test (Lesson 16).

1. Only the test, scratch paper, pencils, and an eraser should be on each student's desk.
2. Steps should be taken to minimize the temptation of dishonesty and the likelihood of accidentally seeing other students' answers. Following are some suggestions.
   a. Desk tops should be level. If the desks are very close to each other, have the students keep their work directly in front of them on the desk.
   b. Students should not look around more than necessary during test time.
   c. No communication should be allowed.
   d. As a rule, students should remain seated during the whole test period. It is a good idea to sharpen a few extra pencils and have them on hand.
   e. Students should hand in their tests before going on to any other work.
3. Encourage the students to do their work carefully and to go back over it if they have time. Do not allow them to hand in their tests too soon. On the other hand, some students are so meticulous that they can hardly finish their tests. If you have this problem, set a time when you will collect all the tests. Once 90% of the tests are completed, the rest of them should generally be finished in the next five or ten minutes. Of course, there are exceptions for slower students.
4. A test is different from homework. Students should realize that they must rely on their own knowledge as they work. The teacher should not help them except to make sure that all instructions are clearly understood.

Evaluating Test Results

1. If you check the tests in class, have students check each others' work. Spot check the corrected tests.
2. Tests are valuable tools in determining what the students have grasped. Are there any places where the class is uniformly weak? If so, reinforcement is needed.
3. One effective way to discover the general performance of the class is to find the class median. This is done by arranging the scores in order from highest to lowest. The middle score is the class median. If there is an even number of students, find the average of the two middle scores.

Disposition of the Corrected Tests

1. As a rule, students should have the privilege to see their tests. Review any weak points and answer any questions about why an answer is wrong.
2. The teacher may use his discretion about whether the students should be allowed to keep their tests permanently. Some teachers prefer to collect them again so that students' younger siblings will have no chance of seeing the tests in later years.

Grading

Homework scores are the basic source of a student's math grade. Quizzes are fewer, but more significant in showing understanding of concepts. Chapter Tests are designed to show what the student has thoroughly mastered and have the greatest influence in determining the final grade.

It is suggested that a Quiz grade is given twice the value of a regular homework score, and that the Chapter Test(s) represent at least half the of the student's report card grade. Speed Tests, if graded, may be handled in the same way as homework scores.

To give a Quiz score double value, enter it twice in the list of addends when calculating an average, and also increase the divisor every time you make a double entry. For example, if you have a series of thirteen homework scores and four Quiz scores, the sum will be the total of those seventeen scores, plus the four Quiz scores repeated. The dividend for finding the average should be 13 plus 4 plus 4, or 21. A simple system for producing this result is to write the entry twice when entering a Quiz score in the gradebook. Then at grade-averaging time the double value is automatically worked in.

To have test scores represent half the value of the math grade, first find the average of all other scores. If there was more than one Chapter Test in the marking period, also average the test scores. Then find the average of those two results (add them and divide by two).

Contents of Teacher Book 1

Chapter 1 Basic Mathematical Operations
Chapter 2 English and Metric Measures
Chapter 3 Factoring and Fractions
Chapter 4 Decimals, Ratios, and Proportions
Chapter 5 Mastering Percents
Chapter 6 Statistics and Graphs

Table of Contents

Numbers given first are pupil page numbers.
Numbers at the far right are actual manual page numbers printed at the bottom of the pages.

To the Teacher 5

Chapter 7   Lines and Planes in Geometry

85.	Geometric Terms	282	16
86.	Working With Angles	286	24
87.	Classifying Triangles	291	35
88.	Constructing Triangles	296	44
89.	Constructing Geometric Figures	300	52
90.	Perimeters of Polygons	305	63
91.	Circles and Circumferences	308	68
92.	Areas of Squares and Rectangles	312	76
93.	Areas of Parallelograms and Triangles	315	83
94.	Areas of Trapezoids	319	91
95.	Areas of Circles	323	99
96.	Areas of Compound Figures	326	104
97.	Reading Problems: Solving Multistep Problems	330	112
98.	Chapter 7 Review	334	120
99.	Chapter 7 Test (in test booklet)		626

Chapter 8   Geometric Solids and the Pythagorean Rule

100.	Geometric Solids and Surface Area of Cubes	340	132
101.	Surface Area of Rectangular Solids	344	140
102.	Surface Area of Cylinders	347	147
103.	Surface Area of Square Pyramids	351	155
104.	Surface Area of Spheres	354	160
105.	Volume of Rectangular Solids and Cubes	357	167
106.	Volume of Cylinders and Cones	360	172
107.	Volume of Square Pyramids	363	179
108.	Volume of Spheres	366	184
109.	Understanding Square Roots	369	191
110.	Extracting Square Roots	373	199
111.	Extracting More Difficult Square Roots	376	204
112.	Understanding the Pythagorean Rule	379	211
113.	Applying the Pythagorean Rule	382	216
114.	Reading Problems: Using Parallel Problems	386	224
115.	Chapter 8 Review	390	232
116.	Chapter 8 Test (in test booklet)		630

Chapter 9   Mathematics and Finances

117.	Checking Accounts: Using Deposit Tickets	394	240
118.	Checking Accounts: Writing Checks	397	247
119.	Checking Accounts: Maintaining Account Balances	401	255
120.	Checking Accounts: Reconciling the Account	405	263
121.	Savings Accounts and Simple Interest	408	268
122.	Calculating Part-year Interest	412	276
123.	Calculating Compound Interest	414	280
124.	Using the Compound Interest Formula	417	287
125.	Calculating Sales Tax and Property Tax	422	296
126.	Calculating Profit	426	304
127.	Calculating Profit as a Percent of Sales	430	312
128.	Reading Problems Relating to Finances	434	320
129.	Mental Math: The Four Basic Operations	437	327
130.	Chapter 9 Review	441	335
131.	Chapter 9 Test (in test booklet)		633

Chapter 10   Introduction to Algebra

132.	Identifying Algebraic Expressions	446	344
133.	Order of Operations	449	351
134.	Evaluating Algebraic Expressions	452	356
135.	Adding and Subtracting Monomials	455	363
136.	Using Addition and Subtraction to Solve Equations	458	368
137.	Using Multiplication and Division to Solve Equations	461	375
138.	Reading Problems: Writing Equations	465	383
139.	Exponents and Literal Numbers	470	392
140.	Exponents in Multiplication	473	399
141.	Exponents in Division	476	404
142.	Chapter 10 Review	479	411
143.	Chapter 10 Test (in test booklet)		637

Chapter 11   Signed Numbers, Tables, and Graphs

144.	Introduction to Signed Numbers	482	416
145.	Adding Signed Numbers	486	424
146.	Subtracting Signed Numbers	490	432
147.	Multiplying and Dividing Signed Numbers	493	439
148.	Evaluating Expressions With Signed Numbers	497	447
149.	Constructing Tables From Formulas	500	452
150.	Constructing Graphs From Formulas	503	459

151.	Polynomials and Signed Numbers	507	467
152.	Reading Problems: More Practice With Equations	510	472
153.	Chapter 11 Review	514	480
154.	Chapter 11 Test (in test booklet)		640

Chapter 12 Other Numeration Systems and Final Reviews

155.	Reading Electric Meters	518	488
156.	Using Scientific Notation	523	499
157.	The Concept of Base in Numeration	527	507
158.	The Duodecimal (Base Twelve) Numeration System	530	512
159.	The Binary (Base Two) Numeration System	533	519
160.	Review of Base Twelve and Base Two Numeration	537	527
161.	Final Review of Basic Mathematical Operations and Measures	540	532
162.	Final Review of Mental Calculation, Factoring, and Fractions	543	539
163.	Final Review of Decimals, Ratios, and Proportions	546	544
164.	Final Review of Percents and Statistics	549	551
165.	Final Review of Plane Geometry and Graphs	552	556
166.	Final Review of Solid Geometry and the Pythagorean Rule	557	567
167.	Final Review of Finances	561	575
168.	Final Review of Algebra	565	583
169.	Review of Chapter 12 and Final Review	568	588
170.	Final Test (in test booklet)		643

Quizzes and Speed Tests 599

Chapter Tests 7–12 625

Index 573 649

Symbols 581 657

Formulas 582 658

Square Roots 584 660

Tables of Measure . . . . .Back Endsheet

Simple geometric principles are an integral part of design. Many aspects of symmetry and balance relate to the 360 degrees of a circle. A regular polygon of any number of sides can be constructed in a circle because the same number of equal angles will share the center vertex, as there are equal sides in the perimeter.

The triangle is another pleasing and versatile form. Triangles are also very useful tools in solving practical problems.

Chapter 7
Lines and Planes in Geometry

The word *geometry* comes from two Greek words meaning "land measure." (*Geo* means "earth" or "land.") Geometry is the branch of mathematics that deals with the measurement and relationships of points, lines, surfaces, and solids.

A geometric point is a specific location with no length or width. The dot which is used to represent a point is actually a small circle, used to show where the point is.

A line has one dimension only. It has no thickness or width. Its length can be specified only as segments, defined by specific points on the line.

Geometric figures that have two dimensions but no thickness are plane figures. Circles, squares, and triangles are examples of figures with length and width that define a certain part of a plane.

You will study geometric forms with three dimensions in Chapter 8.

*Many,
O LORD my God,
are thy wonderful works which thou hast done.
Psalm 40:5*

85. Geometric Terms

Plane geometry is the study of lines, angles, squares, and other two-dimensional figures. Some specific terms of plane geometry are listed below with explanations and examples. If there is a geometric symbol for a term, that symbol is also shown.

Points and Lines

Term	Example	Symbol
Point—A location having no length, width, or thickness.	A	(none)
Line—A straight line that extends without end in two directions. A line has length but no width or thickness.	B C	$\overleftrightarrow{BC}$
Line Segment—A part of a line with two endpoints.	D E	$\overline{DE}$
Ray—A line with only one endpoint.	F G	$\overrightarrow{FG}$
Angle—A figure formed by two rays with the same endpoint, or two line segments that meet at a point. The rays or line segments are the **sides** of the angle, and the meeting point is the **vertex**. An angle is often named by using three letters. Sometimes it is named by a single letter inside the angle. An angle may be considered to show how much a line segment rotates in moving from one side to the other. An arrow is commonly used to show this rotation.	H, I, J (with c inside)	∠HIJ or ∠JIH or ∠c
Right Angle—An angle that has a square corner; a 90° angle.	K, L, M	(This symbol inside an angle means it is a right angle.)

LESSON 85

Objectives

- To review terms relating to points and lines: point, line, line segment, ray, angle, right angle, intersecting lines, parallel lines, perpendicular lines.

- To teach geometric terms relating to geometric planes: plane, circle, polygon, triangle, rectangle, square, quadrilateral, pentagon, hexagon, heptagon, octagon, parallelogram, trapezoid, *polygon, nonagon, decagon.

Review

1. *Use proportions to solve these reading problems.* (Lesson 55)

    a. Aluminum is made from bauxite at the ratio of 4 pounds bauxite to 1 pound aluminum. How much bauxite went into the $6\frac{1}{4}$-pound pyramid-shaped cap on the Washington Monument? (25 lb.)

    $$\frac{\text{bauxite}}{\text{aluminum}} \quad \frac{4}{1} = \frac{n}{6\frac{1}{4}} \quad \frac{\text{bauxite}}{\text{aluminum}}$$

    b. On an afghan 55 inches wide, Marian can add 2 rows in 40 minutes. How long will it take her to crochet 96 rows on the afghan? (32 hr.)

    $$\frac{\text{rows}}{\text{minutes}} \quad \frac{2}{40} = \frac{96}{n} \quad \frac{\text{rows}}{\text{minutes}}$$

2. *Find these equivalents.* (Lesson 23)

    a. 92 l = ___ qt. (97.52)
    b. 514 l = ___ kl (0.514)
    c. 63 gal. = ___ l (239.4)
    d. 24 kl = ___ l (24,000)

Introduction

What does the word *geometry* mean? The *geo* part means "earth or land," and the *metry* part means "measure." So the basic meaning of *geometry* comes from land measurement.

Geometry is the branch of mathematics that deals with points, lines, angles, surfaces, and solids. Geometry, then, is the mathematics of space; and as such, it has gone far beyond measuring land as surveyors and mapmakers do.

Teaching Guide

1. **Plane geometry is the study of points, lines, and other two-dimensional figures.** Following are the plane figures presented in this lesson. Discuss each figure, some examples of the figure, and the geometric symbol (if any) used to represent the figure. The students are expected to memorize the symbols taught in this lesson.

Figures	*Examples*
point	corner of a property corner of a paper place where four states meet
line	axis of the Earth
line segment	lines on a paper boundary of a property
ray	light ray sound wave
angle	hands of a clock brace for a shelf peak of a roof

(In discussing angles, be sure to include the concept of rotation. Clock hands are excellent for illustration.)

right angle
- corner of a room
- corner of a window pane

intersecting lines
- design on a shirt or dress
- lines on a map
- various letters of the alphabet

parallel lines
- center lines on a road
- top and bottom of a wall
- lines on a paper

perpendicular lines
- post in the ground
- various letters of the alphabet
- (Perpendicular lines may intersect, as in the plus sign [+], or they may not, as in a right angle [⌐].)

2. **A geometric plane has length and width but no thickness.** A plane is infinite, like a line, but it extends in two dimensions. In practical experience, we work with portions of planes. Review the common plane figures, and give examples of them.

Figures	*Examples*
plane	*portions of planes:* surface of a desktop, surface of the floor, surface of the chalkboard, a map
circle	clock face, letter O, wheels
polygon	(any closed figure with straight sides)
triangle	yield sign
quadrilateral	(any four-sided polygon)
pentagon	road sign for a school, gable end of a building
hexagon	cells in a honeycomb
heptagon	end of a barn with a gambrel roof
octagon	stop sign
nonagon	(no common example)
decagon	(no common example)

3. **A regular polygon is one in which all sides are equal in length. An irregular polygon has sides of unequal length.**

4. **The quadrilateral is probably the most common kind of polygon.** Quadrilaterals have various specific names as shown below.

parallelogram—Any quadrilateral with parallel sides but usually with corners that are not right angles.

rectangle—A polygon with four parallel sides and four right angles but usually with adjacent sides of unequal length.

rhombus—A regular quadrilateral whose corners are usually not right angles.

square—A regular quadrilateral whose corners are right angles.

trapezoid—A quadrilateral with two parallel sides and two unparallel sides.

Intersecting Lines—Two lines that intersect or cross at a point.

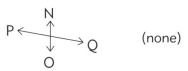

(none)

Parallel Lines—Two lines that are always the same distance apart, no matter how far they extend in either direction.

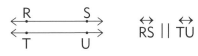

$\overleftrightarrow{RS} \parallel \overleftrightarrow{TU}$

Perpendicular Lines—Lines that form right angles.

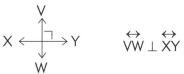

$\overleftrightarrow{VW} \perp \overleftrightarrow{XY}$

Plane Figures

Plane—A flat surface that has width and length, but no thickness. A plane extends infinitely in all directions. The top surface of a desk or table is an example of part of a plane.

Plane Geometry—Geometry that deals with plane figures, that is, points and lines in the same plane.

Circle—A closed curve on which all points are the same distance from the center.

Polygon—Any closed figure that can be drawn with line segments, that is, with straight sides. Following are the names of various polygons, based on the number of sides they have.

Polygons

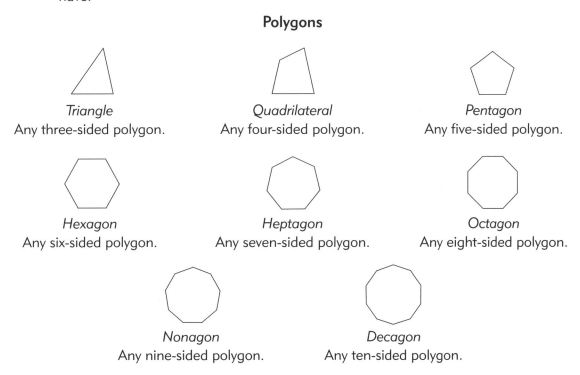

Triangle
Any three-sided polygon.

Quadrilateral
Any four-sided polygon.

Pentagon
Any five-sided polygon.

Hexagon
Any six-sided polygon.

Heptagon
Any seven-sided polygon.

Octagon
Any eight-sided polygon.

Nonagon
Any nine-sided polygon.

Decagon
Any ten-sided polygon.

A **regular polygon** is one in which all sides are equal in length. An **irregular polygon** has sides of unequal length. For example, a regular decagon has ten sides of equal length.

284   Chapter 7   Lines and Planes in Geometry

The quadrilateral is probably the most common kind of polygon. Here is a list of various specific names for specialized quadrilaterals.

Parallelogram—Any quadrilateral with parallel sides. The corners are usually not right angles.

Rectangle—A polygon with four parallel sides and four right angles. Adjacent sides are usually of unequal length.

Rhombus—Any regular quadrilateral. The corners are usually not right angles.

Square—A regular quadrilateral whose corners are right angles.

Trapezoid—A quadrilateral with two parallel sides and two unparallel sides.

CLASS PRACTICE

Write each expression, using symbols.

a. line segment LM $\overline{LM}$

b. line MN is perpendicular to line OP $\overleftrightarrow{MN} \perp \overleftrightarrow{OP}$

Draw these figures.

c. $\overleftrightarrow{YZ}$

d. $\overleftrightarrow{FG} \parallel \overleftrightarrow{HI}$

Name these figures.

e. octagon

f. circle

Label each figure as regular or irregular.

g. regular

h. irregular

WRITTEN EXERCISES

A. Write each expression, using symbols.

1. angle ABC $\angle ABC$
2. line DE $\overleftrightarrow{DE}$
3. ray FG $\overrightarrow{FG}$
4. line segment HI $\overline{HI}$
5. line JK is perpendicular to line LM $\overleftrightarrow{JK} \perp \overleftrightarrow{LM}$
6. line NO is parallel to line PQ $\overleftrightarrow{NO} \parallel \overleftrightarrow{PQ}$

Further Study

In the following list, quadrilaterals are classified in such a way that they become more and more specific and regular.

A *quadrilateral* is any four-sided figure. All the figures named below are quadrilaterals.

A *trapezium* is a quadrilateral with no parallel sides. This quadrilateral is completely irregular.

A *trapezoid* is a quadrilateral with only one set of parallel sides.

An *isosceles* trapezoid is a trapezoid whose unparallel sides are of equal length.

A parallelogram *is any quadrilateral whose opposite sides* are parallel. All the figures named below are parallelograms.

A *rhomboid* is a parallelogram that does not have any right angles and whose adjacent sides are of unequal length.

A *rhombus* is any parallelogram that has four sides of equal length.

A *rectangle* is a parallelogram having four right angles.

A *square* is a rhombus with four right angles.

The following diagram shows the relationships among quadrilaterals. The formula for any given figure will work for all the ones to which the arrows go from that figure. For example, the areas of rhomboids, rhombuses, rectangles, and squares can all be found by using the formula for a parallelogram.

Notice that a square is both a rhombus and a rectangle. The formulas for finding the areas of parallelograms, rhombuses, and rectangles will all work to find the areas of squares. In addition, the areas of the figures in the parallelogram family can all be calculated by using the formula for trapezoids.

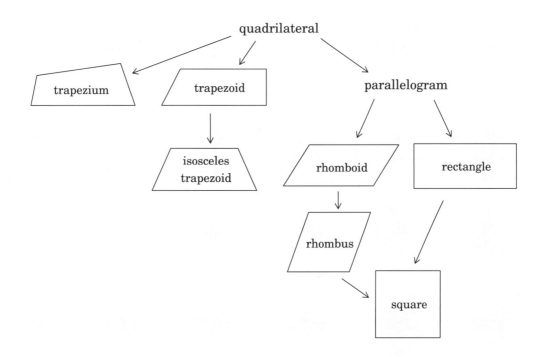

T-285   Chapter 7   *Lines and Planes in Geometry*

Sample Figures for Part B

7.   8.

9.   10.

11.   12.

13.   14.

Sample Figures for Part E

23. ⬡ (7 sides, equal or unequal)   24. ⬠ (5 sides, equal or unequal)

25. △ (3 equal sides)   26. ◇ (4 equal sides)

Proportions for Exercises 29 and 30

29. $\dfrac{\text{butter}}{\text{milk}} \quad \dfrac{1}{20} = \dfrac{n}{88} \quad \dfrac{\text{butter}}{\text{milk}}$ $n = 4\tfrac{2}{5}$ pounds

30. $\dfrac{\text{beats}}{\text{seconds}} \quad \dfrac{65}{60} = \dfrac{n}{84} \quad \dfrac{\text{beats}}{\text{seconds}}$ $n = 91$ times

B. Draw these figures. (See facing page for figures.)

7. $\overline{RS} \perp \overline{TU}$

8. $\overleftrightarrow{VW} \parallel \overleftrightarrow{XY}$

9. $\angle RST$

10. $\overrightarrow{RG}$

11. $\overline{AZ}$

12. $\overleftrightarrow{QR}$

13. intersecting lines ST and UV

14. right angle XYZ

C. Name these figures.

15. rectangle
16. pentagon
17. decagon
18. nonagon

D. Name each figure by using one of the eight terms shown under *Polygons* **in the lesson, including** *regular* **or** *irregular*. **Example:** *regular pentagon.*

19. regular quadrilateral
20. irregular triangle
21. irregular hexagon
22. regular octagon

E. Draw these figures. (See facing page for figures.)

23. heptagon
24. pentagon
25. regular triangle
26. rhombus

F. Solve these reading problems. Use proportions for numbers 29 and 30.

27. What term of plane geometry is associated with a sunbeam?   ray

28. What term of plane geometry is associated with a stop sign?   octagon

29. Butter is made at the approximate ratio of 1 pound butter to 20 pounds raw milk. If a cow produced 88 pounds of milk one day, how much butter could be made from it? $4\frac{2}{5}$ pounds

30. If a heart beats 65 times per minute, how many times does it beat in 84 seconds?   91 times

31. Refining a barrel of crude oil requires about 6.08 kiloliters of water. How many liters of water is that?   6,080 liters

32. An 8,850-mile network of pipes under the streets of London delivers about 1,328,640,300 liters of water per day. If 1 liter = 0.26 gallon, how many gallons of water is that?   345,446,478 gallons

REVIEW EXERCISES

G. Find these equivalents. *(Lesson 23)*

33. 75 *l* = 79.5 qt.

34. 18.5 kl = ____ *l*   18,500

35. 224 gal. = 851.2 *l*

36. 8 kl = 80 hl

286   Chapter 7   *Lines and Planes in Geometry*

86. Working With Angles

An angle is a figure formed by two rays that begin at the same point, or two line segments that meet at a point. The meeting point of the angle is the vertex. The two rays or line segments form the sides of the angle. Angles are classified according to the sharpness of the vertex, as described below.

A **right angle** is made of two perpendicular lines, which form a 90° angle.

An **acute angle** is sharper than a right angle; therefore, it has fewer than 90°. (The word *acute* means "sharp.")

An **obtuse angle** is not as sharp as a right angle; it has between 90° and 180°. (The word *obtuse* means "blunt.")

A **straight angle** has sides extending in opposite directions. A straight angle has 180°.

A **reflex angle** has between 180° and 360°. The arrow shows that it is a reflex angle and not an acute or obtuse angle.

Two angles with a sum of 90° are called **complementary angles**. In the diagram at the right, the 50° angle is the complement of the 40° angle. Together they have 90°.

Two angles with a sum of 180° are called **supplementary angles**. In the diagram at the right, the 110° angle is the supplement of the 70° angle. Together they have 180°.

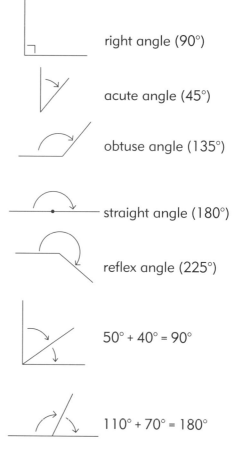

right angle (90°)

acute angle (45°)

obtuse angle (135°)

straight angle (180°)

reflex angle (225°)

50° + 40° = 90°

110° + 70° = 180°

Measuring Angles

The sharpness of an angle is measured by using the degrees of a circle. Because a complete circle has 360°, an angle that represents one-fourth of a circle is a 90° angle.

The degrees of an angle are measured with a protractor. A protractor usually has a semicircular shape, with an arrow indicating the center of the straight edge. One scale begins with 0° on the left side of the protractor and increases to 180° on the right side. The other scale begins with 0° on the right side and increases to 180° on the left side. These two scales allow the measuring of angles that open to either the left or the right.

LESSON 86

Objectives

- To teach angles and the different types of angles: acute, right, obtuse, straight, *reflex, *complementary, *supplementary.
- To review the use of the protractor.
- To review measuring, drawing, and bisecting angles by using the protractor.

Review

1. Give Lesson 86 Speed Test (Basic Math Facts). You need not take grades on this drill, since its main purpose is to improve speed. It can be used for practice over several days.

2. Sketch these figures on the chalkboard. The heptagon may be regular or irregular. *Identify each geometric figure.* (Lesson 85)

   a. (parallelogram)

   b. (heptagon)

3. *Find the distance represented by each measurement.* (Lesson 56)

   Scale: 1 in. = 24 mi.
   a. $2\frac{5}{8}$ in. (63 mi.)
   b. $3\frac{1}{4}$ in. (78 mi.)

   Scale: $\frac{7}{8}$ in. = 14 mi.
   c. $1\frac{1}{2}$ in. (24 mi.)
   d. $7\frac{1}{8}$ in. (114 mi.)

4. *Find each equivalent.* (Lesson 24)
   a. 2.05 km² = ___ ha   (205)
   b. 19 ha = ___ a.   (47.5)

Introduction

What kind of geometric figure has two rays extending from the same point? (an angle) In how many directions can the rays extend? (in any number of directions) Draw a few examples of all kinds of angles, including straight and reflex angles.

Teaching Guide

1. **An angle is a figure formed by two rays that begin at the same point, or two line segments that meet at a point.** The meeting point of the angle is the vertex. The two rays or line segments form the sides of the angle.

    Draw an angle on the board and point out the various parts.

2. **Angles are classified according to the sharpness of the vertex.** They are right angles, acute angles, obtuse angles, straight angles, and reflex angles.

3. **Certain combinations of angles have specific names.**

    a. Complementary angles are two angles with a sum of 90°.

    b. Supplementary angles are two angles with a sum of 180°.

4. **The sharpness of an angle is measured by using the degrees of a circle.** A protractor is used to measure the degrees in an angle.

    Take time to examine your students' protractors. Not all protractors are alike, and the explanation for using one kind of protractor may confuse a student with another kind.

5. **A protractor is used to measure angles.** Discuss the steps in the lesson. Then draw angles on the board for students to measure.

6. **A protractor is used to construct angles of specified sizes.** Discuss the steps in the lesson, and have the students construct the following angles.

    a. 40°   b. 65°
    c. 90°   d. 125°
    e. 175°   f. 225°

To measure an angle, use the following steps. Notice how the right angle is measured in the illustration.

1. If necessary, extend the sides of the angle with a pencil until they are long enough to measure.

2. Lay the protractor on the angle so that the center mark is at the vertex and the straight edge is aligned with one side of the angle. This side is the base from which you are working.

3. Read the number of degrees where the second side of the angle crosses the scale. Be sure to use the scale that has 0° on the line serving as the base. For a reflex angle, measure it as an acute or obtuse angle and subtract the reading from 360°.

4. Check to make sure your answer is logical. If an angle is sharper than a right angle, it will have fewer than 90°. If it is not as sharp as a right angle, it will have more than 90°.

An angle that represents one-fourth of a complete circle is a 90° angle.

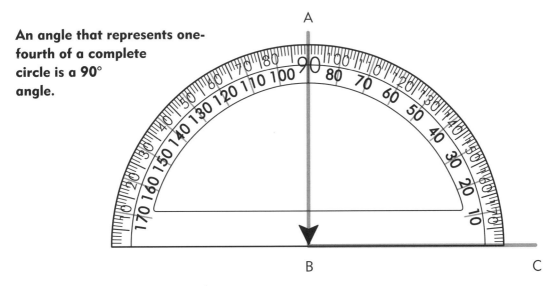

Constructing Angles

To construct an angle with a given number of degrees, use the following steps.

1. Place a dot where you intend the vertex to be. Form the base of the angle by starting at the dot and drawing a line segment in any direction, using a straightedge.

2. Position the straight edge of the protractor on the line segment you drew, with the center mark at the dot that marks the vertex.

3. Find the desired number of degrees on the correct scale of the protractor, and put a dot on the paper at that point.

4. Using a straightedge, draw a line from the vertex through the dot you made in step 3.

5. Draw a curved line between the two sides to show what kind of angle it is.

6. Mark the degrees of the angle.

288 Chapter 7 *Lines and Planes in Geometry*

Bisecting Angles

To **bisect** an angle is to divide it into two equal angles. Do this by following the steps below.

1. Carefully measure the angle with a protractor.

2. Divide the number of degrees by 2.

3. With the protractor in the same position, find the number of degrees on the correct scale of the protractor. Put a dot on the paper at that point.

4. Using a straightedge, draw a line from the vertex through the dot you made in step 3.

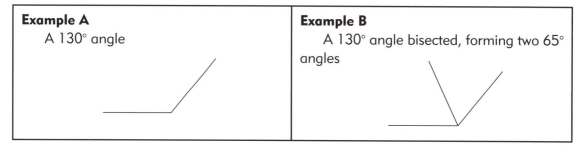

Example A	**Example B**
A 130° angle	A 130° angle bisected, forming two 65° angles

CLASS PRACTICE

Tell what kind of angle each one is.

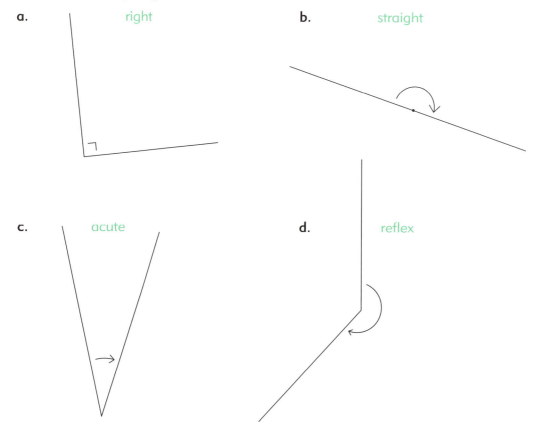

a. right

b. straight

c. acute

d. reflex

Lesson 86   T–288

7. **To bisect an angle is to divide it into two equal angles.** Discuss the steps in the lesson; then have students draw and bisect the following angles.

   a. 80°   b. 120°
   c. 160°   d. 140°

T–289 Chapter 7 Lines and Planes in Geometry

Model Answers for CLASS PRACTICE i–p

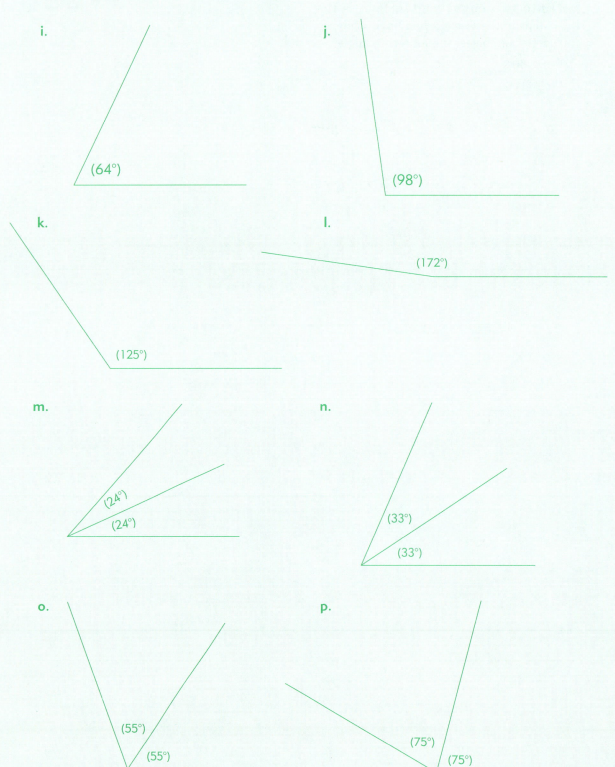

Measure the angles in exercises a–d on page 288. (Allow a tolerance of 1 degree.)

e. angle a = 90°   f. angle b = 180°
g. angle c = 30°   h. angle d = 222°

Draw the following angles. (See facing page for model answers. Allow a tolerance of 1 degree.)

i. 64°   j. 98°   k. 125°   l. 172°

Draw and bisect the following angles.

m. 48°   n. 66°   o. 110°   p. 150°

Identify each pair of angles as complementary or supplementary.

q. 41° and 49°   r. 26° and 64°   s. 92° and 88°   t. 17° and 73°
  complementary   complementary   supplementary   complementary

WRITTEN EXERCISES

A. *Write what kind of angle each one is.*

1. acute

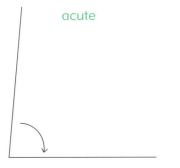

2. reflex

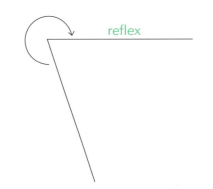

3. straight

4. obtuse

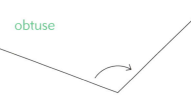

5. right

6. acute

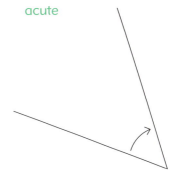

290   Chapter 7   Lines and Planes in Geometry

B. Measure the angles in exercises 1–6 on page 289.   (Allow a tolerance of 1 degree.)

7. angle 1 = 85°
8. angle 2 = 289°
9. angle 3 = 180°
10. angle 4 = 117°
11. angle 5 = 90°
12. angle 6 = 52°

C. Draw these angles, and label each one with the number of degrees.

13. 45°   (See facing page for model answers.)
14. 69°
15. 85°
16. 9°
17. 345°
18. 175°

D. Draw and bisect the following angles.

19. 50°   20. 28°   21. 90°   22. 130°

E. Identify each pair of angles as *complementary* or *supplementary*.

23. 62° and 118°   supplementary
24. 39° and 51°   complementary

F. Solve these reading problems.

25. What type of angle is formed by two adjacent arms of a starfish?   acute
26. What type of angle is formed by a sliding board and the landing at the bottom?   obtuse
27. The scale of a map is $\frac{1}{8}$ inch = 50 miles. If the distance between two cities is 2,600 miles, how far apart should they be on the map? $6\frac{1}{2}$ inches
28. On a map where 1 inch equals $\frac{5}{8}$ mile, the city limits measure $6\frac{3}{4}$ inches north to south and $4\frac{1}{2}$ inches east to west. What is the actual distance from the northern to the southern edge of the city limits? $4\frac{7}{32}$ miles
29. Roy Deitch seeded a wheat field with an area of 76,000 square meters. How large is the field in hectares?   7.6 hectares
30. Lisa bought a throw rug that covered 9,300 square centimeters. How many square meters is that?   0.93 m²

REVIEW EXERCISES

G. Name these geometric figures. *(Lesson 85)*

31.    trapezoid (or quadrilateral)
32. 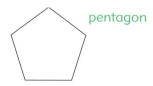   pentagon

H. Find the distance represented by each measurement. Scale: *1 in. = 16 mi.* *(Lesson 56)*

33. $4\frac{1}{4}$ in.   68 mi.
34. $3\frac{5}{8}$ in.   58 mi.

I. Find these equivalents. *(Lesson 24)*

35. 1.6 ha = 0.016 km²
36. 907,000 mm² = 0.907 m²

Lesson 86 T–290

Model Answers for Parts C and D

13.

14.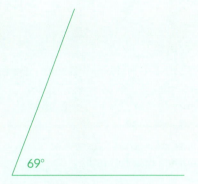

15.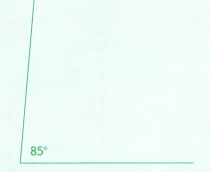

16.

17.

18.

19.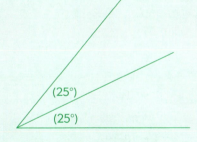

20.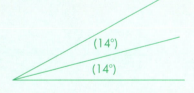

21.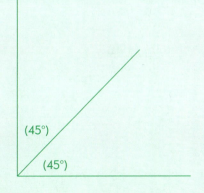

22.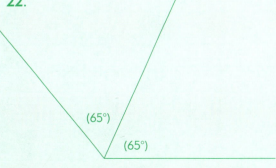

LESSON 87

Objectives

- To review the different types of triangles: acute, obtuse, right, equiangular; and equilateral, isosceles, scalene.
- To review that there are always 180 degrees in a triangle.
- To review similar and congruent triangles.

Review

1. Sketch these angles on the chalkboard. *Identify each angle.* (Lesson 86)

   a.  (reflex)

   b.  (obtuse)

   c. ─────●───→───── (straight)

   d. ◿ (acute)

2. *Find the arithmetic mean of each set, to the nearest whole number.* (Lesson 73)

   a. 27, 26, 34, 30, 24, 28 (28)

   b. 411, 432, 401, 431, 418, 420 (419)

3. *Find the distance represented by each measurement. Scale:* 1 in. = 8 ft. (Lesson 57)

   a. $3\frac{7}{8}$ in. (31 ft.)

   b. $2\frac{3}{4}$ in. (22 ft.)

4. *Add or subtract these compound English measures.* (Lesson 25)

   a. 12 hr. 18 min.
       + 8 hr. 48 min.
       (21 hr. 6 min.)

   b.   8 lb. 2 oz.
       − 5 lb. 14 oz.
       (2 lb. 4 oz.)

87. Classifying Triangles

A triangle is a geometric figure having three straight sides. A triangle can be classified according to the angles that form its corners.

An **acute triangle** has three acute angles. If the angles are equal, it is an **equiangular triangle**.

A **right triangle** has one right angle and two acute angles.

An **obtuse triangle** has one obtuse angle and two acute angles.

Notice that triangles have only obtuse, right, and acute angles. It is not possible to use a straight angle or a reflex angle to make a triangle.

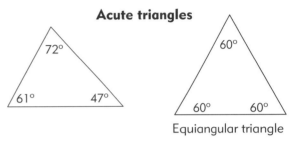

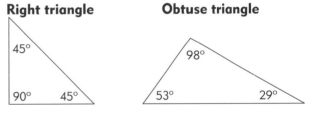

Triangles are also classified according to the lengths of their sides.

An **equilateral triangle** has all sides of equal length. Every equilateral triangle is also an equiangular triangle and vice versa. *Equilateral* means "having equal sides."

An **isosceles triangle** (ī sŏs′ ə lēz′) has two sides of equal length. *Isosceles* means "having equal legs."

A **scalene triangle** (skā′ lēn) has sides that are all different in length. *Scalene* means "uneven."

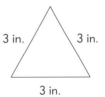

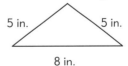

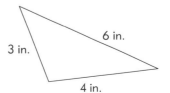

292   Chapter 7   Lines and Planes in Geometry

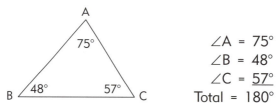

∠A = 75°
∠B = 48°
∠C = <u>57°</u>
Total = 180°

The sum of the angles in any triangle is 180°. If the sizes of two angles are known, the degrees in the third angle can be calculated by subtracting the sum of the two known angles from 180°.

Example		
If two angles of a triangle measure 40° and 60°, what is the size of the third angle?	40° + 60° 100°	180° − 100° 80°

Geometric figures that are the same in shape but not in size are said to be **similar.** Geometric figures that are the same in shape and size are said to be **congruent.** Shape and size alone determine whether figures are similar or congruent. Position does not affect this classification. Two equal triangles may be turned in different directions, but they are still congruent.

The symbol △ means "triangle." The symbol ~ means "is similar to." The symbol ≅ means "is congruent to." (Note the = sign in the last symbol.)

Similar triangles

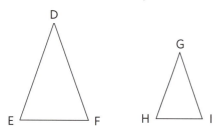

△ DEF ~ △ GHI
Triangle DEF is similar to triangle GHI.

Congruent triangles

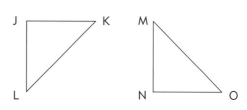

△ JKL ≅ △ MNO
Triangle JKL is congruent to triangle MNO.

CLASS PRACTICE

Identify each triangle as acute, obtuse, *or* right.

a.   acute

b.   right

c. 45°, 15°, 120°   obtuse

d. 75°, 55°, 50°   acute

Lesson 87   T–292

Introduction

Review the names of the polygons. Mention that a triangle is a polygon with three sides and three angles. Then give the following quiz on triangles. If the students have trouble, demonstrate on the board or let them help you demonstrate.

a. Can a triangle have more than one acute angle?
(Yes, all three may be acute.)

b. Can a triangle have more than one right angle?
(No, any figure with two right angles must have at least four sides.)

c. Can a triangle have more than one obtuse angle?
(No, any figure with two obtuse angles must have at least four sides.)

d. Can a triangle have only one acute angle?
(No, at least two of the angles must be acute.)

Teaching Guide

1. **One way to classify triangles is by the angles that form their corners.**

   a. An acute triangle has three acute angles.

   b. An equiangular triangle has three equal angles.

   c. A right triangle has one right angle and two acute angles.

   d. An obtuse triangle has one obtuse angle and two acute angles.

   Notice that only obtuse, right, and acute angles are used to make triangles. It is not possible to use a straight angle or a reflex angle to make a polygon having three straight sides.

2. **Another way to classify triangles is by the lengths of their sides.**

   a. An equilateral triangle has all sides of equal length. Every equilateral triangle is also an equiangular triangle and vice versa.

   b. An isosceles triangle has two sides of equal length. Two of its angles are equal as well.

   c. A scalene triangle has sides that are all different in length.

3. **The sum of the angles in any triangle is 180°.** If the sizes of two angles are known, the degrees in the third angle can be calculated by subtracting the sum of the two known angles from 180°. Have students find the value for the missing angles in the following sets.

   a. 45°, 55°   (80°)

   b. 80°, 60°   (40°)

   c. 50°, 125°   (5°)

   d. 90°, 60°   (30°)

4. **Geometric figures that are the same in shape but not in size are said to be similar. Geometric figures that are the same in shape and size are said to be congruent.** Shape and size alone determine whether figures are similar or congruent. Position does not affect this classification. Two equal triangles may be turned in different directions, but they are still congruent.

The symbol △ means "triangle."

The symbol ~ means "is similar to."

The symbol ≅ means "is congruent to." (Note the = sign in the last symbol.)

Have students identify similar triangles and congruent triangles in *Class Practice*.

Further Study

The word *triangle* merely indicates three angles. In Euclidean geometry, all triangles have three straight sides and the angles equal 180°. Euclidean geometry is named after Euclid, a Greek mathematician whose 13-volume treatise is still a reference guide in mathematics 2,300 years later.

However, there are other definitions of triangles. One definition is "a geometric figure having three vertices and sides that are arbitrary curves." By that definition, a triangle does not need to have straight sides. Neither do the angles necessarily total 180°.

Triangle with sides that are arbitrary curves

In spherical geometry, a triangle is a three-sided portion of the curved surface of a sphere. Each of the three sides is an arc, which is part of a great circle on the sphere. You could cut a spherical triangle by making three "straight" cuts that intersect on the rind of an orange.

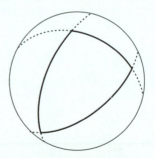

Spherical triangle with sides that are arcs of great circles on a sphere

Identify each triangle as *equilateral, isosceles,* **or** *scalene.*

e.
isosceles

f.
scalene

g.
isosceles

h.
equilateral

i. 2 in., 3 in., 5 in.   scalene

j. 3 in., 3 in., 3 in.   equilateral

Identify each pair of triangles as similar or congruent, using the symbols shown in the lesson.

k.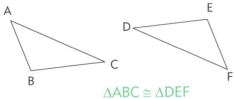
△ABC ≅ △DEF

l.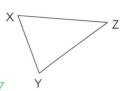
△UVW ~ △XYZ

Each pair gives the degrees in two angles of a triangle. Find the degrees in the third angle.

m. 82°, 65°   33°

n. 25°, 95°   60°

Measure the angles in these triangles. Check your work by making sure the sum of the three angles is 180°.

o.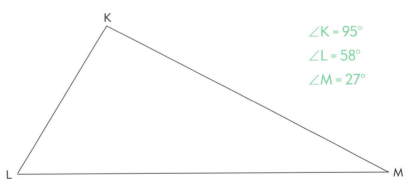
∠K = 95°
∠L = 58°
∠M = 27°

p.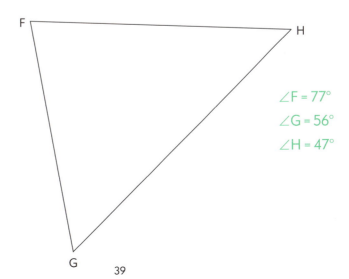
∠F = 77°
∠G = 56°
∠H = 47°

294   Chapter 7   *Lines and Planes in Geometry*

WRITTEN EXERCISES

A. Identify each triangle as *acute,* *obtuse,* **or** *right.*

1. right   2. obtuse

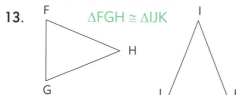

3. 72°, 68°, 40°   4. 27°, 63°, 90°   5. 50°, 35°, 95°   6. 60°, 60°, 60°
   acute   right   obtuse   acute

B. Identify each triangle as *equilateral,* *isosceles,* **or** *scalene.*

7. isosceles   8. scalene

9. 5 in., 5 in., 5 in.   equilateral   10. 7 in., 8 in., 9 in.   scalene
11. 6 in., 8 in., 10 in.   scalene   12. 7 in., 9 in., 7 in.   isosceles

C. Identify each pair of triangles as *similar* **or** *congruent,* **using the symbols shown in the lesson.**

13. △FGH ≅ △IJK

14. △NOP ~ △QRS

D. Each pair gives the degrees in two angles of a triangle. Find the degrees in the third angle.

15. 46°, 68°   66°   16. 38°, 97°   45°   17. 45°, 66°   69°
18. 75°, 75°   30°   19. 37°, 79°   64°   20. 81°, 85°   14°

E. Measure the angles in these triangles. Check your work by making sure the sum of the three angles is 180°.

21. ∠T = 120°
     ∠U = 30°
     ∠V = 30°

22. ∠W = 50°
     ∠X = 110°
     ∠Y = 20°

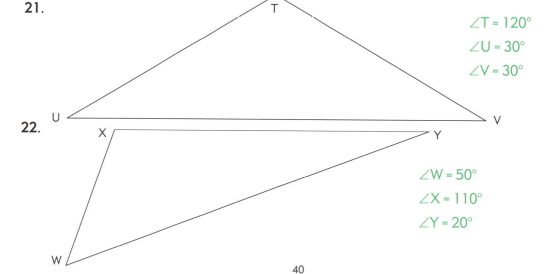

40

T–295   Chapter 7   Lines and Planes in Geometry

Solutions for Part F

23. $180 - (2 \times 22)$

24. $180 - (90 + 57)$

25. $\dfrac{\frac{1}{4}}{1} = \dfrac{5\frac{1}{2}}{n}$ $\dfrac{\frac{1}{4}}{1} = \dfrac{7\frac{1}{4}}{n}$

26. $\dfrac{\frac{1}{8}}{1} = \dfrac{15\frac{1}{2}}{n}$ $\dfrac{\frac{1}{8}}{1} = \dfrac{19}{n}$

27. 6 hr. 30 min. + 1 hr. 35 min. + 3 hr. 10 min. + 4 hr. 55 min. + 5 hr. 43 min.

28. 16 ft. – 12 ft. 9 in.

F. Solve these reading problems.

23. The gable on the end of the Snyder's house has two 22° angles at the eaves. What is the angle at the peak?   136°

24. Lester welded an angle brace across the corner of a gate. In the triangle that was formed by the brace, if one angle was 90° and another was 57°, what was the third angle?   33°

25. On a blueprint of Locust Point School, one classroom measures $5\frac{1}{2}$ inches by $7\frac{1}{4}$ inches. What are the dimensions of the room if $\frac{1}{4}$ inch equals 1 foot?   22 ft. by 29 ft.

26. A kitchen measures $15\frac{1}{2}$ feet wide and 19 feet long. What would be its dimensions on a scale drawing where $\frac{1}{8}$ inch = 1 foot? $1\frac{15}{16}$ in. by $2\frac{3}{8}$ in.

27. Reuben worked 6 hours 30 minutes to dismantle a car engine and replace the head gasket. Machine work and reassembly time over several days took 1 hour 35 minutes, 3 hours 10 minutes, 4 hours 55 minutes, and 5 hours 43 minutes. What was the total time he spent on the engine?   21 hr. 53 min.

28. Gerald used a 16-foot two-by-four to cut two pieces each 6 feet $4\frac{1}{2}$ inches long. How long was the remaining piece? (Disregard any waste from cutting.)   3 ft. 3 in.

REVIEW EXERCISES

G. Classify each angle, and find its size in degrees. *(Lesson 86)*

29. acute; 51°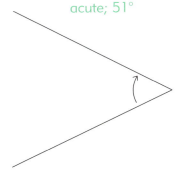

30. obtuse; 127°

H. Find the arithmetic mean of each set, to the nearest whole number. *(Lesson 73)*

31. 64, 53, 48, 55, 59, 50   55

32. 809, 796, 815, 804, 811, 800   806

I. Find the actual distance in meters for each measurement on a scale drawing. Scale: 1:90. *(Lesson 57)*

33. 23.4 cm   21.06 m

34. 16.2 cm   14.58 m

J. Add these English and metric compound measures. *(Lesson 25)*

35. 10 mi. 4,500 ft.
    \+ 2 mi. 1,530 ft.
    13 mi. 750 ft.

36. 21 kg + 32 g = ___ kg   21.032

88. Constructing Triangles

To draw a triangle of a specific size and shape, you must have certain information. Following are three ways in which this information may be stated.

- The lengths of two sides and the measure of the angle between them
- The measure of two angles and the length of the side between them
- The lengths of the three sides

To draw a triangle when the measures of two sides and the angle between are given, use the following steps. (See Example A.)

1. Draw a line segment the length of one of the two sides.
2. Draw the angle from one endpoint of the line segment.
3. Measure along the second side of the angle drawn in step 2. Place a dot at the length given for the second side.
4. Draw the third side of the triangle from the dot placed in step 3 to the other end of the line segment drawn in step 1.
5. Label the triangle with the given dimensions.

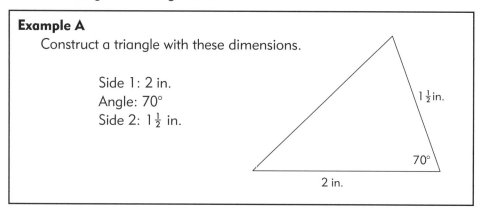

Example A
Construct a triangle with these dimensions.

Side 1: 2 in.
Angle: 70°
Side 2: $1\frac{1}{2}$ in.

To draw a triangle when the measures of two angles and the side between are given, follow the steps below. (See Example B.)

1. Draw a line segment to the length given.
2. Draw the first angle from one endpoint of the line segment.
3. Draw the second angle from the other endpoint of the line segment. This angle should intersect the angle drawn in step 2.
4. Label the triangle with the given dimensions.

LESSON 88

Objectives

- To teach *using a ruler and a protractor to construct a triangle when two sides and the included angle are known.
- To teach *using a ruler and a protractor to construct a triangle when two angles and the included side are known.
- To review using the compass.
- To teach *using a compass to construct a triangle when the three sides are known.

Review

1. Give Lesson 88 Quiz (Geometric Figures).
2. Sketch these triangles on the chalkboard. *Classify each triangle as acute, obtuse, or right. Also tell whether it is equilateral, isosceles, or scalene.* (Lesson 87)

   a. (obtuse, isosceles)

   b. (acute, equilateral)

   c. (right, isosceles)

   d. (obtuse, scalene)

3. *Find the supplementary angle for each of these.* (Lesson 86)

   a. 16° (164°)   b. 28° (152°)
   c. 74° (106°)   d. 131° (49°)

4. *Find the median and the mode of each set.* (Lesson 74)

   a. 5, 6, 5, 7, 9, 11, 7, 8, 3
   (median = 7; mode = 5, 7)

   b. 21, 23, 26, 52, 21, 27, 25, 24, 25, 18
   (median = $24\frac{1}{2}$; mode = 21, 25)

5. *Multiply or divide these compound measures.* (Lesson 26)

   a.

   b. 5 × 18 kg 40 g = __ kg   (90.2)

Introduction

Two sets of three items can be used to draw triangles.

Set 1: a ruler, a compass, and a pencil

Set 2: a ruler, a protractor, and a pencil

T–297   Chapter 7   Lines and Planes in Geometry

Teaching Guide

1. **A specific triangle can be drawn if you know the measure of two of the sides and the angle between them.** Discuss the steps in the lesson; then have students draw the following triangles.

    a. 3 in., 50°, 2 in.

    b. 1 in., 125°, 1 in.

2. **A specific triangle can be drawn if you know the measure of two angles and the side between them.** Discuss the steps in the lesson; then have students draw the following triangles.

    a. 30°, 2 in., 50°

    b. 25°, 2 in., 125°

3. **A specific triangle can be drawn when the lengths of the three sides are given.** Discuss the steps in the lesson; then have students draw the following triangles.

    a. 3 in., 2½ in., 2 in.

    b. 1½ in., 2 in., 2½ in.

Model Triangles for CLASS PRACTICE

b.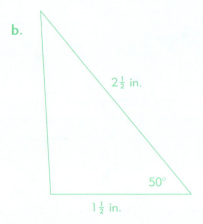

c.

d.

e.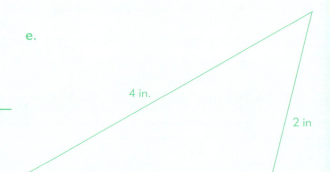

a.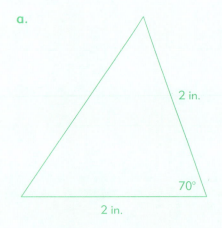

f.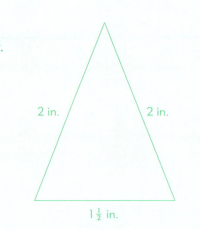

Example B

Construct a triangle with these dimensions.

Angle 1: 40°
Side: 2 in.
Angle 2: 50°

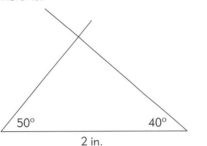

To draw a triangle when the lengths of the three sides are given, follow the steps below. (See Example C.)

1. Draw a line segment the length of the first side.

2. Open a compass to the length of the second side. Place the point of the compass on one endpoint of the line segment, and draw an arc that is sure to cross the peak of the triangle.

3. Open the compass to the length of the third side. Place the point of the compass on the other endpoint of the line segment, and draw an arc that intersects the arc drawn in step 2.

4. Draw the two remaining sides of the triangle from the point where the arcs intersect to the ends of the line segment drawn in step 1.

5. Label the dimensions of the triangle.

Example C

Construct a triangle with these dimensions.

Side 1: $2\frac{1}{2}$ in.
Side 2: 2 in.
Side 3: $1\frac{1}{2}$ in.

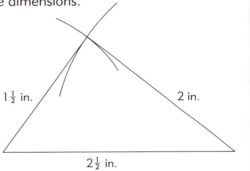

CLASS PRACTICE

Use each set of facts to construct a triangle. Label it with the dimensions given.
(See facing page for model triangles.)

a. 2 in., 70°, 2 in.
b. $1\frac{1}{2}$ in., 50°, $2\frac{1}{2}$ in.
c. 35°, 2 in., 35°
d. 15°, 1 in., 120°
e. 3 in., 4 in., 2 in.
f. $1\frac{1}{2}$ in., 2 in., 2 in.

298   Chapter 7   Lines and Planes in Geometry

WRITTEN EXERCISES

A. *Use each set of facts to construct a triangle. Label it with the dimensions given.*   (See facing page for model triangles.)

1. 1 in., 50°, 2 in.
2. $1\frac{1}{2}$ in., 40°, 2 in.
3. 2 in., 80°, 2 in.
4. 1 in., 90°, 2 in.
5. 2 in., 125°, 1 in.
6. 2 in., 85°, 2 in.
7. 50°, 2 in., 50°
8. 35°, 2 in., 45°
9. 70°, 2 in., 20°
10. 45°, 2 in., 45°
11. 60°, 2 in., 30°
12. 50°, 1 in., 70°
13. 2 in., 3 in., 2 in.
14. 1 in., 2 in., 2 in.
15. 2 in., 2 in., 2 in.
16. $2\frac{1}{2}$ in., $2\frac{1}{2}$ in., 3 in.

B. *Solve these reading problems.*

17. An electric pole and its guy wire form a right triangle with angles of 90° and 60° on the ground. What is the size of the angle at the top of the wire?   30°

18. To make a stairway, Kevin is cutting triangles from a 2-by-12 plank. Each triangle has a 90° angle in the lower back corner and a 37° angle in the lower front corner. What is the angle of the upper corner?   53°

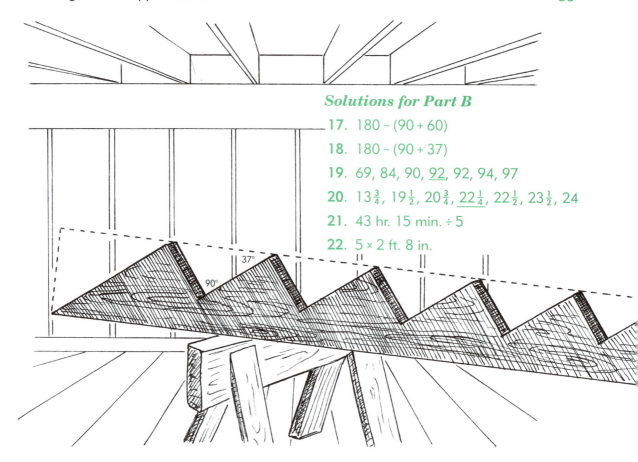

Solutions for Part B

17. 180 − (90 + 60)
18. 180 − (90 + 37)
19. 69, 84, 90, <u>92</u>, 92, 94, 97
20. $13\frac{3}{4}$, $19\frac{1}{2}$, $20\frac{3}{4}$, <u>$22\frac{1}{4}$</u>, $22\frac{1}{2}$, $23\frac{1}{2}$, 24
21. 43 hr. 15 min. ÷ 5
22. 5 × 2 ft. 8 in.

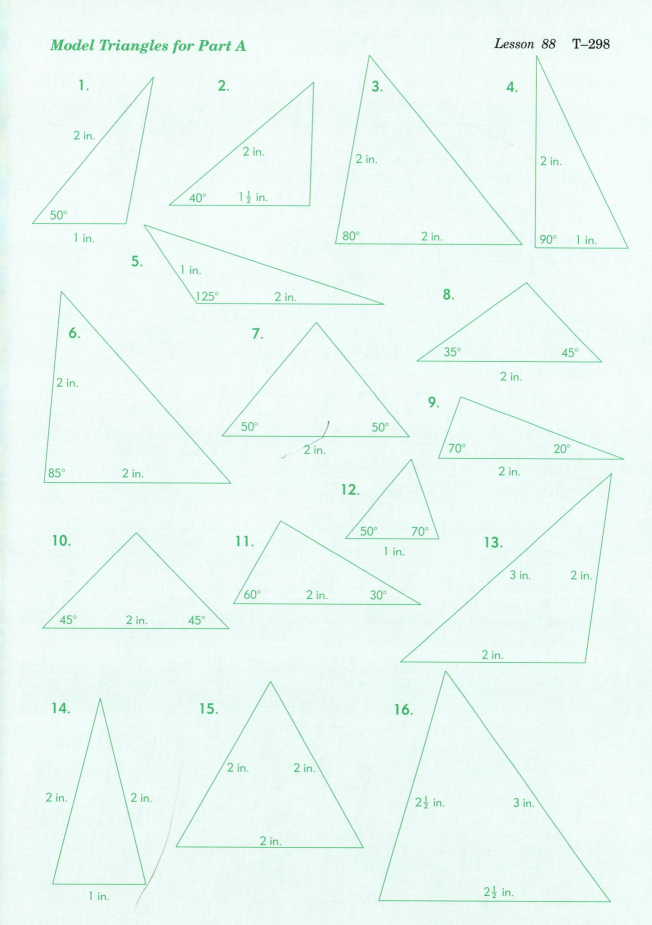

T–299 Chapter 7 Lines and Planes in Geometry

Further Study

Three other sets of criteria for triangles could be considered.

1. Given measures for two angles and one side that is not between them.

    One way to do this construction is to use a trial-and-error method. In the diagram below there is only one base measure that will satisfy the given criteria and form a triangle. That measure could be found by adjusting the length of BC until line CD crosses point A.

    Angle 1: 30°

    Angle 2: 70°

    Side: 3″

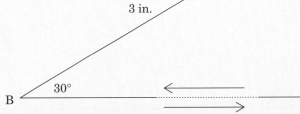

Another way would be to calculate the third angle by subtracting 30 + 70 from 180. Constructing an 80° angle at point A would produce a 70° angle at the correct point on BC.

2. Given measures for two sides and one angle that is not between them.

    This information is not enough to define a particular triangle. As shown below, there are two points where a one-inch line from point A meets the baseline. The triangle can be ABC or ABD.

    Side 1: 3″

    Side 2: 1″

    Angle: 15°

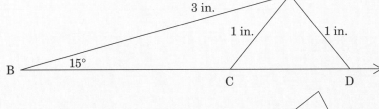

3. Given measures for three angles.

    Having no measure for length of sides allows any number of different-sized triangles, all similar to one another.

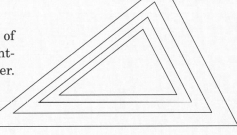

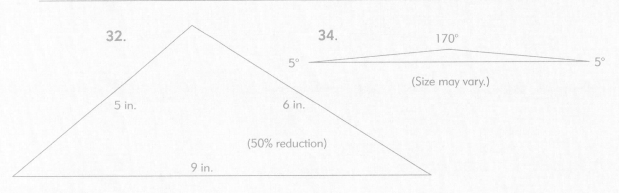

32.

5 in.   5°   6 in.

9 in.

(50% reduction)

34.

170°

5°   5°

(Size may vary.)

50

19. On an eighth grade history test the scores were as follows: 84%, 92%, 97%, 69%, 94%, 90%, 92%. What was the class median?   92%

20. After picking watermelons one day, Allen and Philip recorded the weights of 7 melons as follows: 24 pounds, $20\frac{3}{4}$ pounds, $22\frac{1}{2}$ pounds, $19\frac{1}{2}$ pounds, $23\frac{1}{2}$ pounds, $22\frac{1}{4}$ pounds, and $13\frac{3}{4}$ pounds. What was the median weight? $22\frac{1}{4}$ pounds

21. In 5 days, Carol worked 43 hours 15 minutes at the bakery. How many hours and minutes did she average per day?   8 hours 39 minutes

22. The south side of a church building has 5 windows each measuring 2 feet 8 inches wide. What is the total width of all the windows on that side?   13 feet 4 inches

REVIEW EXERCISES

C. Classify each triangle in two ways: (a) acute, obtuse, **or** right, **and (b)** equilateral, isosceles, **or** scalene. *(Lesson 87)*

23. 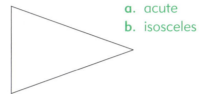   a. acute
     b. isosceles

24. 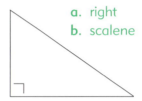   a. right
     b. scalene

D. Classify each angle as acute, obtuse, straight, **or** reflex. *(Lesson 86)*

25. obtuse

26. reflex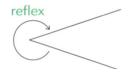

E. Find the median and the mode(s) of each group. *(Lesson 74)*

27. 32, 35, 31, 36, 32, 34, 30   median = 32; mode = 32
28. 99, 102, 95, 84, 96, 95, 105, 99, 98   median = 98; mode = 95, 99

F. Divide these compound English and metric measures. *(Lesson 26)*

29. $4\overline{)18\text{ lb. }14\text{ oz.}}$ 4 lb. $11\frac{1}{2}$ oz.

30. 12 kl 77 l ÷ 5.2 = ____ l 2,322.5

CHALLENGE EXERCISES

G. Is it possible to draw each of these triangles? If so, draw it. If not, explain why.   (See facing page for answers to 32 and 34.)

31. 5 in., 6 in., 14 in.   14 inches is longer than the sum of the other two sides.
32. 9 in., 5 in., 6 in.
33. 65°, 72°, 80°   The sum of the angles exceeds 180°.
34. 170°, 5°, 5°

89. Constructing Geometric Figures

With a compass you can draw precise arcs and circles. These arcs and circles can be used in a multitude of ways. Bisecting line segments, bisecting angles, and constructing perpendicular lines are a few of these uses. The drawing of hexagons will probably stir your imagination to many other artistic uses of arcs and circles.

Bisecting Line Segments

A compass can be used to bisect line segments into two parts of equal length. The letters in the following steps refer to Example A.

1. Open the compass to slightly more than half the length of the line segment.

2. Place the sharp point of the compass on endpoint A. Draw arcs at approximately midpoint above and below AB.

3. With the same compass setting, place the sharp point on endpoint B. Draw arcs above and below AB to intersect with the first pair of arcs.

4. Use a straightedge to bisect AB by drawing a line segment between points C and D.

5. Measure AE and EB to make sure their lengths are equal.

Example A

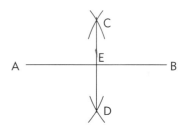

Bisecting Angles

A compass can be used to bisect an angle into two equal angles. In Lesson 86 you bisected angles with a protractor. The following steps tell how to do it with a compass.

1. Place the sharp point of the compass on point G. With the same compass setting, draw arcs that intersect FG and GH at points I and J.

2. Open the compass to more than half the distance between I and J. Place the sharp point on point I and draw an arc in the middle of the angle.

3. With the same setting, place the compass point on point J and draw an arc that intersects the arc inside the angle at K.

4. Use a straightedge to draw a line from G through point K.

Example B

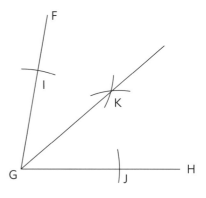

5. Use a protractor to make sure angle FGK is equal to angle KGH.

LESSON 89

Objectives

- To teach using a compass *to bisect line segments, *to bisect angles, and *to draw perpendicular lines.
- To review using a compass and a straightedge to draw hexagons, equilateral triangles, and artistic designs.

Review

1. *Use each set of facts to construct a triangle. Label it with the dimensions given.* (Lesson 88)

    a. $1\frac{1}{2}$ in., 60°, $1\frac{1}{2}$ in.

    b. 1 in., 130°, 2 in.

2. *Classify each triangle as acute, obtuse, or right. Also tell whether it is equilateral, isosceles, or scalene.* (Lesson 87)

    a. 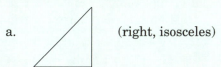 (right, isosceles)

    b. 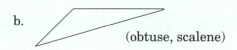 (obtuse, scalene)

3. *Find each equivalent, using the tables.* (Lesson 27)

    a. 18 cm = ___ in. (7.02)

    b. 14 tons = ___ kg (12,698)

    c. 55 *l* = ___ gal. (14.3)

    d. 12 sq. mi. = ___ km² (31.08)

    e. $20\frac{1}{2}$ in. = ___ cm (52.07)

    f. 150 MT = ___ tons (165)

Introduction

Have the students get out their compasses. Then see if they can guess reasonable answers to the following questions.

(1) Why does a compass have its name? Why not call it a circlescriber or an arcdrawer or something like that? (*Compass* comes from two Latin elements: *com-* [completely] and *passus* [step, pace]. To compass something is to walk completely around it. When we draw a circle, the pencil of the compass "walks" completely around the center.)

(2) If I gave you a pair of compasses, how many would you have? (You would have one. A pair of compasses is one item in the same way as a pair of scissors or a pair of trousers.)

Teaching Guide

1. **A compass can be used to bisect line segments into two parts of equal length.** The letters in the following steps refer to Example A in the lesson text.

    (1) Open the compass to slightly more than half the length of the line segment.

    (2) Place the sharp point of the compass on endpoint A. Draw arcs at approximately midpoint above and below AB.

    (3) With the same compass setting, place the sharp point on endpoint B. Draw arcs above and below AB to intersect with the first pair of arcs.

T–301   *Chapter 7   Lines and Planes in Geometry*

(4) Use a straightedge to bisect AB by drawing a line segment between points C and D.

(5) Measure AE and EB to make sure their lengths are equal.

Have the students do exercises *a* and *b* of *Class Practice*.

2. **A compass can be used to bisect an angle.** The letters in the following steps refer to Example B in the lesson text.

   (1) Place the sharp point of the compass on point G. With the same compass setting, draw arcs that intersect FG and GH at points I and J.

   (2) Open the compass to more than half the distance between I and J. Place the sharp point on I, and draw an arc in the middle of the angle.

   (3) With the same setting, place the compass point on point J and draw an arc that intersects the arc inside the angle at K.

   (4) Use a straightedge to draw a line from G through point K.

   (5) Use a protractor to make sure angle FGK is equal to angle KGH.

Have the students do exercises *c* and *d* of *Class Practice*.

3. **A compass can be used to construct perpendicular lines.** The bisecting line in number 1 is also perpendicular to the original line, but this exercise allows the perpendicular to be placed at any chosen point along the line. The letters in the following steps refer to Example C in the lesson text.

   (1) Mark a point (N) on LM where you want the perpendicular line to be.

   (2) Place the point of the compass on point N. With one compass setting, draw arcs that intersect LM on both sides of point N.

   (3) Open the compass farther. Place its point on point O and draw an arc above point N.

   (4) With the same setting, place the compass point on P and draw an arc that intersects the arc above at Q.

   (5) Use a straightedge to draw a line from N through point Q.

   (6) Use a protractor to make sure that QN is perpendicular to LM.

Have the students do exercises *e* and *f* of *Class Practice*.

4. **A compass can be used to draw many geometric figures other than circles.** Following are the steps for using a compass and a straightedge to draw a regular hexagon or triangle.

   (1) Use a compass to draw a circle the size of the regular hexagon that you want.

   (2) Mark a starting point anywhere on the circle.

   (3) With the compass setting still the same as the radius, place the compass point on the starting point and draw a short arc that intersects the circle.

   (4) Move the compass point to this point of intersection, and draw another arc farther around the circle. Keep moving the compass point to the next arc and drawing another one until the last arc intersects the starting point.

Constructing Perpendicular Lines

A perpendicular line forms a right angle with a base line. Use the following steps to construct a perpendicular line with a compass. The letters in these steps refer to Example C.

1. Mark a point (N) on LM where you want the perpendicular line to be.

2. Place the point of the compass on point N. With one compass setting, draw arcs that intersect LM on both sides of point N.

3. Open the compass farther. Place its point on point O, and draw an arc above point N.

4. With the same setting, place the compass point on point P and draw an arc that intersects the arc above at Q.

5. Use a straightedge to draw a line from N through point Q.

6. Use a protractor to make sure that QN is perpendicular to LM.

Example C

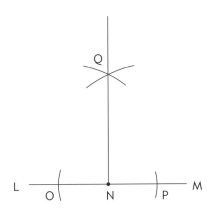

Drawing Hexagons and Equilateral Triangles

A compass can be used to draw many geometric figures other than circles. Following are the steps for using a compass and a straightedge to draw a regular hexagon or triangle.

1. Use a compass to draw a circle the size of the regular hexagon that you want.

2. Mark a starting point anywhere on the circle.

3. With the same radius setting on the compass, place the compass point on your starting point and draw a short arc that intersects the circle.

4. Move the compass point to this point of intersection and draw another arc farther around the circle. Keep moving the compass point to the next arc and drawing another one until the last arc intersects the starting point.

5. To draw a hexagon, use a straightedge to draw lines between the points of intersection and form the six sides of the hexagon. To draw an equilateral triangle, connect every other point of intersection.

6. Erase the circle.

Example D

1-inch radius

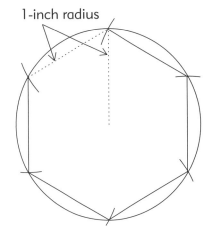

302   Chapter 7   Lines and Planes in Geometry

With a bit of artistic imagination, you can use the same method to draw many other attractive designs. The Star of David below (Example E) was drawn by connecting every other intersection to complete one equilateral triangle, and then doing the same thing with the remaining intersections to form an overlapping triangle.

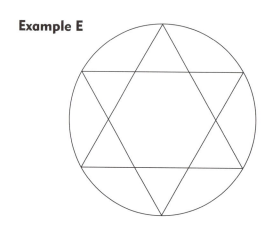

Example E

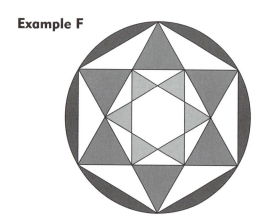

Example F

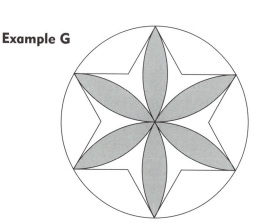

Example G

CLASS PRACTICE   (See facing page for model answers.)

Draw line segments of these lengths, and use a compass to bisect them.

   a. 2 in.   b. 7 cm

Draw angles with the sizes given, and use a compass to bisect them.

   c. 90°   d. 32°

Draw line segments of these lengths, and use a compass to draw lines perpendicular to them.

   e. 3 in.   f. 10 cm

Draw circles having these radii.

   g. 0.5 in.   h. 3 cm

(5) To draw a hexagon, use a straightedge to draw lines between the points of intersection and form the six sides of the hexagon. To draw an equilateral triangle, connect every other point of intersection.

(6) Erase the circle.

Have the students do exercises g–j of *Class Practice*.

With a bit of artistic imagination, the same method can be used to draw many other attractive designs. If you have class time, stimulate the students' interest by drawing one of the sample designs, or devise an original one with your class. You may wish to have the students construct full-page designs in an art class.

Model Answers for CLASS PRACTICE

c.

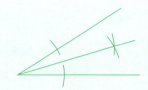

d.

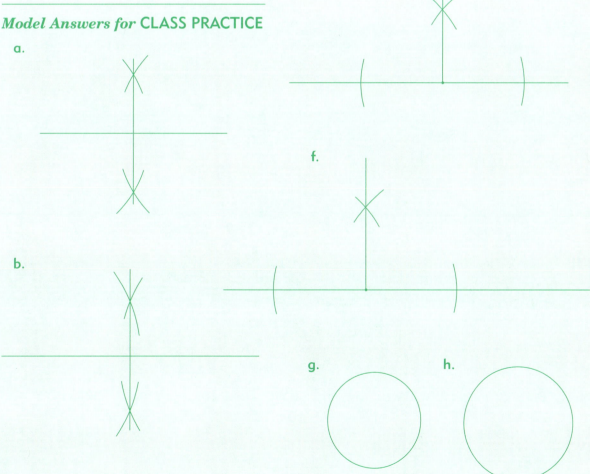

T–303 Chapter 7 *Lines and Planes in Geometry*

i.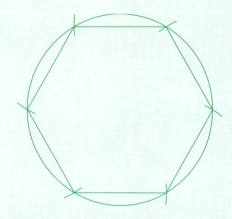
j.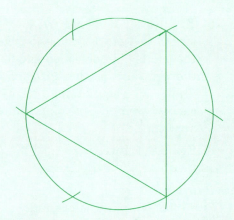

Model Answers for WRITTEN EXERCISES

1.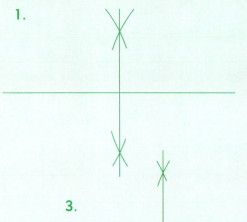
2.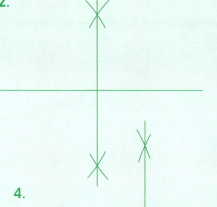

3.
4.

5.
6.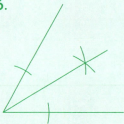
7.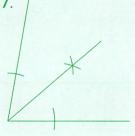

8.
9.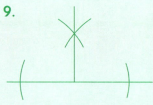
10.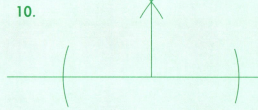

(Continued on page T–304.)

Use a compass and a ruler to do these exercises.

   i. Draw a regular hexagon in a circle with a 2-inch diameter.

   j. Draw an equilateral triangle in a circle with a 2-inch diameter.

WRITTEN EXERCISES

A. *Draw line segments of the following lengths, and use a compass to bisect them. Your work will need to be accurate within $\frac{1}{16}$ inch (for English units) or 1 millimeter (for metric units).*

1. $2\frac{1}{2}$ in.
2. $2\frac{1}{4}$ in.
3. 6 cm
4. 8 cm

B. *Draw the following angles, and use a compass to bisect them. Your work will be need to be accurate within 1 degree.*

5. 70°
6. 60°
7. 80°
8. 50°

C. *Draw line segments of these lengths, and use a compass to draw lines perpendicular to them.*

9. 4 cm
10. 7 cm
11. 1 in.
12. $2\frac{1}{2}$ in.

D. *Use a compass and a ruler to draw these hexagons.*

13. Draw a regular hexagon, beginning with a 2-inch diameter.
14. Draw a regular hexagon, beginning with a 3-inch diameter.
15. Draw an equilateral triangle, beginning with a 2-inch diameter.
16. Draw an equilateral triangle, beginning with a 3-inch diameter.

E. *Draw a copy of Example F or Example G, or one of the figures shown below. If you wish, you may make a design of your own.*

17–20. (4 points for one drawing)   (Check students' designs.)

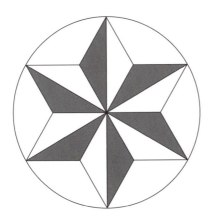

304 Chapter 7 Lines and Planes in Geometry

F. Solve these reading problems.

21. By recess time, Nancy had completed $68\frac{3}{4}\%$ of her 32 math problems. How many problems did she still have to do?
10 problems

22. In November a dozen eggs cost $0.97. In March the price was $0.88 per dozen. What was the rate of decrease, to the nearest whole percent?
9%

23. A triangular 5-acre field has one corner of 50° and another of 45°. How many degrees are in the third corner?
85°

24. The A-frame of a swing set forms an isosceles triangle. The angle at the peak is 28°. What is the measure of each of the angles at the ground? (An isosceles triangle has two equal angles as well as two equal sides.)
76°

25. Is the opening of a half-inch wrench larger or smaller than the head of a 13-millimeter bolt?
smaller

26. Brymesser Dairy buys milk replacer in 25-kilogram bags. How many pounds are in each bag?
55 pounds

REVIEW EXERCISES

G. Construct a triangle with each set of dimensions. Label the given dimensions. *(Lesson 88)*
(See facing page for model triangles.)

27. 2 in., 50°, 3 in. 28. $1\frac{1}{2}$ in., 60°, $1\frac{1}{2}$ in.

H. Classify each triangle in two ways: (a) *acute, obtuse,* or *right,* and (b) *equilateral, isosceles,* or *scalene.* (Lesson 87)

29.  a. obtuse
b. scalene

30.  a. acute
b. isosceles

I. Find these equivalents, using the tables. *(Lesson 27)*

31. 14 tbsp. = 210 ml   32. 9 km = 5.58 mi.

Solutions for Part F

21. 32 − (0.6875 × 32)
22. 97 − 89 = 9; 9 = __% of 97
23. 180 − (50 + 45)
24. (180 − 28) ÷ 2
25. 13 mm = 1.3 cm; 0.5 × 2.54 < 1.3
26. 25 × 2.2

Lesson 89 T-304

11. **12.** **13.**
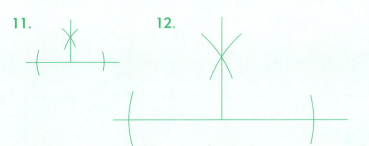
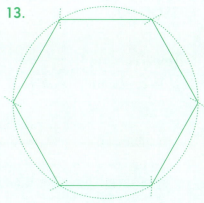

14. **15.**
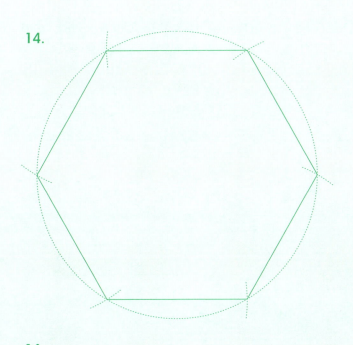
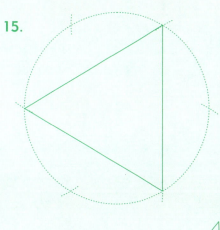

16. **27.**
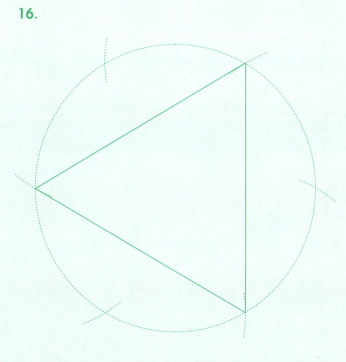
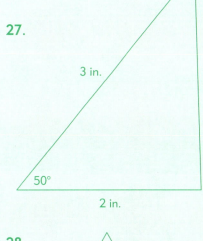

28.
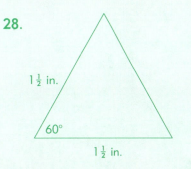

T–305 Chapter 7 Lines and Planes in Geometry

LESSON 90

Objectives

- To review finding the perimeter of any polygon.
- To review calculating the perimeter of a square by using the formula $p = 4s$.
- To review calculating the perimeter of a rectangle by using the formula $p = 2(l + w)$.

Review

1. Use each set of facts to construct a triangle. Label it with the dimensions given. (Lesson 88)
   a. $1\frac{1}{2}$ in., 3 in., $2\frac{1}{2}$ in.
   b. 6 cm, 3 cm, 6 cm
2. Change these decimals, ratios, and fractions to percents. (Lesson 60)
   a. $\frac{1}{4}$ (25%)
   b. 0.02 (2%)
   c. 12:25 (48%)
3. Find each temperature equivalent, to the nearest degree. (Lesson 28)
   a. 9°C = ___°F (48)
   b. 44°F = ___°C (7)

90. Perimeters of Polygons

The word *perimeter* comes from two Greek words that mean "around" and "measure." Thus, perimeter is the measure or distance around a plane figure. To find the perimeter of any polygon, measure all the sides and find the sum of the measurements.

Example A
Find the perimeter of this figure.
$p = 3 + 2 + 2 + 1 + 3 + 2 + 2 + 1 = 16$ ft.

Example B
Find the perimeter of this square.
$p = 4s$
$p = 4 \times 5\frac{1}{2}$ ft.
$p = 22$ ft.

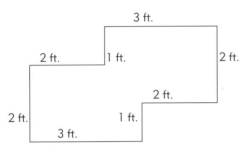

Special formulas are used to find the perimeters of common polygons. Because every square has four sides of equal length, the formula is *perimeter = 4 times the length of one side*, or $p = 4s$. In Example B, the perimeter of the square is 22 feet because $4 \times 5\frac{1}{2} = 22$.

A rectangle has two pairs of sides with equal lengths. To find the perimeter of a rectangle, add the length to the width and multiply the sum by 2. The formula is *perimeter = 2 times the sum of the length and the width*, or $p = 2(l + w)$.

Example C
Find the perimeter of this rectangle.
$p = 2(l + w)$
$p = 2(10 + 2)$
$p = 2 \times 12$ ft.
$p = 24$ ft.

Memorize these formulas for perimeters.
Square: perimeter = 4 × side, or $p = 4s$
Rectangle: perimeter = 2 × (length + width), or $p = 2(l + w)$

CLASS PRACTICE

Give the formulas for finding these perimeters. Be sure you know them by memory.

a. Perimeter of a square $p = 4s$
b. Perimeter of a rectangle $p = 2(l + w)$

Find the perimeter of polygons that have sides with these dimensions.

c. 4 in., 1 in., 5 in., 2 in.   12 in.   d. 8 in., 9 in., 11 in.   28 in.
e. 6 in., 4 in., 6 in., 7 in., 8 in., 5 in.   36 in.   f. 11 in., 12 in., 10 in., 12 in., 11 in., 10 in.   66 in.

Find the perimeter of each square. Write the formula and solve.

g. $s = 18$ cm $p = 4s$ 72 cm   h. $s = 20$ ft. 3 in. $p = 4s$ 81 ft.
i. $s = 4$ ft. 6 in. $p = 4s$ 18 ft.   j. $s = 2.3$ km $p = 4s$ 9.2 km

Find the perimeter of each rectangle. Write the formula and solve.

k. 6 in. by 5 in. $p = 2(l + w)$ 22 in.   l. $4\frac{1}{2}$ ft. by 5 ft. $p = 2(l + w)$ 19 ft.
m. 14 cm by 11 cm $p = 2(l + w)$ 50 cm   n. 12 m by 8.2 m $p = 2(l + w)$ 40.4 m

WRITTEN EXERCISES

A. Write the formulas for finding these perimeters. Be sure you know them by memory.

1. Perimeter of a square $p = 4s$
2. Perimeter of a rectangle $p = 2(l + w)$

B. Find the perimeter of polygons that have sides with these dimensions.

3. 5 in., 4 in., 3 in., 9 in.   21 in.   4. 6 in., 5 in., 7 in.   18 in.
5. 9 in., 8 in., 8 in.   25 in.   6. 10 in., 9 in., 8 in., 6 in.   33 in.
7. 11 in., 9 in., 7 in., 6 in., 9 in., 8 in.   50 in.   8. 6 in., 7 in., 5 in., 7 in., 9 in., 4 in.   38 in.

C. Find the perimeter of each square. Write the formula and solve.

9. $s = 12$ cm $p = 4s$ 48 cm   10. $s = 9$ ft. 2 in. $p = 4s$ 36 ft. 8 in.
11. $s = 3$ ft. 9 in. $p = 4s$ 15 ft.   12. $s = 5$ yd. 9 in. $p = 4s$ 21 yd.
13. $s = 2.2$ m $p = 4s$ 8.8 m   14. $s = 0.8$ km $p = 4s$ 3.2 km

D. Find the perimeter of each rectangle. Write the formula and solve.

15. 3 in. by 4 in. $p = 2(l + w)$ 14 in.   16. 9 in. by 8 in. $p = 2(l + w)$ 34 in.
17. 9 in. by 12 in. $p = 2(l + w)$ 42 in.   18. $8\frac{1}{2}$ in. by 11 in. $p = 2(l + w)$ 39 in.
19. 2.5 cm by 3.3 cm $p = 2(l + w)$ 11.6 cm   20. 6.7 m by 2 m $p = 2(l + w)$ 17.4 m

Lesson 90 T–306

Introduction

Ask the students if they know what a periscope is. A periscope is an instrument that is used to look around corners. *Peri-* comes from a word meaning "around," and *scope* comes from a word meaning "to look."

Then what does *perimeter* mean? *Peri-* means "around" and *meter* means "measure." So the word *perimeter* means "the measure around" something.

Teaching Guide

1. **Perimeter is the distance around a plane figure.** Have the students point out some perimeters in the classroom.

2. **To find the perimeter of any polygon, measure all the sides and find the sum of the measurements.** Measure the perimeters of some polygons such as a desktop or your classroom floor. Also find the perimeter of polygons with sides of the lengths below.

   a. 7 in., 13 in., 6 in., 7 in. (33 in.)
   b. 16 in., 12 in., 19 in., 25 in., 9 in. (81 in.)

3. **To find the perimeter of a square, use the following formula: perimeter = 4 × side, or $p = 4s$.** Students are to memorize this formula. Emphasize using the formula, replacing s with the length of one side of the square.

   a. $s = 5$ in. (20 in.)
   b. $s = 11$ cm (44 cm)
   c. $s = 13$ ft. 4 in. (53 ft.)
   d. $s = 6$ yd. 2 ft. (26 yd. 2 ft.)

4. **To find the perimeter of a rectangle, use the following formula: perimeter = 2 × the sum of the length and width, or $p = 2(l + w)$.** Again use the formula to find several perimeters.

   a. $l = 8$ in. $w = 6$ in. (28 in.)
   b. $l = 22.3$ cm $w = 16.9$ cm (78.4 cm)
   c. $l = 15.3$ m $w = 11.8$ m (54.2 m)
   d. $l = 3\frac{1}{3}$ yd. $w = 2\frac{3}{4}$ yd. ($12\frac{1}{6}$ yd.)

T-307 Chapter 7 Lines and Planes in Geometry

Solutions for Part E

21. 2(100 + 50) − 20
22. 2(10 + 13)
23. 46 ÷ 4
24. 4 × 26
25. $F = \frac{9}{5}(357 + 32)$
26. $F = \frac{9}{5}(80 + 32)$

Model Triangles for Part F

27.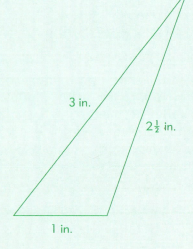

28.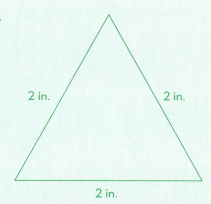

E. Solve these reading problems. Be careful, for some of them contain extra information.

21. The tabernacle had a courtyard 100 cubits long and 50 cubits wide, with a gate 20 cubits wide on the east side (Exodus 27:9–16). How many cubits of fine white linen were needed to enclose the courtyard, not including the gate? 280 cubits

22. Ezekiel 40–42 describes Ezekiel's vision of the new temple. The dimensions of the temple gate are recorded in Ezekiel 40:11 as 10 cubits by 13 cubits. What was the perimeter of the temple gate? 46 cubits

23. Brother David poured a square concrete slab for a feed bin. The 4 planks he used as forms had a total length of 46 feet. How long was one side of the square? $11\frac{1}{2}$ feet

24. A cylindrical pressure tank with a 26-inch diameter is packaged inside a square wooden crate. What is the inside perimeter of the crate? 104 inches

25. Mercury boils at 357°C. To the nearest whole degree, what is the Fahrenheit equivalent? 675°F

26. Gasoline evaporates before water because it boils at 80°C. What is the Fahrenheit equivalent? 176°F

REVIEW EXERCISES

F. Construct a triangle with each set of dimensions. Label the dimensions given. *(Lesson 88)* (See facing page for model triangles.)

27. 1 in., 3 in., $2\frac{1}{2}$ in.

28. 2 in., 2 in., 2 in.

G. Change these decimals, ratios, and fractions to percents. *(Lesson 60)*

29. $\frac{3}{4}$ 75%

30. $\frac{9}{25}$ 36%

31. 11:20   55%

32. 0.09   9%

H. Find each temperature equivalent, to the nearest degree. *(Lesson 28)*

33. 14°C = __57__ °F

34. 200°F = __93__ °C

91. Circles and Circumferences

A circle is a closed curve on which all points are the same distance from the center. The diagram below gives the names of the various parts of a circle.

Line segment AB is a **diameter** of the circle. It begins at one edge, passes through the center, and extends to the opposite edge.

Line segment CD is a **radius** of the circle. It extends from the center to the edge.

The perimeter of a circle is called its **circumference**.

Line segment EF is an **arc**. It is a part of the curve of the circle.

Arc AEFB is a **semicircle**. It is one-half of the circle.

Line segment GH is a **chord**; it extends from one point on the circumference to another. The diameter is a special chord that passes through the center of the circle.

A pie-shaped piece such as BCD is a **sector** of the circle. Its edges are two radii and an arc. The arc AGHD and the two radii AC and DC enclose another sector. The half circle enclosed by arc AEFB and radii AC and BC is also a sector.

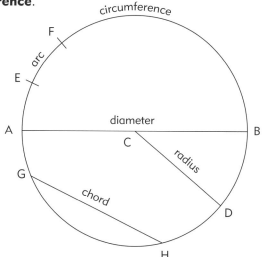

Finding the perimeter of a polygon is a simple matter of adding the lengths of the sides. But the circumference of a circle is harder to measure because it is curved. And there is no readily noticeable relationship between the diameter and circumference of a circle.

Thousands of years ago, the builders of Solomon's temple knew that the circumference of the molten sea was about three times its diameter. Several centuries later, a Greek mathematician named Archimedes calculated the circumference of a circle at about $3\frac{1}{7}$ times the length of its diameter.

Today the ratio between the diameter and circumference of a circle is known as **pi** (pronounced like *pie*). The Greek letter for *p*, π, is the symbol for pi.

Mathematicians have found that the value of pi is an **irrational number.** This means that its exact value cannot be stated as either a common fraction or a decimal.

To find the circumference of a circle, multiply its diameter by pi. As a formula, this is stated as circumference = pi times diameter or $c = \pi d$. If the radius is given, use $c = 2\pi r$.

> **Memorize these formulas for the circumference of a circle.**
> circumference = π × diameter, or $c = \pi d$
> circumference = 2 × π × radius, or $c = 2\pi r$

LESSON 91

Objectives

- To review the terms relating to a circle: diameter, radius, circumference, arc, semicircle, sector, and chord.
- To memorize the value of pi and recognize its symbol, π.
- To teach that *pi is an irrational number and that a more accurate value for pi is 3.1416.
- To memorize the formulas for calculating the circumference of a circle when the diameter or radius is known: $c = \pi d$ and *$c = 2\pi r$.
- To teach *finding the diameter of a circle when the circumference is known, using the formula $d = c \div \pi$.

Review

1. Give Lesson 91 Quiz (Geometric Symbols and Figures).

2. *Give the formula for finding each fact.* (Lesson 90)
   a. Perimeter of a square ($p = 4s$)
   b. Perimeter of a rectangle
      ($p = 2[l + w]$)

3. *Find the perimeter of each figure.* (Lesson 90)

   Squares
   a. $s = 12.5$ cm (50 cm)
   b. $s = 8$ in. (32 in.)
   c. $s = 0.4$ km (1.6 km)

   Rectangles
   d. $2\frac{1}{2}$ ft. by 3 ft. (11 ft.)
   e. 14 cm by 15 cm (58 cm)
   f. 10 in. by 13 in. (46 in.)

4. *Write each expression, using symbols.* (Lesson 85)
   a. line segment OP ($\overline{OP}$)
   b. line AB is parallel to line CD
      ($\overleftrightarrow{AB} \parallel \overleftrightarrow{CD}$)
   c. angle FGH ($\angle FGH$)
   d. line UV is perpendicular to line WX
      ($\overleftrightarrow{UV} \perp \overleftrightarrow{WX}$)

5. *Find each percentage.* (Lesson 61)
   a. 40% of 45 (18)
   b. 21% of 85 (17.85)
   c. 92% of 125 (115)
   d. 63% of 63 (39.69)

6. *Find the missing parts. Express any remainder as a fraction.* (Lesson 29)

Distance	Rate	Time
a. 530 mi.	60 m.p.h.	___ hr. ($8\frac{5}{6}$)
b. 471 mi.	___ m.p.h.	5 hr. ($94\frac{1}{5}$)
c. ___ mi.	175 m.p.h.	4 hr. (700)

T–309   Chapter 7   Lines and Planes in Geometry

Introduction

Ask the students what we call the perimeter of a circle. It is usually known as the circumference.

Ask the students for suggestions on the best way to find the circumference of a circle. A rigid ruler is not very accurate. A flexible seamstress's tape would do better. Depending on the object, perhaps rolling it on the floor and measuring the distance it takes to make one rotation would be most accurate.

Now measure the diameter and circumference of a cylindrical or spherical object. Write a ratio by dividing the circumference by the diameter, and round the result to the nearest hundredth. Your answer should be close to 3.14.

Teaching Guide

1. **A circle is a closed curve on which all points are the same distance from the center.** Review this definition from Lesson 85. Also review the terms relating to a circle: diameter, radius, circumference, arc, semicircle, sector, and chord.

2. **The ratio of the circumference of a circle to its diameter is known as pi.** This ratio is indicated by the Greek letter π and is equal to about 3.14 or $3\frac{1}{7}$. Students are to memorize these values of pi.

3. **To find the circumference of a circle, multiply pi times the diameter. The formula is $c = \pi d$.** Use $3\frac{1}{7}$ for pi when the diameter or radius is given in fraction form or when it is a multiple of 7. Otherwise, use 3.14 as the value of pi.

   Have students use the formula to find several circumferences. Use both values of pi for at least one set to show how nearly alike the results are. See *d–g* below.

   Use 3.14 as the value of pi.
   a. $d = 8$ in.   (25.12 in.)
   b. $d = 12$ ft. (37.68 ft.)
   c. $d = 28$ cm   (87.92 cm)

   Use values of pi as indicated.
   d. $\pi = 3\frac{1}{7}$ $d = 21$ ft.   (66 ft.)
   e. $\pi = 3.14$ $d = 21$ ft.   (65.94 ft.)
   f. $\pi = 3\frac{1}{7}$ $d = 14$ in.   (44 in.)
   g. $\pi = 3.14$ $d = 14$ in.   (43.96 in.)

4. **To find the circumference when the radius is known, double the radius to find the diameter before multiplying by pi. This formula is $c = 2\pi r$.** Alert the students to carefully observe whether the diameter or the radius is given.

   a. $r = 5$ in.   (31.4 in.)
   b. $r = 7.8$ cm (48.984 cm)

For most purposes, $3\frac{1}{7}$ ($\frac{22}{7}$) or 3.14 is a satisfactory value of pi. Use 3.1416 when you need a more precise answer. **In this course, use $3\frac{1}{7}$ for pi when working with fractions or with multiples of 7. Use 3.14 for other problems unless you are directed otherwise.**

> **Memorize these values of pi.**
> $\pi = 3.14$ or $3\frac{1}{7}$

Example A

$d = 4.5$ in.; $\pi = 3.14$; $c = $ ___

$c = \pi d$

$c = 3.14 \times 4.5$

$c = 14.13$ in.

Example B

$d = 56$ in.; $\pi = 3\frac{1}{7}$; $c = $ ___

$c = \pi d$

$c = 3\frac{1}{7} \times 56$

$c = \frac{22}{7} \times \frac{\overset{8}{56}}{1} = \frac{176}{1}$

$c = 176$ in.

Example C

$r = 10.5$ cm; $\pi = 3.14$; $c = $ ___

$c = 2\pi r$

$c = 2 \times 3.14 \times 10.5$

$c = 65.94$ cm

Example D

$r = 4\frac{3}{8}$ in.; $\pi = 3\frac{1}{7}$; $c = $ ___

$c = 2\pi r$

$c = 2 \times 3\frac{1}{7} \times 4\frac{3}{8}$

$c = \frac{\overset{1}{2}}{1} \times \frac{\overset{11}{22}}{7} \times \frac{\overset{5}{35}}{8} = \frac{55}{2}$

$c = 27\frac{1}{2}$ in.

When the circumference of a circle is known, its diameter can be found by dividing the circumference by pi. Use the following formula.

> diameter = circumference ÷ π, or $d = \frac{c}{\pi}$

Example E

$c = 11$ in.; $\pi = 3\frac{1}{7}$; $d = $ ___

$d = \frac{c}{\pi}$

$d = 11 \div 3\frac{1}{7}$

$d = \frac{\overset{1}{11}}{1} \times \frac{7}{\underset{2}{22}} = \frac{7}{2}$

$d = 3\frac{1}{2}$ in.

Example F

$c = 15$ in.; $\pi = 3.14$;

$d = $ ___ to the nearest hundredth

$d = \frac{c}{\pi}$

$d = 15 \div 3.14$

$d = 4.78$ in. (rounded)

310   Chapter 7   Lines and Planes in Geometry

CLASS PRACTICE

Give the formulas for finding these facts. Be sure you know them by memory.

a. Circumference of a circle when the radius is known $c = 2\pi r$
b. Circumference of a circle when the diameter is known $c = \pi d$

Find the circumference of each circle, using $3\frac{1}{7}$ for pi.

c. $d = 35$ in.   110 in.   d. $d = 11\frac{1}{2}$ in. $36\frac{1}{7}$ in.   e. $d = 14$ cm   44 cm

Find the circumference of each circle, using 3.14 for pi.

f. $d = 2.5$ m   7.85 m   g. $d = 9$ ft.   28.26 ft.   h. $d = 5.8$ cm   18.212 cm
i. $r = 1.3$ m   8.164 m   j. $r = 7$ in.   43.96 in.   k. $r = 12\frac{1}{2}$ ft.   78.5 ft.

Find the diameter of each circle, using $3\frac{1}{7}$ for pi.

l. $c = 44$ in.   14 in.   m. $c = 38\frac{1}{2}$ in. $12\frac{1}{4}$ in.
n. $c = 55$ in. $17\frac{1}{2}$ in.   o. $c = 27\frac{1}{2}$ in. $8\frac{3}{4}$ in.

WRITTEN EXERCISES

A. *Write the correct words to complete these sentences.*

1. The distance around the outside of a circle is its ___.   circumference
2. The distance through the center of a circle, from edge to edge, is its ___.   diameter
3. The ratio of the circumference of a circle to its diameter is known as ___.   pi
4. Any part of the edge of a circle is called a(n) ___.   arc
5. One-half of a circle is called a(n) ___.   semicircle
6. The distance from the center of a circle to the outside edge is its ___.   radius
7. The approximate value of pi in fraction form is ___. $3\frac{1}{7}$
8. The approximate value of pi in decimal form is ___ to the nearest hundredth.   3.14

B. *Write the formulas for finding these facts. Be sure you know them by memory.*

9. Circumference of a circle when the radius is known $c = 2\pi r$
10. Circumference of a circle when the diameter is known $c = \pi d$

C. *Find the circumference of each circle, using $3\frac{1}{7}$ for pi.*

11. $d = 7$ in.   22 in.   12. $d = 14$ in.   44 in.   13. $d = 21$ cm   66 cm
14. $d = 28$ cm   88 cm   15. $d = 10\frac{1}{2}$ in.   33 in.   16. $d = 17\frac{1}{2}$ yd.   55 yd.

D. *Find the circumference of each circle, using 3.14 for pi.*

17. $d = 5$ m   15.7 m   18. $d = 6.4$ cm   20.096 cm   19. $d = 3.8$ cm   11.932 cm
20. $d = 14.75$ ft.   46.315 ft.   21. $r = 6$ in.   37.68 in.   22. $r = 1.5$ m   9.42 m

E. *Find the diameter of each circle, using $3\frac{1}{7}$ for pi.*

23. $c = 16\frac{1}{2}$ in. $5\frac{1}{4}$ in.   24. $c = 33$ in. $10\frac{1}{2}$ in.

5. **To find the diameter when the circumference is known, divide the circumference by pi. The formula is $d = c \div \pi$.**

   Use $3\frac{1}{7}$ as the value of pi.
   a. $c = 33$ in. ($10\frac{1}{2}$ in.)
   b. $c = 16$ ft. ($5\frac{1}{11}$ ft.)

Further Study

The ratio of the circumference to the diameter of a circle is represented by the Greek letter π. This letter stands for the Greek word *periphereia*, which is the source of the English word *periphery* (a synonym of *circumference*).

Saying that π is equal to 3.14, 3.1416, or $3\frac{1}{7}$ is accurate enough for many calculations. However, all these values are approximate. The exact value of π has long intrigued mathematicians, and modern computers have calculated π to more than 100 million decimal places. But no exact value has ever been found, neither is there any pattern of repeating digits. Therefore, π is an irrational number; it cannot be stated with any numbers in the form $a:b$.

In 1 Kings 7:23, the value of π is considered as 3. The Greek mathematician Archimedes correctly stated that the value is between $3\frac{10}{70}$ and $3\frac{10}{71}$. For most practical purposes, however, the value 3.14 or $3\frac{1}{7}$ will serve the students well throughout life. The degree of accuracy required for π varies according to the degree of accuracy required in the answer. A highly precise value of π is obviously more critical for putting a satellite into orbit than for finding the circumference of a tire.

T–311   Chapter 7   Lines and Planes in Geometry

Solutions for Part F

25. pad diameter = $\frac{3}{4}$ft. + 36ft. + $\frac{3}{4}$ft.
    $3\frac{1}{7} \times 37\frac{1}{2} \div 3 = 39\frac{2}{7}$

26. $\frac{1}{2} \times 3.14 \times 15 \times 2$

27. $18 \div 3\frac{1}{7}$

28. $7{,}500 \div (3\frac{1}{7} \times 4 \times 2)$

29. 0.0625×112

30. $0.065 \times \$339.89$

F. Solve these reading problems.

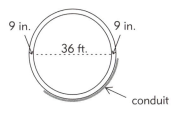

conduit

25. Ray Dhiel poured a concrete pad for a circular grain bin 36 feet in diameter. The pad extended 9 inches outside the circumference of the bin. How many feet of electrical conduit will he need to bury along the edge of the pad to reach a fan $\frac{1}{3}$ of the way around the bin? $39\frac{2}{7}$ feet

26. A dog is tied to a porch post with a 15-foot chain. What is the length of the semicircle at the outer edge of his living space?   47.1 feet

27. According to 1 Kings 7:15, two brass pillars were cast for the temple. The compass (circumference) of the pillars was 12 cubits. If the circumference was 18 feet, what was the diameter of the pillars? (Use $3\frac{1}{7}$ for pi.) $5\frac{8}{11}$ feet

28. Brother Daniel buys 7,500-foot rolls of net wrap for his baler, which makes round bales 4 feet in diameter. If each bale is wrapped twice, how many round bales can he figure per roll of net wrap? (Round to the nearest 10 bales.)   300 bales

29. On the Statue of Liberty, the distance from head to toe is 112 feet. What is the distance from head to toe on the model that is $6\frac{1}{4}\%$ of the actual size?   7 feet

30. Leonard bought a box spring and mattress set for $339.89. How much sales tax did he need to pay if the rate was $6\frac{1}{2}\%$? $22.09

REVIEW EXERCISES

G. Write the formulas for finding these facts. *(Lesson 90)*

31. Perimeter of a square $p = 4s$
32. Perimeter of a rectangle $p = 2(l + w)$

H. Find the perimeters of squares and rectangles having these dimensions. *(Lesson 90)*

33. $s = 11.5$ cm   46 cm
34. $l = 5.5$ ft.; $w = 4$ ft.   19 ft.

I. Write the symbols for these expressions. *(Lesson 85)*

35. line segment YZ $\overline{YZ}$
36. line NO is perpendicular to line PQ $\overleftrightarrow{NO} \perp \overleftrightarrow{PQ}$

J. Find these percentages. *(Lesson 61)*

37. 31% of 52   16.12
38. 35% of 184   64.4

K. Find the missing parts. *(Lesson 29)*

	Distance	Rate	Time
39.	330 mi.	55 m.p.h.	6 hr.
40.	440 mi.	80 m.p.h.	$5\frac{1}{2}$ hr.

312 Chapter 7 *Lines and Planes in Geometry*

92. Areas of Squares and Rectangles

Area is the amount of surface covered by an object. A sheet of regular typing paper, for example, covers an area of over 90 square inches. Both the circle and the square below have an area of one square inch.

To calculate the area of a rectangle, multiply its length by its width.

Memorize this formula for the area of a rectangle.
area = length times width, or $a = lw$

A square is a rectangle whose length and width are equal. To find the area of a square, multiply the length of one side by itself: *area = side × side*. For a 4-inch square, multiply 4 × 4 to find the area. This can be stated as $4^2 = 16$ square inches.

Memorize this formula for the area of a square.
area = side × side, or $a = s^2$

Example A
Find the area of a rectangle with these dimensions.
$l = 3$ ft. 6 in.; $w = 2$ ft. 4 in.; $a =$ ___
$a = lw$
$a = 3\frac{1}{2} \times 2\frac{1}{3} = 8\frac{1}{6}$ sq. ft.

Example B
Find the area of this square.
$s = 14$ cm; $a =$ ___
$a = s^2$
$a = 14 \times 14 = 196$ cm^2

Remember that 1 square foot = 144 square inches and 1 square yard = 9 square feet. Also, be sure to include the word *square* when you label units of area, since area has two dimensions. In metric abbreviations, the exponent 2 is used to express *square*. No symbol for *square* is used with *acre* and *hectare*.

LESSON 92

Objectives

- To review finding the area of a rectangle by using the formula $a = lw$.
- To review the use of exponents in metric abbreviations of area.
- To review finding the area of a square by using the formula $a = s^2$.

Review

1. *Give the formula for finding each fact.* (Lessons 90, 91)

    a. Circumference of a circle
    $\qquad (c = \pi d \text{ or } c = 2\pi r)$

    b. Perimeter of a square $(p = 4s)$

    c. Perimeter of a rectangle
    $\qquad (p = 2[l + w])$

2. *Find the circumference of each circle.* (Lesson 91)

    a. $d = 5$ ft. (15.7 ft.)

    b. $d = 15$ in. (47.1 in.)

    c. $r = 4$ in. (25.12 in.)

    d. $r = 6.2$ cm (38.936 cm)

3. *Find the perimeter of each figure.* (Lesson 90)

    Squares

    a. $s = 9$ ft. (36 ft.)

    b. $s = 2.1$ m (8.4 m)

    Rectangles

    c. $l = 11$ in.; $w = 10$ in. (42 in.)

    d. $l = 5.2$ cm; $w = 4$ cm (18.4 cm)

4. *Identify each angle as acute, obtuse, straight, or reflex.* (Lesson 86)

    a. 180° (straight)

    b. 225° (reflex)

    c. 15° (acute)

    d. 105° (obtuse)

5. *Identify each pair of angles as complementary or supplementary.* (Lesson 86)

    a. 25° and 155° (supplementary)

    b. 78° and 12° (complementary)

6. *Find the amount of each change, to the nearest cent. Also find the new amount.* (Lesson 63)

    a. $14.56 increased by 4%
    $\qquad$ ($0.58; $15.14)

    b. $245.65 decreased by 84%
    $\qquad$ ($206.35; $39.30)

7. *Find these equivalents.* (Lesson 30)

    a. 60 talents = ___ lb. (4,500)

    b. 14 homers = ___ gal. (1,120)

T–313   *Chapter 7   Lines and Planes in Geometry*

Introduction

Ask the students how many yards of carpet it would take for the floor of a room measuring 5 yards by 4 yards. If the answer is given as 20 yards, ask them if a 20-yard piece of thread will do! How about a 20-yard piece of carpet 1 inch wide?

Perhaps your students answered the question correctly. The amount of carpet needed to cover a floor measuring 5 yards by 4 yards cannot be found in yards because the yard is a linear unit with only one dimension. A unit of area is needed; in this case, the square yard. Twenty square yards of carpet would cover the floor.

Teaching Guide

1. **Area is the amount of surface covered by a geometric figure.** Two dimensions are indicated in area: length and width.

2. **To find the area of a rectangle, multiply the length by the width. The formula is $a = lw$.** Students are to memorize this formula.

   a. $l = 7$ ft. $w = 6$ ft.   (42 sq. ft.)
   b. $l = 3$ ft. 9 in. $w = 7$ in.   ($2\frac{3}{16}$ sq. ft.)
   c. $l = 12$ cm $w = 11$ cm   (132 cm²)
   d. $l = 15$ km $w = 5$ km   (75 km²)

3. **To find the area of a square, multiply the length of one side by itself. The formula is $a = s^2$.** Students are to memorize this formula.

   a. $s = 10$ in.   (100 sq. in.)
   b. $s = 12$ ft.   (144 sq. ft.)
   c. $s = 15$ cm   (225 cm²)
   d. $s = 19$ m   (361 m²)

4. **The word *square* must be included in labels for units of area, since area has two dimensions.** In metric abbreviations, the exponent 2 is used to express *square*. No symbol for *square* is used with *acre* and *hectare*.

Solutions for Part D

21. $42 \times 19\frac{1}{2} \times 2$
22. $10\frac{1}{2} \times 10\frac{1}{2} \times \frac{1}{4}$
23. $52 \times 34\frac{1}{2}$

Lesson 92

CLASS PRACTICE

Give the formulas for finding these areas. Be sure you know them by memory.

a. Area of a square $a = s^2$
b. Area of a rectangle $a = lw$

Find the areas of rectangles with these dimensions.

c. l = 6 ft. 9 in.
 w = 4 ft.   27 sq. ft.
d. l = 9 in.
 w = 7 in.   63 sq. in.
e. l = 12 cm
 w = 15 cm   180 cm²
f. l = 1.2 m
 w = 4.1 m   4.92 m²

Find the areas of these squares.

g. s = 2 ft. 8 in. $7\frac{1}{9}$ sq. ft.
h. $s = 9\frac{1}{4}$ in. $85\frac{9}{16}$ sq. in.
i. s = 17 cm   289 cm²
j. s = 5.2 m   27.04 m²

WRITTEN EXERCISES

A. *Write the formulas for finding these areas. Be sure you know them by memory.*

1. Area of a square $a = s^2$
2. Area of a rectangle $a = lw$

B. *Find the areas of rectangles with these dimensions.*

3. l = 5 ft.
 w = 3 ft.   15 sq. ft.
4. l = 13 cm
 w = 11 cm   143 cm²
5. l = 16 m
 w = 13 m   208 m²
6. l = 5.4 m
 w = 3.75 m   20.25 m²
7. l = 7.2 km
 w = 5.6 km   40.32 km²
8. l = 8.2 m
 w = 6.8 m   55.76 m²
9. l = 7 ft. 4 in.
 w = 4 ft. 6 in.   33 sq. ft.
10. l = 18 ft.
 w = 6 ft. 8 in.   120 sq. ft.
11. l = 31 ft.
 w = 24 ft.   744 sq. ft.

C. *Find the areas of these squares.*

12. s = 3 in.   9 sq. in.
13. s = 12 cm   144 cm²
14. s = 16 m   256 m²
15. s = 12.5 m   156.25 m²
16. s = 5.5 mi.   30.25 sq. mi.
17. s = 6.8 m   46.24 m²
18. s = 3 ft. 9 in. $14\frac{1}{16}$ sq. ft.
19. s = 8 yd. 1 ft. $69\frac{4}{9}$ sq. yd.
20. s = 18 in.   324 sq. in.

D. *Solve these reading problems.*

21. A house has a roof that measures 42 feet by $19\frac{1}{2}$ feet on each side of the peak. How many square feet of plywood will it take to cover the roof?   1,638 square feet

22. The block for playing foursquare at Oak Grove Christian School measures $10\frac{1}{2}$ feet on each side. How many square feet are in each of the 4 smaller squares? $27\frac{9}{16}$ square feet

23. The play area in the basement at Center Point Mennonite School is 52 feet long and $34\frac{1}{2}$ feet wide. What is its area?   1,794 square feet

314 Chapter 7 Lines and Planes in Geometry

24. The inner covering of the tabernacle consisted of 10 curtains of fine-twined linen, each measuring 28 cubits by 4 cubits (Exodus 26:1, 2). What was the total area of the 10 curtains? See if you can solve this problem mentally.   1,120 square cubits

25. According to Esther 3:9, Haman promised to pay 10,000 talents of silver to those in charge of destroying the Jews. If silver is worth $560 per pound today, how many dollars' worth of silver did Haman promise? $420,000,000

26. Each year Solomon gave Hiram 20,000 measures (homers) of wheat (1 Kings 5:11). If this continued for the 7 years of building the temple, how many bushels of wheat did Solomon send to Hiram?   1,120,000 bushels

REVIEW EXERCISES

E. Write these formulas. *(Lessons 90, 91)*

27. Circumference of a circle $c = \pi d$ or $c = 2\pi r$
28. Perimeter of a square $p = 4s$
29. Perimeter of a rectangle $p = 2(l + w)$

F. Find the circumference of each circle. *(Lesson 91)*

30. $d = 15$ cm   47.1 cm
31. $d = 8.5$ cm   26.69 cm
32. $r = 3\frac{1}{2}$ ft.   22 ft.

G. Find the perimeter of each square. *(Lesson 90)*

33. $s = 50$ ft.   200 ft.
34. $s = 11\frac{1}{2}$ in.   46 in.

H. Identify each pair of angles as *complementary* **or** *supplementary.* *(Lesson 86)*

35. 48° and 132°   supplementary
36. 27° and 63°   complementary

I. Identify each angle as *acute, obtuse, straight,* **or** *reflex.* *(Lesson 86)*

37. 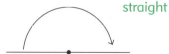   straight
38.   acute

J. Find the amount of each change, to the nearest cent. Also find the new amount. *(Lesson 63)*

39. $1,300 increased by 36% $468 $1,768
40. $184.62 decreased by 68% $125.54 $59.08

K. Find these equivalents. *(Lesson 30)*

41. 1,500 shekels = 600 oz.
42. 34 cubits = 1,564 cm

24. 28 × 4 × 10
25. 10,000 × 75 × 560
26. 7 × 20,000 × 8

T–315   *Chapter 7   Lines and Planes in Geometry*

LESSON 93

Objectives

- To review finding the area of a parallelogram by using the formula $a = bh$.
- To review finding the area of a triangle by using the formula $a = \frac{1}{2}bh$.

Review

1. Give Lesson 93 Quiz (Perimeter and Circumference).

2. *Give the formula for finding each fact.* (Lessons 90–92)
   a. Area of a square   ($a = s^2$)
   b. Area of a rectangle   ($a = lw$)
   c. Circumference of a circle
      ($c = \pi d$ or $c = 2\pi r$)
   d. Perimeter of a square   ($p = 4s$)
   e. Perimeter of a rectangle
      ($p = 2[l + w]$)

3. *Find the area of each square or rectangle.* (Lesson 92)
   a. $s = 13$ cm   (169 cm²)
   b. $l = 9$ in.; $w = 6$ in.   (54 sq. in.)

4. *Find the circumference of each circle.* (Lesson 91)
   a. $d = 5$ mi.   (15.7 mi.)
   b. $r = 14$ cm   (88 cm)

5. *Tell whether each triangle is equilateral, isosceles, or scalene.* (Lesson 87)
   a. 3 ft., 4 ft., 4 ft.   (isosceles)
   b. $2\frac{1}{2}$ in., $2\frac{3}{4}$ in., $2\frac{5}{8}$ in.   (scalene)

6. *Find these percentages.* (Lesson 62)
   a. 160% of 18   (28.8)
   b. $\frac{4}{5}$% of 200   (1.6)

7. *Choose the facts needed to solve each problem; then find the answer.* (Lesson 31)
   a. A 134-ton blue whale has about 8 tons of blood, which contains about 8 trillion red blood cells. How many gallons of blood is that if 1 gallon equals 8.6 pounds? Round your answer to the nearest thousandth.

   (8 tons; 8.6 lb. per gal.; 1,860.465 gal.)

   b. Of the blue whale's 134 tons, a useful oil accounts for about 20% of the weight. Its total amount is about 162 barrels of 42 gallons per barrel. What is this amount in gallons?

   (162 barrels; 42 gal. per barrel; 6,804 gal.)

93. Areas of Parallelograms and Triangles

A parallelogram is a quadrilateral with parallel sides, but its corners are usually not 90° angles. Study the parallelogram at the right. Notice that if the triangular part on the left end is moved to the other end, the figure becomes a rectangle measuring 1 inch by 3 inches.

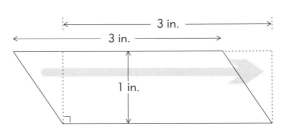

To find the area of a parallelogram, multiply the length of the base times the height. Note carefully that the height is not the length of the slanted side. Rather, it is the perpendicular distance between the two bases.

Memorize this formula for the area of a parallelogram.
area = base × height, or $a = bh$

Example A: Parallelogram
$b = 3$ in.; $h = 2$ in.; $a = $ ___
$a = bh$
$a = 3 \times 2 = 6$ sq. in.

Example B: Parallelogram
$b = 11$ cm; $h = 9$ cm; $a = $ ___
$a = bh$
$a = 11 \times 9 = 99$ cm²

Now look at the diagram below. It shows two congruent triangles put together to form a parallelogram. This can be done with any triangle: place it beside another triangle of the same shape and size, and the result is a parallelogram. Thus, a triangle has one-half the area of a parallelogram with the same base and height. So the area of a triangle can be calculated by finding half of bh.

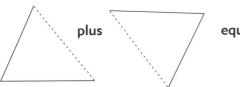

Memorize this formula for the area of a triangle.
area = ½ × base × height, or $a = \frac{1}{2}bh$

Example C
Find the area of a triangle with these dimensions.
$b = 7$ ft. 4 in.; $h = 7$ ft.; $a = $ ___
$a = \frac{1}{2}bh$
$a = \frac{1}{2} \times 7\frac{1}{3} \times 7 = 25\frac{2}{3}$ sq. ft.

Example D
Find the area of a triangle with these dimensions.
$b = 25$ cm; $h = 16$ cm; $a = $ ___
$a = \frac{1}{2}bh$
$a = \frac{1}{2} \times 25 \times 16 = 200$ cm²

316   Chapter 7   *Lines and Planes in Geometry*

CLASS PRACTICE

Give the formulas for finding these areas. Be sure you know them by memory.

a. Area of a parallelogram $a = bh$

b. Area of a triangle $a = \frac{1}{2}bh$

Find the areas of parallelograms with these dimensions.

c. $b = 5$ in.   d. $b = 5$ ft. 6 in.   e. $b = 10$ cm   f. $b = 8.5$ m
   $h = 3$ in. 15 sq. in. $h = 2$ ft. 9 in. $h = 12$ cm 120 cm² $h = 6.2$ m 52.7 m²
                        $15\frac{1}{8}$ sq. ft.

Find the area of parallelogram g in square inches and h in square centimeters.

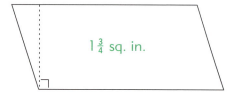

g. $1\frac{3}{4}$ sq. in.   h. 12.5 cm²

Find the areas of triangles with these dimensions.

i. $b = 8$ ft.   j. $b = 2$ ft. 4 in.   k. $b = 16$ cm   l. $b = 6.8$ m
   $h = 4$ ft. 16 sq. ft. $h = 3$ ft. $3\frac{1}{2}$ sq. ft. $h = 13$ cm 104 cm² $h = 3.1$ m 10.54 m²

Find the area of triangle m in square inches and n in square centimeters.

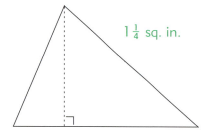 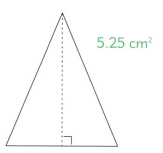

m. $1\frac{1}{4}$ sq. in.   n. 5.25 cm²

WRITTEN EXERCISES

A. *Write the formulas for finding these areas. Be sure you know them by memory.*

1. Area of a parallelogram $a = bh$
2. Area of a triangle $a = \frac{1}{2}bh$

B. *Find the areas of parallelograms with these dimensions.*

3. $b = 6$ ft.   4. $b = 16$ cm   5. $b = 19$ m
   $h = 4$ ft. 24 sq. ft. $h = 12$ cm 192 cm² $h = 15$ m 285 m²
6. $b = 13$ ft. 6 in.   7. $b = 4$ yd. 2 ft.   8. $b = 7.2$ m
   $h = 7$ ft. 4 in. 99 sq. ft. $h = 1$ yd. 2 ft. $7\frac{7}{9}$ sq. yd. $h = 5.8$ m 41.76 m²

Introduction

Draw a parallelogram on the board and ask the students to explain in their own words what a parallelogram is. Some of the following explanations may be given.

A parallelogram looks like a rectangle that was carrying too much weight.

A parallelogram has sides of equal length and bases of equal length.

The opposite sides of a parallelogram are parallel lines.

Say, "Suppose we were to cut off one end of the parallelogram to make a 90° angle at that end, and we moved that piece to the other end to form a 90° angle there. What shape would we have?" As shown by the diagram in the lesson, the result would be a rectangle with a length equal to the base of the parallelogram and a width equal to the height of a parallelogram.

Teaching Guide

1. **A parallelogram is a quadrilateral with parallel sides, but its corners are usually not 90° angles.** Every rectangle is also a parallelogram, but not every parallelogram is a rectangle. Point out that since the opposite sides are parallel, they are of equal length.

2. **To find the area of a parallelogram, multiply the length of the base times the height. The formula is $a = bh$.** Emphasize that the height is not the length of the slanted side. It is rather the perpendicular distance between the two bases.

   a. $b = 7$ in. $h = 6$ in.   (42 sq. in.)
   b. $b = 12$ ft. $h = 7$ ft.   (84 sq. ft.)
   c. $b = 15$ cm $h = 13$ cm   (195 cm^2)
   d. $b = 18$ m $h = 14$ m   (252 m^2)

3. **To find the area of a triangle, multiply $\frac{1}{2}$ times the base times the height. The formula is $a = \frac{1}{2}bh$.** This formula is based on the fact that a triangle has one-half the area of a parallelogram with the same base and height.

   a. $b = 4$ ft. $h = 2$ ft.   (4 sq. ft.)
   b. $b = 8$ in. $h = 6$ in.   (24 sq. in.)
   c. $b = 16$ cm $h = 12$ cm   (96 cm^2)
   d. $b = 24$ m $h = 15$ m   (180 m^2)

T–317 *Chapter 7 Lines and Planes in Geometry*

Solutions for Part F

19. 7×12
20. $\frac{1}{2} \times 8\frac{1}{4} \times 5$

C. Find the area of parallelogram 9 in square inches and 10 in square centimeters.

9. $3\frac{7}{16}$ sq. in.

D. Find the areas of triangles with these dimensions.

11. b = 9 ft.
    h = 6 ft.   27 sq. ft.
12. b = 14 cm
    h = 14 cm   98 cm²
13. b = 18 m
    h = 15 m   135 m²
14. b = 8 yd.
    h = 8 yd. 9 in.   33 sq. yd.
15. b = 48 ft.
    h = 5 ft. 5 in.   130 sq. ft.
16. b = 9.6 m
    h = 4.5 m   21.6 m²

10. 15 cm²

E. Find the area of triangle 17 in square inches and 18 in square centimeters.

17. $1\frac{3}{4}$ sq. in.

18. 18 cm²

F. Solve these reading problems.

19. The Ebys are putting wallpaper on the parallelogram-shaped wall along an open stairway. The wall is 12 feet wide, and the distance between the steps and the ceiling is 7 feet. How many square feet will they cover with wallpaper?   84 sq. ft.

20. Below the stairs on the open side, the Ebys will put a triangle of paneling. The height is 5 feet, and the width is 8 feet 3 inches. How many square feet of paneling will they use? $20\frac{5}{8}$ sq. ft.

318   Chapter 7   *Lines and Planes in Geometry*

21. A building committee plans to use siding for the gable ends of a new school, and brick for the rest of the exterior. The end wall is 48 feet wide, and the peak is 8 feet above the brick. How many square feet of siding in all will they need for the two ends?   **384 sq. ft.**

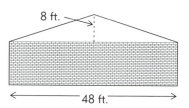

22. Dorcas is cutting triangular patches of different sizes to make a quilt. For 80 of the triangles, the base is 6 inches and the height is 3 inches. What is the area of these 80 triangles in square feet?   **5 sq. ft.**

23. Moses was instructed to use a double piece of linen to make a breastplate for the high priest's robe (Exodus 28:16). The piece was 1 span wide and 2 spans long. If 1 span is 9 inches and 1 cubit is 18 inches, how many square inches were in the linen cloth? Write the facts needed for solution, then solve.   **Facts: 1 span, 2 spans, 9 inches**
   Answer: 162 square inches

24. Deuteronomy 3:11 states that the bed of Og, king of Bashan, was 9 cubits long and 4 cubits wide. These dimensions were $13\frac{1}{2}$ feet by 6 feet in English units and and 4.14 meters by 1.84 meters in metric units. What was the area of the bed in square feet? Write the facts needed for solution, then solve.   **Facts: $13\frac{1}{2}$ feet by 6 feet**
   Answer: 81 square feet

REVIEW EXERCISES

G. *Write the formula for each fact.* (Lessons 90–92)

25. Area of a square $a = s^2$
26. Area of a rectangle $a = lw$
27. Perimeter of a square $p = 4s$
28. Perimeter of a rectangle $p = 2(l + w)$
29. Circumference of a circle $c = \pi d$ or $c = 2\pi r$

H. *Find the area of each square or rectangle.* (Lesson 92)

30. $s = 4.6$ m   **21.16 m²**   31. $s = 19$ in.   **361 sq. in.**   32. $l = 6$ ft.; $w = 3$ ft.   **18 sq. ft.**

I. *Find the circumference of each circle.* (Lesson 91)

33. $d = 12$ m   **37.68 m**   34. $r = 6\frac{1}{2}$ in.   **$40\frac{6}{7}$ in.**

J. *Identify each pair of triangles as similar or congruent, using the symbols you have learned. Also tell whether they are* equilateral, isosceles, *or* scalene. (Lesson 87)

35. 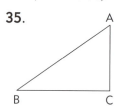   △ABC ≅ △DEF
   scalene

36. 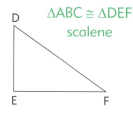   △GHI ~ △JKL
   equilateral

K. *Solve these percent problems.* (Lesson 62)

37. 115% of 42   **48.3**   38. 255% of $518   **$1,320.90**
39. $\frac{1}{2}$% of $3,000   **$15**   40. $\frac{5}{8}$% of 840   **5.25**

21. ($\frac{1}{2}$ × 48 × 8) × 2
22. ($\frac{1}{2}$ × $\frac{1}{2}$ ft. × $\frac{1}{4}$ ft.) × 80
23. 9 × 18
24. 13$\frac{1}{2}$ × 6

T-319 Chapter 7 *Lines and Planes in Geometry*

LESSON 94

Objectives

- To review the shape of a trapezoid.
- To review finding the area of a trapezoid by using the formula $a = \frac{1}{2}h(b_1 + b_2)$.

Review

1. *Give the formula for finding each fact.* (Lessons 90–93)
   a. Area of a parallelogram $(a = bh)$
   b. Area of a triangle $(a = \frac{1}{2}bh)$
   c. Area of a square $(a = s^2)$
   d. Area of a rectangle $(a = lw)$
   e. Circumference of a circle
      $(c = \pi d \text{ or } c = 2\pi r)$
   f. Perimeter of a square $(p = 4s)$
   g. Perimeter of a rectangle
      $(p = 2[l + w])$

2. *Find the area of each figure.* (Lessons 92, 93)

   Square and rectangle
   a. $s = 21$ cm (441 cm²)
   b. $l = 8\frac{3}{4}$ in.; $w = 4\frac{1}{2}$ in. ($39\frac{3}{8}$ sq. in.)

   Parallelograms
   c. $b = 14$ cm; $h = 11$ cm (154 cm²)
   d. $b = 5\frac{1}{2}$ ft.; $h = 2\frac{3}{4}$ ft. ($15\frac{1}{8}$ sq. ft.)

   Triangles
   e. $b = 10$ cm; $h = 9$ cm (45 cm²)
   f. $b = 6\frac{1}{3}$ ft.; $h = 2\frac{1}{2}$ ft. ($7\frac{11}{12}$ sq. in.)

3. *Use each set of facts to construct a triangle. Label it with the facts given.* (Lesson 88)
   a. 8 cm, 65°, 9 cm
   b. 1 in., 115°, 2 in.
   c. 2 in., 90°, 2 in.

4. *Find the new amount after each change.* (Lesson 64)
   a. $25.20 increased by 45% ($36.42)
   b. $875 decreased by 18% ($717.50)

94. Areas of Trapezoids

A **trapezoid** (trăp′ ə zoid) is a quadrilateral that has one set of parallel sides, called the bases. The diagram below shows how this odd-shaped figure can be rearranged to form a parallelogram. If a trapezoid is cut exactly halfway between the bases, and parallel to them, the upper part can be inverted and placed against the lower part to form a perfect parallelogram. This parallelogram has the same area as the original trapezoid. What is the base of the parallelogram? What is its height?

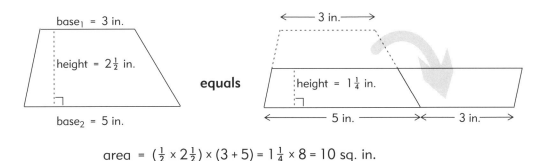

$$\text{area} = (\tfrac{1}{2} \times 2\tfrac{1}{2}) \times (3 + 5) = 1\tfrac{1}{4} \times 8 = 10 \text{ sq. in.}$$

Memorize this formula for the area of a trapezoid.
area = $\tfrac{1}{2}$ × height × (base$_1$ + base$_2$), or $a = \tfrac{1}{2}h(b_1 + b_2)$

Example A	**Example B**
Find the area of this trapezoid.	Find the area of this trapezoid.
h = 7 yd. 1 ft.	h = 25 cm
b_1 = 2 ft.	b_1 = 14 cm
b_2 = 10 yd. 1 ft.	b_2 = 22 cm
$a = \tfrac{1}{2}h(b_1 + b_2)$	$a = \tfrac{1}{2}h(b_1 + b_2)$
$a = \tfrac{1}{2} \times 7\tfrac{1}{3} \times (\tfrac{2}{3} + 10\tfrac{1}{3})$	$a = \tfrac{1}{2} \times 25 \times (14 + 22)$
$a = \tfrac{1}{2} \times 7\tfrac{1}{3} \times 11 = 40\tfrac{1}{3}$ sq. yd.	$a = (\tfrac{1}{2} \times 36) \times 25 = 450 \text{ cm}^2$

In the last step of Example B, the rearranging and grouping of factors is permitted by the commutative and associative laws of multiplication.

320   Chapter 7   Lines and Planes in Geometry

CLASS PRACTICE

Give the formula for finding this area. Be sure you know it by memory.

a. Area of a trapezoid $a = \frac{1}{2}h(b_1 + b_2)$

Find the areas of trapezoids with these dimensions.

b. h = 4 ft. 10 in.
b_1 = 4 ft. 6 in.
b_2 = 9 in. $12\frac{11}{16}$ sq. ft.

c. h = 10 ft.
b_1 = 12 ft.
b_2 = 8 ft.   100 sq. ft.

d. h = 7 cm
b_1 = 14 cm
b_2 = 11 cm   87.5 cm²

e. h = 12.1 m
b_1 = 18.5 m
b_2 = 16 m   208.725 m²

Find the area of trapezoid f in square inches and g in square centimeters.

f.

$2\frac{1}{8}$ sq. in.

g.

15.75 cm²

WRITTEN EXERCISES

A. Write the formula for finding this area. Be sure you know it by memory.

1. Area of a trapezoid $a = \frac{1}{2}h(b_1 + b_2)$

B. Find the areas of trapezoids with these dimensions.

2. h = 4 in.
b_1 = 3 in.
b_2 = 5 in.   16 sq. in.

3. h = 14 cm
b_1 = 6 cm
b_2 = 20 cm   182 cm²

4. h = 6 m
b_1 = 10 m
b_2 = 4 m   42 m²

5. h = 12 ft. 8 in.
b_1 = 8 ft.
b_2 = 10 ft.   114 sq. ft.

6. h = 15 ft.
b_1 = 8 ft. 8 in.
b_2 = 12 ft.   155 sq. ft.

7. h = 11 m
b_1 = 9 m
b_2 = 17 m   143 m²

8. h = 28 cm
b_1 = 16 cm
b_2 = 19 cm   490 cm²

9. h = 32 ft.
b_1 = 12 ft.
b_2 = 21 ft.   528 sq. ft.

10. h = 26 in.
b_1 = 12 in.
b_2 = 17 in.   377 sq. in.

Introduction

Draw a trapezoid on the board and ask the class, "Why is this not a parallelogram?" A parallelogram has two sets of parallel sides, whereas a trapezoid has only one set of parallel sides. A four-sided figure with only one set of parallel sides is a trapezoid.

Teaching Guide

1. **A trapezoid is a quadrilateral with only one set of parallel sides.** Draw several trapezoids to help the students recognize them in different positions. One set of parallel sides is the criterion. The parallel sides are always the two bases, and the height is always the perpendicular measure of distance between them.

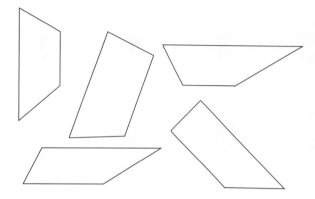

2. **To find the area of a trapezoid, multiply $\frac{1}{2}$ times the height times the sum of the two bases. The formula is $a = \frac{1}{2}h(b_1 + b_2)$**

   a. $h = 8$ cm
      $b_1 = 6$ cm
      $b_2 = 4$ cm (40 cm²)

   b. $h = 12$ cm
      $b_1 = 8$ cm
      $b_2 = 10$ cm (108 cm²)

   c. $h = 30$ ft.
      $b_1 = 14$ ft.
      $b_2 = 21$ ft. (525 sq. ft.)

   d. $h = 28$ in.
      $b_1 = 15$ in.
      $b_2 = 16$ in. (434 sq. in.)

T–321   Chapter 7   *Lines and Planes in Geometry*

Solutions for Part D

13. $\frac{1}{2} \times 14 \times (40 + 32)$
14. $\frac{1}{2} \times 446 \times (418 + 169) = 130{,}901$ sq. ft.
    $130{,}901 \div 43{,}560$
15. $25\% \times \$24.95$
16. $132\% \times \$49.23$
17. $(2 \times 26 \times 38) \div 100$
18. $3 \times 4 \times 603{,}550 \div 43{,}560$

C. Find the area of trapezoid 11 in square inches and 12 in square centimeters.

11.

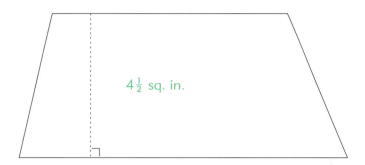

$4\frac{1}{2}$ sq. in.

12.

15 cm²

D. Solve these reading problems.

13. The end of a steel frame implement shed has the shape of a trapezoid. It is 40 feet wide at the base, 32 feet wide at the roof, and 14 feet high. How many square feet of sheet metal will it take to cover one end?   504 square feet

14. The east boundary of a school property is 418 feet. A parallel boundary 446 feet to the west is 169 feet long. The other two boundaries are straight lines from corner to corner. What is the size of the property in acres? (Drop any remainder.)   3 acres

15. On the last day of a clearance sale, Allen's Hardware discounted everything 75%. What was the sale price on a 60-amp load center if the regular price was $24.95? $6.24

16. West Shore Tire figures a 32% markup on new tires. What is the selling price of a tire with a cost of $49.23? $64.98

17. Brother Lawrence is replacing the shingles on his house roof. If the roof measures 26 feet by 38 feet on each side of the peak, how many squares of shingles will he need? (One square is 100 square feet. For a fraction of a square, raise the answer to the next whole square.)   20 squares

18. Israel's men of war numbered 603,550 at one time (Numbers 1:46). If each man was given an area measuring 3 feet by 4 feet, how many acres would this army have covered, to the nearest tenth of an acre?   166.3 acres

322   Chapter 7 Lines and Planes in Geometry

REVIEW EXERCISES

E. Write the formula for each fact. *(Lessons 90–93)*

19. Area of a parallelogram $a = bh$
20. Area of a triangle $a = \frac{1}{2}bh$
21. Area of a square $a = s^2$
22. Area of a rectangle $a = lw$
23. Perimeter of a square $p = 4s$
24. Perimeter of a rectangle $p = 2(l + w)$
25. Circumference of a circle $c = \pi d$ or $c = 2\pi r$

F. Find the areas of the figures indicated. *(Lessons 92, 93)*

26. Square: s = 15.2 cm   231.04 cm²   27. Square: s = 115 ft.   13,225 sq. ft.
28. Rectangle: $l = 10\frac{1}{2}$ in.   29. Rectangle: l = 8.4 m
    w = 7 in. $73\frac{1}{2}$ sq. in. w = 2.65 m   22.26 m²
30. Parallelogram: b = 28 cm   31. Parallelogram: $b = 6\frac{3}{4}$ ft.
    h = 15 cm   420 cm² h = 2 ft. $13\frac{1}{2}$ sq. ft.
32. Triangle: b = 16 cm; h = 8 cm   64 cm²

G. Construct a triangle with each set of dimensions. Label the given dimensions. *(Lesson 88)*   (See facing page for model triangles.)

33. 7 cm, 30°, 12 cm   34. $1\frac{1}{4}$ in., 50°, $1\frac{1}{2}$ in.

H. Find the new amount after each change, to the nearest cent or whole number. *(Lesson 64)*

35. $36.50 increased by 54% $56.21   36. 912 increased by 8%   985

Lesson 94   T–322

Model Triangles for Part G

33.

34.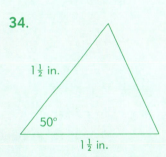

T–323 Chapter 7 *Lines and Planes in Geometry*

LESSON 95

Objective

- To review finding the area of a circle by using the formula $a = \pi r^2$.

Review

1. *Give the formula for finding each fact. (Lessons 90–94)*
   a. Area of a trapezoid $(a = \frac{1}{2}h[b_1 + b_2])$
   b. Area of a parallelogram $(a = bh)$
   c. Area of a triangle $(a = \frac{1}{2}bh)$
   d. Area of a square $(a = s^2)$
   e. Area of a rectangle $(a = lw)$
   f. Circumference of a circle $(c = \pi d \text{ or } c = 2\pi r)$
   g. Perimeter of a square $(p = 4s)$
   h. Perimeter of a rectangle $(p = 2[l + w])$

2. *Find the area of each trapezoid. (Lesson 94)*
   a. $h = 3$ in.
      $b_1 = 4\frac{1}{2}$ in.
      $b_2 = 5$ in. $(14\frac{1}{4}$ sq. in.$)$
   b. $h = 30$ ft.
      $b_1 = 28$ ft.
      $b_2 = 36$ ft. $(960$ sq. ft.$)$

3. *Find the area of each figure. (Lesson 93)*
   a. Parallelogram
      $b = 5.3$ m
      $h = 2.6$ m $(13.78$ m$^2)$
   b. Triangle
      $b = 8.3$ cm
      $h = 6.4$ cm $(26.56$ cm$^2)$

4. *Use each set of facts to construct a triangle. Label it with the given dimensions. (Lesson 88)*
   a. 5 cm, 4 cm, 5 cm
   b. $1\frac{1}{2}$ in., 1 in., $\frac{3}{4}$ in.
   c. 2 in., 1 in., $1\frac{1}{2}$ in.

5. *Find each percent. Express remainders as fractions. (Lesson 65)*
   a. 5 is ___ of 6 $(83\frac{1}{3}\%)$
   b. 11 is ___ of 18 $(61\frac{1}{9}\%)$

6. *Write these numbers using words. (Lesson 1)*
   a. 15,290,400
      (Fifteen million, two hundred ninety thousand, four hundred)
   b. 8,000,456,000 (Eight billion, four hundred fifty-six thousand)

95. Areas of Circles

A dog is chained to a post in the middle of a lawn. If his chain is 7 feet long, he can move about in a circular area extending 7 feet in all directions from the post. How many square feet are in his living space?

To find the area of a circle, multiply pi times the square of the radius. The radius in the example above is 7 feet. To find the area over which the dog can move, multiply $3\frac{1}{7} \times 7 \times 7$. The dog has a living space of 154 square feet.

A circle with a 1-inch radius covers $3\frac{1}{7}$ square inches ($3\frac{1}{7} \times 1 \times 1 = 3\frac{1}{7}$). The diagram at the right shows that this is a reasonable answer.

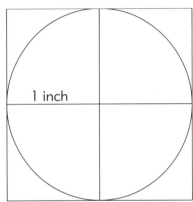

1 inch

Memorize this formula for the area of a circle.

area = pi × radius × radius, or $a = \pi r^2$

Remember to use 3.14 as the value of pi, or $3\frac{1}{7}$ when working with fractions or multiples of 7, unless you are directed otherwise.

If the diameter of a circle is given, divide it by 2 to find the radius. Then find the area in the usual manner.

Example A	**Example B**
Find the area of this circle.	Find the area of this circle.
radius = 21 ft.; π = $3\frac{1}{7}$; area = ___	diameter = 13 in.; π = 3.14; area = ___
$a = \pi r^2$	radius = 13 ÷ 2 = 6.5 in.
$a = 3\frac{1}{7} \times 21 \times 21$	$a = \pi r^2$
$a = 1{,}386$ sq. ft.	$a = 3.14 \times 6.5 \times 6.5$
	$a = 132.665$ sq. in.

CLASS PRACTICE

Give the formula for finding this area. Be sure you know it by memory.

a. Area of a circle $a = \pi r^2$

Find the areas of circles with these radii. Use $3\frac{1}{7}$ for pi.

b. $r = 4\frac{2}{3}$ in. $68\frac{4}{9}$ sq. in.   c. $r = 10\frac{1}{2}$ in. $346\frac{1}{2}$ sq. in.   d. $r = 7$ in.   154 sq. in.

324   Chapter 7   *Lines and Planes in Geometry*

Find the areas of circles with these radii. Use 3.14 for pi.

e. $r = 11$ in.   379.94 sq. in.   f. $r = 3.1$ cm   30.1754 cm²   g. $r = 6.5$ m   132.665 m²
h. $r = 15$ in.   706.5 sq. in.   i. $r = 8.4$ cm   221.5584 cm²   j. $r = 4.1$ m   52.7834 m²

Find the areas of circles with these diameters. Use 3.14 for pi.

k. $d = 16$ in.   200.96 sq. in.   l. $d = 24$ cm   452.16 cm²   m. $d = 27$ in.   572.265 sq. in.

WRITTEN EXERCISES

A. *Write the formula for finding this area. Be sure you know it by memory.*

1. Area of a circle $a = \pi r^2$

B. *Find the areas of circles with these radii. Use $3\frac{1}{7}$ for pi.*

2. $r = 14$ in.   616 sq. in.   3. $r = 1\frac{1}{2}$ in. $7\frac{1}{14}$ sq. in.   4. $r = 2\frac{1}{3}$ in. $17\frac{1}{9}$ sq. in.

C. *Find the areas of circles with these radii. Use 3.14 for pi.*

5. $r = 5$ in.   78.5 sq. in.   6. $r = 8$ in.   200.96 sq. in.   7. $r = 12$ in.   452.16 sq. in.
8. $r = 14$ cm   615.44 cm²   9. $r = 8.5$ cm   226.865 cm²   10. $r = 9.7$ m   295.4426 m²

D. *Find the areas of circles with these diameters. Use 3.14 for pi.*

11. $d = 12$ ft.   113.04 sq. ft.   12. $d = 18$ in.   254.34 sq. in.
13. $d = 35$ cm   961.625 cm²   14. $d = 11$ m   94.985 m²

E. *Solve these reading problems.*

15. Martha bought a lawn sprinkler that can spray water 20 feet in all directions. How many square feet can the sprinkler water?   1,256 square feet

16. Mr. Shultz planted a circular flower bed around a large rock about 4 feet in diameter. The flower bed is 12 feet in diameter. If he spreads mulch over the open soil in the flower bed, how many square feet will he cover? (Hint: The last step is subtraction.)   100.48 square feet

17. According to Numbers 35:4, 5, each city given to the Levites included a certain amount of land around it. If the radius of a city and its surrounding land was 5,000 feet, how many acres belonged to that city? (Round to the nearest whole number.)   1,802 acres

18. How many square inches are in the opening of a pipe (a) with a 4-inch diameter? (b) with a 2-inch diameter? (c) The larger area is how many times greater than the smaller area?   a. 12.56 sq. in.   b. 3.14 sq. in.   c. 4 times   (Doubling diameter will quadruple capacity.)

19. Mr. Richards lost 25 hogs out of a group of 1,100. What was the mortality rate to the nearest tenth of a percent?   2.3%

20. Leah had 2 wrong answers on a Bible test with 36 questions. What was her percent score?   94%

Introduction

Draw a circle on the board, and ask the students to suggest a way to find its area without using the formula in the lesson. Show them how by first dividing the circle into sectors as shown.

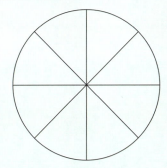

Point out that the sectors are roughly in the shape of triangles. If we were to make them smaller and smaller, we would finally have many thin triangular pieces with two straight sides and one side very nearly straight. These triangles could be arranged in the form of a parallelogram as shown below, with half of the short bases at the bottom and half at the top.

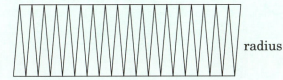

The height of this parallelogram is equal to the radius of the circle. Its base is equal to half of the circumference (which is equal to $\pi \times $ radius). So for the area of a circle, we can rewrite the parallelogram formula ($a = b \times h$) as $a = \pi \times r \times r$. In simplest form, this is stated as $a = \pi r^2$.

Teaching Guide

To find the area of a circle, multiply pi times the square of the radius. The formula is $a = \pi r^2$.

a. $r = 7$ in. $\pi = 3\frac{1}{7}$ (154 sq. in.)

b. $r = 11\frac{2}{3}$ ft. $\pi = 3\frac{1}{7}$ ($427\frac{7}{9}$ sq. ft.)

c. $r = 9$ cm $\pi = 3.14$ (254.34 cm^2)

d. $r = 19$ m $\pi = 3.14$ (1,133.54 m^2)

Solutions for Part E

15. 3.14×20^2

16. $(3.14 \times 6^2) - (3.14 \times 2^2)$

17. $(3.14 \times 5,000^2) \div 43,560$

18. 3.14×22
    3.14×12
    $12.56 \div 3.14$

19. $25 \div 1,100$

20. 34 correct ÷ 36

Model Triangles for Part I

33.
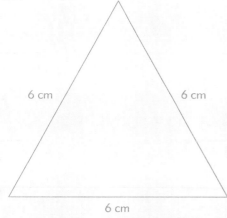

34.
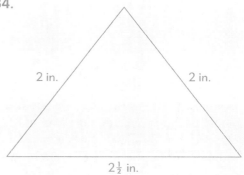

REVIEW EXERCISES

F. Write the formula for each fact. *(Lessons 90–94)*

21. Area of a parallelogram $a = bh$
22. Area of a triangle $a = \frac{1}{2}bh$
23. Area of a square $a = s^2$
24. Area of a rectangle $a = lw$
25. Perimeter of a square $p = 4s$
26. Perimeter of a rectangle $p = 2(l + w)$
27. Area of a trapezoid $a = \frac{1}{2}h(b_1 + b_2)$
28. Circumference of a circle $c = \pi d$ or $c = 2\pi r$

G. Find the areas of trapezoids with these dimensions. *(Lesson 94)*

29. $h = 2\frac{1}{2}$ in.
    $b_1 = 3\frac{3}{4}$ in.
    $b_2 = 3$ in. $8\frac{7}{16}$ sq. in.

30. $h = 8.4$ cm
    $b_1 = 15.6$ cm
    $b_2 = 17$ cm 136.92 cm²

H. Find the areas of triangles with these dimensions. *(Lesson 93)*

31. $b = 12$ cm
    $h = 8.9$ cm 53.4 cm²

32. $b = 11$ ft.
    $h = 4\frac{3}{4}$ ft. $26\frac{1}{8}$ sq. ft.

I. Construct a triangle with each set of dimensions. Label the given dimensions. *(Lesson 88)* (See facing page for model triangles.)

33. 6 cm, 6 cm, 6 cm
34. $2\frac{1}{2}$ in., 2 in., 2 in.

J. Find each percent to the nearest whole number. *(Lesson 65)*

35. 8 is 73% of 11
36. 14 is 47% of 30

K. Write these numbers, using words. *(Lesson 1)*

37. 15,300,000,000,000 Fifteen trillion, three hundred billion
38. 158,000,000,000,000,000 One hundred fifty-eight quadrillion

326   Chapter 7 Lines and Planes in Geometry

96. Areas of Compound Figures

Many geometric forms are combinations of the basic figures that you have studied in this chapter. Find the area of such a figure by combining the areas of its various parts. Following are the steps for calculating the area of a compound figure.

1. Divide the compound figure into sections that are simple geometric shapes. See Examples A and B.
2. Calculate the area of each individual section.
3. Add the areas of the individual sections. The result is the total area of the figure.

If the compound figure includes semicircles, find the area of a full circle and divide by 2 (Example A).

Example A
   Find the area of the figure below.
   Think: A rectangle and a semicircle.

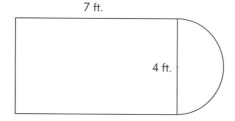

Rectangle:
   7 × 4 = 28 sq. ft.
Semicircle:
   radius = 2 ft.
   (3.14 × 2 × 2) ÷ 2 = 6.28 sq. ft.
28 + 6.28 = 34.28 sq. ft.

Example B
   Find the area of the figure below.
   Think: A rectangle and two triangles.

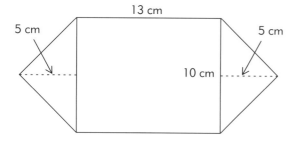

Rectangle:
   13 × 10 = 130 cm²
Triangles:
   base = 10 cm
   height = 5 cm
   ($\frac{1}{2}$ × 10 × 5) × 2 = 50 cm²
130 + 50 = 180 cm²

For sections not to be included in the total area, such as doors and windows in a wall, find the area of those sections and subtract them from the total area (Examples C and D).

LESSON 96

Objective
- To teach *finding the area of a compound figure such as a semicircle attached to a square.

Review

1. Give the formula for finding each fact. (Lessons 90–95)
   a. Area of a circle $(a = \pi r^2)$
   b. Area of a trapezoid $(a = \frac{1}{2}h[b_1 + b_2])$
   c. Area of a parallelogram $(a = bh)$
   d. Area of a triangle $(a = \frac{1}{2}bh)$
   e. Area of a square $(a = s^2)$
   f. Area of a rectangle $(a = lw)$
   g. Circumference of a circle $(c = \pi d$ or $c = 2\pi r)$
   h. Perimeter of a square $(p = 4s)$
   i. Perimeter of a rectangle $(p = 2[l + w])$

2. Find the area of each figure. (Lessons 94, 95)

   Circles
   a. $r = 7$ in. $\pi = 3\frac{1}{7}$ (154 sq. in.)
   b. $r = 9$ cm $\pi = 3.14$ (254.34 cm²)

   Trapezoids
   c. $h = 5$ in.
   $b_1 = 6\frac{1}{2}$ in.
   $b_2 = 7$ in.   ($33\frac{3}{4}$ sq. in.)

   d. $h = 14.2$ cm
   $b_1 = 19$ cm
   $b_2 = 20.5$ cm (280.45 cm²)

3. Find the perimeter of each rectangle or square. (Lesson 90)
   a. $s = 4\frac{1}{4}$ in.   (17 in.)
   b. $l = 5.6$ m $w = 3.5$ m   (18.2 m)

4. Find the rate of each change, to the nearest whole percent. Label your answer as an increase or a decrease. (Lesson 66)
   a. From $14.15 to $16.23
   (15% increase)
   b. From $214.50 to $225.00
   (5% increase)

5. Identify each number as prime or composite. (Lesson 34)
   a. 72   (composite)
   b. 74   (composite)

6. Round each number as indicated. (Lesson 2)
   a. 1,596 to the nearest ten   (1,600)
   b. 5,387,249,421,000,000 to the nearest ten trillion
   (5,390,000,000,000,000)

Introduction

On the board, draw the compound figure shown below, including the dimensions. Ask the students if they can find its area.

Large semicircle: $\frac{1}{2} \times 3.14 \times 10 \times 10 = 157.00$
Small semicircle: $\frac{1}{2} \times 3.14 \times 5 \times 5 = 39.25$
Rectangle: $20 \times 10 = 200.00$
Left triangle: $\frac{1}{2} \times 10 \times 15 = 75.00$
Lower triangle: $\frac{1}{2} \times 20 \times 10 = \underline{100.00}$
Total: 571.25 sq. ft.

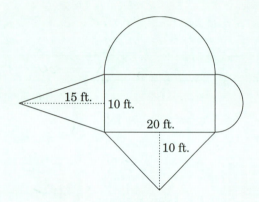

Teaching Guide

1. **To calculate the area of a compound figure, use the following steps.**

   (1) Divide the compound figure into sections that are simple geometric shapes.

   (2) Calculate the area of each individual section.

   (3) Add the areas of the individual sections. The result is the total area of the figure.

   Use the figures in *Class Practice* as specified in points 2 and 3 below.

2. **If the compound figure includes semicircles, find the area of a full circle and divide by 2.** Use exercise *a* of *Class Practice* to demonstrate this point.

3. **For sections not to be included in the total area, such as doors and**

Solutions for CLASS PRACTICE

a. Rectangle: $5 \times 2 = 10$ sq. ft.
Large semicircle: $(3.14 \times 2.5^2) \times \frac{1}{2} = 9.8125$ sq. ft.
Small semicircle: $(3.14 \times 1^2) \times \frac{1}{2} = 1.57$ sq. ft.
Large triangle: $\frac{1}{2} \times 5 \times 2 = 5$ sq. ft.
Small triangle: $\frac{1}{2} \times 2 \times 1 = 1$ sq. ft.
Combined area: $10 + 9.8125 + 1.57 + 5 + 1 = 27.3825$ sq. ft.

b. Rectangle: $30 \times 14 = 420$ m²
Semicircle: $(3\frac{1}{7} \times 72) \times \frac{1}{2} = 77$ m²
Triangle subtracted: $\frac{1}{2} \times 14 \times 10 = 70$ m²
Rectangle subtracted: $7 \times 9 = 63$ m²
Combined area: $(420 + 77) - (70 + 63) = 364$ m²

Solutions for WRITTEN EXERCISES

1. Square: $4 \times 4 = 16$ m²
   Semicircle: $(3.14 \times 2^2) \times \frac{1}{2} = 6.28$ m²
   Combined area: $16 + 6.28 = 22.28$ m²

2. Square: $7 \times 7 = 49$ sq. ft.
   Semicircle subtracted: $(3\frac{1}{7} \times 3\frac{1}{2}^2) \times \frac{1}{2} = 19\frac{1}{4}$ sq. ft.
   Combined area: $49 - 19\frac{1}{4} = 29\frac{3}{4}$ sq. ft.

Example C Find the area of the unshaded part of this figure.

Think: Rectangle; two sections excluded.
Rectangle:
  31 × 21 = 651 sq. ft.
Section 1:
  8 × 11 = 88 sq. ft.
Section 2:
  4 × 5 = 20 sq. ft.
Total excluded: 88 + 20 = 108 sq. ft.
Total area: 651 − 108 = 543 sq. ft.

Example D Find the area of the wall below, not including the windows.

Think: Rectangle; four sections excluded.
Rectangle:
  25 × 20 = 500 sq. ft.
Sections 1, 2:
  (3 × 5) × 2 = 30 sq. ft.
Sections 3, 4:
  (5 × 5) × 2 = 50 sq. ft.
Total excluded: 30 + 50 = 80 sq. ft.
Total area: 500 − 80 = 420 sq. ft.

CLASS PRACTICE

Find the area of the unshaded part of each figure below. In exercise b, use $3\frac{1}{7}$ for pi.
(See facing page for solutions.)

a. 27.3825 sq. ft.
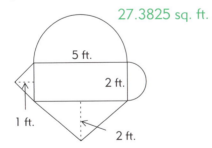

b. 364 m²
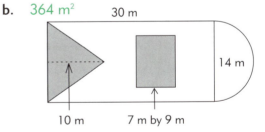

WRITTEN EXERCISES

A. *Find the area of the unshaded part of each figure below. In exercise 2, use $3\frac{1}{7}$ for pi.*
(See facing page for solutions.)

1. 22.28 m²

2. $29\frac{3}{4}$ sq. ft.

328 Chapter 7 Lines and Planes in Geometry

3. 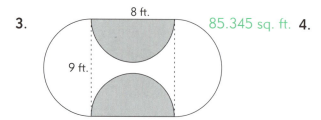 85.345 sq. ft.   4. 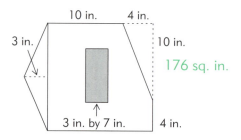 176 sq. in.

5. 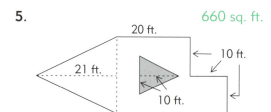 660 sq. ft.   6. 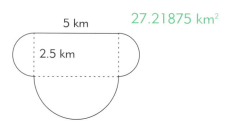 27.21875 km²

7. 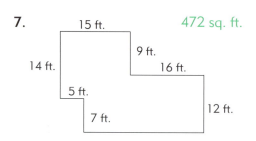 472 sq. ft.   8. 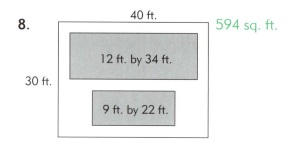 594 sq. ft.

B. Solve these reading problems.

9. Sister Eunice is teaching her students to make 4-inch by 7-inch envelopes. The two overlapping triangular side flaps have a base of 4 inches and a height of $3\frac{1}{2}$ inches. The top and bottom triangular flaps have a base of 7 inches and a height of $2\frac{1}{2}$ in. How many square inches of paper are in each envelope? $59\frac{1}{2}$ square inches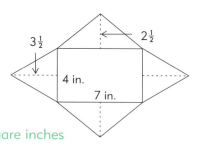

10. Brother Neil covered the two ends of his greenhouse with plastic. Each end is a semicircle with an 8-foot radius. On one end he cut an opening for a door measuring 80 inches by 36 inches and an opening for a furnace measuring $2\frac{1}{2}$ feet by 4 feet. How many square feet of plastic are on the two ends of his greenhouse?
    170.96 square feet

11. South Mountain Body Shop is replacing the side panel on the body of a moving truck. The panel is 20 feet long and 8 feet high, and it has an extension over the cab that measures $3\frac{1}{2}$ feet by 5 feet. What is the area of the side panel? $177\frac{1}{2}$ square feet

12. A church auditorium is 50 feet long, 40 feet wide, and 8 feet high at the side walls. Since it has a cathedral ceiling, the end walls have the shape of the figure shown at the right. The peak of the ceiling is 4 feet higher than the side walls. What is the area of the four walls of the auditorium? (Ignore the doors and windows for this calculation.) 1,600 square feet

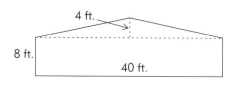

windows in a wall, find the area of those sections and subtract them from the total area. Use exercise *b* of *Class Practice* to demonstrate this point.

---

3. Square: 8 × 9 = 72 sq. ft.
   Two semicircles added: 3.14 × 4.5² = 63.585 sq. ft.
   Two semicircles subtracted: 3.14 × 4² = 50.24 sq. ft.
   Combined area: (72 + 63.585) − 50.24 = 85.345 sq. ft.

4. Full square: 14 × 14 = 196 sq. in.
   Triangle added: ½ × 14 × 3 = 21 sq. in.
   Triangle corner subtracted: ½ × 10 × 4 = 20 sq. in.
   Rectangle subtracted: 3 × 7 = 21 sq. in.
   Combined area: (196 + 21) − (20 + 21) = 176 sq. in.

5. Full rectangle: 30 × 20 = 600 sq. ft.
   Triangle added: ½ × 20 × 21 = 210 sq. ft.
   Square corner subtracted: 10 × 10 = 100 sq. ft.
   Triangle subtracted: ½ × 10 × 10 = 50 sq. ft.
   Combined area: (600 + 210) − (100 + 50) = 660 sq. ft.

6. Rectangle: 5 × 2.5 = 12.5 km²
   Large semicircle: (3.14 × 2.5²) × ½ = 9.8125 km²
   Two small semicircles: 3.14 × 1.25² = 4.90625 km²
   Combined area: 12.5 + 9.8125 + 4.90625 = 27.21875 km²

7. Full rectangle: 31 × 21 = 651 sq. ft.
   Rectangle subtracted: 9 × 16 = 144 sq. ft.
   Rectangle subtracted: 5 × 7 = 35 sq. ft.
   Combined area: 651 − (144 + 35) = 472 sq. ft.

8. Full rectangle: 40 × 30 = 1,200 sq. ft.
   Rectangle subtracted: 12 × 34 = 408 sq. ft.
   Rectangle subtracted: 9 × 22 = 198 sq. ft.
   Combined area: 1,200 − (408 + 198) = 594 sq. ft.

Solutions for Part B

9. $(4 \times 7) + 2(\frac{1}{2} \times 4 \times 3\frac{1}{2}) + 2(\frac{1}{2} \times 7 \times 2\frac{1}{2})$
10. $2 \times \frac{1}{2}(3.14 \times 8^2) - [(6\frac{2}{3} \times 3) + (2\frac{1}{2} \times 4)]$
11. $(20 \times 8) + (3\frac{1}{2} \times 5)$
12. $2(50 \times 8) + 2(40 \times 8) + 2(\frac{1}{2} \times 40 \times 4)$

T–329 *Chapter 7 Lines and Planes in Geometry*

13. 65 – 35 = 30; 30 = __% of 32,000
14. 32,000 – 300 = 31,700; 31,700 = __% of 32,000

13. Carlisle Milk Products produces 35 million pounds of butter per year. The company plans to expand to 65 million pounds per year. What is the rate of the proposed increase, to the nearest whole percent?   86%
14. In Judges 7, Gideon's army shrank from 32,000 men to 300 men. To the nearest whole percent, what was the rate of decrease in the size of the army?   99%

REVIEW EXERCISES

C. *Write the formula for each fact.* (Lessons 90–95)

15. Area of a circle $a = \pi r^2$
16. Circumference of a circle $c = \pi d$ or $c = 2\pi r$
17. Perimeter of a square $p = 4s$
18. Perimeter of a rectangle $p = 2(l + w)$
19. Area of a square $a = s^2$
20. Area of a rectangle $a = lw$
21. Area of a parallelogram $a = bh$
22. Area of a triangle $a = \frac{1}{2}bh$
23. Area of a trapezoid $a = \frac{1}{2}h(b_1 + b_2)$

D. *Find the area of each circle.* (Lesson 95)

24. $r = 6$ in.   113.04 sq. in.
25. $r = 16$ in.   803.84 sq. in.
26. $r = 18$ cm   1,017.36 cm²

E. *Find the area of each trapezoid.* (Lesson 94)

27. $h = 4\frac{1}{2}$ in.
    $b_1 = 6\frac{1}{4}$ in.
    $b_2 = 8\frac{1}{2}$ in. $33\frac{3}{16}$ sq. in.
28. $h = 11.5$ cm
    $b_1 = 23.8$ cm
    $b_2 = 18.2$ cm   241.5 cm²

F. *Find the perimeter of each rectangle.* (Lesson 90)

29. $l = 9.15$ m; $w = 6.28$ m   30.86 m
30. $l = 12\frac{1}{4}$ in.; $w = 12$ in. $48\frac{1}{2}$ in.

G. *Find the rate of each change, to the nearest whole percent. Include the label* increase *or* decrease. (Lesson 66)

31. From $56.12 to $38.50   31% decrease
32. From $513.25 to $628.12   22% increase

H. *Identify each number as* prime *or* composite. (Lesson 34)

33. 91   composite
34. 86   composite

I. *Round each number to the place indicated.* (Lesson 2)

35. 2,014,160 (hundred thousands)   2,000,000
36. 128,314,000,000 (billions)   128,000,000,000

330 *Chapter 7 Lines and Planes in Geometry*

97. Reading Problems: Solving Multistep Problems

Many reading problems, like other challenges in life, require more than one step to solve them. Also, the solutions require different operations in various orders.

The secret to solving any reading problem is to gain a clear understanding of it, just as a mechanic understands the many devices that make a vehicle operate. After you understand a reading problem, you must follow an orderly method to solve it. Below are four steps for solving a multistep reading problem.

1. Carefully read the problem. Note the facts that are given, and pay special attention to the question that is asked.

2. Decide which mathematical operation or operations are required. Do the clue words indicate addition, subtraction, multiplication, or division? Must you find some facts by addition before you can multiply?

3. Solve the problem by an orderly method. Begin with the first step and then do the second, the third, and so on.

4. Check your answer to make sure it is logical.

Here is a reading problem that clearly cannot be solved in one quick operation. Carefully read the problem, and then note how it is solved by using the steps above.

> At a glass plant, a ribbon of glass moves continuously at a rate of 635 tons per 24-hour day. One linear foot of glass 15 feet wide weighs about 24 pounds. How fast is the glass moving in feet per minute, to the nearest whole number?

Step 1: Note the following facts.
    (a) The glass ribbon moves at the rate of 635 tons per 24 hours.
    (b) One linear foot of glass weighs 24 pounds.
    (c) The question asks, "What is the rate of speed in feet per minute?"

Steps 2 and 3:
    (a) Multiply to change tons per day to pounds per day.
        635 tons per day × 2,000 lb. = 1,270,000 lb. per day
    (b) Multiply to change 24 hours to minutes.
        24 hr. × 60 min. = 1,440 min. per day
    (c) Divide to change pounds per day to pounds per minute.
        1,270,000 lb. per day ÷ 1,440 min. per day = 882 lb. per min. (rounded)
    (d) Divide to change pounds per minute to feet per minute.
        882 lb. per min. ÷ 24 lb. per ft. = 36.75 ft. per min.

LESSON 97

Objective
- To give practice with solving multistep reading problems.

Review
1. Give Lesson 97 Quiz (Area).
2. *Give the formula for finding each fact.* (Lessons 90–95)
   a. Area of a circle $(a = \pi r^2)$
   b. Area of a trapezoid $(a = \frac{1}{2}h[b_1 + b_2])$
   c. Area of a parallelogram $(a = bh)$
   d. Area of a triangle $(a = \frac{1}{2}bh)$
   e. Area of a square $(a = s^2)$
   f. Area of a rectangle $(a = lw)$
   g. Circumference of a circle $(c = \pi d \text{ or } c = 2\pi r)$
   h. Perimeter of a square $(p = 4s)$
   i. Perimeter of a rectangle $(p = 2[l + w])$
3. *Find the area of the unshaded part of each figure.* (Lesson 96)

   a. (332.96 sq. ft.)

   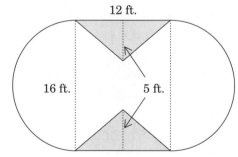

   b. (1,393 m²)

   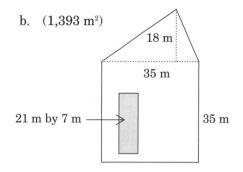

4. *Find the area of each circle.* (Lesson 95)
   a. $r = 2\frac{1}{2}$ in. ($19\frac{5}{8}$ sq. in.)
   b. $d = 26$ in. (530.66 sq. in.)
5. *Find the circumferences of each circle.* (Lesson 91)
   a. $r = 12$ in. (75.36 sq. in.)
   b. $d = 55$ ft. (172.7 sq. ft.)
6. *Find the base in each problem.* (Lesson 67)
   a. 70 is 28% of ___ (250)
   b. 21 is 35% of ___ (60)
7. *Find the greatest common factor of each pair.* (Lesson 35)
   a. 16, 36 (4)
   b. 96, 128 (32)
8. *Find the lowest common multiple of each pair.* (Lesson 35)
   a. 12, 16 (48)
   b. 12, 27 (108)

T–331 *Chapter 7 Lines and Planes in Geometry*

Introduction

Read together the multistep reading problem in the lesson text. Ask for volunteers to solve the problem mentally. When no one accepts the challenge, ask why this problem cannot be solved mentally. Following are two reasons.

1. The numbers are too large to calculate mentally.
2. The problem requires a number of steps that would be difficult to follow mentally.

All reading problems require some concentration and organization to solve. However, multistep reading problems require an extra degree of both.

Teaching Guide

To solve a multistep reading problem, use the following steps.

(1) Carefully read the problem. Note the facts that are given, and pay special attention to the question that is asked.

(2) Decide which mathematical operation or operations are required. Do the clue words indicate addition, subtraction, multiplication, or division? Must you find some facts by addition before you can multiply?

(3) Solve the problem by an orderly method. Begin with the first step and then do the second, the third, and so on.

(4) Check your answer to make sure it is logical.

Use the multistep reading problem in the lesson text and those in *Class Practice* to demonstrate these steps.

Class Practice a.

If you have sufficient time and chalkboard space, thoroughly explore various solution routes.

Solution 1

(1) Multiply to find how many pounds the neighbor bought.
$12 \times 68 = 816$ pounds

(2) Set up a direct proportion with the price per ton.

$$\frac{\text{price}}{\text{pounds}} \quad \frac{\$170}{2{,}000} = \frac{n}{816} \quad \frac{\text{price}}{\text{pounds}}$$

(3) Multiply the extremes.
$170 \times 816 = 138{,}720$

(4) Divide to find n.
$138{,}720 \div 2{,}000 = \$69.36$

Step 4: The answer is logical. 37 ft. per min. is 2,220 ft. per hr., or 53,280 ft. per day. At 24 lb. per ft., this is 1,278,720 lb. per day, or 639 tons per day. This is near the given information, but differs because of rounding.

CLASS PRACTICE
(See *Teaching Guide* for solutions.)

Solve each problem in a step-by-step manner, showing the steps.

a. King Farms sells alfalfa hay for $170 per ton. A neighbor bought 12 bales that averaged 68 pounds each. What was his bill? $69.36

b. A certain dairy can process 350,000 gallons of milk per day. If the average cow produces 70 pounds per day, how many cows does it take to supply the dairy? (Milk weighs 8.6 pounds per gallon.) 43,000 cows

c. Witmer Diesel purchased a rebuilt starter for $146. The business added $12\frac{1}{2}$% for overhead; and then on the cost plus overhead, it added another 12% for its profit. What was the total charge for the starter, including a 6% sales tax? $195.00

d. On a 900-mile trip, the Rissers averaged 25 miles per gallon of gasoline. They left home with the 16-gallon tank full of fuel. On the trip, they put in 14.5 gallons at the first refueling and 12 gallons at the second refueling. How many gallons were left in the fuel tank at the end of the trip? 6.5 gallons

WRITTEN EXERCISES

A. Solve each problem in a step-by-step manner, showing your work. Numbers 1–4 have the steps listed for you.

1. Matthew wants to make pens for nursery pigs in a barn 55 feet long and 38 feet wide. Aisles occupy 165 square feet. Each pen will hold 25 pigs and allow each pig $3\frac{1}{2}$ square feet of floor space. How many pens can he put in the barn? 22 pens
   a. Multiply to find the total area of the barn.   55 × 38 = 2,090 sq. ft.
   b. Subtract the area of the aisles.   2,090 – 165 = 1,925 sq. ft.
   c. Multiply to find how much area each pen must have.   25 × $3\frac{1}{2}$ = $87\frac{1}{2}$ sq. ft.
   d. Divide to find how many pens can be put in the barn.   1,925 ÷ $87\frac{1}{2}$ = 22 pens

2. Roger needs to replace the linoleum on the kitchen floor, which measures $10\frac{1}{2}$ feet wide and 18 feet long. He will use 9-inch square tiles. How many tiles will he need? 336 tiles
   a. Multiply to find the area of the floor in square feet. $10\frac{1}{2}$ × 18 = 189 sq. ft.
   b. Multiply to find the area of the floor in square inches.   189 × 144 = 27,216 sq. in.
   c. Multiply to find the area of one tile in square inches.   9 × 9 = 81 sq. in.
   d. Divide to find the number of tiles needed.   27,216 ÷ 81 = 336 tiles

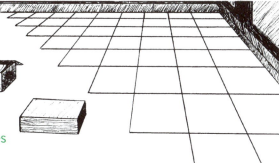

332 Chapter 7 *Lines and Planes in Geometry*

3. Noah covered the inside and the outside of the ark with pitch. The ark was 300 cubits long and 50 cubits wide. If one gallon of pitch covered 50 square feet, how many gallons would it have taken to cover the inside and outside of just the bottom of the ark? *1,350 gallons*
   a. Multiply to change cubits to feet.   *300 × 1.5 = 450 ft.   50 × 1.5 = 75 ft.*
   b. Multiply to find the area of both sides of the floor.   *2(450 × 75) = 67,500 sq. ft.*
   c. Divide to find how many gallons of pitch.   *67,000 ÷ 50 = 1,350 gallons*

4. In one recipe for float glass, 75% of the weight is sand. A glass factory has 2 lines each producing 625 tons of this glass per day. If a dump truck hauls 25 tons per load, how many truckloads of sand does the factory use in one day?   *$37\frac{1}{2}$ truckloads*
   a. Multiply to find the number of tons for 2 lines of glass.   *2 × 625 = 1,250 tons*
   b. Multiply to find what part of the total tons is sand.   *0.75 × 1,250 = 937.5 tons*
   c. Divide to find the number of truckloads used.   *937.5 ÷ 25 = $37\frac{1}{2}$ truckloads*

5. Glass weighs about 1.6 pounds per square foot at 3 millimeters thick. If a truck can haul 22 tons of glass, how many square feet of glass is that?   *27,500 square feet*

6. One ribbon of glass in the factory is 1,400 feet long and 15 feet wide. How many window panes can be cut from this ribbon if each pane measures 24 by 30 inches?   *4,200 panes*

7. A recipe for regular window glass includes 14% broken glass. On a window pane that measures 22 by $28\frac{1}{2}$ inches, the recycled part represents how many square inches?   *87.78 square inches*

8. Materials for the ceremony at the completion of the transcontinental railroad included a golden spike that cost $415.24. The gold in the spike was worth $400. The rest of the cost was for the engraved lettering on the four sides of the spike. How many letters were there if the cost of each letter was 4 cents?   *381 letters*

B. Solve these reading problems.

9. A salesman for Brenner Furniture receives a $7\frac{1}{2}$% commission. What was the selling price of a bedroom suite if his commission was $140.70?   *$1,876*

10. A dry goods outlet store sells fabrics at 65% off the regular price. What was the regular price if Ann's bill was $58.80? (Hint: What percent of the regular price was Ann's bill?)   *$168*

11. Arlene is creasing the rim of a fresh pie crust in a 9-inch pan. To the nearest whole number, how many creases must she make to have 2 per inch?   *57 creases*

12. Katherine has a round glass motto with a $4\frac{1}{2}$-inch radius. If she surrounds the motto with a flat chain, what length of chain will she use?   *28.26 inches*

13. The enrollment at Green Springs School is the lowest common multiple of the 7 students in sixth grade and the 6 students in seventh grade. What is the enrollment?   *42 students*

14. James has two nephews whose ages are 6 and 8. James calculated that he is 3 years older than the lowest common multiple of their ages. How old is James?   *27 years*

Solution 2

(1) Divide to find the price per pound.
$170 ÷ 2,000 = $0.085

(2) Multiply to find how many pounds the neighbor bought.
12 × 68 = 816 pounds

(3) Multiply to find the price.
816 × $0.085 = $69.36

Solution 3

(1) Multiply to find how many pounds the neighbor bought.
12 × 68 = 816 pounds

(2) Divide to find what part of a ton the neighbor bought ($\frac{816}{2000}$).
816 ÷ 2,000 = 0.408

(3) Multiply to find the price.
0.408 × $170 = $69.36

Class Practice b.

Solution 1

(1) Multiply to find how many pounds are processed per day.
350,000 × 8.6 = 3,010,000 pounds

(2) Divide to find how many cows are needed.
3,010,000 ÷ 70 = 43,000 cows

Solution 2

(1) Divide to find how many gallons one cow produces.
70 ÷ 8.6 = 8.14 (rounded)

(2) Divide to find how many cows are needed.
350,000 ÷ 8.14 = 42,998 (rounded)

Different levels of rounding produce different answers. If the first step of solution 2 is calculated to the nearest ten-thousandth, the answers for the two solutions will match (to the nearest whole cow). Solution 1 is clearly the more satisfactory way of solving this problem.

Solutions for Exercises 5–8

5. 22 × 2,000 = 44,000 pounds in a truckload
44,000 ÷ 1.6 = 27,500 sq. ft.

6. 1,400 × 15 = 21,000 sq. ft. in the ribbon
2 × 2½ = 5 sq. ft. in one pane
21,000 ÷ 5 = 4,200 panes

7. 22 × 28½ = 627 sq. in. in the window
14% of 627 = 87.78 sq. in.

8. $415.24 − $400 = $15.24 for lettering
$15.24 ÷ $0.04 = 381 letters

Solutions for Part B

9. 75% of ___ = $140.70; $140.70 ÷ 0.75

10. 35% of ___ = $58.80; $58.80 ÷ 0.35

11. 3.14 × 9 × 2

12. 3.14 × 9

14. 24 + 3

T-333 Chapter 7 *Lines and Planes in Geometry*

Class Practice c.

Are there alternate solution routes for this problem? Do the students see why it would be incorrect to add all the percents and then find $130\frac{1}{2}\%$ of $146?

(1) Multiply to find the cost increased by $12\frac{1}{2}\%$.
 146 × 1.125 = $164.25

(2) Multiply to find cost plus overhead increased by 12%.
 164.25 × 1.12 = $183.96

(3) Multiply to find the selling price increased by 6%.
 183.96 × 1.06 = $195.00

Class Practice d.

This problem allows the option of which of these two steps you do first:

Find how many gallons consumed
$$(900 \div 25 = 36).$$

Find the total number of gallons in and added to the tank
$$(16 + 14.5 + 12 = 42.5).$$

Either way, you must find these two facts and then subtract.

An Ounce of Prevention

Solutions given with the answer key are not intended for showing the process that students need to follow. Many reading problems can be solved by various routes. These solutions are provided to give direction for the teacher who needs to help students find a way of solution.

REVIEW EXERCISES

C. Write the formula for each fact. *(Lessons 90–95)*

15. Area of a circle $a = \pi r^2$
16. Circumference of a circle $c = \pi d$ or $c = 2\pi r$
17. Perimeter of a square $p = 4s$
18. Perimeter of a rectangle $p = 2(l + w)$
19. Area of a square $a = s^2$
20. Area of a rectangle $a = lw$
21. Area of a parallelogram $a = bh$
22. Area of a triangle $a = \frac{1}{2}bh$
23. Area of a trapezoid $a = \frac{1}{2}h(b_1 + b_2)$

D. Find the areas of circles with these radii. Use 3.14 for pi. *(Lesson 95)*

24. $r = 24$ in.   1,808.64 sq. in.   25. $r = 11.4$ m   408.0744 m²   26. $r = 71$ cm   15,828.74 cm²

E. Find the circumference of each circle. Use $3\frac{1}{7}$ for pi. *(Lesson 91)*

27. $d = 49$ ft.   154 ft.   28. $d = 28$ km   88 km

F. Find the missing numbers. *(Lesson 67)*

29. 126 is 70% of 180   30. 48 is 32% of 150

G. Find the greatest common factor of each pair. *(Lesson 35)*

31. 48, 72   24   32. 21, 91   7

H. Find the lowest common multiple of each pair. *(Lesson 35)*

33. 26, 39   78   34. 42, 63   126

98. Chapter 7 Review

A. Write the formulas for these facts. Be sure you know them all by memory. *(Lessons 90–95)*

1. Perimeter of a square — $p = 4s$
2. Perimeter of a rectangle — $p = 2(l + w)$
3. Circumference of a circle when the radius is known — $c = 2\pi r$
4. Circumference of a circle when the diameter is known — $c = \pi d$
5. Area of a square — $a = s^2$
6. Area of a rectangle — $a = lw$
7. Area of a parallelogram — $a = bh$
8. Area of a triangle — $a = \frac{1}{2}bh$
9. Area of a trapezoid — $a = \frac{1}{2}h(b_1 + b_2)$
10. Area of a circle — $a = \pi r^2$

B. Do these exercises with geometric symbols, terms, and figures. *(Lessons 85–87, 91)*

11. Use symbols to write "line segment YZ." $\overline{YZ}$
12. Use symbols to write "line WX is perpendicular to line YZ." $\overleftrightarrow{WX} \perp \overleftrightarrow{YZ}$
13. Draw the figure indicated by this expression: $\overleftrightarrow{AB} \parallel \overleftrightarrow{CD}$.
14. Draw the figure indicated by this symbol: ∠EFG.
15. A polygon with nine straight sides is a(n) <u>nonagon</u>
16. A polygon with ten straight sides is a(n) <u>decagon</u>
17. This is a(n) ___ angle.   reflex

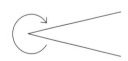

18. This is a(n) ___ angle.   obtuse

19. Draw a 55° angle.   (See facing page for model answers.)
20. Draw a 130° angle.

Lesson 98   T-334

LESSON 98

Objectives

- To review the material taught in Chapter 7 (Lessons 85–97).

Teaching Guide

1. Lesson 98 reviews the material taught in Lessons 85–97. Be sure to discuss any parts with which your students had special difficulty. The problems in each lesson are arranged so that each odd-numbered problem is of the same type and difficulty as the next even-numbered problem. Thus 1 and 2 are a pair, 3 and 4 are a pair, and so on. This is especially useful in review lessons because you can assign either the even-numbered or the odd-numbered problems, and use the others for class practice. If you do assign all the exercises, page through the lessons in Chapter 1 and select problems for class review.

   This odd-even pattern applies for all the review lessons in this text.

2. **Math problems lend themselves well to board work.** Review lessons are especially good for board work because you usually do not need to demonstrate how to do the problems on the board yourself.

3. **Be sure to review the new concepts introduced in this chapter.** The main ones are shown on page T-335, along with the exercises in this lesson that review those concepts.

Model Answers for Exercises 19 and 20

(Allow a tolerance of 1°.)

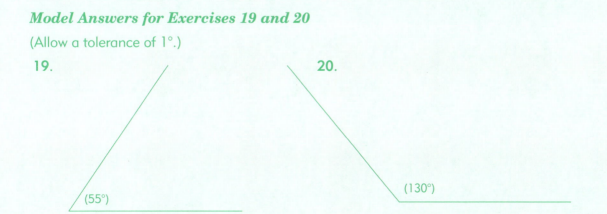

19.   (55°)

20.   (130°)

T-335 Chapter 7 Lines and Planes in Geometry

Lesson number and new concept	Exercises in Lesson 98
85—Identifying various polygons.	15, 16
86—Identifying reflex angles.	18
86—Identifying complementary and supplementary angles.	21, 22
88—Constructing triangles from different sets of information.	None
89—Bisecting line segments and angles, and drawing perpendicular lines by using a compass.	None
91—Finding the diameter of a circle when the circumference is known.	None
91—Understanding that pi cannot be expressed as an exact value and that a more precise value is 3.1416.	None
96—Finding the areas of compound figures.	62, 63

Lesson 98   335

21. The complementary angle to a 47° angle has ____ degrees.   43
22. The supplementary angle to a 48° angle has ____ degrees.   132
23. If a triangle has angles of 43°, 48° and 89°, it is (obtuse, acute, right).   acute
24. If a triangle has angles of 59°, 22°, and 99°, it is (obtuse, acute, right).   obtuse
25. If two angles of a triangle have 15° and 35°, the third angle has ____ degrees.   130
26. If two angles of a triangle have 80° and 90°, the third angle has ____ degrees.   10
27. If a triangle has sides each 6 inches long, it is (scalene, equilateral, isosceles).   equilateral
28. The ___ of a circle is the distance from the outer edge to the center.   radius
29. Half the circumference of a circle is a ___.   semicircle
30. Identify these triangles as similar or congruent, using symbols.   ΔEFG ≅ ΔKLM

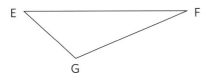
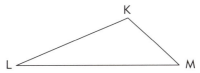

31. Identify these triangles as similar or congruent, using symbols.   ΔRST ~ ΔUVW

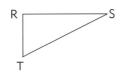
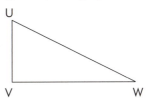

C. **Measure these angles.** *(Lesson 86)*   (Allow a tolerance of 1 degree.)

32.   322°

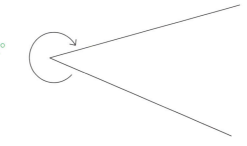

33.   46°

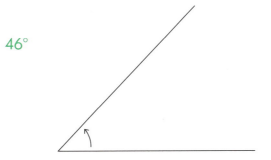

336   Chapter 7   Lines and Planes in Geometry

D. Measure the angles in these triangles. Check your work by making sure the sum of the three angles is 180°. *(Lesson 87)*

(Measurements may vary by 1 degree.)

34.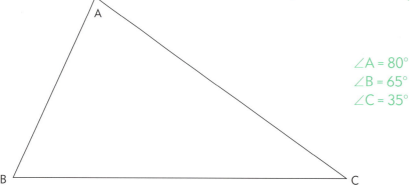

∠A = 80°
∠B = 65°
∠C = 35°

35.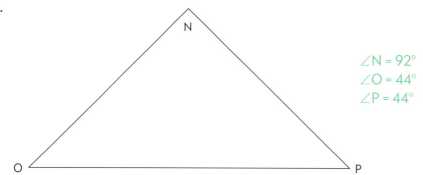

∠N = 92°
∠O = 44°
∠P = 44°

E. Find the perimeter or circumference of each geometric figure. *(Lessons 90, 91)*

Irregular polygons

36. 12 in., 8 in., 9 in., 7 in., 5 in., 8 in., 7 in.   56 in.
37. 9 in., 6 in., 7 in., 8 in., 8 in., 5 in., 3 in., 7 in.   53 in.

Squares

38. s = 15 in.   60 in.   39. s = 22 cm   88 cm

Rectangles

40. l = 15 cm
    w = 13 cm   56 cm   41. l = 14 ft.
                                            w = 11 ft.   50 ft.

Circles

42. r = 18 in.   43. r = $3\frac{1}{2}$ ft.   44. d = 25 cm   45. d = $12\frac{1}{2}$ yd.
    π = 3.14   113.04 in. π = $3\frac{1}{7}$ 22 ft. π = 3.14   78.5 cm π = $3\frac{1}{7}$ $39\frac{2}{7}$ yd.

F. Find the area of each geometric figure. *(Lessons 92–95)*

Rectangles

46. l = 6.5 m
    w = 4.75 m   30.875 m²   47. l = $3\frac{1}{2}$ in.
                                             w = 7 in. $24\frac{1}{2}$ sq. in.

T–337   *Chapter 7   Lines and Planes in Geometry*

Solutions for Part G

62. Rectangle: 44 × 22 = 968 sq. ft.
Triangle added: $\frac{1}{2}$ × 44 × 16 = 352 sq. ft.
Triangle subtracted: $\frac{1}{2}$ × 44 × 8 = 176 sq. ft.
Combined area: (968 + 352) − 176 = 1,144 sq. ft.

63. Rectangle: 80 × 56 = 4,480 m²
Semicircle added: (3.14 × 40²) × $\frac{1}{2}$ − 2,512 m²
Semicircle subtracted: (3$\frac{1}{7}$ × 14²) × $\frac{1}{2}$ = 308 m²
Combined area: (4,480 + 2,512) − 308 = 6,684 m²

Solutions for Part H

64. $\frac{1}{2}$(3.14 × 800²) = 1,004,800 sq. ft. irrigated
1,004,800 ÷ 43,560 = 23 acres

65. 4 × 3 × 9 = 108 hours burned
120 − 108 = 12 hours left

Squares

48. $s = 11.1$ m   123.21 m²   **49.** $s = 2\frac{1}{2}$ mi. $6\frac{1}{4}$ sq. mi.

Parallelograms

50. $b = 5.5$ m   **51.** $b = 6\frac{1}{2}$ ft.
$h = 3.75$ m   20.625 m² $h = 4\frac{2}{3}$ ft. $30\frac{1}{3}$ sq. ft.

Triangles

52. $b = 12$ ft.   **53.** $b = 2.4$ m
$h = 10$ ft.   60 sq. ft. $h = 3.5$ m   4.2 m²

Trapezoids

54. $h = 6$ in.   **55.** $h = 8$ cm
$b_1 = 4$ in. $b_1 = 4$ cm
$b_2 = 6$ in.   30 sq. in. $b_2 = 7$ cm   44 cm²

Circles

56. $r = 49$ in.   **57.** $r = 56$ in.
$\pi = 3\frac{1}{7}$ 7,546 sq. in. $\pi = 3\frac{1}{7}$ 9,856 sq. in.

58. $r = 15.7$ m   **59.** $r = 12.5$ cm
$\pi = 3.14$ 773.9786 m² $\pi = 3.14$ 490.625 cm²

60. $d = 14$ in.   **61.** $d = 15$ in.
$\pi = 3.14$ 153.86 sq. in. $\pi = 3.14$ 176.625 sq. in.

G. Find the area of the unshaded part of each figure below. *(Lesson 96)*

62. 1,144 sq. ft.

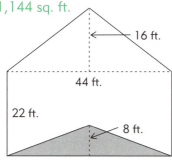

63. 6,684 m²

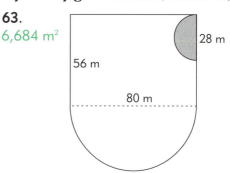

H. Solve these multistep reading problems. *(Lesson 97)*

64. An irrigation system has an 800-foot "arm" moving in a circular path around a central pivot. To the nearest whole number, how many acres does it water when it completes half of one circuit?   23 acres

65. Mabel lights her long-burning candle 4 times a week for 3 hours each time. If the candle's total burning time is 120 hours, how many of those hours are left after 9 weeks of use?   12 hours

338   Chapter 7   Lines and Planes in Geometry

66. Brother Nathan needed straw for his heifers. He paid $60 apiece for 24 large square bales weighing 1,200 pounds each. What was the cost per ton of straw? $100

67. Father is planning to paint both ends and one side wall of the garage, which is 24 feet long and 20 feet wide. It has 8-foot side walls, and the peak at each gable end is 5 feet higher than the side walls. If one gallon of paint covers 250 square feet, how many gallons should he buy? (Round to the next higher gallon.)   3 gallons

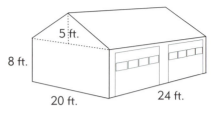

68. Matthew helped his brothers fill the haymow, which measured 20 feet by $31\frac{1}{2}$ feet. There were 9 layers of bales with 175 bales per layer, and the bales weighed an average of 65 pounds each. What was the weight of the hay on each square foot of the floor? $162\frac{1}{2}$ pounds

69. A fuel supplier sells heating fuel at $0.789 per gallon for 150 gallons or more; otherwise, the price is $0.849 per gallon. How much cheaper is 150 gallons at the lower price than at the higher price? $9.00

66. 24 × $60 = $1,440 total cost
    24 × 1,200 = 28,800 pounds purchased
    28,800 ÷ 2,000 = 14.4 tons purchased
    $1,440 ÷ 14.4 = $100 per ton

67. 24 × 8 = 192 sq. ft. on side wall
    2[(20 × 8)+($\frac{1}{2}$ × 20 × 5)] = 420 sq. ft. on two end walls
    192 + 420 = 612 total sq. ft. to paint
    612 ÷ 250 = 2.448 gallons

68. 9 × 175 × 65 = 102,375 pounds of hay
    20 × $31\frac{1}{2}$ = 630 sq. ft. of floor
    102,375 ÷ 630 = $162\frac{1}{2}$ pounds per square foot

69. 0.849 − 0.789 = $0.06 saved per gallon
    150 × 0.06 = $9.00 saved

99. Chapter 7 Test

LESSON 99

Objective

- To test the students' mastery of the concepts in Chapter 7.

Teaching Guide

1. Correct Lesson 98.
2. Review any areas of special difficulty.
3. Administer the test.

Following are a few pointers for testing:

1. Only the test, scratch paper, pencils, and an eraser should be on each student's desk.
2. Steps should be taken to minimize the temptation for dishonesty and the likelihood of accidentally seeing other students' answers. Following are some suggestions.
   a. Desk tops should be level. If the desks are very close to each other, have the students keep their work directly in front of them on the desk.
   b. Students should not look around more than necessary during test time.
   c. No communication should be allowed.
   d. As a rule, students should remain seated during the whole test period. It is a good idea to sharpen a few extra pencils and have them on hand.
   e. Students should hand in their tests before going on to any other work.
3. Encourage the students to do their work carefully and to go back over it if they have time. Do not allow them to hand in their tests too soon. On the other hand, some students are so meticulous that they can hardly finish their tests. If you have this problem, set a time when you will collect all the tests. Once 90 percent of the tests are completed, the rest of them should be finished in the next five or ten minutes. Of course, there are exceptions for slower students.
4. A test is different from homework. Students should realize that they must rely on their own knowledge as they work. The teacher should not help them except to make sure they understand all instructions.

Evaluating the Results

1. If you check the tests in class, have the students check each other's work. Spot check the corrected tests.
2. Tests are valuable tools in determining what the students have grasped. Are there any places where the class is uniformly weak? If so, reinforcement is needed.
3. One effective way to discover the general performance of the class is to find the class median. This is done by arranging all the scores in order from highest to lowest and finding the middle score. If there are an even number of students, find the average of the two middle scores.

Value of the Test Score

Test scores should carry considerable weight in determining report card grades. It is suggested that the test grade average have a value at least equal to that of the homework grade average. That is, the report card grade would be the homework grade average plus the test grade average, divided by 2.

Disposition of the Corrected Tests

1. As a rule, students should have the privilege to see their tests. Review any weak points, and answer any questions about why an answer is wrong.

2. The teacher may use his discretion about whether the student should be allowed to keep his test permanently. Some teachers prefer to collect them again so that the students' younger siblings will have no chance of seeing the tests in later years.

Pythagoras was a man whose labor was in wisdom and in knowledge. Many other mathematicians have also done great work through their studies. Pythagoras, Eratosthenes, and Isaac Newton have left their labors for our portion. Many of the things in our books are there because of others' work. We have not labored therein, but we can benefit in learning and using these rules because someone else gave them to us.

Chapter 8
Geometric Solids and the Pythagorean Rule

The figures of plane geometry have two dimensions: length and width. Solid geometry deals with figures having three dimensions: length, width, and height.

In the days when Daniel was a captive and then a prince in Babylon, there lived a Greek philosopher and mathematician named Pythagoras. This man's name is associated with a principle that he stated about right triangles, though the principle was understood before his time. The Pythagorean rule describes the consistent relationship between the sides of a right triangle. Solutions to some special problems can be found by applying this principle.

There is a man whose labour is in wisdom, and in knowledge, . . . yet to a man that hath not laboured therein shall he leave it for his portion.
Ecclesiastes 2:21

340 Chapter 8 *Geometric Solids and the Pythagorean Rule*

100. Geometric Solids and Surface Area of Cubes

Solid geometry is the study of three-dimensional objects. This chapter deals with finding the surface area and the volume of the regular geometric solids listed below.

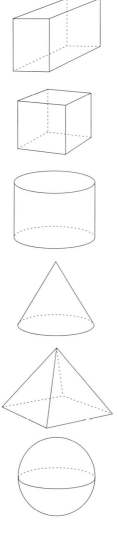

Rectangular Solid—A box-shaped object having six rectangular sides. The six sides are three pairs of congruent rectangles. Another name for a rectangular solid is a rectangular prism.

Cube—A rectangular solid whose length, width, and height are equal. The surface of any cube consists of six congruent square sides.

Cylinder—A solid shaped like a can, with two parallel circular bases. The lateral surface (side) of a cylinder is a rectangle if it is rolled out flat.

Cone—A solid shaped like some ice cream cones, with a circular base and a lateral surface that comes to a point.

Pyramid—A solid shaped like the pyramids in Egypt, with a polygon as its base and triangular sides that meet at a point (the apex). A square pyramid has a square as its base.

Sphere—A ball-shaped solid, with every point on its surface the same distance from the center.

Finding the Surface Area of a Cube

The **surface area** of a geometric solid is the total area of all its faces (sides). A cube has 6 square faces. Because all the faces are congruent, the surface area of a cube is 6 times the area of one face. The formula is written as $a_s = 6e^2$. The symbol a_s means "area (surface)," and e represents the length of one edge of the cube.

LESSON 100

Objectives

- To review that objects having three dimensions are geometric solids.
- To teach the shapes and terms used for the following geometric figures: cube, prism, cylinder, cone, pyramid, *sphere.
- To review that surface area is the area on the surface of a geometric solid.
- To review finding the surface area of a cube by using the formula $a_s = 6e^2$. (Students will be expected to memorize most of the formulas taught in this chapter.)
- To teach that *lateral area is the area of the vertical faces of a geometric solid.

Review

1. Give Lesson 100 Quiz (Finding Area).
2. *Find the area of each square or rectangle.* (Lesson 92)
   a. $s = 1.4$ m (1.96 m²)
   b. $l = 9$ ft.; $w = 4\frac{1}{2}$ ft. ($40\frac{1}{2}$ sq. ft.)
3. *Find the missing part in each commission problem.* (Lesson 68)

	Sales	Rate	Commission	
a.	$4,540	—	$181.60	(4%)
b.	$216.00	$2\frac{1}{2}$%	—	($5.40)
c.	—	3%	$32.16	($1,072)

4. *Change these mixed numbers to improper fractions.* (Lesson 36)
   a. $12\frac{4}{5}$ ($\frac{64}{5}$)
   b. $1\frac{5}{11}$ ($\frac{16}{11}$)
   c. $7\frac{3}{4}$ ($\frac{31}{4}$)

T–341   Chapter 8   *Geometric Solids and the Pythagorean Rule*

Introduction

Ask, "What are the geometric similarities between a classroom, a textbook, and a piece of paper?" All three are geometric solids because they have length, width, and height. Depending on the shape of the classroom, they may all be rectangular solids.

Perhaps the students question whether a piece of paper actually is a geometric solid. Does it have height? Yes, although we usually call it thickness. Anything with physical substance is a geometric solid.

Teaching Guide

1. **A geometric solid has three dimensions: length, width, and height.** Examples of geometric solids are cubes, rectangular solids, and square pyramids. Identify a number of geometric solids, including some in your classroom.

    Rectangular solid: classroom, book, lunch box
    Cube: box, child's block
    Cylinder: fluorescent light bulb, Thermos, jar
    Cone: paper cup, sharpened end of a pencil
    Square pyramid: feed hopper, hip roof on a square building
    Sphere: softball, globe, basketball

2. **The surface area of a geometric solid is the total area of all its faces (sides).** List the plane figures on each of the geometric solids in the lesson.

    Rectangular solid: 6 rectangles (some may be squares)
    Cube: 6 squares
    Cylinder: 2 circles, 1 rectangle
    Square pyramid: 1 square, 4 triangles

3. **To find the surface area of a cube, multiply 6 times the area of one side.** Because the area of one side is found by multiplying edge × edge, the formula is $a_s = 6e^2$.

    a. $e = 2$ in.   ($a_s = 24$ sq. in.)
    b. $e = 23$ in. ($a_s = 3{,}174$ sq. in.)
    c. $e = 3$ cm   ($a_s = 54$ cm^2)
    d. $e = 27$ cm ($a_s = 4{,}374$ cm^2)

> **Memorize this formula for the area of a cube.**
> surface area = 6 times edge squared, or $a_s = 6e^2$

Sometimes we need to find only the **lateral area** (the area of the vertical faces) of a cube, such as when painting the walls of a room. The word *lateral* means "side." Since the vertical faces of a cube consist of 4 congruent squares, the lateral area is found by multiplying 4 times the square of one edge. This is written as $a_\ell = 4e^2$. The symbol a_ℓ in this formula means "area (lateral)."

Example A	Example B
Find the surface area of a cube with edges measuring 5 inches. $a_s = 6e^2$ $a_s = 6 \times 5 \times 5$ $a_s = 150$ sq. in.	Find the lateral area of a cube with edges measuring 13 centimeters. $a_\ell = 4e^2$ $a_\ell = 4 \times 13 \times 13$ $a_\ell = 676$ cm^2

The faces of a cube are surfaces; therefore, they are measured with units of area such as square inches or square centimeters. In metric abbreviations, the exponent 2 is used to designate square measure.

CLASS PRACTICE

Find the surface area of each cube.

a. $e = 6$ in.
   216 sq. in.
b. $e = 8$ in.
   384 sq. in.
c. $e = 12$ m
   864 m^2
d. $e = 11$ ft.
   726 sq. ft.

Find the lateral area of each cube.

e. $e = 4$ in.
   64 sq. in.
f. $e = 15$ cm
   900 cm^2
g. $e = 23$ in.
   2,116 sq. in.
h. $e = 25$ cm
   2,500 cm^2

Name the plane figures that make up the faces of each geometric solid. Then give the total number of those plane figures on that geometric solid.

Geometric solids	Plane figures	Number of those plane figures
i. cube	squares	6
j. square pyramid (base)	square	1
k. square pyramid (sides)	triangles	4

WRITTEN EXERCISES

A. Write the name of each geometric solid.

1.
   cylinder

2.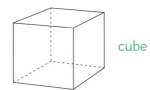
   cube

342 Chapter 8 Geometric Solids and the Pythagorean Rule

3. 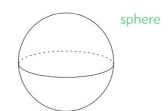 sphere

4. 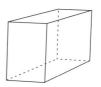 rectangular solid (or rectangular prism)

5. 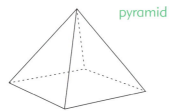 pyramid

6.  cone

B. Name the plane figures that make up the faces of each geometric solid. Then give the total number of those plane figures on that geometric solid.

Geometric solids	Plane figures	Number of those plane figures
7. square pyramid (base)	square	1
8. cylinder (bases)	circles	2
9. cylinder (side)	rectangle	1
10. rectangular solid (Assume that no face is square.)	rectangles	6

C. Find the surface area of each cube.

11. e = 4 in.   96 sq. in.
12. e = 7 in.   294 sq. in.
13. e = 10 in.   600 sq. in.
14. e = 25 mm   3,750 mm^2
15. e = 21 m   2,646 m^2
16. e = 18 cm   1,944 cm^2

D. Find the lateral area of each cube.

17. e = 9 in.   324 sq. in.
18. e = 20 in.   1,600 sq. in.
19. e = 35 in.   4,900 sq. in.
20. e = 19 cm   1,444 cm^2
21. e = 32 cm   4,096 cm^2
22. e = 55 cm   12,100 cm^2

E. Solve these reading problems.

23. Jeremiah 52:21 describes the pillars of the temple that Nebuchadnezzar destroyed. What kind of geometric solid is a pillar?   cylinder

24. What kind of geometric solid is a classroom globe?   sphere

25. Solomon overlaid the entire interior of the most holy place of the temple with gold (1 Kings 6:20). The most holy place was in the shape of a cube with edges of 20 cubits. If the floor, the ceiling, and the walls were all overlaid with gold, how many square feet were covered? (First change cubits to feet; 1 cubit = $1\frac{1}{2}$ feet.)   5,400 square feet

26. Rodney is varnishing a cubical toy box that he built. If each edge measures 22 inches, how much inside surface area must he varnish? (Include the lid.)   2,904 square inches

4. **To find the lateral area of a cube, multiply 4 times the area of one side.** The word *lateral* means "side"; finding lateral area is useful, for example, when one needs only the area of the walls of a room. This formula is written as $a_\ell = 4e^2$.

   a. $e = 4$ in. ($a_\ell = 64$ sq. in.)
   b. $e = 30$ ft. ($a_\ell = 3{,}600$ sq. ft.)
   c. $e = 5$ m   ($a_\ell = 100$ m²)
   d. $e = 40$ cm ($a_\ell = 6{,}400$ cm²)

Solutions for Exercises 25 and 26

25. 6×30^2
26. 6×22^2

T–343 Chapter 8 Geometric Solids and the Pythagorean Rule

Solutions for Exercises 27 and 28

27. $(4 \times 12) + 2(\frac{1}{2} \times 3.14 \times 2^2)$

28. $\$85 = 5\% \times \underline{}$

27. James is mulching the ground below a small grape arbor in the back yard. If the area consists of a rectangle 4 feet wide by 12 feet long, plus a semicircle at each end, how many square feet must he mulch? (Use 3.14 for pi.) 60.56 square feet

28. A salesman received $85 for selling a lawn mower at Harder's Lawn and Garden. If his commission is 5% of the selling price, what was the price of the lawn mower? $1,700

REVIEW EXERCISES

F. Find the area of each square or rectangle. *(Lesson 92)*

29. $s = 8.6$ cm   73.96 cm²
30. $s = 3\frac{1}{2}$ mi. $12\frac{1}{4}$ sq. mi.
31. $l = 12$ in.
    $w = 5$ in.   60 sq. in.
32. $l = 18$ ft.
    $w = 9\frac{1}{2}$ ft.   171 sq. ft.

G. Find the missing part in each commission problem. *(Lesson 68)*

	Sales	Rate	Commission		Sales	Rate	Commission
33.	$3,610	3%	$108.30	34.	$464	$7\frac{1}{2}$%	$34.80

H. Change these mixed numbers to improper fractions. *(Lesson 36)*

35. $8\frac{3}{4}$ $\frac{35}{4}$
36. $6\frac{11}{12}$ $\frac{83}{12}$

We have thought of thy lovingkindness,
O God, in the midst of thy temple.
Psalm 48:9

101. Surface Area of Rectangular Solids

The surface area of a rectangular solid is the combined area of three pairs of congruent rectangles. The first pair is the upper and lower faces, whose dimensions are the length and width (lw). The second pair is the end faces, whose dimensions are the width and height (wh). The third pair is the two side faces, whose dimensions are the length and height (lh).

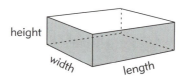

The total surface area is found by using the following formula.

$$a_s = \underset{\text{area of top and bottom faces}}{2lw} + \underset{\text{area of two end faces}}{2wh} + \underset{\text{area of front and back faces}}{2lh}$$

Memorize this formula for the surface area of a rectangular solid.
$$a_s = 2lw + 2wh + 2lh$$

The lateral area is the area of the vertical sides (shaded in the diagram above). To find the lateral area of a rectangular solid, use this formula: $a_\ell = 2wh + 2lh$.

Example A	**Example B**
Find the surface area of a rectangular solid with the following dimensions.	Find the lateral area of a rectangular solid with the following dimensions.
$l = 10$ in.; $w = 8$ in.; $h = 6$ in.	$l = 18$ cm; $w = 10$ cm; $h = 5$ cm
$a_s = 2lw + 2wh + 2lh$	$a_\ell = 2wh + 2lh$
$a_s = 2(10 \times 8) + 2(8 \times 6) + 2(10 \times 6)$	$a_\ell = 2(10 \times 5) + 2(18 \times 5)$
$a_s = 160 + 96 + 120$	$a_\ell = 100 + 180$
$a_s = 376$ sq. in.	$a_\ell = 280$ cm²

CLASS PRACTICE

Find the surface area of each rectangular solid.

a. $l = 7$ ft.   188 sq. ft.
$w = 6$ ft.
$h = 4$ ft.

b. $l = 11$ in.   366 sq. in.
$w = 8$ in.
$h = 5$ in.

c. $l = 14$ m   666 m²
$w = 9$ m
$h = 9$ m

LESSON 101

Objectives

- To review finding the surface area of a rectangular solid by using the formula $a_s = 2lw + 2wh + 2lh$.
- To teach *finding the lateral area of a rectangular solid by using the formula $a_\ell = 2wh + 2lh$. (Students are not required to memorize this formula.)

Review

1. *Find the surface area of cubes with these dimensions.* (Lesson 100)

    a. 6 in. (216 sq. in.)

    b. 13 ft. (1,014 sq. ft.)

2. *Solve these multistep reading problems.* (Lesson 97)

    a. A recipe for chocolate chip cookies calls for $1\frac{1}{4}$ cups of brown sugar. One batch makes 4 dozen cookies. If Laura has 3 cups of brown sugar, by how much is she short for making 10 dozen cookies?

    ($\frac{1}{8}$ cup)

    $\dfrac{\text{cups brown sugar}}{\text{dozen cookies}} \ \dfrac{1\frac{1}{4}}{4} = \dfrac{n}{10} \ \dfrac{\text{cups brown sugar}}{\text{dozen cookies}}$

    $n = 3\frac{1}{8} \quad 3\frac{1}{8} - 3 = \frac{1}{8}$

    b. A farmer has his forage equipment for sale. He is asking $4,000 for the harvester, $500 for the corn head, $2,200 each for 3 forage wagons, and $700 for the blower. If someone wants to buy all the pieces together, he will sell them for $10,000. Buying all the pieces together is how much cheaper than buying them separately?

    ($1,800)

    $4,000 + $500 + (3 × $2,200) + $700 = $11,800

    $11,800 − $10,000 = $1,800

3. *Find the area of parallelograms with these dimensions.* (Lesson 93)

    a. $b = 4$ in.
       $h = 2\frac{1}{2}$ in. (10 sq. in.)

    b. $b = 250$ ft.
       $h = 140$ ft. (35,000 sq. ft.)

4. *Write these expressions, using symbols.* (Lesson 85)

    a. ray RS ($\overrightarrow{RS}$)

    b. line DE is parallel to line MN ($\overleftrightarrow{DE} \parallel \overleftrightarrow{MN}$)

5. *Solve these percent problems mentally.* (Lesson 69)

    a. 50% of 14 = ___ (7)

    b. 9 is ___ of 12 (75%)

6. *Solve these fraction problems, and express the answers in simplest form.* (Lesson 37)

    a. $\frac{7}{8}$
       $+ \frac{1}{4}$
       ($1\frac{1}{8}$)

    b. $\frac{11}{12}$
       $- \frac{2}{3}$
       ($\frac{1}{4}$)

T–345 Chapter 8 *Geometric Solids and the Pythagorean Rule*

Introduction

Find the surface area of your classroom. For simple calculations, use approximate dimensions. Have students decide which measurements to use to find the area of each rectangle.

Teaching Guide

1. **The formula for finding the surface area of a rectangular solid is $a_s = 2lw + 2wh + 2lh$.** Students are to memorize this formula.

    a. $l = 6$ in.
        $w = 3$ in.
        $h = 3$ in. (90 sq. in.)

    b. $l = 8$ ft.
        $w = 7$ ft.
        $h = 5$ ft. (262 sq. ft.)

    c. $l = 4$ ft.
        $w = 2$ ft. 6 in.
        $h = 3$ ft. 9 in. ($68\frac{3}{4}$ sq. ft.)

2. **The formula for finding the lateral area of a rectangular solid is $a_\ell = 2wh + 2lh$.**

    a. $l = 7$ in.
        $w = 4$ in.
        $h = 4$ in. (88 sq. in.)

    b. $l = 9$ cm
        $w = 8$ cm
        $h = 6$ cm (204 cm^2)

    c. $l = 5\frac{1}{4}$ in.
        $w = 1\frac{1}{2}$ in.
        $h = 2\frac{1}{4}$ ($30\frac{3}{8}$ sq. in.)

Find the lateral area of each rectangular solid.

d. l = 8 ft. 3 in.   152½ sq. ft.
   w = 7 ft.
   h = 5 ft.

e. l = 7 yd.   86 sq. yd.
   w = 7 yd. 1 ft.
   h = 3 yd.

f. l = 15 cm   400 cm²
   w = 10 cm
   h = 8 cm

WRITTEN EXERCISES

A. *Write the formulas for finding these areas. Be sure you know the first one by memory.*

1. Surface area of a rectangular solid $a_s = 2lw + 2wh + 2lh$
2. Lateral area of a rectangular solid $a_\ell = 2wh + 2lh$

B. *Find the surface area of each rectangular solid.*

3. l = 5 in.   62 sq. in.
   w = 3 in.
   h = 2 in.

4. l = 10 in.   344 sq. in.
   w = 7 in.
   h = 6 in.

5. l = 12 ft.   552 sq. ft.
   w = 9 ft.
   h = 8 ft.

6. l = 9 cm   254 cm²
   w = 7 cm
   h = 4 cm

7. l = 14 cm   1,012 cm²
   w = 13 cm
   h = 12 cm

8. l = 15 m   850 m²
   w = 11 m
   h = 10 m

C. *Find the lateral area of each rectangular solid.*

9. l = 6 yd. 1 ft.   64 sq. yd.
   w = 4 yd. 1 ft.
   h = 3 yd.

10. l = 12 ft.   418 sq. ft.
    w = 10 ft.
    h = 9 ft. 6 in.

D. *Find the surface area of each rectangular solid, and answer the question after each pair.*

11. **a.** l = 4 cm; w = 3 cm; h = 2 cm   52 cm²
    b. l = 8 cm; w = 6 cm; h = 4 cm   208 cm²
    c. Look at your answers for a and b. When all the dimensions of a rectangular solid are doubled, the surface area is multiplied by __4__.

12. **a.** l = 3 cm; w = 2 cm; h = 1 cm   22 cm²
    b. l = 9 cm; w = 6 cm; h = 3 cm   198 cm²
    c. Look at your answers for a and b. When all the dimensions of a rectangular solid are tripled, the surface area is multiplied by __9__.

E. *Solve these multistep reading problems. Use formulas from this lesson for numbers 13–16.*

13. The Israelites were instructed that if a house was smitten with leprosy, they were to scrape the walls and replaster them (Leviticus 14). Suppose a room was 14 feet long, 12 feet wide, and 7 feet high, and it had a door measuring 3 feet by 6 feet and a window measuring 1 foot by 1 foot. What was the lateral area that would have needed replastering? (The door and window would be excluded.)   345 square feet

346 Chapter 8 Geometric Solids and the Pythagorean Rule

14. A large fish aquarium is 4 feet long, 2 feet wide, and $1\frac{1}{2}$ feet high. How many square feet of glass is needed for the four sides?   18 square feet

15. Benjamin is painting the four plywood walls on his sister's playhouse, inside and out. The walls are each 5 feet high and 7 feet long. There is a door 2 feet by 4 feet on one side and a window 1 foot square on another side. How many square feet must Benjamin paint? (Hint: Subtract the area of the door and window before doubling the surface area.)   262 square feet

16. Mother is sewing new covers for the sofa cushions. If each cushion measures 18 by 24 inches and is 6 inches thick, how many square feet of material will she need to cover 3 cushions? (Disregard seam allowance.) $28\frac{1}{2}$ square feet

17. At Landon's Consignment Shop, people bring in used clothing and furniture to be sold on commission. For one person, the shop sold a pair of shoes for $15, a coat for $30, and a living room suite for $395. How much commission did the shop receive if the rate is 50% for clothing and 30% for furniture? $141.00

18. Jay has agreed to raise 2,000 sunflower plants for a nearby produce stand. A 1-ounce seed packet contains about 360 seeds. If the germination rate is 85%, how many packets should he buy to be reasonably sure that he will have enough plants?   7 packets

REVIEW EXERCISES

F. Write the formula for finding each fact. *(Lesson 100)*

19. Surface area of a cube $a_s = 6e^2$
20. Lateral area of a cube $a_\ell = 4e^2$

G. Find the surface area of each cube. *(Lesson 100)*

21. $e = 7$ in.   294 sq. in.   22. $e = 11$ ft.   726 sq. ft.

H. Find the area of each parallelogram. *(Lesson 93)*

23. $b = 5$ in. $22\frac{1}{2}$ sq. in.   24. $b = 100$ ft.   7,500 sq. ft.
    $h = 4\frac{1}{2}$ in. $h = 75$ ft.

I. Write these expressions, using symbols. *(Lesson 85)*

25. Line segment ST $\overline{ST}$
26. Line HI is perpendicular to line JK $\overleftrightarrow{HI} \perp \overleftrightarrow{JK}$

J. Solve these percent problems mentally. *(Lesson 69)*

27. 40% of 15 = __6__   28. 3 is $16\frac{2}{3}$% of 18

K. Solve these fraction problems, and express the answers in simplest form. *(Lesson 37)*

29. $\frac{7}{9} + \frac{1}{3} = 1\frac{1}{9}$

30. $\frac{5}{8} - \frac{1}{2} = \frac{1}{8}$

31. $\frac{3}{4} - \frac{1}{6} = \frac{7}{12}$

32. $\frac{2}{5} + \frac{3}{4} = 1\frac{3}{20}$

An Ounce of Prevention

1. Students tend to use these formulas by rote without understanding what they are really doing. Help them to grasp the principle behind the calculations by making frequent use of sketches or examples of the geometric objects under discussion.
2. Bring a can with a label to class for Lesson 102.

Further Study

For most practical purposes, lateral area is found by multiplying perimeter times height. This is the method used by people who do painting and insulating; when there is a gable end, they find an average height. The method in the lesson is taught because it is closely related to the method for finding total surface area. You may wish to mention the other method to the students, or allow them to discover it on their own.

Exercises 11 and 12 show how doubling or tripling the dimensions of a rectangular solid affects its surface area. In mathematical terms, surface area varies as the square of the multiple by which the dimensions are increased.

Solutions for Part E

13. $a = 2(14 \times 7) + 2(12 \times 7) = 364$; $364 - (18 + 1) = 345$
14. $a = 2(4 \times 1\frac{1}{2}) + 2(2 \times 1\frac{1}{2})$
15. $a = 4(5 \times 7) = 140$; $140 - (8 + 1) = 131$; $2 \times 131 = 262$
16. $a = 2(1\frac{1}{2} \times 2) + 2(1\frac{1}{2} \times \frac{1}{2}) + 2(2 \times \frac{1}{2}) = 9\frac{1}{2}$; $3 \times 9\frac{1}{2} = 28\frac{1}{2}$
17. $0.5(15 + 30) + 0.3 \times 395 = 141$
18. $2{,}000 \div (0.85 \times 360) = 6.5$

T-347 Chapter 8 Geometric Solids and the Pythagorean Rule

LESSON 102

Objectives

- To review finding the surface area of a cylinder by using the formula $a_s = 2\pi r^2 + 2\pi rh$.

- To teach *finding the lateral area of a cylinder by using the formula $a_\ell = 2\pi rh$. (Students are not required to memorize this formula.)

Review

1. *Find surface areas of rectangular solids with these dimensions.* (Lesson 101)

    a. $l = 14$ in.
       $w = 11$ in.
       $h = 7$ in.   (658 sq. in.)

    b. $l = 4\frac{1}{4}$ ft.
       $w = 3$ ft.
       $h = 2\frac{1}{2}$ ft.   (61.75 sq. ft.)

2. *Find the surface areas of cubes with these dimensions.* (Lesson 100)

    a. $e = 12.2$ cm   (893.04 cm^2)

    b. $e = 15$ ft.   (1,350 sq. ft.)

3. *Find the areas of trapezoids with these dimensions.* (Lesson 94)

    a. $h = 9$ in.
       $b_1 = 9$ in.
       $b_2 = 7\frac{1}{2}$ in.   ($74\frac{1}{4}$ sq. in.)

    b. $h = 15$ cm
       $b_1 = 16$ cm
       $b_2 = 14$ cm   (225 cm^2)

4. *Use sketches to solve these problems.* (Lesson 70)

    a. Donald replaced the rotting boards on the picnic table top. He used 4 pieces of treated lumber that were each $7\frac{1}{2}$ inches wide, and he allowed a $\frac{1}{2}$-inch gap between the boards. How wide was the new table top?   ($31\frac{1}{2}$ in.)

    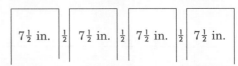

    b. Mother makes pancakes on her griddle that is 12 inches wide and 20 inches long. She can fry 6 pancakes at a time if the pancakes are 6 inches in diameter. While 6 pancakes are frying, what is the area of the griddle space left unused? (Use 3.14 for pi.)
       (70.44 sq. in.)

    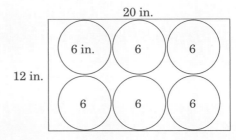

5. *Multiply these fractions.* (Lesson 38)

    a. $19 \times \frac{1}{4}$ ($4\frac{3}{4}$)

    b. $\frac{4}{9}$ of 65   ($28\frac{8}{9}$)

102. Surface Area of Cylinders

Finding the surface area of a geometric solid is simplified by thinking of the parts as separate plane figures. A cylinder has two circles for bases. The rest of the cylinder (lateral surface), if cut vertically and rolled out flat, would be a rectangle. Its length is equal to the circumference of the circular base, and its width is equal to the height of the cylinder.

surface area equals   circle   plus   circle   plus   rectangle (circumference of cylinder × height of cylinder)

The surface area of a cylinder is found by combining the areas of the various parts of the cylinder. The formula is given below.

$$a_s = \underbrace{2\pi r^2}_{\text{area of the two bases}} + \underbrace{2\pi rh}_{\text{area of the lateral surface}}$$

Memorize this formula for the surface area of a cylinder.
$$a_s = 2\pi r^2 + 2\pi rh$$

The lateral area of a cylinder is the area of a rectangle. To find the lateral area, use this formula: $a_\ell = 2\pi rh$.

Example A	Example B
Find the surface area of a cylinder with the following dimensions. Use 3.14 for pi. $r = 1.5$ in. $h = 4$ in. $a_s = 2\pi r^2 + 2\pi rh$ $a_s = (2 \times 3.14 \times 1.5 \times 1.5)$ $\quad + (2 \times 3.14 \times 1.5 \times 4)$ $a_s = 14.13 + 37.68$ $a_s = 51.81$ sq. in.	Find the lateral area of a cylinder with the following dimensions. Use 3.14 for pi. $r = 4$ m $h = 5$ m $a_\ell = 2\pi rh$ $a_\ell = 2 \times 3.14 \times 4 \times 5$ $a_\ell = 125.6$ m²

CLASS PRACTICE

Give the formulas for finding these facts.

a. Surface area of a cylinder $a_s = 2\pi r^2 + 2\pi rh$

b. Lateral area of a cylinder $a_\ell = 2\pi rh$

Find the surface area of each cylinder. Use 3.14 for pi.

c. $r = 2$ in.   113.04 sq. in.   d. $r = 3$ in.   207.24 sq. in.   e. $r = 6$ cm   602.88 cm²
   $h = 7$ in. $h = 8$ in. $h = 10$ cm

Find the lateral area of each cylinder.

f. $r = 2$ in.   62.8 sq. in.   g. $r = 14$ m   2,640 m²   h. $r = 11$ ft.   2,210.56 sq. ft.
   $h = 5$ in. $h = 30$ m $h = 32$ ft.
   $\pi = 3.14$ $\pi = 3\frac{1}{7}$ $\pi = 3.14$

WRITTEN EXERCISES

A. Write the formulas for finding these facts. Be sure you know the first one by memory.

1. Surface area of a cylinder $a_s = 2\pi r^2 + 2\pi rh$

2. Lateral area of a cylinder $a_\ell = 2\pi rh$

B. Find the surface area of each cylinder. Use 3.14 for pi.

3. $r = 8$ in.   854.08 sq. in.   4. $r = 13$ cm   4,816.76 cm²
   $h = 9$ in. $h = 46$ cm

5. $r = 4$ in.   251.2 sq. in.   6. $r = 5$ ft.   408.2 sq. ft.
   $h = 6$ in. $h = 8$ ft.

7. $r = 10$ cm   1,256 cm²   8. $r = 7$ cm   1,186.92 cm²
   $h = 10$ cm $h = 20$ cm

C. Find the surface area of each cylinder, and answer the question after each pair. Use 3.14 for pi.

9. a. $r = 3$ in.; $h = 4$ in.   131.88 sq. in.
   b. $r = 6$ in.; $h = 8$ in.   527.52 sq. in.
   c. Look at your answers for a and b. When the dimensions of a cylinder are doubled, the surface area is multiplied by __4__.

10. a. $r = 2$ in.; $h = 4$ in.   75.36 sq. in.
    b. $r = 8$ in.; $h = 16$ in.   1,205.76 sq. in.
    c. Look at your answers for a and b. When the dimensions of a cylinder are quadrupled, the surface area is multiplied by __16__.

Introduction

Bring a can with a label to class, and have the students identify it as a cylinder. Ask the following questions about the cylinder.

1. What are the geometric shapes of the top and bottom of a cylinder? (circles)
2. What is the formula for the area of one base? ($a = \pi r^2$)
3. What formula can we use for the area of the two bases? ($a = 2\pi r^2$)
4. (Cut vertically across the label and pull it off.) What figure is the lateral surface of the cylinder? (rectangle)
5. The length of the rectangle is equal to what? (the circumference of the base)
6. How can we find the length of the rectangle, since it equals the circumference of a circle? ($c = \pi d$ or $c = 2\pi r$)
7. The width of the rectangle is equal to what? (the height of the can)
8. How can we find the area of the rectangle? (by multiplying the length times the width: $2\pi r \times h$)
9. How can we combine these formulas for the surface area of the entire cylinder?

$$(a_s = 2\pi r^2 + 2\pi rh)$$

Teaching Guide

1. **The surface of a cylinder consists of two circles and a rectangle.** The two bases of a cylinder are circles; and its side, cut vertically and rolled flat, forms a rectangle.

2. **The formula for the surface area of a cylinder is $a_s = 2\pi r^2 + 2\pi rh$.** This is a combination of the formula for finding the area of the bases (two circles) and the rectangular side.

   a. $r = 8$ in.
      $h = 4$ in. (602.88 sq. in.)

   b. $r = 4$ m
      $h = 25$ m (728.48 m²)

3. **The formula for the lateral area of a cylinder is $a_\ell = 2\pi rh$.** Show that the circumference of the can (πd) is what yields the length of the rectangle. This might be hard for students to visualize.

   a. $r = 12$ in.
      $h = 48$ in. (3,617.28 sq. in.)

   b. $r = 14.6$ cm
      $h = 105$ cm (9,627.24 cm²)

Note: At the end of the class period, hand out the sheets *Lesson 102 Speed Test Preparation* (Addition), to help students prepare for the Lesson 103 Speed Test.

An Ounce of Prevention

Warn the students that some of the reading problems give the diameter or circumference instead of the radius.

Further Study

The lateral area of a cylinder can be found by multiplying perimeter times height, the same as with rectangular solids. Of course, the perimeter in this case is the circumference, which is found by multiplying πd or $2\pi r$. The formula taught in this lesson uses $2\pi r$ because this makes it simpler to use (its parts are more similar and have fewer different letters). Otherwise the formula would be $a_s = 2\pi r^2 + \pi dh$.

Solutions and Sketches for Part E

13. $2(3.14 \times 1.75^2) + (2 \times 3.14 \times 1.75 \times 4.5)$
14. $2 \times 3.14 \times 12 \times 80$
15. $2 \times 27 \times 18$
16. $2 \times 3.14 \times 1.5 \times 6$

17.
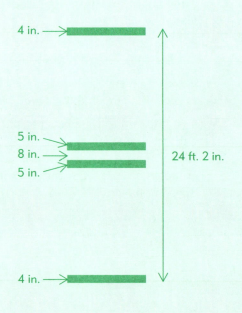

18.
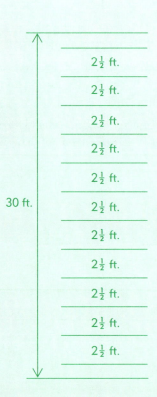

D. Find the lateral area of each cylinder. Use 3.14 for pi.

11. $r = 2$ in.　　37.68 sq. in.
　　$h = 3$ in.

12. $r = 7$ ft.　　395.64 sq. ft.
　　$h = 9$ ft.

E. Solve these reading problems. Use 3.14 for pi. Draw sketches for numbers 17 and 18.

13. A tin can has a height of $4\frac{1}{2}$ inches and a radius of $1\frac{3}{4}$ inches. How many square inches of metal were used to make this can? (Answer to the nearest hundredth.) 68.69 sq. in.

14. A silo is 80 feet high and 24 feet in diameter. What is the outside lateral area of the silo?　6,028.8 sq. ft.

15. King Hezekiah overlaid with gold the two brass pillars that Solomon made for the temple porch (1 Kings 7:15–22; 2 Kings 18:16). The pillars had a height of about 27 feet and a circumference of about 18 feet. What was the lateral area of the two pillars together?　972 sq. ft.

16. The front roller on a paving machine is a cylinder with a radius of $1\frac{1}{2}$ feet and a height of 6 feet. How much area does this roller cover each time it makes one revolution?　56.52 sq. ft.

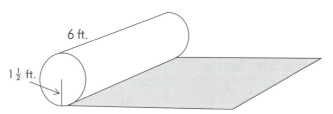

17. The two-lane road adjacent to the Shultz's property has a total width of 24 feet 2 inches. The white line along each edge is 4 inches wide, and each of the two solid yellow lines in the center is 5 inches wide. A space of 8 inches separates the two yellow lines. How wide is each driving lane?　11 ft.

18. Marla would like to plant 12 rows of vegetables in a garden 30 feet wide. The rows will be spaced $2\frac{1}{2}$ feet apart. How close to the edge of the garden must she plant the outside rows?　$1\frac{1}{4}$ ft.

REVIEW EXERCISES

F. Write the formulas for finding these areas. *(Lessons 100, 101)*

19. Surface area of a cube　　$a_s = 6e^2$

20. Surface area of a rectangular solid　　$a_s = 2lw + 2wh + 2lh$

G. Find the surface area of each rectangular solid. *(Lesson 101)*

21. $l = 10$ in.　　412 sq. in.
　　$w = 8$ in.
　　$h = 7$ in.

22. $l = 5$ ft.　　94 sq. ft.
　　$w = 4$ ft.
　　$h = 3$ ft.

H. Find the surface area of each cube. *(Lesson 100)*

23. $e = 36$ m　　7,776 m²

24. $e = 18$ ft.　　1,944 sq. ft.

350 Chapter 8 *Geometric Solids and the Pythagorean Rule*

I. Find the area of each trapezoid. *(Lesson 94)*

25. h = 8 in.   66 sq. in.   **26.** h = 16 cm   168 cm²
b_1 = 10 in. b_1 = 7 cm
b_2 = 6½ in. b_2 = 14 cm

J. Measure these angles. *(Lesson 86)*

27.   69°

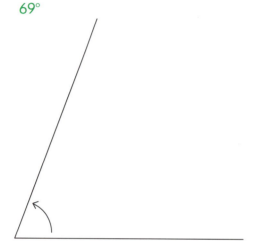

28.   82°

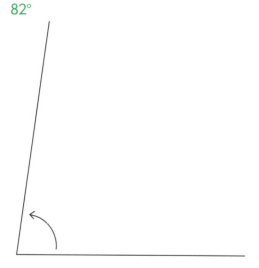

K. Multiply these fractions. *(Lesson 38)*

29. 21 × ⅕   4⅕   **30.** ⅝ of 57   35⅝

T–351   Chapter 8   Geometric Solids and the Pythagorean Rule

LESSON 103

Objectives

- To teach *finding the surface area of a square pyramid by using the formula $a_s = 4(\frac{1}{2}b\ell) + b^2$ or its simpler form $a_s = 2b\ell + b^2$.

- To teach *finding the lateral area of a square pyramid by using the formula $a_\ell = 4(\frac{1}{2}b\ell)$ or its simpler form $a_\ell = 2b\ell$. (Students are not required to memorize this formula.)

Review

1. Give Lesson 103 Speed Test (Addition). *Note:* You may want to repeat this speed test over several days to sharpen the pupils' addition skills.

2. *Find the surface area of cylinders with these dimensions.* (Lesson 102)

   a. $r = 5$ m
      $h = 16$ m   (659.4 m²)

   b. $r = 11$ in.
      $h = 29$ in.   (2,763.2 sq. in.)

3. *Find the surface area of rectangular solids with these dimensions.* (Lesson 101)

   a. $l = 12$ ft.
      $w = 8$ ft.
      $h = 5$ ft.   (392 sq. ft.)

   b. $l = 45$ in.
      $w = 30$ in.
      $h = 25$ in.   (6,450 sq. in.)

4. *Find the area of circles with these dimensions.* (Lesson 95)

   a. $r = 16$ cm
      $\pi = 3.14$ (803.84 cm²)

   b. $r = 3\frac{1}{2}$ mi.
      $\pi = 3\frac{1}{7}$ ($38\frac{1}{2}$ sq. mi.)

5. *Each pair gives the degrees in two angles of a triangle. Find the degrees in the third angle.* (Lesson 87)

   a. 45°, 63°   (72°)

   b. 15°, 123°   (42°)

6. *Solve these division problems.* (Lesson 39)

   a. $16 \div \frac{2}{5}$ (40)

   b. $\frac{4}{5} \div \frac{1}{2}$ ($1\frac{3}{5}$)

103. Surface Area of Square Pyramids

A square pyramid has a square base and four triangular faces. The four triangular sides meet at an apex or tip like the pyramids in Egypt. To calculate the surface area of a square pyramid, use the formulas for the area of a square ($a = s^2$) and the area of a triangle ($a = \frac{1}{2}bh$).

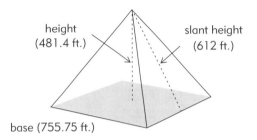

Great Pyramid of Egypt

In the sketch at the right, notice that the height of a triangular face is not the same as the height of the pyramid. This is because the triangles are slanted, not vertical. To avoid confusion, the cursive ℓ is used for the lateral, or slant, height of the triangle to distinguish from the height of the pyramid. So the formula for the area of a triangle changes from $\frac{1}{2}bh$ to $\frac{1}{2}b\ell$ when the triangle is a face of a pyramid.

To find the area of all four triangular faces, multiply $\frac{1}{2}b\ell$ by 4; this is written as $4(\frac{1}{2}b\ell)$.

To find the area of the bottom face of the pyramid, use the formula for the area of a square, $a = s^2$. The side of this square is also the base of a triangular face. That measure is already designated as b in the formula for the triangular faces. It is helpful to use the same symbol for the same measure. Therefore, the area of this square is b^2.

By combining these formulas, we obtain the following formula for the surface area of a square pyramid.

$$a_s = \underset{\substack{\text{area of four} \\ \text{triangular sides}}}{4(\tfrac{1}{2}b\ell)} + \underset{\substack{\text{area of} \\ \text{bottom face}}}{b^2}$$

> **Memorize this formula for the surface area of a square pyramid.**
> $$a_s = 4(\tfrac{1}{2}b\ell) + b^2$$

The formula given above is meaningful as it relates to the different parts of the pyramid. This is the formula that you are expected to memorize. But according to the associative law of multiplication, $4 \times (\frac{1}{2} \times b \times \ell) = (4 \times \frac{1}{2}) \times b \times \ell$. Therefore the formula above can be simplified to $a_s = 2b\ell + b^2$. This form is more efficient in actual use because it requires fewer calculations.

The lateral area of a square pyramid is the area of the four triangular faces. To find this area, use only the first part of the formula above. This can be stated as $a_\ell = 4(\frac{1}{2}b\ell)$, or simplified to $a_\ell = 2b\ell$.

352 Chapter 8 *Geometric Solids and the Pythagorean Rule*

> **Example A**
> Find the surface area of a square pyramid having a base of 10 feet and a slant height of 8 feet.
> $a_s = 4(\frac{1}{2}b\ell) + b^2$
> $a_s = 2b\ell + b^2$
> $a_s = 2(10 \times 8) + (10 \times 10)$
> $a_s = 160 + 100$
> $a_s = 260$ sq. ft.

> **Example B**
> Find the lateral area of the Great Pyramid of Egypt from the diagram on page 351.
> $a_\ell = 4(\frac{1}{2}b\ell)$
> $a_\ell = 2b\ell$
> $a_\ell = 2 \times 755.75 \times 612$
> $a_\ell = 925{,}038$ sq. ft.

CLASS PRACTICE

Find the surface area of each square pyramid.

a. $b = 4$ cm   88 cm²   b. $b = 12$ ft.   936 sq. ft.   c. $b = 11$ in.   341 sq. in.
   $\ell = 9$ cm $\ell = 33$ ft. $\ell = 10$ in.

Find the lateral area of each square pyramid.

d. $b = 8$ cm   128 cm²   e. $b = 11$ m   352 m²   f. $b = 20$ in.   720 sq. in.
   $\ell = 8$ cm $\ell = 16$ m $\ell = 18$ in.

WRITTEN EXERCISES

A. *Write these formulas. Be sure you know the first one by memory.*

1. Surface area of a square pyramid $a_s = 4(\frac{1}{2}b\ell) + b^2$
2. Lateral area of a square pyramid $a_\ell = 4(\frac{1}{2}b\ell)$

B. *Find the surface area of each square pyramid.*

3. $b = 5$ in.   65 sq. in.   4. $b = 14$ cm   504 cm²   5. $b = 3$ in.   27 sq. in.
   $\ell = 4$ in. $\ell = 11$ cm $\ell = 3$ in.

6. $b = 5$ ft.   85 sq. ft.   7. $b = 8$ in.   224 sq. in.   8. $b = 9$ in.   207 sq. in.
   $\ell = 6$ ft. $\ell = 10$ in. $\ell = 7$ in.

9. $b = 12$ cm   360 cm²   10. $b = 15$ cm   585 cm²   11. $b = 18$ in.   1,224 sq. in.
   $\ell = 9$ cm $\ell = 12$ cm $\ell = 25$ in.

C. *Find the lateral area of each square pyramid.*

12. $b = 8$ cm   13. $b = 9$ cm   14. $b = 15$ in.   15. $b = 100$ ft.
    $\ell = 7$ cm 112 cm² $\ell = 8$ cm 144 cm² $\ell = 22$ in. 660 sq. in. $\ell = 60$ ft.
                                                                             12,000 sq. ft.

D. *Solve these reading problems.*

16. Carl set up a small tent in the shape of a square pyramid. Each side was 8 feet wide, and the slant height was 7 feet. How many square feet of plastic were needed for the sides of the tent?   112 square feet

Lesson 103 T–352

Introduction

Refer to the sketch of the Great Pyramid of Egypt, and ask the following questions.

1. Why is this pyramid called a square pyramid? (It has a square base.)
2. What geometric shapes are on the faces of a square pyramid? (one square and four triangles)
3. What formula can we use to find the area of the base? ($a = s^2$)
4. What formula can we use to find the area of one of the triangular sides? ($a = \frac{1}{2}bh$; but since h is the slant height, use ℓ for that part: $a = \frac{1}{2}b\ell$)
5. What formula can we write to find the area of all four triangles? ($a = 4[\frac{1}{2}b\ell]$)
6. How can we combine these formulas for the surface area of the entire square pyramid?
   (Since $s = b$, we can write the formula $a_s = 4(\frac{1}{2}b\ell) + b^2$.)
7. How can we simplify this formula?
   (Since $4 \times \frac{1}{2} = 2$, we can use $a = 2b\ell$.)

Teaching Guide

1. **The surface of a square pyramid consists of a square base and four triangular sides.**

2. **The following letters are used for the dimensions of a pyramid.**

   b = base of triangular side, which is also one side of the square base of the pyramid

   ℓ = slant height of triangular side

   The symbol ℓ is used for the height of the triangle because this is different from the height of the pyramid itself.

3. **The formula for finding the surface area of a square pyramid is $a_s = 4(\frac{1}{2}b\ell) + b^2$. This can be simplified to $a_s = 2b\ell + b^2$.**

   a. $b = 11$ cm
      $\ell = 8$ cm (297 cm²)

   b. $b = 14$ cm
      $\ell = 11$ cm (504 cm²)

4. **The formula for finding the lateral area of a square pyramid is $a_\ell = 4(\frac{1}{2}b\ell)$. This can be simplified to $a_\ell = 2b\ell$.**

   a. $b = 10$ in.
      $\ell = 7$ in. (140 sq. in.)

   b. $b = 13$ cm
      $\ell = 10$ cm (260 cm²)

T–353   Chapter 8   Geometric Solids and the Pythagorean Rule

Further Study

A pyramid is any geometric solid whose base is a polygon and whose sides are triangles that meet at one point. The pyramid is named according to the shape of its base; therefore, a square pyramid has a square base. The base can also be a triangle, a pentagon, or any other polygon. A regular pyramid is one whose base is a regular polygon and whose sides are all congruent isosceles triangles.

The formula for a triangular pyramid requires a few adjustments from the one used for square pyramids. Instead of finding the area of a square base, one must find the area of a triangle ($\frac{1}{2}bh$). Also, a triangular pyramid has three sides instead of four. Thus the formula for the surface area of a triangular pyramid is as = $3(\frac{1}{2}b\ell) + \frac{1}{2}bh_b$ (where h_b represents the height of the triangular base).

The Egyptian pyramids confirm that God created man with the full intellect that he has today. Though built shortly after the Flood, the pyramids reveal amazing architectural skill. For example, the 755-foot sides of the Great Pyramid are accurate within $\frac{1}{2}$ inch.

Solutions for Part D

16. $4 \times \frac{1}{2} \times 8 \times 7$
17. $4 \times \frac{1}{2} \times 9 \times 6$
18. $4 \times \frac{1}{2} \times 708 \times 471$
19. $4 \times \frac{1}{2} \times 357 \times 218$
20. $3\frac{1}{7} \times 140^2$
21. 3.14×2^2

17. A steel grain bin with an open top is in the shape of an inverted square pyramid. Each side measures 9 feet, and the slant height is 6 feet. How many square feet of steel were used to make this bin? 108 square feet

18. Through centuries of weathering, the pyramids in Egypt have lost some of their original size. The middle pyramid originally had a base of 708 feet. The height of each triangular face was 471 feet. What was the lateral area of this pyramid? 666,936 square feet

19. The southernmost pyramid originally had a base of 357 feet and a slant height of 218 feet. What was the lateral area of this pyramid? 155,652 square feet

20. The Nolts have a sprinkler that can shoot water 140 feet out from a rotating nozzle. If it irrigates a full circle, what is the area that it can water from a single location? (Use $3\frac{1}{7}$ for pi.) 61,600 square feet

21. Find the area of a round tabletop 4 feet in diameter. 12.56 square feet

REVIEW EXERCISES

E. Write the formulas for finding these areas. *(Lessons 100–102)*

22. Surface area of a cube $a_s = 6e^2$

23. Surface area of a rectangular solid $a_s = 2lw + 2wh + 2lh$

24. Surface area of a cylinder $a_s = 2\pi r^2 + 2\pi rh$

F. Find the surface area of each cylinder. *(Lesson 102)*

25. $r = 2$ m   100.48 m²
    $h = 6$ m
    $\pi = 3.14$

26. $r = 7$ in.   1,408 sq. in.
    $h = 25$ in.
    $\pi = 3\frac{1}{7}$

G. Find the surface area of each rectangular solid. *(Lesson 101)*

27. $l = 16$ ft.   788 sq. ft.
    $w = 10$ ft.
    $h = 9$ ft.

28. $l = 3$ yd.   32 sq. yd.
    $w = 2$ yd.
    $h = 2$ yd.

H. Find the area of each circle. *(Lesson 95)*

29. $r = 21$ cm   1,384.74 cm²
    $\pi = 3.14$

30. $r = 14$ yd.   616 sq. yd.
    $\pi = 3\frac{1}{7}$

I. Find the measure of the third angle for each triangle. Two angles are given. *(Lesson 87)*

31. 54°, 21°   105°

32. 18°, 118°   44°

J. Solve these division problems. *(Lesson 39)*

33. $14 \div \frac{1}{3}$ 42

34. $\frac{4}{9} \div \frac{4}{5}$ $\frac{5}{9}$

354 Chapter 8 Geometric Solids and the Pythagorean Rule

104. Surface Area of Spheres

A sphere is a geometric solid that does not have any polygonal faces. However, the surface area of a sphere has been found to equal 4 times the area of a circle with the same diameter. Therefore, the surface area of a sphere can be calculated with the formula $a_s = 4\pi r^2$.

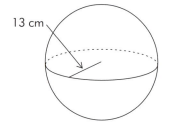

Memorize this formula for the surface area of a sphere.
$$a_s = 4\pi r^2$$

Example A
Find the surface area of a sphere with a radius of 13 centimeters. Use 3.14 for pi.
$a_s = 4\pi r^2$
$a_s = 4 \times 3.14 \times 13 \times 13$
$a_s = 2{,}122.64$ cm²

Example B
Find the surface area of sphere with a radius of $\frac{7}{16}$ inch.
$a_s = 4\pi r^2$
$a_s = \frac{\overset{1}{4}}{1} \times \frac{\overset{11}{22}}{\underset{1}{7}} \times \frac{\overset{1}{7}}{\underset{4}{16}} \times \frac{7}{\underset{8}{16}} = \frac{77}{32}$
$a_s = 2\frac{13}{32}$ sq. in.

CLASS PRACTICE

Find the surface area of each sphere.

a. $r = \frac{3}{4}$ in.
 $\pi = 3\frac{1}{7}$ $7\frac{1}{14}$ sq. in.

b. $r = 5$ cm
 $\pi = 3.14$ 314 cm²

c. $r = 21$ in.
 $\pi = 3\frac{1}{7}$ 5,544 sq. in.

d. $r = 42$ in.
 $\pi = 3\frac{1}{7}$
 22,176 sq. in.

WRITTEN EXERCISES

A. Write this formula. Be sure you know it by memory.

1. Surface area of a sphere $a_s = 4\pi r^2$

B. Find the surface area of each sphere. Use 3.14 for pi.

2. $r = 2$ in.   50.24 sq. in.
3. $r = 6$ in.   452.16 sq. in.
4. $r = 18$ in. 4,069.44 sq. in.
5. $r = 10$ cm   1,256 cm²
6. $r = 25$ cm   7,850 cm²
7. $r = 75$ cm   70,650 cm²

C. Find the surface area of each sphere. Use $3\frac{1}{7}$ for pi.

8. $r = 17\frac{1}{2}$ ft.   3,850 sq. ft.
9. $r = 28$ m   9,856 m²
10. $r = 5\frac{1}{4}$ in. $346\frac{1}{2}$ sq. in.

LESSON 104

Objective

- To teach *finding the surface area of a sphere by using the formula $a_s = 4\pi r^2$.

Review

1. *Find the surface area of square pyramids with these dimensions.* (Lesson 103)

    a. $b = 6$ cm
       $\ell = 9$ cm   (144 cm²)

    b. $b = 25$ in.
       $\ell = 26$ in.   (1,925 sq. in.)

2. *Find the surface area of cylinders with these dimensions.* (Lesson 102)

    a. $r = 3$ in.
       $h = 12$ in.   (282.6 sq. in.)

    b. $r = 5$ ft.
       $h = 65$ ft.   (2,198 sq. ft.)

3. *Find the surface area of cubes with these dimensions.* (Lesson 100)

    a. $e = 19$ cm   (2,166 cm²)

    b. $e = 15$ ft.   (1,350 sq. ft.)

4. *Use each set of facts to construct a triangle. Label it with the facts given.* (Lesson 88)

    a. 45°, $1\frac{1}{2}$ in., 50°

    b. $1\frac{1}{2}$ in., 45°, 2 in.

5. *Solve these division problems.* (Lesson 40)

    a. $2\frac{2}{3} \div 4\frac{4}{5}$ ($\frac{5}{9}$)

    b. $5\frac{1}{2} \div 11$ ($\frac{1}{2}$)

T-355 Chapter 8 Geometric Solids and the Pythagorean Rule

Introduction

Bring an orange or a grapefruit to class. Review the facts about the sphere, a ball-shaped geometric solid. What is unique about this figure?

a. The sphere is the only geometric solid on which every point on the surface is the same distance from the center.

b. The sphere is the only geometric solid studied so far that has no face consisting of a polygon.

c. An infinite number of circles pass through both poles of a sphere.

Cut the orange in half. Find the area of the cut side of one half of the orange. The surface area of the complete sphere is directly related to the area of this circle. Can the students guess how many times greater it is? The surface area is 4 times the area of the circle.

Teaching Guide

The formula for finding the surface area of a sphere is $a_s = 4\pi r^2$.

a. $r = 9$ in. (1,017.36 sq. in.)
b. $r = 11$ cm (1,519.76 cm^2)
c. $r = \frac{1}{2}$ in. ($23\frac{1}{7}$ sq. in.)
d. $r = \frac{1}{4}$ in. ($\frac{11}{14}$ sq. in.)

Solutions for Part E

13. $4 \times 3.14 \times 6^2$
14. $4 \times 3.14 \times 1,080^2$
15. $(4 \times 3.14 \times 3^2) \div 3$
16. $4 \times \frac{22}{7} \times \frac{7}{4} \times \frac{7}{4}$
17. $(5 \times 8^2) \div 32$
18. $(2 \times 3.14 \times 1.5^2) + (12 \times 3.14 \times 3) = 127.17$; $127.17 \div 5$

D. Find the surface area of each sphere, and answer the question after each pair. Use 3.14 for pi.

11. a. r = 4 in.   200.96 sq. in.
    b. r = 8 in.   803.84 sq. in.
    c. Look at your answers for a and b. When the radius of a sphere is doubled, the surface area is multiplied by __4__.

12. a. r = 3 cm   113.04 cm²
    b. r = 12 cm   1,808.64 cm²
    c. Look at your answers for a and b. When the radius of a sphere is quadrupled, the surface area is multiplied by __16__.

E. Solve these reading problems. Note that numbers 13–16 give the diameter, not the radius.

13. A typical classroom globe measures 12 inches in diameter. What is the surface area of such a globe?   452.16 square inches

14. The average diameter of the moon has been calculated to be 2,160 miles. (It is not a perfect sphere.) What is the surface area of the moon, to the nearest thousand square miles?   14,650,000 square miles

15. James uses ammonia to disinfect his 6-foot spherical spray tank. If he uses 1 ounce of ammonia per 3 square feet of surface area, how much ammonia does he need for each cleaning? (Answer to the nearest whole ounce.)   38 ounces

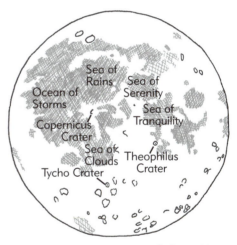

Prominent Features of the Moon

16. How many square inches of leather does it take to cover a softball that has a diameter of $3\frac{1}{2}$ inches? (Use $3\frac{1}{7}$ for pi.) $38\frac{1}{2}$ square inches

17. The Torkelsons have a walk-in refrigerator whose interior is an 8-foot cube. To improve its efficiency, they plan to cover the walls and ceiling with sheets of insulation measuring 4 feet by 8 feet. How many sheets will they need?   10 sheets

18. Marlene is painting a cylindrical tank that is 12 feet long and has a diameter of 3 feet. If she paints at the rate of 5 square feet per minute, how long should it take her to paint the tank? (Round to the nearest 5 minutes.)   25 minutes

356 Chapter 8 *Geometric Solids and the Pythagorean Rule*

REVIEW EXERCISES

F. Write these formulas. *(Lessons 100–103)*

19. Surface area of a cube $\quad a_s = 6e^2$
20. Surface area of a rectangular solid $\quad a_s = 2lw + 2wh + 2lh$
21. Surface area of a cylinder $\quad a_s = 2\pi r^2 + 2\pi rh$
22. Surface area of a square pyramid $\quad a_s = 4(\frac{1}{2}b\ell) + b^2$

G. Find the surface area of each square pyramid. *(Lesson 103)*

23. $b = 7$ cm $\quad$ 301 cm² $\qquad$ 24. $b = 18$ in. $\quad$ 1,188 sq. in.
    $l = 18$ cm $\qquad\qquad\qquad\qquad$ $l = 24$ in.

H. Find the surface area of each cylinder. *(Lesson 102)*

25. $r = 5$ in. $\quad$ 628 sq. in. $\qquad$ 26. $r = 25$ ft. $\quad$ 13,345 sq. ft.
    $h = 15$ in. $\qquad\qquad\qquad\qquad$ $h = 60$ ft.
    $\pi = 3.14$ $\qquad\qquad\qquad\qquad$ $\pi = 3.14$

I. Find the surface area of each cube. *(Lesson 100)*

27. $e = 8$ m $\quad$ 384 m² $\qquad$ 28. $e = 21$ ft. $\quad$ 2,646 sq. ft.

J. Construct a triangle with each set of dimensions. Label the given dimensions. *(Lesson 88)* $\qquad$ (See facing page for model triangles.)

29. 35°, $2\frac{1}{2}$ in., 35° $\qquad$ 30. 1 in., 55°, $2\frac{1}{2}$ in.

K. Solve these division problems. *(Lesson 40)*

31. $\frac{5}{9} \div \frac{5}{6}$ $\quad$ $\frac{2}{3}$ $\qquad$ 32. $3\frac{1}{3} \div 25$ $\quad$ $\frac{2}{15}$

Further Study

It was the ancient Greeks who discovered how to calculate the surface area of a sphere. They devised a way to "square" the circle and the sphere—that is, to define their surface areas in terms of square units—and then they proved that the surface area equals the diameter times the circumference ($a_s = d \times \pi d$ or $2r \times 2\pi r$). This can be simplified as either $a_s = \pi d^2$ or $a_s = 4\pi r^2$, but the latter is preferred because it is easily associated with the formula for the area of a circle ($a = \pi r^2$).

Model Triangles for Part J

29.

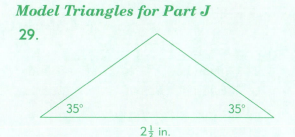

30.

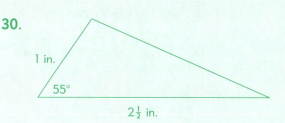

T–357 Chapter 8 Geometric Solids and the Pythagorean Rule

LESSON 105

Objectives

- To review that volume is the space occupied by a three-dimensional object.
- To review that in abbreviations, *cubic* is indicated by *cu.* with English units and by the exponent 3 with metric units.
- To review finding the volume of a rectangular solid by using the formula $v = lwh$.
- To review finding the volume of a cube by using the formula $v = e^3$.

Review

1. *Find the surface area of spheres with these radii.* (Lesson 104)
   a. $r = 7$ in. (615.44 sq. in.)
   b. $r = 23$ mm (6,644.24 mm²)

2. *Find the surface area of square pyramids having these dimensions.* (Lesson 103)
   a. $b = 11$ m
      $\ell = 16$ m (473 m²)
   b. $b = 18$ in.
      $\ell = 28$ in. (1,332 sq. in.)

3. *Find the surface area of rectangular solids with these dimensions.* (Lesson 101)
   a. $l = 6$ ft.
      $w = 5$ ft.
      $h = 6$ ft. (192 sq. ft.)
   b. $l = 19$ in.
      $w = 15$ in.
      $h = 8$ in. (1,114 sq. in.)

4. *Solve these multistep reading problems.* (Lesson 97)
   a. Morgan Hardware is preparing to pour concrete 6 inches thick over an area that measures 90 feet by 150 feet. At $70 per cubic yard, what will be the cost of the concrete? (One cubic yard equals 27 cubic feet.)
      ($17,500)
      $\frac{1}{2} \times 90 \times 150 = 6{,}750$ cu. ft.
      $6{,}750 \div 27 \times \$70 = \$17{,}500$
   b. The potato is a nutritious, fat-free food. It takes 7 pounds of potatoes to provide a total of 2,500 calories, the approximate daily requirement for adults. If the average potato weighs about $\frac{3}{10}$ pound, how many calories are in a kettle of 5 potatoes?
      ($535\frac{5}{7}$ calories)
      $5 \times \frac{3}{10} = 1\frac{1}{2}$
      $\dfrac{\text{pounds}}{\text{calories}} \quad \dfrac{7}{2{,}500} = \dfrac{1\frac{1}{2}}{n} \quad \dfrac{\text{pounds}}{\text{calories}}$

5. *Find the arithmetic mean of each set.* (Lesson 73)
   a. 42, 56, 48, 43, 85, 38, 61 ($53\frac{2}{7}$)
   b. 325, 316, 295, 308, 330, 315
      ($314\frac{5}{6}$)

6. *Solve these fraction problems.* (Lesson 41)
   a. 8 is $\frac{2}{3}$ of ___ (12)
   b. $\frac{5}{16}$ is $\frac{1}{6}$ of ___ ($1\frac{7}{8}$)

105. Volume of Rectangular Solids and Cubes

Volume is the capacity or amount of space contained in a geometric solid. Visualize a geometric solid as a hollow object. Your classroom represents a geometric solid even though it is filled mostly with air rather than being a solid form.

To find the volume of any solid with vertical sides, multiply the area of its base times its height. This can be stated as *volume = area of base × height*, or $v = Bh$. (Use a capital B to represent the square measure of the base, as distinguished from b for a linear base measure.) This is the general formula for the volume of any geometric solid with vertical sides.

There are also specific formulas for finding volumes. For a rectangular solid, the first step is to find the area of the base by multiplying length times width ($a = lw$). The second step is to multiply that product by the height. The formula is *volume = length × width × height*, or $v = lwh$.

Memorize this formula for the volume of a rectangular solid. $v = lwh$

Rectangular solid
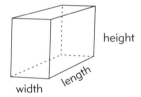
height, length, width

Cube
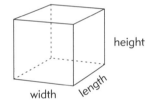
height, length, width

The length, width, and height of a cube are equal. So the formula *volume = length × width × height* can be restated as *volume = edge × edge × edge*, or $v = e^3$.

Memorize this formula for the volume of a cube. $v = e^3$

Example A	**Example B**
Find the volume of a rectangular solid with the following dimensions. $l = 6$ ft.; $w = 4$ ft.; $h = 8$ in. $v = lwh$ $v = 6 \times 4 \times \frac{2}{3}$ $v = 16$ cu. ft.	Find the volume of a cube with edges measuring 8 centimeters. $v = e^3$ $v = 8 \times 8 \times 8$ $v = 512$ cm³

In Example A, notice that 8 inches is changed to feet so that all three factors are stated as feet. In abbreviations for cubic units, *cu.* is used with English units (Example A answer), and the exponent 3 is used with metric units (Example B answer).

358 Chapter 8 Geometric Solids and the Pythagorean Rule

CLASS PRACTICE

Find the volume of each rectangular solid.

a. $l = 6$ m   120 m³   b. $l = 20$ ft. 80 cu. ft. c. $l = 9$ yd. 42 cu. yd. d. $l = 15$ in. 3,780 cu. in.
   $w = 4$ m $w = 12$ ft. $w = 7$ yd. $w = 12$ in.
   $h = 5$ m $h = 4$ in. $h = 2$ ft. $h = 21$ in.

Find the volume of each cube.

e. $e = 5$ ft.   f. $e = 18$ m   g. $e = 31$ cm   h. $e = 12$ in.
   125 cu. ft.   5,832 m³   29,791 cm³   1,728 cu. in.

WRITTEN EXERCISES

A. *Write these formulas. Be sure you know them by memory.*

1. Volume of a rectangular solid $v = lwh$

2. Volume of a cube $v = e^3$

B. *Find the volume of each rectangular solid.*

3. $l = 4$ in.   384 cu. in.   4. $l = 36$ ft.   1,620 cu. ft.   5. $l = 21$ ft.   70 cu. ft.
   $w = 8$ in. $w = 18$ ft. $w = 5$ in.
   $h = 12$ in. $h = 2$ ft. 6 in. $h = 8$ ft.

6. $l = 5$ cm   280 cm³   7. $l = 15$ cm   5,250 cm³   8. $l = 18$ cm   5,940 cm³
   $w = 8$ cm $w = 14$ cm $w = 15$ cm
   $h = 7$ cm $h = 25$ cm $h = 22$ cm

C. *Find the volume of each cube.*

9. $e = 7$ in.   10. $e = 23$ cm   11. $e = 19$ m   12. $e = 30$ ft.
   343 cu. in.   12,167 cm³   6,859 m³   27,000 cu. ft.

D. *Find the volume of each rectangular solid, and answer the question after each pair.*

13. a. $l = 3$ ft.; $w = 8$ ft.; $h = 10$ ft.   240 cu. ft.

    b. $l = 6$ ft.; $w = 16$ ft.; $h = 20$ ft.   1,920 cu. ft.

    c. Look at your answers for *a* and *b*. When the dimensions of a rectangular solid are doubled, the volume is multiplied by __8__.

14. a. $e = 2$ yd.   8 cu. yd.

    b. $e = 4$ yd.   64 cu. yd.

    c. Look at your answers for *a* and *b*. When the dimensions of a cube are doubled, the volume is multiplied by __8__.

E. *Solve these reading problems. Numbers 19 and 20 are multistep problems.*

15. Brother Fred is building a chicken house 500 feet long and 40 feet wide. He plans to blow in a 9-inch layer of insulation above the ceiling. How many cubic feet of insulation will he need?   15,000 cubic feet

Lesson 105 T–358

Introduction

Ask the students how finding the area of a rectangle, a square, and a parallelogram are similar. They all involve multiplying the two dimensions. The only difference is in the terminology: *length* and *width* are used for the rectangle, *side* is used for the square, and *base* and *height* are used for the parallelogram.

Likewise, the same general formula is used for finding the volume of any geometric solid with a flat base and vertical sides. The principle is that volume equals the area of the base times the height, or $v = Bh$.

Rectangular solid: Find the area of the base ($a = lw$), and then multiply by the height.

Cube: Find the area of base ($a = s^2$), and then multiply by the height. But the term *edge* is used since all the dimensions are equal.

Cylinder: Find the area of the circular base ($a = \pi r^2$), and then multiply by the height.

Teaching Guide

1. **Volume is the amount of space contained in a geometric solid.** The volume of the classroom is the amount of space within it.

2. **The basic principle for finding volume is to multiply the area of the base times the height.** In other words, volume = length × width × height.

3. **The formula for finding the volume of a rectangular solid is $v = lwh$.**

    a. $l = 5$ yd.
        $w = 3$ yd.
        $h = 4$ in.   ($1\frac{2}{3}$ cu. yd.)

    b. $l = 14$ cm
        $w = 12$ cm
        $h = 20$ cm   (3,360 cm³)

4. **The formula for finding the volume of a cube is $v = e^3$.**

    a. $e = 25$ cm   (15,625 cm³)

    b. $e = 4\frac{1}{2}$ in.   ($91\frac{1}{8}$ cu. in.)

Chapter 8 Geometric Solids and the Pythagorean Rule

Further Study

Exercises 13 and 14 show that volume varies as the cube of the multiple by which the dimensions are increased.

Solutions for Part E

15. $500 \times 40 \times \frac{3}{4}$
16. $10 \times 8 \times \frac{1}{9}$
17. $1 \times 3 \times \frac{5}{6} \times 7\frac{1}{2}$
18. $(26 \times 30 \times 7\frac{1}{2}) \div 18$
19. $12 \div (2 \times 2\frac{1}{2}) = 2\frac{2}{5}$; 3×15
20. $2 \times 3 \times 6 \times 85$

16. Father is preparing to concrete the garage floor, which is 10 yards long and 8 yards wide. If the concrete will have an average thickness of 4 inches, how many cubic yards should he order? (Round to the nearest whole number.)   9 cubic yards

17. Anthony filled the fish aquarium in the living room with water to a depth of 10 inches. The aquarium is 1 foot wide and 3 feet long. If 1 cubic foot of water equals $7\frac{1}{2}$ gallons, how many gallons did it take to fill the aquarium? $18\frac{3}{4}$ gallons

18. The classroom for grades 7–10 is 26 feet wide and 30 feet long, and the ceiling is $7\frac{1}{2}$ feet high. How many cubic feet of air are there for each of the 18 people in the room?   325 cubic feet

19. The oven in Sister Katrina's kitchen has room for 2 cookie sheets that each hold $2\frac{1}{2}$ dozen peanut butter cookies. If the cookies must bake for 15 minutes, how long will it take Katrina to bake 12 dozen peanut butter cookies? (A partial oven load needs to bake the full 15 minutes.)   45 minutes

20. Vernon has 85 blocks each 2 inches thick, 3 inches wide, and 6 inches long. What is the total volume of the blocks?   3,060 cubic inches

REVIEW EXERCISES

F. Write these formulas. *(Lessons 100–104)*

21. Surface area of a cube $a_s = 6e^2$

22. Surface area of a rectangular solid $a_s = 2lw + 2wh + 2lh$

23. Surface area of a cylinder $a_s = 2\pi r^2 + 2\pi rh$

24. Surface area of a square pyramid $a_s = 4(\frac{1}{2}b\ell) + b^2$

25. Surface area of a sphere $a_s = 4\pi r^2$

G. Find the surface area of each sphere. *(Lesson 104)*

26. r = 6 in.   452.16 sq. in.   27. r = 25 cm   7,850 cm²

H. Find the surface area of each square pyramid. *(Lesson 103)*

28. b = 21 m   1,155 m²   29. b = 15 in.   1,185 sq. in.
     ℓ = 17 m ℓ = 32 in.

I. Find the surface area of each rectangular solid. *(Lesson 101)*

30. l = 25 ft.   3,160 sq. ft.   31. l = 250 ft.   19,100 sq. ft.
     w = 20 ft. w = 25 ft.
     h = 24 ft. h = 12 ft.

J. Find the arithmetic mean of each set. *(Lesson 73)*

32. 15, 21, 35, 19, 10, 26, 21   21

33. 925, 879, 914, 908, 899, 894 $903\frac{1}{6}$

K. Solve these fraction problems. *(Lesson 41)*

34. 6 is $\frac{8}{9}$ of $6\frac{3}{4}$ 35. 24 is $\frac{3}{5}$ of   40   36. $\frac{2}{3}$ is $\frac{1}{7}$ of $4\frac{2}{3}$

360 Chapter 8 Geometric Solids and the Pythagorean Rule

106. Volume of Cylinders and Cones

To find the volume of any solid with vertical sides, multiply the area of its base times its height ($v = Bh$). The base of a cylinder is a circle, so the volume of a cylinder is found by multiplying $\pi r^2 \times$ height. The result is the volume of the cylinder in cubic units.

Diameter = 21 in.; radius = $10\frac{1}{2}$ in.
Height = 20 in.
Area of base = $3\frac{1}{7} \times 10\frac{1}{2} \times 10\frac{1}{2} = 346\frac{1}{2}$ sq. in.
Volume of cylinder = $346\frac{1}{2} \times 20 = 6{,}930$ cu. in.

$v = \pi r^2 h$ **Memorize this formula for the volume of a cylinder.**

A cone has a circular base, like a cylinder. But instead of having vertical sides, its lateral surface slopes to a point. It has been found that a cone always has $\frac{1}{3}$ the volume of a cylinder with the same base and height. The general formula for such a figure is $v = \frac{1}{3}Bh$.

In the specific formula for the volume of a cone, πr^2 replaces B for the area of the base. Multiply πr^2 by the height (h) to find the volume of a cylinder of equal height. Then multiply by $\frac{1}{3}$ to find the volume of the cone. The formula is $v = \frac{1}{3}\pi r^2 h$.

The height of a cone must be measured along a line perpendicular to the base. It is not measured along the sloping side.

 height

Diameter = 21 in.; radius = $10\frac{1}{2}$ in.
Height = 20 in.
Area of base = $3\frac{1}{7} \times 10\frac{1}{2} \times 10\frac{1}{2} = 346\frac{1}{2}$ sq. in.
Volume of cone = $\frac{1}{3} \times 346\frac{1}{2} \times 20 = 2{,}310$ cu. in.

$v = \frac{1}{3}\pi r^2 h$ **Memorize this formula for the volume of a cone.**

Example A
Find the volume of a cylinder with a radius of 6 centimeters and a height of 8 centimeters. Use 3.14 for pi.

$v = \pi r^2 h$
$v = 3.14 \times 6 \times 6 \times 8$
$v = 904.32$ cm³

Example B
Find the volume of a cone with a radius of 10 yards 18 inches and a height of 13 yards 1 foot. Use $3\frac{1}{7}$ for pi.

$v = \frac{1}{3}\pi r^2 h$
$v = \frac{1}{3} \times 3\frac{1}{7} \times 10\frac{1}{2} \times 10\frac{1}{2} \times 13\frac{1}{3}$
$v = \frac{1}{3} \times \frac{\overset{11}{\cancel{22}}}{\underset{1}{\cancel{7}}} \times \frac{\overset{3}{\cancel{21}}}{\underset{1}{\cancel{2}}} \times \frac{\overset{7}{\cancel{21}}}{\underset{1}{\cancel{2}}} \times \frac{\overset{20}{\cancel{40}}}{\underset{1}{\cancel{3}}}$
$v = 1{,}540$ cu. yd.

LESSON 106

Objectives

- To review finding the volume of a cylinder by using the formula $v = \pi r^2 h$.
- To review finding the volume of a cone by using the formula $v = \frac{1}{3}\pi r^2 h$.

Review

1. Give Lesson 106 Quiz (Finding Surface Area).

2. *Find the volume of each cube or rectangular solid.* (Lesson 105)
   a. $e = 35$ cm   (42,875 cm³)
   b. $l = 11$ ft.
      $w = 8$ ft.
      $h = 8$ ft.   (704 cu. ft.)

3. *Find the surface area of each cylinder or sphere.* (Lessons 102, 104)
   (Use $3\frac{1}{7}$ for pi with multiples of 7.)
   a. $r = 28$ in.
      $h = 25$ in.   (9,328 sq. in.)
   b. $r = 35$ in.   (15,400 sq. in.)

4. *Find the perimeter of each square or rectangle.* (Lesson 90)
   a. $l = 95$ ft.
      $w = 63$ ft.   (316 ft.)
   b. $s = 63$ ft.   (252 ft.)

5. *Find the median of each set.* (Lesson 74)
   a. 26, 28, 32, 18, 26, 24, 25   (26)
   b. 85, 75, 94, 82, 80, 70, 93, 85, 86, 82, 90, 70, 93   (85)

6. *Find the mode of each set.* (Lesson 74)
   a. 6, 8, 7, $6\frac{1}{2}$, $7\frac{1}{2}$, 6, 7, $7\frac{1}{2}$, 8, $6\frac{1}{2}$, 7, 7, $7\frac{1}{2}$ (7)
   b. 26, 32, 28, 29, 27, 31, 32, 30, 26, 28, 30, 31, 28, 29, 31, 33, 37, 21   (28, 31)

7. *Do these divisions mentally.* (Lesson 42)
   a. $25 \div \frac{5}{7}$ (35)
   b. $12 \div \frac{3}{7}$ (28)

T-361   Chapter 8   Geometric Solids and the Pythagorean Rule

Introduction

A large fuel tank has a diameter of 14 feet and a length (height) of 24 feet. How many gallons of fuel can it hold? (1 cu. ft. = $7\frac{1}{2}$ gal.)

$r = 7$ ft.; $h = 24$ ft.
$v = \pi r^2 h = 3\frac{1}{7} \times 7 \times 7 \times 24 = 3{,}696$ cu. ft.
$3{,}696 \times 7\frac{1}{2} = 27{,}720$ gal.

Teaching Guide

1. **The formula for finding the volume of a cylinder is $v = \pi r^2 h$.** Remind students to use $3\frac{1}{7}$ for pi if the problem contains fractions or if one of the factors is a multiple of 7.

   a. $r = 9$ in.
       $h = 15$ in.   (3,815.1 cu. in.)

   b. $r = 12$ cm
       $h = 23$ cm   (10,399.68 cm³)

   c. $r = 21$ in.
       $h = 12$ in.
       $\pi = 3\frac{1}{7}$ (16,632 cu. in.)

2. **A cone has one lateral surface that slopes from the circular base to a point.** It has $\frac{1}{3}$ the volume of a cylinder with the same radius and height.

3. **The formula for finding the volume of a cone is $v = \frac{1}{3}\pi r^2 h$.**

   a. $r = 1$ ft. 2 in.
       $h = 12$ ft.   ($17\frac{1}{9}$ cu. ft.)

   b. $r = 11$ cm
       $h = 24$ cm   (3,039.52 cm³)

Solutions for Part E

11. $(\frac{22}{7} \times \frac{7}{2} \times \frac{7}{2} \times 12) \times 7\frac{1}{2}$

12. $\frac{22}{7} \times 17\frac{1}{2} \times 17\frac{1}{2} \times 33$

13. $(\frac{1}{3} \times 3.14 \times 15^2 \times 12) \div 1.25$

In general, use 3.14 for pi with decimal problems, and $3\frac{1}{7}$ when working with fractions or when one of the factors is a multiple of 7.

CLASS PRACTICE

Find the volume of each cylinder. In exercises c and d, first find the radius. Use 3.14 for pi.

a. $r = 2$ in.
   $h = 4$ in.
   50.24 cu. in.

b. $r = 3$ ft. 6 in.
   $h = 8$ ft.
   307.72 cu. ft.

c. $d = 8$ m
   $h = 5$ m
   251.2 m³

d. $d = 13$ cm
   $h = 16$ cm
   2,122.64 cm³

Find the volume of each cone. Use $3\frac{1}{7}$ for pi.

e. $r = 4$ ft. 6 in.
   $h = 7$ ft.
   $148\frac{1}{2}$ cu. ft.

f. $r = 9$ yd.
   $h = 3$ yd. 18 in.
   297 cu. yd.

g. $r = 21$ cm
   $h = 18$ cm
   8,316 cm³

h. $r = 9$ ft.
   $h = 27$ ft.
   $2,291\frac{1}{7}$ cu. ft.

WRITTEN EXERCISES

A. *Write these formulas. Be sure you know them by memory.*

1. Volume of a cylinder $v = \pi r^2 h$

2. Volume of a cone $v = \frac{1}{3}\pi r^2 h$

B. *Find the volume of each cylinder. Use 3.14 for pi.*

3. $r = 2$ mm   113.04 mm³
   $h = 9$ mm

4. $r = 2$ cm   75.36 cm³
   $h = 6$ cm

5. $r = 25$ m   15,700 m³
   $h = 8$ m

C. *Find the volume of each cone. Use $3\frac{1}{7}$ for pi.*

6. $r = 3$ ft.
   $h = 7$ ft.
   66 cu. ft.

7. $r = 4$ yd. 18 in.
   $h = 9$ yd. 1 ft.
   198 cu. yd.

8. $r = 9$ in.
   $h = 10\frac{1}{2}$ in.
   891 cu. in.

9. $r = 4\frac{1}{5}$ ft.
   $h = 8\frac{3}{4}$ ft.
   $161\frac{7}{10}$ cu. ft.

D. *Find the volume of each cylinder, and answer the question after each pair. Use 3.14 for pi.*

10. a. $r = 5$ ft.; $h = 10$ ft.   785 cu. ft.

    b. $r = 10$ ft.; $h = 20$ ft.   6,280 cu. ft.

    c. Look at your answers for a and b. When the dimensions of a cylinder are doubled, its volume is multiplied by __8__.

E. *Solve these reading problems.*

11. A cylindrical milk tank is 12 feet long and has a radius of $3\frac{1}{2}$ feet. One cubic foot equals $7\frac{1}{2}$ gallons. How many gallons can the tank hold?   3,465 gallons

12. The soldiers of Abner and Joab held a contest by the pool of Gibeon (2 Samuel 2:12–17). In 1956–57, debris was cleared from what is believed to be this pool. It is roughly in the shape of a cylinder with a diameter of 35 feet and a depth of 33 feet. What is its volume? (First find the radius. Use $3\frac{1}{7}$ for pi.) $31,762\frac{1}{2}$ cubic feet

13. A cone-shaped pile of grain is 12 feet high and has a radius of 15 feet. One bushel of grain is 1.25 cubic feet. How many bushels of grain are in the pile, to the nearest bushel?   2,261 bushels

362   Chapter 8   *Geometric Solids and the Pythagorean Rule*

14. The Rohrers' grain bin is in the shape of a cylinder 20 feet high and 16 feet in diameter. If one bushel is 1.25 cubic feet, how much grain does the bin hold? (Round to the nearest bushel.)   3,215 bushels

15. A stack of hay is 20 feet wide, 12 feet high, and 60 feet long. If the bales have an average volume of $4\frac{1}{2}$ cubic feet, how many bales are in the stack?   3,200 bales

16. A tin can has a radius of 2 inches and a height of 6 inches. What is the area of a label that covers the sides of the can? (Include an overlap of $\frac{1}{4}$ inch where the ends are joined.)   76.86 square inches

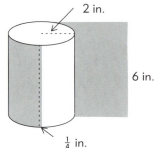

REVIEW EXERCISES

F. Write these formulas. *(Lessons 100–105)*

17. Surface area of a cube $a_s = 6e^2$

18. Surface area of a rectangular solid $a_s = 2lw + 2wh + 2lh$

19. Surface area of a cylinder $a_s = 2\pi r^2 + 2\pi rh$

20. Surface area of a square pyramid $a_s = 4(\frac{1}{2}b\ell) + b^2$

21. Surface area of a sphere $a_s = 4\pi r^2$

22. Volume of a rectangular solid $v = lwh$

23. Volume of a cube $v = e^3$

G. Find the volume of each cube or rectangular solid. *(Lesson 105)*

24. l = 21 ft.   2,835 cu. ft.   25. e = 14 ft.   2,744 cu. ft.
    w = 15 ft.
    h = 9 ft.

H. Find the surface area of each figure. Use $3\frac{1}{7}$ for pi. *(Lessons 102, 104)*

26. Cylinder: r = 2 m $113\frac{1}{7}$ m²   27. Sphere: r = 7 mm   616 mm²
    h = 7 m

I. Find the perimeter of each square or rectangle. *(Lesson 90)*

28. l = 88 ft.   266 ft.   29. s = 26.8 m   107.2 m
    w = 45 ft.

J. Find the median of each set. *(Lesson 74)*

30. 126, 123, 132, 128, 136, 137   130   31. 64, 68, 62, 69, 67, 61, 65, 66 $65\frac{1}{2}$

K. Find the mode of each set. *(Lesson 74)*

32. 45, 48, 51, 53, 54, 47, 48, 50   48   33. 9, 6, 2, 5, 11, 10, 8, 7, 4   no mode

L. Do these divisions mentally. *(Lesson 42)*

34. $28 \div \frac{4}{5}$ 35   35. $14 \div \frac{2}{7}$ 49   36. $35 \div \frac{5}{8}$ 56

14. $(3.14 \times 8^2 \times 20) \div 1.25$
15. $(20 \times 12 \times 60) \div 4\frac{1}{2}$
16. $(3.14 \times 4 + \frac{1}{4}) \times 6$

LESSON 107

Objective

- To teach *finding the volume of a square pyramid by using the formula $v = \frac{1}{3}lwh$.

Review

1. Find the volume of each geometric solid. Use $3\frac{1}{7}$ for pi. (Lessons 105, 106)

    a. Cube
    $e = 18$ m
    ($5,832$ m³)

    b. Rectangular solid
    $l = 24$ in.
    $w = 18$ in.
    $h = 9$ in. ($3,888$ cu. in.)

    c. Cylinder
    $r = 14$ cm
    $h = 45$ cm ($27,720$ cm³)

    d. Cone
    $r = 3$ in.
    $h = 7$ in. (66 cu. in.)

2. Find the surface area of square pyramids with these dimensions. (Lesson 103)

    a. $b = 11$ ft.
    $\ell = 15$ ft. (451 sq. ft.)

    b. $b = 25$ ft.
    $\ell = 36$ ft. ($2,425$ sq. ft.)

3. Find the circumference of each circle. (Lesson 91)

    a. $d = 16$ mm
    $\pi = 3.14$ (50.24 mm)

    b. $r = 17\frac{1}{2}$ in.
    $\pi = 3\frac{1}{7}$ (110 in.)

107. Volume of Square Pyramids

A square pyramid has a square base, but it does not have vertical sides like a rectangular solid. Instead, its sides taper to a point or apex. As with the cone and cylinder, a pyramid has $\frac{1}{3}$ the volume of a rectangular solid with the same base and height. The general formula is $v = \frac{1}{3}Bh$.

The specific formula for the volume of a pyramid is written by inserting $\frac{1}{3}$ into the $v = lwh$ formula for the volume of a rectangular solid. The volume of a square pyramid is $v = \frac{1}{3}lwh$.

As with a cone, the height of a pyramid is measured along a line perpendicular to the base. It is not the slant height.

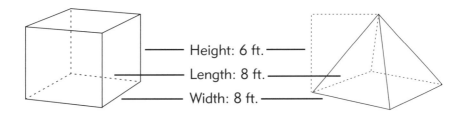

Height: 6 ft.
Length: 8 ft.
Width: 8 ft.

Memorize this formula for the volume of a pyramid.

$$v = \tfrac{1}{3}lwh$$

The formula above works for pyramids that have square or rectangular bases. This lesson deals with square pyramids, since they are the most common. Therefore, the letter *s* (for one *side* of the square base) is used when the dimensions of square pyramids are given.

Example A	Example B
Find the volume of a square pyramid with the following dimensions.	Find the volume of a square pyramid with the following dimensions.
$s = 8$ ft.; $h = 6$ ft.	$s = 12$ cm; $h = 8$ cm 9 mm
$v = \tfrac{1}{3}lwh$	$v = \tfrac{1}{3}lwh$
$v = \tfrac{1}{3} \times 8 \times 8 \times 6$	$v = \tfrac{1}{3} \times 12 \times 12 \times 8.9$
$v = 128$ cu. ft.	$v = 48 \times 8.9$
	$v = 427.2$ cm³

364 Chapter 8 Geometric Solids and the Pythagorean Rule

CLASS PRACTICE

Find the volume of each square pyramid. Round decimals in your answers to the nearest hundredth.

a. $s = 5$ in.
 $h = 8$ in. $66\frac{2}{3}$ cu. in.

b. $s = 12$ ft.
 $h = 10$ ft.   480 cu. ft.

c. $s = 1$ ft. 6 in.
 $h = 3$ ft. 4 in. $2\frac{1}{2}$ cu. ft.

d. $s = 8.8$ m
 $h = 5.6$ m   144.55 m³

WRITTEN EXERCISES

A. Write this formula. Be sure you know it by memory.

1. Volume of a square pyramid $v = \frac{1}{3}lwh$

B. Find the volume of each square pyramid.

2. $s = 3$ in.   12 cu. in.
   $h = 4$ in.

3. $s = 5$ m   65 m³
   $h = 7$ m 80 cm

4. $s = 16$ cm   1,024 cm³
   $h = 12$ cm

5. $s = 15$ ft.   1,500 cu. ft.
   $h = 20$ ft.

6. $s = 3$ ft. 9 in. $93\frac{3}{4}$ cu. ft.
   $h = 20$ ft.

7. $s = 75$ ft.   112,500 cu. ft.
   $h = 60$ ft.

C. Find the volume of each square pyramid, and answer the question after each pair.

8. a. $s = 4$ in.   48 cu. in.
    $h = 9$ in.
   b. $s = 4$ in.   96 cu. in.
    $h = 18$ in.
   c. Look at your answers for *a* and *b*. When one dimension (height) of a square pyramid is doubled, the volume is multiplied by __2__.

9. a. $s = 5$ in.   50 cu. in.
    $h = 6$ in.
   b. $s = 10$ in.   200 cu. in.
    $h = 6$ in.
   c. Look at your answers for *a* and *b*. When two dimensions of a square pyramid (length and width) are doubled, the volume is multiplied by __4__.

10. a. $s = 3$ in.   21 cu. in.
     $h = 7$ in.
    b. $s = 6$ in.   168 cu. in.
     $h = 14$ in.
    c. Look at your answers for *a* and *b*. When all three dimensions of a square pyramid are doubled, the volume is multiplied by __8__.

D. Solve these reading problems. Round decimals in your answers to the nearest hundredth.

11. A grain hopper is shaped like an inverted square pyramid. It has a height of 5 feet and sides measuring 6 feet. How many bushels of grain can it hold? (1 bushel = $1\frac{1}{4}$ cubic feet)   48 bushels

12. A grain bin shaped like an inverted square pyramid has a height of 9 feet and a square top with 7-foot sides. How many bushels of grain can it hold? (1 bushel = 1.25 cubic feet)   117.6 bushels

13. The Middle Pyramid in Egypt has sides measuring 708 feet and a height of 471 feet. What is its volume?   78,698,448 cubic feet

Lesson 107 T–364

Introduction

Call attention to the pyramid in the lesson text, and use the following questions to draw out the formula taught in this lesson.

1. What is the general formula for finding volume? ($v = Bh$)

2. How is this general formula modified when the sides of a geometric solid narrow from a polygon to a point? (Include the factor $\frac{1}{3}$ in the formula.)

3. How can we find the area of the base of the pyramid in the diagram? ($a = s^2$, or in more general terms, $a = lw$)

4. Since the base is $a = lw$, how can we find the volume of the pyramid? (by multiplying $\frac{1}{3}$ times the area of the base times the height: $v = \frac{1}{3}lwh$)

Teaching Guide

1. **A square pyramid has a square base and sides that taper to a point.** As with the cone and cylinder, a pyramid has a volume $\frac{1}{3}$ as great as that of a rectangular solid with the same base and height.

2. **The formula for the volume of a pyramid is $v = \frac{1}{3}lwh$.** Since this lesson deals only with square pyramids, the letter s (for one side of the square base) is used when the dimensions of pyramids are given.

   a. $s = 6$ in.
   $h = 9$ in. (108 cu. in.)

   b. $s = 4\frac{1}{2}$ in.
   $h = 6\frac{1}{2}$ in. ($43\frac{7}{8}$ cu. in.)

Solutions for Part D

11. ($\frac{1}{3} \times 6 \times 6 \times 5$) ÷ $1\frac{1}{4}$

12. ($\frac{1}{3} \times 7 \times 7 \times 9$) ÷ 1.25

13. $\frac{1}{3} \times 708 \times 708 \times 471$

T–365 Chapter 8 Geometric Solids and the Pythagorean Rule

14. $\frac{1}{3} \times 357 \times 357 \times 218$
15. $3.14 \times 1^2 \times 20$
16. $3.14 \times 0.5^2 \times 5$

14. The southernmost pyramid in Egypt has sides measuring 357 feet and a height of 218 feet. What is its volume? 9,261,294 cubic feet

15. A piece of sausage is 20 inches long and 2 inches in diameter. How many cubic inches are in it? (Note that the diameter is given. Use 3.14 for pi.) 62.8 cubic inches

16. What is the volume of a hypodermic syringe 1 inch in diameter and 5 inches long? (First find the radius.) 3.93 cubic inches

REVIEW EXERCISES

E. Write these formulas. *(Lessons 100–106)*

17. Surface area of a cube $a_s = 6e^2$
18. Surface area of a rectangular solid $a_s = 2lw + 2wh + 2lh$
19. Surface area of a cylinder $a_s = 2\pi r^2 + 2\pi rh$
20. Surface area of a square pyramid $a_s = 4(\frac{1}{2}b\ell) + b^2$
21. Surface area of a sphere $a_s = 4\pi r^2$
22. Volume of a rectangular solid $v = lwh$
23. Volume of a cube $v = e^3$
24. Volume of a cylinder $v = \pi r^2 h$
25. Volume of a cone $v = \frac{1}{3}\pi r^2 h$

F. Find the volume of each geometric solid. Use 3.14 for pi. *(Lessons 105, 106)*

26. Cube: $e = 9$ in. 729 cu. in.
27. Cube: $e = 11$ m 1,331 m³
28. Rectangular solid
 $l = 15$ ft. 810 cu. ft.
 $w = 9$ ft.
 $h = 6$ ft.
29. Cylinder
 $r = 6$ cm 791.28 cm³
 $h = 7$ cm
30. Cone
 $r = 4$ in. 150.72 cu. in.
 $h = 9$ in.

G. Find the surface area of each square pyramid. *(Lesson 103)*

31. $b = 20$ ft. 1,400 sq. ft.
 $\ell = 25$ ft.
32. $b = 40$ ft. 8,000 sq. ft.
 $\ell = 80$ ft.

H. Find the circumference of each circle. *(Lesson 91)*

33. $d = 6.5$ m; $\pi = 3.14$ 20.41 m
34. $r = 21$ ft.; $\pi = 3\frac{1}{7}$ 132 ft.

CHALLENGE EXERCISES

I. Find the volumes of these geometric solids.

35. Rectangular pyramid
 448 cu. ft.

36. Triangular pyramid
 84 m³

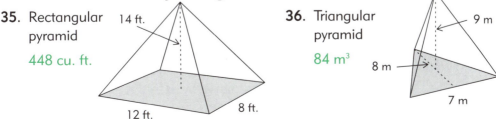

108. Volume of Spheres

It is difficult to comprehend the volume of a sphere as the product of length, width, and height (*lwh*) because we cannot use these terms to describe a sphere. Rather, we use the radius or diameter.

The volume of a sphere is found to be $1\frac{1}{3}$ times as great as the product of pi and the radius cubed. The formula is usually written as $v = \frac{4}{3}\pi r^3$.

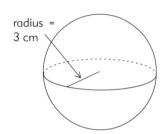

radius = 3 cm

Memorize this formula for the volume of a sphere.
$$v = \frac{4}{3}\pi r^3$$

The factors $\frac{4}{3}$ and π in the formula may be combined by the associative law of multiplication. If pi = 3.14, multiply 1.33 × 3.14 to obtain 4.18 (to the nearest hundredth). If pi = $3\frac{1}{7}$, multiply $\frac{4}{3} \times 3\frac{1}{7}$ to obtain $4\frac{4}{21}$. Therefore, the volume of a sphere may be calculated by multiplying 4.18 or $4\frac{4}{21}$ ($\frac{88}{21}$) times the radius cubed.

Example A
Find the volume of a sphere with a radius of 5 centimeters. Use 3.14 for pi.
$v = \frac{4}{3}\pi r^3 = 4.18 \times r^3$
$v = 4.18 \times 5 \times 5 \times 5$
$v = 4.18 \times 125$
$v = 522.5$ cm³

Example B
Find the volume of a sphere with a radius of $1\frac{3}{4}$ feet. Use $3\frac{1}{7}$ for pi.
$v = \frac{4}{3}\pi r^3 = \frac{88}{21} \times r^3$
$v = \frac{88}{21} \times \frac{7}{4} \times \frac{7}{4} \times \frac{7}{4} = \frac{539}{24}$
$v = 22\frac{11}{24}$ cu. ft.

CLASS PRACTICE

Find the volume of each sphere. Use the value of pi as indicated.

a. $r = 8$ in.
$\pi = 3.14$
2,140.16 cu. in.

b. $r = 11$ m
$\pi = 3.14$
5,563.58 m³

c. $r = 3\frac{1}{2}$ ft.
$\pi = 3\frac{1}{7}$
$179\frac{2}{3}$ cu. ft.

d. $r = 5\frac{1}{4}$ in.
$\pi = 3\frac{1}{7}$
$606\frac{3}{8}$ cu. in.

WRITTEN EXERCISES

A. Write this formula. Be sure you know it by memory.

1. Volume of a sphere $\quad v = \frac{4}{3}\pi r^3$

B. Find the volume of each sphere. Use 3.14 for pi.

2. $r = 10$ in.   4,180 cu. in.   3. $r = 12$ in.   7,223.04 cu. in.   4. $r = 6$ cm   902.88 cm³

LESSON 108

Objective

- To teach *finding the volume of a sphere by using the formula $v = \frac{4}{3}\pi r^3$.

Review

1. *Find the volume of each figure.* (Lessons 106, 107)

    a. Square pyramid
    $s = 21$ ft.
    $h = 25$ ft. (3,675 cu. ft.)

    b. Cylinder
    $r = 10$ cm
    $h = 32$ cm (10,048 cm³)

    c. Cone
    $r = 8$ in.
    $h = 9$ in. (602.88 cu. in.)

2. *Find the surface area of each figure.* (Lessons 100, 104)

    a. Sphere: $r = 24$ ft. (7,234.56 sq. ft.)

    b. Cube: $e = 12.1$ m (878.46 m²)

3. *Find the area of each rectangle or square.* (Lesson 92)

    a. $l = 15$ ft.
    $w = 12$ ft. (180 sq. ft.)

    b. $s = 54$ cm
    (2,916 cm²)

4. *Tell what missing information is needed to solve each problem.* (Lesson 44)

    a. Rodney weeded the six bean rows in the garden. Two of the rows were ten feet shorter than the other four. What was the total length of the rows he weeded?
    (length of two shorter rows or four longer rows)

    b. A giant grain conveyer can load corn onto trucks at the rate of 75 bushels per minute. At that rate, how many trucks can it load in one hour?
    (capacity of one truck)

Introduction

Find the volume of a softball with a circumference of 11 inches. Use $3\frac{1}{7}$ for pi and $\frac{4}{3}$ as the multiplier. Note that this exercise gives the circumference, not the radius.

It is difficult to apply the formula $v = lwh$ to a sphere because the sphere has no obvious base. However, there is a formula for finding the volume of a sphere.

Teaching Guide

1. **The formula for the volume of a sphere is $v = \frac{4}{3}\pi r^3$.**

2. **Finding the volume of a sphere can be simplified because the first two factors are always $\frac{4}{3}$ and π.** If pi = 3.14, we can multiply 1.33 × 3.14 to obtain 4.18 (to the nearest hundredth). If pi = $3\frac{1}{7}$, we can multiply $\frac{4}{3} \times 3\frac{1}{7}$ to obtain $4\frac{4}{21}$. So the volume of a sphere can be found by multiplying 4.18 or $4\frac{4}{21}$ ($\frac{88}{21}$) times the radius cubed.

   a. $r = 9$ in. (3,047.22 cu. in.)

   b. $r = 7\frac{1}{2}$ ft. (1,767$\frac{6}{7}$ cu. ft.)

Solutions for Part E

10. $\frac{88}{21} \times \frac{9}{2} \times \frac{9}{2} \times \frac{9}{2}$

11. $(4.18 \times 9^3) \div 231$

12. $4.18 \times 3^3 \times 7.5$

13. $\frac{88}{21} \times \frac{7}{2} \times \frac{7}{2} \times \frac{7}{2} \times \frac{15}{2}$

C. Find the volume of each sphere. Use $3\frac{1}{7}$ for pi.

5. $r = 1\frac{1}{2}$ in. $14\frac{1}{7}$ cu. in.
6. $r = 2\frac{1}{2}$ in. $65\frac{10}{21}$ cu. in.
7. $r = 7$ in. $1{,}437\frac{1}{3}$ cu. in.

D. Find the volume of these spheres, using 3.14 for pi. Then answer the question after each pair.

8. a. $r = 2$ in.; $v =$ <u>33.44</u> cu. in.   b. $r = 4$ in.; $v =$ <u>267.52</u> cu. in.

   c. Look at your answers for *a* and *b*. When the radius of a sphere is doubled, its volume is multiplied by <u>8</u>.

9. a. $r = 1$ cm; $v =$ <u>4.18</u> cm³   b. $r = 3$ cm; $v =$ <u>112.86</u> cm³

   c. Look at your answers for *a* and *b*. When the radius of a sphere is tripled, its volume is multiplied by <u>27</u>.

E. Solve these reading problems. Round decimals in your answers to the nearest hundredth. For numbers 14 and 15, write what additional information is needed to solve the problems.

10. What is the volume of a kickball with a radius of $4\frac{1}{2}$ inches? $381\frac{6}{7}$ cubic inches

11. The Hursts have a spherical receiver jar in their milking parlor. If the jar has a radius of 9 inches, how many gallons of milk can it hold? Answer to the nearest whole number. (One gallon equals 231 cubic inches.)   13 gallons

12. Brother Paul does custom spraying for farmers in his area. The tank on his truck is a sphere 6 feet in diameter. How many gallons of water can it hold? (One cubic foot equals 7.5 gallons.)   846.45 gallons

13. Suppose Brother Paul uses a spherical tank with a 7-foot diameter. How many gallons could that tank hold? (Use pi = $3\frac{1}{7}$ and 1 cubic foot = $7\frac{1}{2}$ gallons.) $1{,}347\frac{1}{2}$ gallons

14. A wooden table top is 40 inches wide and 7 feet long. How many board feet are in it? (One board foot is the amount of wood in a piece 12 inches square and 1 inch thick.)
    Needed: thickness of the table top

15. The ice cream piled on top of David's cone was 4 inches high. If that amount was enough to fill a cylinder 3 inches high, how many cubic inches of ice cream were there?
    Needed: diameter of the cone, or radius of the cylinder

REVIEW EXERCISES

F. Write these formulas. *(Lessons 100–107)*

16. Surface area of a cube $a_s = 6e^2$
17. Surface area of a rectangular solid $a_s = 2lw + 2wh + 2lh$
18. Surface area of a cylinder $a_s = 2\pi r^2 + 2\pi rh$
19. Surface area of a square pyramid $a_s = 4(\frac{1}{2}b\ell) + b^2$
20. Surface area of a sphere $a_s = 4\pi r^2$
21. Volume of a rectangular solid $v = lwh$
22. Volume of a cube $v = e^3$

368 *Chapter 8 Geometric Solids and the Pythagorean Rule*

23. Volume of a cylinder $v = \pi r^2 h$
24. Volume of a cone $v = \frac{1}{3}\pi r^2 h$
25. Volume of a square pyramid $v = \frac{1}{3}lwh$

G. Find the volume of each geometric solid. *(Lessons 106, 107)*

26. Cylinder 4,019.2 cm³
    $r = 8$ cm
    $h = 20$ cm
    $\pi = 3.14$

27. Cone $564\frac{2}{3}$ cu. in.
    $r = 7$ in.
    $h = 11$ in.
    $\pi = 3\frac{1}{7}$

28. Square pyramid $2,389\frac{1}{3}$ cu. ft.
    $s = 16$ ft.
    $h = 28$ ft.

29. Square pyramid 2.7 m³
    $s = 1.5$ m
    $h = 3.6$ m

H. Find the surface area of each geometric solid. *(Lessons 100, 104)*

30. Cube 153,600 m²
    $e = 160$ m

31. Sphere 5,544 sq. ft.
    $r = 21$ ft.
    $\pi = 3\frac{1}{7}$

32. Sphere 3,215.36 sq. in.
    $r = 16$ in.
    $\pi = 3.14$

I. Find the area of each rectangle or square. *(Lesson 92)*

33. $l = 38$ ft. 1,064 sq. ft.
    $w = 28$ ft.

34. $s = 61$ cm 3,721 cm²

An Ounce of Prevention

Beware of the frustration of "wrong" answers resulting from different calculation routes.

Both the decimal and the fraction form of pi are rounded values. This produces answers that differ slightly when the different forms are used. Compare these answers for the area of a circle with a 7-foot radius calculated with the two approximations of pi.

$\frac{22}{7} \times r^2 = 154$ sq. ft.

$3.14 \times r^2 = 153.86$ sq. ft.

(which rounds off to 154)

The difference increases as the number of factors increase in a formula. Compare these answers for the volume of a sphere with a 7-foot radius.

$\frac{88}{21} \times r^3 = 1437\frac{1}{3}$ cu. ft.

$4.18 \times r^3 = 1433.74$ cu. ft.

Should a student work from the memorized formula, $v = \frac{4}{3}\pi r^3$, and use the fraction $\frac{4}{3}$ and the decimal 3.14, he will have still another correct answer.

$\frac{4}{3} \times 3.14 \times r^3 = 1436.03$ cu. ft.

These answers will not all agree unless you round to the nearest hundred.

Further Study

1. Finding the volume of a sphere is based on finding the surface area of a sphere. Imagine the surface of a sphere being marked off into many small squares. Each square is the base of a pyramid that tapers to a point at the center of the sphere. For a pyramid, volume = $\frac{1}{3}$ × area of base × height. Hence, the volume of all the pyramids that make up the sphere equals the total surface area × $\frac{1}{3}$ × the height of each pyramid (which equals the radius of the sphere).

    The formula for finding the surface area of a sphere is $v = 4\pi r^2$. So the volume of a sphere can be found by multiplying $4\pi r^2 \times \frac{1}{3}r$. In simplest form, $4 \times \frac{1}{3} \times \pi \times r \times r \times r$ is written as $\frac{4}{3}\pi r^3$.

2. Exercises 8 and 9 show that the volume of a sphere varies as the cube of the multiple by which the radius is increased.

LESSON 109

Objectives

- To review the concept of square roots.
- To review finding square roots by using a table.
- To review finding square roots by estimation.

Review

1. Give Lesson 109 Quiz (Finding Volume).

2. *Find the volume of each geometric solid.* (Lessons 105, 107, 108)

    a. Sphere: $r = 11$ in.
       (5,563.58 cu. in.)

    b. Cube: $e = 7$ in.
       (343 cu. in.)

    c. Rectangular solid
       $l = 26$ ft.
       $w = 18$ ft.
       $h = 15$ ft. (7,020 cu. ft.)

    d. Square pyramid
       $s = 18$ m
       $h = 26$ m (2,808 m^3)

3. *Find the surface area of each rectangular solid.* (Lesson 101)

    a. $l = 27$ ft.
       $w = 19$ ft.
       $h = 15$ ft. (2,406 sq. ft.)

    b. $l = 15$ yd.
       $w = 12$ yd.
       $h = 4$ yd. (576 sq. yd.)

4. *Find the area of each parallelogram.* (Lesson 93)

    a. $b = 63$ cm
       $h = 45$ cm (2,835 cm^2)

    b. $b = 18$ in.
       $h = 11$ in. (198 sq. in.)

109. Understanding Square Roots

Mother's sewing room is a square room with 144 square feet of floor space. How long is each side of the room? Since the area of a square is side times side, you can find the side of a square when given the area by finding what number times itself equals that area. What number times itself equals 144? Mother's sewing room measures 12 feet on each side.

In the example above, the **square root** of 144 is 12 because 12 times itself (or 12^2) equals 144. The **radical sign** ($\sqrt{}$) is used to indicate square root. The **radicand** is the number under the radical sign.

Memorize the square roots of these perfect squares.

$\sqrt{1} = 1$ $\sqrt{16} = 4$ $\sqrt{49} = 7$ $\sqrt{100} = 10$
$\sqrt{4} = 2$ $\sqrt{25} = 5$ $\sqrt{64} = 8$ $\sqrt{121} = 11$ $\sqrt{225} = 15$
$\sqrt{9} = 3$ $\sqrt{36} = 6$ $\sqrt{81} = 9$ $\sqrt{144} = 12$ $\sqrt{625} = 25$

Every number has a square root. A number is a **perfect square** if its square root is a whole number. When a square root is not a whole number, it is usually stated as a decimal rounded to the nearest thousandth.

Any number between 0 and 10,000 has a square root less than 100. To find the roots of perfect squares up to 10,000, you can use estimation as shown in the following steps.

Table for Step 1

Square	Square root
100	10
400	20
900	30
1,600	40
2,500	50
3,600	60
4,900	70
6,400	80
8,100	90
10,000	100

Table for Step 2

Ones' digits of perfect squares	Ones' digits of square roots
1	1 or 9
4	2 or 8
5	5
6	4 or 6
9	3 or 7

1. Determine the tens' group in which the square root will be. For the perfect square 3,364, find its place in the first column in the table at the left. Since 3,364 falls between 2,500 and 3,600, its square root will be in the 50s.

2. Look at the digit in the ones' place of the perfect square, and estimate the ones' digit of the square root according to the second table at the left. Since the last digit of 3,364 is 4, the square root must end with 2 or 8. (This is because $2^2 = 4$ and $8^2 = 64$.) Thus the square root must be either 52 or 58.

3. Use trial and error to find which root is correct. The square of 52 is 2,704. The square of 58 is 3,364. So the square root of 3,364 is 58.

Chapter 8 Geometric Solids and the Pythagorean Rule

The table for step 2 is based on the fact that a certain digit in the ones' place of a square root will always produce the same result in the ones' place of the square. The numerals in the shaded column at the right are the only possible ones' digits for a perfect square. A perfect square cannot end with 2, 3, 7, or 8 because no digit multiplied by itself yields a product ending with 2, 3, 7, or 8. Zero is not needed in the table because if a perfect square ends with 0, the square root appears in the table for step 1.

$1^2 = 1$
$2^2 = 4$
$3^2 = 9$
$4^2 = 16$
$5^2 = 25$
$6^2 = 36$
$7^2 = 49$
$8^2 = 64$
$9^2 = 81$

Example
Find the square root of 2,401 by estimation.
Think: 2,401 is between 1,600 and 2,500. The square root will be in the 40s.
2,401 ends with 1. The square root must end with 1 or 9.
The square root is either 41 or 49.
$41^2 = 1,681$ $49^2 = 2,401$
$\sqrt{2,401} = 49$

After finding a square root, you can check your work by multiplying the square root times itself. You can also cast out nines as shown below.

$\sqrt{2,401} = 49 \rightarrow 4$
$\downarrow \qquad\qquad \times 4$
$7 \quad\leftarrow\quad 16$

When you have cast out nines from the root, square the result and cast out nines from that product. The number left should match the result of casting out nines from the radicand.

On page 584 is a table of square roots expressed to the nearest thousandth. (Most square roots are irrational numbers.) If a number from 1 to 200 is not a perfect square, you can find its square root by using the table.

CLASS PRACTICE

Give the square root of each number by using the table of square roots.

a. $\sqrt{115}$ 10.724
b. $\sqrt{125}$ 11.180
c. $\sqrt{135}$ 11.619
d. $\sqrt{145}$ 12.042
e. $\sqrt{155}$ 12.450
f. $\sqrt{160}$ 12.649

Name the tens' group in which the square root of each number will be.

g. $\sqrt{4,362}$ 60s
h. $\sqrt{1,873}$ 40s
i. $\sqrt{3,615}$ 60s
j. $\sqrt{428}$ 20s

Use estimation to find the square roots of these perfect squares.

k. $\sqrt{4,225}$ 65
l. $\sqrt{7,056}$ 84
m. $\sqrt{841}$ 29
n. $\sqrt{5,329}$ 73

Introduction

Ask the students how the following sets of numbers are related.

169, 13   144, 12   121, 11   64, 8

The students should readily see that the second number squared equals the first number. Review the fact that the square root of a number is the number that, when multiplied by itself, equals the original number.

Ask the students to list the numbers and their square roots that they know by memory. Following is a list of combinations that are familiar or should become familiar.

4, 2	9, 3	16, 4
25, 5	36, 6	49, 7
64, 8	81, 9	100, 10
121, 11	144, 12	225, 15
400, 20	625, 25	900, 30
1,600; 40	2,500; 50	3,600; 60
4,900; 70	6,400; 80	

Teaching Guide

1. **The square root of a number is the number whose square is the original number.** The following terms relate to finding square roots.

   a. The symbol for *square root* is $\sqrt{}$ which is known as the radical sign.

   b. The radicand is the number for which the square root is to be found.

   c. A perfect square is a number whose square root is a whole number. The squares listed in the introduction are all perfect squares.

2. **The square root table lists the square roots of the numbers 1–200.** Use the following numbers for practice with finding square roots on the chart.

   a. $\sqrt{179}$ (13.379)
   b. $\sqrt{159}$ (12.610)
   c. $\sqrt{97}$ (9.849)
   d. $\sqrt{145}$ (12.042)
   e. $\sqrt{137}$ (11.705)
   f. $\sqrt{153}$ (12.369)

3. **The square roots of perfect squares between 1 and 10,000 can be found by estimation.** Use the following steps.

   (1) Determine the tens' group in which the square root will be.

   (2) Estimate the digit in the ones' place of the square root.

   (3) Use trial and error to determine which root is correct.

   Be sure to discuss step 2. Emphasize the importance of the ones' digit in deciding what the square root must be. Use the following numbers for practice with finding square roots by estimation.

   a. $\sqrt{1,369}$ (37)
   b. $\sqrt{2,704}$ (52)
   c. $\sqrt{5,184}$ (72)

T–371   Chapter 8   Geometric Solids and the Pythagorean Rule

Extractions for Lesson 110, CLASS PRACTICE

a.
$$\begin{array}{r} 6\ 3 \\ \sqrt{39_\wedge 69} \\ \underline{36} \end{array}$$
(20 × 6) 120   3 69
3 × 123 = 3 69
                0

b.
$$\begin{array}{r} 7\ 1 \\ \sqrt{50_\wedge 41} \\ \underline{49} \end{array}$$
(20 × 7) 140   1 41
1 × 141 = 1 41
                0

c.
$$\begin{array}{r} 1\ 7\ 2 \\ \sqrt{2_\wedge 95_\wedge 84} \\ \underline{1} \end{array}$$
(20 × 1) 20   1 95
7 × 27 = 1 89
(20 × 17) 340   6 84
2 × 342 = 6 84
                     0

d.
$$\begin{array}{r} 2\ 3\ 7 \\ \sqrt{5_\wedge 61_\wedge 69} \\ \underline{4} \end{array}$$
(20 × 2) 40   1 61
3 × 43 = 1 29
(20 × 23) 460   32 69
7 × 467 = 32 69
                     0

Extractions for Lesson 110, Part A

1.
$$\begin{array}{r} 6\ 2 \\ \sqrt{38_\wedge 44} \\ \underline{36} \end{array}$$
(20 × 6) 120   2 44
2 × 122 = 2 44
                0

2.
$$\begin{array}{r} 5\ 6 \\ \sqrt{31_\wedge 36} \\ \underline{25} \end{array}$$
(20 × 5) 100   6 36
6 × 106 = 6 36
                0

3.
$$\begin{array}{r} 8\ 8 \\ \sqrt{77_\wedge 44} \\ \underline{64} \end{array}$$
(20 × 8) 160   13 44
8 × 168 = 13 44
                 0

4.
$$\begin{array}{r} 2\ 2 \\ \sqrt{4_\wedge 84} \\ \underline{4} \end{array}$$
(20 × 2) 40   0 84
2 × 42 = 84
                0

5.
$$\begin{array}{r} 6\ 1 \\ \sqrt{37_\wedge 21} \\ \underline{36} \end{array}$$
(20 × 6) 120   1 21
1 × 121 = 1 21
                0

6.
$$\begin{array}{r} 9\ 5 \\ \sqrt{90_\wedge 25} \\ \underline{81} \end{array}$$
(20 × 9) 180   9 25
5 × 185 = 9 25
                0

Continued on page T–372

Solutions for Exercises 23 and 24

23. $4(7 \times 4) + 4(12 \times 4)$

24. $8 \div 2\frac{2}{3}$

Lesson 109

WRITTEN EXERCISES

A. Give the square root of each number by using the table of square roots.

1. $\sqrt{120}$ 10.954
2. $\sqrt{150}$ 12.247
3. $\sqrt{95}$ 9.747
4. $\sqrt{1}$ 1
5. $\sqrt{196}$ 14
6. $\sqrt{187}$ 13.675

B. Name the tens' group in which the square root of each number will be.

7. $\sqrt{1,255}$ 30s
8. $\sqrt{6,799}$ 80s
9. $\sqrt{978}$ 30s
10. $\sqrt{3,717}$ 60s
11. $\sqrt{7,699}$ 80s
12. $\sqrt{8,512}$ 90s

C. Use estimation to find the square roots of these perfect squares.

13. $\sqrt{6,241}$ 79
14. $\sqrt{529}$ 23
15. $\sqrt{8,281}$ 91
16. $\sqrt{4,356}$ 66
17. $\sqrt{7,396}$ 86
18. $\sqrt{1,369}$ 37

D. Solve these reading problems. The numbers in exercises 19–22 are perfect squares.

19. Grandfather's age times itself equals 4,096. Find his age.   64 years
20. A large square carpet covers an area of 576 square feet. What is the length of its sides?   24 feet
21. Grandmother has a square flower bed with an area of 484 square feet. What is the length of its sides?   22 feet
22. The Zimmermans have a square garage with an area of 784 square feet. What is the length of its sides?   28 feet
23. Kenton painted the wooden sides and tailgate of a dump truck. The truck bed is 7 feet wide and 12 feet long. The sides are 4 feet high. How many square feet will he cover if he paints the inside and the outside of the four sides?   304 square feet
24. The area inside a large picture frame is 8 square feet. If the width of the frame is 32 inches, how long is it?   36 inches (or 3 feet)

372   Chapter 8   Geometric Solids and the Pythagorean Rule

REVIEW EXERCISES

E. Write these formulas. *(Lessons 100–108)*

25. Surface area of a cube $a_s = 6e^2$
26. Surface area of a rectangular solid $a_s = 2lw + 2wh + 2lh$
27. Surface area of a cylinder $a_s = 2\pi r^2 + 2\pi rh$
28. Surface area of a square pyramid $a_s = 4(\frac{1}{2}b\ell) + b^2$
29. Surface area of a sphere $a_s = 4\pi r^2$
30. Volume of a rectangular solid $v = lwh$
31. Volume of a cube $v = e^3$
32. Volume of a cylinder $v = \pi r^2 h$
33. Volume of a cone $v = \frac{1}{3}\pi r^2 h$
34. Volume of a square pyramid $v = \frac{1}{3}lwh$
35. Volume of a sphere $v = \frac{4}{3}\pi r^3$

F. Find the volume of each sphere. Use 4.18 in your calculations. *(Lesson 108)*

36. r = 12 in.   7,223.04 cu. in.   37. r = 20 ft.   33,440 cu. ft.

G. Find the volume of each geometric solid. *(Lessons 105, 107)*

38. Rectangular solid   39. Cube   40. Square pyramid
    l = 14 ft.   1,540 cu. ft. e = 34 cm   39,304 cm³ s = 8 m   512 m³
    w = 11 ft. h = 24 m
    h = 10 ft.

H. Find the surface area of each rectangular solid. *(Lesson 101)*

41. l = 8 ft.   472 sq. ft.   42. l = 11 yd.   268 sq. yd.
    w = 7 ft. w = 6 yd.
    h = 12 ft. h = 4 yd.

I. Find the area of each parallelogram. *(Lesson 93)*

43. b = 55 cm   2,310 cm²   44. b = 26 in.   390 sq. in.
    h = 42 cm h = 15 in.

Extractions for Lesson 110, continued

7.
$$\begin{array}{r}1\ 8\ 6\\\hline\sqrt{3{,}45{,}96}\end{array}$$
(20 × 1) 20 | 2 45 ; 1
8 × 28 = 2 24
(20 × 18) 360 | 21 96
6 × 366 = 21 96
0

8.
$$\begin{array}{r}1\ 8\ 7\\\hline\sqrt{3{,}49{,}69}\end{array}$$
(20 × 1) 20 | 2 49 ; 1
8 × 28 = 2 24
(20 × 18) 360 | 25 69
7 × 367 = 25 69
0

9.
$$\begin{array}{r}1\ 7\ 5\\\hline\sqrt{3{,}06{,}25}\end{array}$$
(20 × 1) 20 | 2 06 ; 1
7 × 27 = 1 89
(20 × 17) 340 | 17 25
5 × 345 = 17 25
0

10.
$$\begin{array}{r}1\ 8\ 2\\\hline\sqrt{3{,}31{,}24}\end{array}$$
(20 × 1) 20 | 2 31 ; 1
8 × 28 = 2 24
(20 × 18) 360 | 7 24
2 × 362 = 7 24
0

11.
$$\begin{array}{r}4\ 3\\\hline\sqrt{18{,}49}\end{array}$$
 ; 16
(20 × 4) 80 | 2 49
3 × 83 = 2 49
0

12.
$$\begin{array}{r}4\ 5\\\hline\sqrt{20{,}25}\end{array}$$
 ; 16
(20 × 4) 80 | 4 25
5 × 85 = 4 25
0

13.
$$\begin{array}{r}4\ 5\ 5\\\hline\sqrt{20{,}70{,}25}\end{array}$$
 ; 16
(20 × 4) 80 | 4 70
5 × 85 = 4 25
(20 × 45) 900 | 45 25
5 × 905 = 45 25
0

14.
$$\begin{array}{r}7\ 5\\\hline\sqrt{56{,}25}\end{array}$$
 ; 49
(20 × 7) 140 | 7 25
5 × 145 = 7 25
0

15.
$$\begin{array}{r}2\ 8\\\hline\sqrt{7{,}84}\end{array}$$
 ; 4
(20 × 2) 40 | 3 84
8 × 48 = 3 84
0

16.
$$\begin{array}{r}1\ 6\ 0\\\hline\sqrt{2{,}56{,}00}\end{array}$$
 ; 1
20 | 1 56
6 × 26 = 1 56
0 00

T–373   Chapter 8   *Geometric Solids and the Pythagorean Rule*

LESSON 110

Objective

- To teach *extracting square roots by calculation.

Review

1. *Find the square root of each number by using the table.* (Lesson 109)
   a. $\sqrt{18}$ (4.243)
   b. $\sqrt{38}$ (6.164)
   c. $\sqrt{153}$ (12.369)

2. *Find the volume of each figure.* (Lessons 106, 108)
   a. Sphere
      $r = 15$ ft.   (14,107.5 cu. ft.)
   b. Cylinder
      $r = 2$ in.
      $h = 8$ in.   (100.48 cu. in.)
   c. Cone
      $r = 18$ ft.
      $h = 18$ ft.   (6,104.16 cu. ft.)

3. *Find the surface area of each cylinder.* (Lesson 102)
   a. $r = 14$ ft. (multiple of 7)
      $h = 36$ ft.   (4,400 sq. ft.)
   b. $r = 6$ in.
      $h = 14$ in.   (753.6 sq. in.)

4. *Find the area of each trapezoid.* (Lesson 94)
   a. $h = 9$ in.
      $b_1 = 6$ in.
      $b_2 = 8$ in.   (63 sq. in.)
   b. $h = 8.9$ m
      $b_1 = 3.7$ m
      $b_2 = 5.1$ m   (39.16 m²)

5. *Select the necessary information, and solve each reading problem.* (Lesson 44)
   a. A mountain lion can cover 23 feet in a single bound. It can leap as much as 18 feet into the air, and it can drop 65 feet to the ground without injury. If a mountain lion could run at full speed for 1 mile, how many times would it touch the ground in that distance? Round to the nearest whole number.

   (23 ft., 1 mi.; 230 times)

   b. The martial eagle of Africa has a wingspread up to 8 feet. Its nest may be as much as $6\frac{1}{2}$ feet in diameter and $3\frac{1}{2}$ feet deep. What would be the circumference of such a nest?   ($6\frac{1}{2}$ ft.; $20\frac{3}{7}$ ft.)

110. Extracting Square Roots

The square root of a number that is not on the table can be found through the process known as extracting the square root. This process is somewhat similar to division.

To extract the square root of a number, follow the steps below.

1. Starting at the right, break the number into periods of two digits each. Use a caret (∧) to mark each separation. (The period farthest to the left may have only one digit.)

   $\sqrt{6{,}81{,}21}$

2. Choose the largest single digit whose square is equal to or less than the number(s) in the first period on the left. Write that number above the first period as a quotient figure. Write its square below the first period and subtract.

   $$\begin{array}{r} 2 \\ \sqrt{6{,}81{,}21} \\ 4 \\ \hline 2 \end{array}$$

   (Squares of 1-digit numbers: 1, 4, 9, 16, 25, 36, 49, 64, 81)

3. Bring down the next period to form a new radicand.

   $$\begin{array}{r} 2 \\ \sqrt{6{,}81{,}21} \\ 4 \\ \hline 2\ 81 \end{array}$$

4. Multiply the partial square root (the quotient obtained so far) by 20. This product is a trial divisor for the new radicand.

5. Estimate a single-digit quotient for the new radicand divided by the trial divisor. Add that digit to the trial divisor to form the real divisor. Also write that digit in the quotient position over the next period.

   (20 × 2) 40 $\begin{array}{r} 2 \\ \sqrt{6{,}81{,}21} \\ 4 \\ \hline 2\ 81 \end{array}$

6. Multiply the real divisor by the last digit in the square root. Write the product below the new radicand and subtract. (If your product is larger than the new radicand, lower the quotient digit and also the ones' digit of the divisor.)

   (20 × 2) 40 $\begin{array}{r} 2\ 6 \\ \sqrt{6{,}81{,}21} \\ 4 \\ \hline 2\ 81 \end{array}$

   6 × 46 = $\underline{2\ 76}$

   5

7. Repeat steps 3–6 until the problem is complete. In step 4, multiply all the digits of the partial square root, not just by the last digit obtained. After doing the division in step 5, remember to add the new quotient figure to the divisor before multiplying in step 6.

8. Check the answer by multiplying the square root by itself and comparing the product with the radicand. You can also check by casting out nines as shown in Lesson 109.

   $\begin{array}{r} 2\ 6\ 1 \\ \sqrt{6{,}81{,}21} \\ 4 \\ \hline 2\ 81 \\ 2\ 76 \\ \hline 5\ 21 \end{array}$

   (20 × 26) 520

   1 × 521 = $\underline{5\ 21}$

199

374　Chapter 8　Geometric Solids and the Pythagorean Rule

Example A
Extract the square root of 841.

$$\begin{array}{r} 2\ 9 \leftarrow \text{square root} \\ \sqrt{8{,}41} \leftarrow \text{radicand} \\ 4 \\ \overline{4\ 41} \leftarrow \text{new radicand} \end{array}$$

$20 \times 2 = 40$
$+\ 9$
$9 \times 49 = 4\ 41$

Check
$\begin{array}{r} 29 \\ \times\ 29 \\ \hline 841 \end{array}$

Example B
Extract the square root of 15,876.

$$\begin{array}{r} 1\ 2\ 6 \\ \sqrt{1{,}58{,}76} \\ 1 \\ \overline{58} \\ 44 \\ \overline{14\ 76} \\ \underline{14\ 76} \end{array}$$

$20 \times 1 = 20$
$2 \times 22 = 44$
$20 \times 12 = 240$
$6 \times 246 = 14\ 76$

Check
$\begin{array}{r} 126 \\ \times\ 126 \\ \hline 15{,}876 \end{array}$

CLASS PRACTICE　　(See page T–371 for extractions worked out.)

Extract the square roots of these perfect squares. Show your work.

a. $\sqrt{3{,}969}$ = 63　　b. $\sqrt{5{,}041}$ = 71　　c. $\sqrt{29{,}584}$ = 172　　d. $\sqrt{56{,}169}$ = 237

WRITTEN EXERCISES　　(See page T–371 for extractions worked out.)

A. *Extract the square roots of these perfect squares. Show your work.*

1. $\sqrt{3{,}844}$ = 62　　2. $\sqrt{3{,}136}$ = 56　　3. $\sqrt{7{,}744}$ = 88　　4. $\sqrt{484}$ = 22

5. $\sqrt{3{,}721}$ = 61　　6. $\sqrt{9{,}025}$ = 95　　7. $\sqrt{34{,}596}$ = 186　　8. $\sqrt{34{,}969}$ = 187

9. $\sqrt{30{,}625}$ = 175　　10. $\sqrt{33{,}124}$ = 182　　11. $\sqrt{1{,}849}$ = 43　　12. $\sqrt{2{,}025}$ = 45

B. *Solve these reading problems.*

13. A square field has an area of 207,025 square feet. Find the length of one of its sides. **455 feet**

14. Josiah is fencing off a newly planted orchard in the meadow to keep the cows away from the trees. The orchard is a square plot with an area of 5,625 square feet. How many feet of wire will he need to fence off this area? **300 feet**

15. In calculating the area of a circle, James squared the radius and obtained a product of 784. What was the radius with which he started? **28**

16. Musser's Building Supply erected a large warehouse for storing lumber. It was a square building with an area of 25,600 square feet. What was the length of one side? **160 feet**

17. In 1947, the Dead Sea Scrolls were found in pottery jars in caves of the Judean foothills. Some of the jars ranged from 25 to 29 inches high and were about 10 inches wide. Suppose one covered jar was in the shape of a cylinder 28 inches high and 10 inches in diameter. What would be the surface area of this jar? **1,036.2 square inches**

18. A spherical water tower has a radius of 21 feet. What is its volume in gallons at $7\frac{1}{2}$ gallons per cubic foot? (Use $\frac{4}{3}$ and $3\frac{1}{7}$.) **291,060 gallons**

Introduction

Work out this division problem on the board.

```
          124
    124)15,376
        12 4
         2 97
         2 48
           496
           496
```

The process of extracting a square root is somewhat like division, but you do not have a divisor to start with. Write the number 15,376 on the board, and extract the square root by following the steps in the lesson.

Compare the division with the extraction. Notice that the basic steps are the same: estimate, multiply, and subtract. An estimate cannot be greater than 9, and a subtraction answer cannot be greater than the real divisor.

$$
\begin{array}{r}
1\ 2\ 4 \\
\sqrt{1{}_\wedge 53{}_\wedge 76}
\end{array}
$$

```
                        1
20 × 1 = 20            53
             2 × 22 = 44
20 × 12 = 240         9 76
            4 × 244 = 9 76
```

Solutions for Part B

13. $\sqrt{207{,}025}$

14. $4 \times \sqrt{5{,}625}$

15. $\sqrt{784}$

16. $\sqrt{25{,}600}$

17. $2(3.14 \times 5^2) + (28 \times 3.14 \times 10)$

18. $\frac{88}{21} \times 21^3 \times 7\frac{1}{2}$

Teaching Guide

The square root of a number can be calculated by a process called extracting the square root. To use this process, follow the steps in the lesson.

(1) Starting at the right, break the number into periods of two digits each. Use a caret ($_\wedge$) to mark each separation. (The period farthest to the left may have only one digit.)

(2) Choose the largest single digit whose square is equal to or less than the number(s) in the first period on the left. Write that number above the first period as a quotient figure. Write its square below the first period and subtract.

(3) Bring down the next period to form a new radicand.

(4) Multiply the partial square root (the quotient obtained so far) by 20. This product is a trial divisor for the new radicand.

(5) Estimate a single-digit quotient for the new radicand divided by the trial divisor. Add that digit to the trial divisor to form the real divisor. Also write that digit in the quotient position over the next period.

(6) Multiply the real divisor by the last digit in the square root. Write the product below the new radicand and subtract.

(7) Repeat steps 3–6 until the problem is complete. In step 4, multiply all the digits of the partial square root, not just the last digit obtained. After doing the division in step 5, remember to add the new quotient figure to the divisor before multiplying in step 6.

T–375 Chapter 8 Geometric Solids and the Pythagorean Rule

(8) Check the answer by multiplying the square root by itself and comparing the product with the radicand.

a. $\sqrt{8{,}464}$ (92)

$$\begin{array}{r}9\ \ 2\\ \sqrt{84{,}64}\\ \underline{81}\\ (20\times 9)\ 180\quad 3\ 64\\ 2\times 182 = \underline{3\ 64}\\ 0\end{array}$$

b. $\sqrt{5{,}776}$ (76)

$$\begin{array}{r}7\ \ 6\\ \sqrt{57{,}76}\\ \underline{49}\\ (20\times 7)\ 140\quad 8\ 76\\ 6\times 146 = \underline{8\ 76}\\ 0\end{array}$$

c. $\sqrt{31{,}684}$ (178)

$$\begin{array}{r}1\ \ 7\ \ 8\\ \sqrt{3{,}16{,}84}\\ \underline{1}\\ (20\times 1)\ 20\quad 2\ 16\\ 7\times 27 = \underline{1\ 89}\\ (20\times 17)\ 340\quad 27\ 84\\ 8\times 348 = \underline{27\ 84}\\ 0\end{array}$$

d. $\sqrt{45{,}796}$ (214)

$$\begin{array}{r}2\ \ 1\ \ 4\\ \sqrt{4{,}57{,}96}\\ \underline{4}\\ (20\times 2)\ 40\quad 57\\ 1\times 41 = \underline{41}\\ (20\times 21)\ 420\quad 16\ 96\\ 4\times 424 = \underline{16\ 96}\\ 0\end{array}$$

An Ounce of Prevention

Be careful not to assign too much material in this lesson. The material is completely new to the students, as new as multiplication and division are to second and third graders. Be sure the students grasp how to do the work before assigning the lesson. Once they gain confidence, they may even consider it fun. It gives a sense of satisfaction to have learned how to extract square roots.

REVIEW EXERCISES

C. Give the square root of each number by using the table of square roots. *(Lesson 109)*

19. $\sqrt{61}$ 7.810

20. $\sqrt{194}$ 13.928

D. Find the volume of each sphere. Use 4.18 in your calculations. *(Lesson 108)*

21. $r = 11$ ft.   5,563.58 cu. ft.

22. $r = 13$ cm   9,183.46 cm³

E. Find the volume of each geometric solid. *(Lesson 106)*

23. Cylinder   254.34 cu. in.
    $r = 3$ in.
    $h = 9$ in.
    $\pi = 3.14$

24. Cone $975\frac{1}{3}$ cu. ft.
    $r = 7$ ft.
    $h = 19$ ft.
    $\pi = 3\frac{1}{7}$

F. Find the surface area of each cylinder. *(Lesson 102)*

25. $r = 21$ ft.   12,276 sq. ft.
    $h = 72$ ft.
    $\pi = 3\frac{1}{7}$

26. $r = 4$ in.   653.12 sq. in.
    $h = 22$ in.
    $\pi = 3.14$

G. Find the area of each trapezoid. *(Lesson 94)*

27. $h = 5$ in.   40 sq. in.
    $b_1 = 7$ in.; $b_2 = 9$ in.

28. $h = 9.9$ m   51.48 m²
    $b_1 = 5.6$ m; $b_2 = 4.8$ m

Hints for Finding Trouble Spots in Extracting Square Roots

1. Multiply all of the accumulated square root (quotient) by 20 for each new trial divisor.

2. Remember to change the trial divisor to the real divisor before multiplying by the new quotient digit.

3. If a quotient digit times the real divisor is larger than the radicand for that step, use a smaller number for your quotient digit.

4. If the remainder in a subtraction step is larger than the real divisor, use a larger number for your quotient digit.

5. If you adjust the new quotient digit, be sure to adjust the ones' digit of the real divisor to match it.

376 Chapter 8 Geometric Solids and the Pythagorean Rule

111. Extracting More Difficult Square Roots

The process of extracting square roots can be used when the radicand is not a perfect square. A square root can be calculated to as many decimal places as desired, the same as the quotient of a division problem.

Here are the steps from the last lesson in condensed form.

1. Break the number into periods of two digits each.

2. Choose the largest single digit whose square is equal to or less than the numbers in the first period. Write the digit above the period. Write the square below, and subtract.

3. Bring down the next period to form a new radicand.

4. Multiply the partial square root by 20. This product is the trial divisor for the new radicand.

5. Estimate a single-digit quotient for the new radicand divided by the trial divisor. Write that digit in the quotient and add it to the trial divisor for the real divisor.

6. Multiply the real divisor by the last digit in the square root. Write the product below the new radicand and subtract.

7. Repeat steps 3–6 until the problem is complete. If it does not work out evenly by the last period, place a decimal point in both radicand and square root. Annex a period of two zeroes in the radicand for each decimal place in the square root.

When the square root is a nonterminating decimal, you will need to round it as instructed.

Example A
Find the square root of 3,271 to the nearest whole number.

$$57\,.\,1 = 57$$
$$\sqrt{32{,}71.00}\text{(rounded)}$$

```
                    5  7 . 1 = 57
                 √32ˏ71.00   (rounded)
                   25
20 × 5 = 100       7 71
                7 × 107 = 7 49
20 × 57 = 1,140        22 00
                1 × 1,141 = 11 41
                           10 59
```

Example B
Find the square root of 4,689 to the nearest tenth.

```
                    6  8 . 4  7 = 68.5
                 √46ˏ89.00ˏ00   (rounded)
                   36
20 × 6 = 120       10 89
                8 × 128 = 10 24
20 × 68 = 1,360        65 00
                4 × 1,364 = 54 56
20 × 684 = 13,680       10 44 00
                7 × 13,687 = 9 58 09
                                85 91
```

LESSON 111

Objectives

- To give more practice with extracting square roots.
- To teach *extracting the square roots of numbers that are not perfect squares.

Review

1. *Find the square root of each number by using a table.* (Lesson 109)

    a. $\sqrt{48}$ (6.928)

    b. $\sqrt{78}$ (8.832)

2. *Extract the square root of each number.* (Lesson 110)

    a. $\sqrt{576}$ (24)

    $$\sqrt{5\wedge 76} \quad 2\ 4$$
    $$\underline{4}$$
    $$(20 \times 2)\ 40\ \overline{1\ 76}$$
    $$4 \times 44 = \underline{1\ 76}$$
    $$0$$

    b. $\sqrt{4{,}096}$ (64)

    $$\sqrt{40\wedge 96} \quad 6\ 4$$
    $$\underline{36}$$
    $$(20 \times 6)\ 120\ \overline{4\ 96}$$
    $$4 \times 124 = \underline{4\ 96}$$
    $$0$$

3. *Find the volume of each square pyramid, to the nearest hundredth.* (Lesson 107)

    a. $b = 26$ cm
    $h = 35$ cm   (7886.67 cm³)

    b. $b = 115$ ft.
    $h = 98$ ft.   (432,016.67 cu. ft.)

4. *Find the surface area of each square pyramid.* (Lesson 103)

    a. $b = 32$ in.
    $\ell = 28$ in.   (2,816 sq. in.)

    b. $b = 19$ ft.
    $\ell = 25$ ft.   (1,311 sq. ft.)

5. *Find the area of this circle.* (Lesson 95)

    $r = 31\frac{1}{2}$ ft.
    $\pi = 3\frac{1}{7}$ ($3{,}118\frac{1}{2}$ sq. ft.)

6. *Write these decimals, using words.* (Lesson 47)

    a. 3.025   (Three and twenty-five thousandths)

    b. 16.8079
    (Sixteen and eight thousand seventy-nine ten-thousandths)

Chapter 8 Geometric Solids and the Pythagorean Rule

Teaching Guide

The process of extracting square roots can be used when the radicand is not a perfect square. A square root can be calculated to as many decimal places as desired, the same as the quotient of a division problem.

Review the steps for extracting square roots; then work through a few problems with the students. Be sure they annex two zeroes at a time to the right side of the radicand. Watch also for accuracy in placing the decimal point in the square root, and in rounding the answer.

Find the square root of each number to the place indicated.

a. $\sqrt{5,712}$ to the nearest whole number (76)

```
           7 5 . 5
       √57̭12.00
         49
(20 × 7) 140    8 12
    5 × 145 = 7 25
(20 × 75) 1500   87 00
   5 × 1505 = 75 25
                 11 75
```

b. $\sqrt{4,778}$ to the nearest tenth (69.1)

```
             6 9 . 1 2
         √47̭78.00̭00
           36
(20 × 6) 120    11 78
    9 × 129 =   11 61
(20 × 69) 1380     17 00
   1 × 1381 =     13 81
(20 × 691) 13820    3 19 00
   2 × 13822 =      2 76 44
                      42 56
```

c. $\sqrt{7,442}$ to the nearest tenth (86.3)

```
             8 6 . 2 6
         √74̭42.00̭00
           64
(20 × 8) 160    10 42
    6 × 166 =    9 96
(20 × 86) 1720     46 00
   2 × 1722 =      34 44
(20 × 862) 17240    11 56 00
   6 × 17246 =      10 34 76
                     1 21 24
```

An Ounce of Prevention

As in the last lesson, be sure the students understand their work. Also be careful not to assign too much homework. Extracting square roots takes time.

Consider checking the students' first two or three answers after about fifteen minutes of work. If these answers are correct, they can keep on going. If they are wrong, have the students work on them until they are correct.

Extractions for CLASS PRACTICE f–i

f.
```
              6 1 . 5
         √37̭89.00
           36
(20 × 6) 120    1 89
   1 × 121 =   1 21
(20 × 61) 1220    68 00
   5 × 1225 =    61 25
                  6 75
```

g.
```
              7 8 . 8
         √62̭15.00
           49
(20 × 7) 140   13 15
   8 × 148 =   11 84
(20 × 78) 1560    1 31 00
   8 × 1568 =    1 25 44
                    5 56
```

h.
```
              6 7 . 4 1
         √45̭45.00̭00
           36
(20 × 6) 120    9 45
   7 × 127 =   8 89
(20 × 67) 1340    56 00
   4 × 1344 =    53 76
(20 × 674) 13480    2 24 00
   1 × 13481 =     1 34 81
                      89 19
```

i.
```
              9 3 . 0 2
         √86̭54.00̭00
           81
(20 × 9) 180    5 54
   3 × 183 =   5 49
(20 × 93) 1860     5 00
                   0
(20 × 930) 18600    5 00 00
   2 × 18602 =     3 72 04
                   1 27 96
```

These answers too can be checked by casting out nines. Disregard any decimal point in the answer, and treat the last subtraction answer as a remainder.

Check of Example A:
Answer = 57.1 R 1,059
57.1 → 4; 4 × 4 = 16 → 7
1,059 → 6; 6 + 7 = 13 → 4
Radicand = 3,271 → 4

Check of Example B:
Answer = 68.47 R 8,591
68.47 → 7; 7 × 7 = 49 → 4
8,591 → 5; 5 + 4 = 9 → 0
Radicand = 4,689 → 0

CLASS PRACTICE

Name the tens' group in which the square root of each number will be.

a. $\sqrt{3,789}$ 60s
b. $\sqrt{6,215}$ 70s
c. $\sqrt{4,545}$ 60s
d. $\sqrt{8,654}$ 90s
e. $\sqrt{928}$ 30s

Find each square root to the place indicated.

f. $\sqrt{3,789}$ (whole number) 62
g. $\sqrt{6,215}$ (whole number) 79
h. $\sqrt{4,545}$ (tenths) 67.4
i. $\sqrt{8,654}$ (tenths) 93.0

WRITTEN EXERCISES

A. *Name the tens' group in which the square root of each number will be.*

1. $\sqrt{2,799}$ 50s
2. $\sqrt{5,555}$ 70s
3. $\sqrt{798}$ 20s
4. $\sqrt{9,899}$ 90s
5. $\sqrt{4,850}$ 60s
6. $\sqrt{6,543}$ 80s

B. *Find each square root to the place indicated.* (See page T–378 for extractions.)

7. $\sqrt{2,799}$ (whole number) 53
8. $\sqrt{8,787}$ (whole number) 94
9. $\sqrt{5,555}$ (whole number) 75
10. $\sqrt{7,777}$ (whole number) 88
11. $\sqrt{798}$ (tenths) 28.2
12. $\sqrt{9,899}$ (tenths) 99.5
13. $\sqrt{4,011}$ (tenths) 63.3
14. $\sqrt{6,543}$ (tenths) 80.9

C. *Solve these reading problems.*

15. Raymond has found that the square root of 185 is very near to his exact age. Find his age to the nearest tenth of a year.   (May be found on table.)   13.6 years

16. Glenn is building a square enclosure for his sheep. If it has an area of 5,476 square feet, how long is each side of the enclosure?   (May be found by estimation.)   74 feet

378 Chapter 8 Geometric Solids and the Pythagorean Rule

17. The basement of Brother Jesse's house is a square with an area of nearly 562 square feet. How long is each side of the basement? (Calculate to the nearest tenth.)
    23.7 feet

18. The Millers' new washing machine came in a square cardboard box whose base had an area of nearly 745 square inches. Find the length of each side to the nearest tenth of an inch.
    27.3 inches

19. A small pyramid has a square base with sides of 14 inches and a height of 20 inches. What is its volume?
    $1{,}306\tfrac{2}{3}$ cubic inches

20. Marcus is constructing a bin to store ground corn. The bin is an inverted pyramid with a slant height of 30 inches and with sides 34 inches long. He will also make a square lid for the bin. How much plywood will he use to build it?
    3,196 square inches

REVIEW EXERCISES

D. Give the square root of each number by using the table of square roots. *(Lesson 109)*

21. $\sqrt{52}$ 7.211
22. $\sqrt{105}$ 10.247

E. Find the volume of each square pyramid. *(Lesson 107)*

23. $s = 30$ cm
    $h = 24$ cm
    7,200 cm³

24. $s = 100$ ft.
    $h = 90$ ft.
    300,000 cu. ft.

F. Find the surface area of each square pyramid. *(Lesson 103)*

25. $b = 15$ in.
    $\ell = 14$ in.
    645 sq. in.

26. $b = 20$ ft.
    $\ell = 12$ ft.
    880 sq. ft.

G. Find the area of each circle. *(Lesson 95)*

27. $r = 42$ ft.
    $\pi = 3\tfrac{1}{7}$
    5,544 sq. ft.

28. $r = 104$ m
    $\pi = 3.14$
    33,962.24 m²

H. Write these decimals, using words. *(Lesson 47)*

29. 5.108 Five and one hundred eight thousandths
30. 11.0851 Eleven and eight hundred fifty-one ten-thousandths

Solutions for Exercises 17–20

17.
```
         2 3 . 7  0
      √5̄,6̄2̄.0̄0̄,̄0̄0̄
        4
(20 × 2) 40    1 62
         3 × 43 = 1 29
(20 × 23) 460    33 00
         7 × 467 = 32 69
(20 × 237) 4740      31 00
```

18.
```
         2 7 . 2  9
      √7̄,4̄5̄.0̄0̄,̄0̄0̄
        4
(20 × 2) 40    3 45
         7 × 47 = 3 29
(20 × 27) 540    16 00
         2 × 542 = 10 84
(20 × 272) 5440    5 16 00
         9 × 5449 = 4 90 41
                    25 59
```

19. $\tfrac{1}{3} \times 14^2 \times 20$

20. $4(\tfrac{1}{2} \times 34 \times 30) + 34^2$

Lesson 111 T–378

Further Study

Typically, square roots are calculated to the thousandths' place. They may be computed further if greater precision is required. However, in this lesson students are finding square roots only to the nearest tenth so that they will learn the process without being burdened with laborious work. The divisors become larger and larger with each new digit in the square root.

Extractions for Part B

7.
$$\begin{array}{r}5\ 2.9\\ \sqrt{27{,}99.00}\\ \underline{25}\end{array}$$
(20 ×5) 100   2 99
2 × 102 = $\underline{2\ 04}$
(20 × 52) 1040   95 00
9 × 1049 = $\underline{94\ 41}$
59

8.
$$\begin{array}{r}9\ 3.7\\ \sqrt{87{,}87.00}\\ \underline{81}\end{array}$$
(20 × 9) 180   6 87
3 × 183 = $\underline{5\ 49}$
(20 × 93) 1860   1 38 00
7 × 1867 = $\underline{1\ 30\ 69}$
7 31

9.
$$\begin{array}{r}7\ 4.5\\ \sqrt{55{,}55.00}\\ \underline{49}\end{array}$$
(20 × 7) 140   6 55
4 × 144 = $\underline{5\ 76}$
(20 × 74) 1480   79 00
5 × 1485 = $\underline{74\ 25}$
4 75

10.
$$\begin{array}{r}8\ 8.1\\ \sqrt{77{,}77.00}\\ \underline{64}\end{array}$$
(20 × 8) 160   13 77
8 × 168 = $\underline{13\ 44}$
(20 × 88) 1760   33 00
1 × 1761 = $\underline{17\ 61}$
15 39

11.
$$\begin{array}{r}2\ 8.2\ 4\\ \sqrt{7{,}98.00{,}00}\\ \underline{4}\end{array}$$
(20 × 2) 40   3 98
8 × 48 = $\underline{3\ 84}$
(20 × 28) 560   14 00
2 × 562 = $\underline{11\ 24}$
(20 × 282) 5640   2 76 00
4 × 5644 = $\underline{2\ 25\ 76}$
50 24

12.
$$\begin{array}{r}9\ 9.4\ 9\\ \sqrt{98{,}99.00{,}00}\\ \underline{81}\end{array}$$
(20 × 9) 180   17 99
9 × 189 = $\underline{17\ 01}$
(20 × 99) 1980   98 00
4 × 1984 = $\underline{79\ 36}$
(20 × 994) 19880   18 64 00
9 × 19889 = $\underline{17\ 90\ 01}$
73 99

13.
$$\begin{array}{r}6\ 3.3\ 3\\ \sqrt{40{,}11.00{,}00}\\ \underline{36}\end{array}$$
(20 × 6) 120   4 11
3 × 123 = $\underline{3\ 69}$
(20 × 63) 1260   42 00
3 × 1263 = $\underline{37\ 89}$
(20 × 633) 12660   4 11 00
3 × 12663 = $\underline{3\ 79\ 89}$
31 11

14.
$$\begin{array}{r}8\ 0.8\ 8\\ \sqrt{65{,}43.00{,}00}\\ \underline{64}\end{array}$$
(20 × 8) 160   1 43
$\underline{0}$
(20 × 80) 1600   1 43 00
8 × 1608 = $\underline{1\ 28\ 64}$
(20 × 808) 16160   14 36 00
8 × 16168 = $\underline{12\ 93\ 44}$
1 42 56

T-379 Chapter 8 Geometric Solids and the Pythagorean Rule

LESSON 112

Objectives

- To teach that *the hypotenuse of a right triangle is the side opposite the right angle.
- To teach *the Pythagorean rule, which states that in a right triangle, the square of the hypotenuse equals the sum of the squares of the other two sides ($c^2 = a^2 + b^2$). This is also called the right triangle rule.
- To teach *using the Pythagorean rule to find the hypotenuse when the lengths of both legs of a right triangle are known.

Review

1. *Find the square root of each number to the place indicated.* (Lessons 110, 111).

    a. $\sqrt{5{,}024}$ to the nearest whole number (71)

    $$\begin{array}{r} 7\ \ 0\,.\,8 \\ \sqrt{50{,}24.00} \\ \underline{49} \end{array}$$

    (20 × 7) 140   1 24
    $\underline{0}$
    (20 × 70) 1400   1 24 00
    8 × 1408 = $\underline{1\ 12\ 64}$
    11 36

    b. $\sqrt{14{,}723}$ to the nearest tenth

    (121.3)

    $$\sqrt{1{,}47{,}23.00{,}00}$$
    $\phantom{\sqrt{1{,}}}1$
    20   47
    2 × 22 = $\underline{44}$
    (20 × 12) 240   3 23
    1 × 241 = $\underline{2\ 41}$
    (20 × 121) 2420   82 00
    3 × 2423 = $\underline{72\ 69}$
    (20 × 1213) 24260   9 31 00
    3 × 24263 = $\underline{7\ 27\ 89}$
    2 03 11

2. *Give the formula for each fact.* (Lessons 100–108)

    a. Surface area of a cube   ($a_s = 6e^2$)
    b. Surface area of a rectangular solid
       ($a_s = 2lw + 2wh + 2lh$)
    c. Surface area of a cylinder
       ($a_s = 2\pi r^2 + 2\pi rh$)
    d. Surface area of a square pyramid
       ($a_s = 4(\frac{1}{2}b\ell) + b^2$)
    e. Surface area of a sphere
       ($a_s = 4\pi r^2$)
    f. Volume of a rectangular solid
       ($v = lwh$)
    g. Volume of a cube   ($v = e^3$)
    h. Volume of a cylinder ($v = \pi r^2 h$)
    i. Volume of a cone   ($v = \frac{1}{3}\pi r^2 h$)
    j. Volume of a square pyramid
       ($v = \frac{1}{3}lwh$)
    k. Volume of a sphere   ($v = \frac{4}{3}\pi r^3$)

3. *Find the volume of each sphere.* (Lesson 108)

    a. $r = 8$ in. (2,140.16 cu. in.)
       (Calculated with 4.18 for $\frac{4}{3}\pi$.)
    b. $r = 14$ ft. (11,498$\frac{2}{3}$ cu. ft.)
       (Calculated with $\frac{88}{21}$ for $\frac{4}{3}\pi$.)

4. *Find the surface area of each sphere.* (Lesson 104)

    a. $r = 8$ in. (803.84 sq. in.)
    b. $r = 36$ ft. (16,277.76 sq. ft.)

5. *Multiply these decimals.* (Lesson 48)

    a. 6.8   b. 30.29
       × 2.8   × 0.094
       (19.04)   (2.84726)

112. Understanding the Pythagorean Rule

A right triangle has one right angle. The two sides that form the right angle are called the legs of the triangle. The diagonal side opposite the right angle is called the **hypotenuse** (hī pŏt′ ən ūs′).

There is a fixed relationship between the length of the hypotenuse and the lengths of the two legs. This relationship is named the **Pythagorean rule** (pĭ thăg′ ə rē′ ən) after Pythagoras (pĭ thăg′ ər əs), a Greek mathematician of the 500s B.C. who was the first to state it. The Pythagorean rule says that the square of the hypotenuse on a right triangle equals the sum of the squares of the two legs. As a formula, the rule is expressed in the statement $c^2 = a^2 + b^2$. The letters a and b represent the two legs, and c represents the hypotenuse.

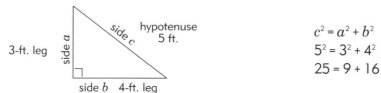

$$c^2 = a^2 + b^2$$
$$5^2 = 3^2 + 4^2$$
$$25 = 9 + 16$$

The Pythagorean rule is often called the 3-4-5 rule. A right triangle that has one 3-foot leg and one 4-foot leg will have a 5-foot hypotenuse. This calculation is shown with the illustration above.

Carpenters often use the Pythagorean rule to see if the walls of a building are square. If they measure 3 feet along one side of a right angle and 4 feet along the other side, the hypotenuse will measure 5 feet if the corner is square. But in actual use, carpenters usually multiply the 3-4-5 combination by 2, 3, or some other number. This yields combinations such as 6-8-10 and 9-12-15, which can be used in the same way to see if a corner is square.

To find the hypotenuse when the two legs are known, follow these steps.

1. Substitute a and b in the formula with the measurements of the legs.
2. Find the squares of the leg measurements.
3. Add the squares found in step 2.
4. Find the square root of the sum obtained in step 3. Use the table of square roots whenever you can.

Example A
Find the hypotenuse of this right triangle: leg a = 5 ft. leg b = 12 ft.
$$c^2 = a^2 + b^2$$
$$c^2 = 5^2 + 12^2$$
$$c^2 = 25 + 144 = 169$$
$$\sqrt{c^2} = \sqrt{169}$$
$$c = 13 \text{ ft.}$$

Example B
Find the hypotenuse of this right triangle: leg a = 7 ft. leg b = 9 ft.
$$c^2 = a^2 + b^2$$
$$c^2 = 7^2 + 9^2$$
$$c^2 = 49 + 81 = 130$$
$$\sqrt{c^2} = \sqrt{130}$$
$$c = 11.402 \text{ ft.}$$

380   Chapter 8   Geometric Solids and the Pythagorean Rule

> **In a right triangle,
> the square of the hypotenuse equals the sum of the squares of the other two sides.**
> $c^2 = a^2 + b^2$

CLASS PRACTICE

Find the hypotenuse of each right triangle. Use the table of square roots.

a. leg a = 3 ft. b. leg a = 4 ft. c. leg a = 5 m d. leg a = 11 m
   leg b = 3 ft.   leg b = 6 ft.   leg b = 7 m   leg b = 6 m
   4.243 ft.   7.211 ft.   8.602 m   12.530 m

WRITTEN EXERCISES

A. *Find the hypotenuse of each right triangle. Use the table of square roots.*

1. leg a = 2 ft.   2.828 ft.   2. leg a = 2 ft.   3.606 ft.   3. leg a = 3 m   5.831 m
   leg b = 2 ft.   leg b = 3 ft.   leg b = 5 m

4. leg a = 2 m   6.325 m   5. leg a = 7 ft.   9.899 ft.   6. leg a = 8 ft.   11.314 ft.
   leg b = 6 m   leg b = 7 ft.   leg b = 8 ft.

7. leg a = 9 m   10.817 m   8. leg a = 8 m   12.042 m   9. leg a = 12 ft.   12.649 ft.
   leg b = 6 m   leg b = 9 m   leg b = 4 ft.

10. leg a = 11 ft.   13.038 ft.   11. leg a = 10 m   13.454 m   12. leg a = 12 m   13.416 m
    leg b = 7 ft.   leg b = 9 m   leg b = 6 m

B. *Solve these reading problems. Draw sketches for numbers 17 and 18.*

13. The supporting cables for a tower are anchored 60 feet above the ground and 40 feet away from its base. How long is each cable, to the nearest foot?   72 feet

14. A guy wire is fastened to a pole, with the upper end 25 feet above the ground and the lower end 12 feet from the base of the pole. How long is the wire, to the nearest tenth of a foot?   27.7 feet

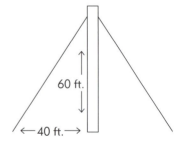

15. What is the diagonal measurement of a rectangular book 8 inches wide by 10 inches long? (Calculate to the nearest tenth.)   12.8 inches

16. When Brother Millard started building a garage onto one side of the house, he made sure the foundation was square with the existing structure. From the corner where the house and the garage met, he measured along the house wall and made a mark at 24 feet. Then from the same corner, he measured along the garage foundation and made a mark at 18 feet. What was the distance between the two marks if the corner was square?   30 feet

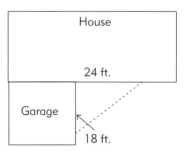

212

Lesson 112   T–380

Introduction

Ask the students to identify a large right angle in the room. (It may be a corner of the room or the chalkboard.) Tell them there is a way to prove whether it is truly a right angle. Measure 12 units along one leg of the angle and 16 units along the other leg. Now ask the students if they can identify the hypotenuse of this angle. The hypotenuse is a diagonal line extending from the end of one leg of the right angle to the end of the other leg.

Ask the students to guess the length of the hypotenuse. Tell them without measuring that its length will be 20 units if the angle is square.

Assuming the hypotenuse is 20 feet, write the following problems on the board. Can the students see how the squares of the three numbers are related?

```
     12              16              20
   × 12            × 16            × 20
   ----            ----            ----
    144    +       256     =       400
```

The sum of the squares of the two legs of the triangle are equal to the square of the hypotenuse. This principle is used to solve various problems relating to measuring.

Perhaps the students can share times their fathers have used this principle. Builders use it for things such as squaring a foundation, finding the length of a staircase, and determining how long the slope of a roof will be from the peak to the eaves.

Teaching Guide

1. **In a right triangle, the two sides that form the right angle are called the legs of the triangle, and the diagonal side opposite the right angle is called the hypotenuse.**

2. **The Pythagorean rule states that the square of the hypotenuse equals the sum of the squares of the two legs, or $c^2 = a^2 + b^2$.** Review your findings from the lesson introduction. Point out that the letters a and b always represent the legs, and c always represents the hypotenuse.

3. **To find the hypotenuse when the two legs are known, follow these steps.**

    (1) Substitute a and b in the formula with the measurements of the legs.

    (2) Find the squares of the leg measurements.

    (3) Add the squares found in step 2.

    (4) Find the square root of the sum obtained in step 3. Use the table of square roots whenever you can.

    a. $a = 5$ ft.
       $b = 6$ ft.
       ($c = 7.8$ ft.)

    b. $a = 13$ ft.
       $b = 5$ ft.
       ($c = 13.9$ ft.)

    c. $a = 25$ ft.
       $b = 15$ ft.
       ($c = 29.2$ ft.)

Solutions for Part B

13. $60^2 + 40^2 = 5{,}200$; $\sqrt{5{,}200}$

14. $25^2 + 12^2 = 769$; $\sqrt{769}$

15. $8^2 + 10^2 = 164$; $\sqrt{164}$

16. $24^2 \times 18^2 = 900$ $\sqrt{900}$;
       (3-4-5 rule multiplied by 6)

T–381 Chapter 8 Geometric Solids and the Pythagorean Rule

Further Study

The principle of the Pythagorean rule was first recognized by the ancient Egyptians, who noticed the 3-4-5 pattern when they laid out their fields. The Greeks learned it from the Egyptians, and Pythagoras was the first to study it thoroughly and state the rule. Pythagoras is considered today as the first true mathematician.

Sketches for Exercises 17 and 18

17.

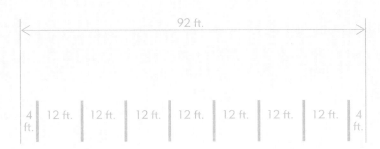

18.

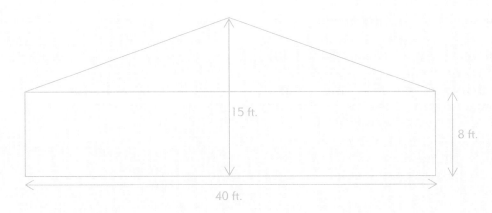

Extractions for Part C

19.
```
                8 2.7 8
              √68ˬ54.00ˬ00
                64
    (20 × 8) 160    4 54
      2 × 162 = 3 24
    (20 × 82) 1640    1 30 00
      7 × 1647 = 1 15 29
   (20 × 827) 16540    14 71 00
      8 × 16548 = 13 23 84
                     1 47 16
```

20.
```
                6 1.1 9
              √37ˬ45.00ˬ00
                36
    (20 × 6) 120    1 45
      1 × 121 = 1 21
    (20 × 61) 1220    24 00
      1 × 1221 = 12 21
   (20 × 611) 12220    11 79 00
      9 × 12229 = 11 00 61
                       78 39
```

(See facing page for sketches.)

17. Brother David wants to build a fence from his barn to a shed that stands 92 feet away. He plans to set the posts 12 feet apart and to leave a 4-foot opening at each end. How many posts will he need?   **8 posts**

18. Brother Scott plans to replace the siding at one gable end of his house. (A gable end has the shape of a pentagon.) The end wall measures 40 feet wide, 8 feet high to the eaves, and 15 feet from the ground to the peak. If one square of siding equals 100 square feet, how many squares must he buy? (Count a partial square as another full square.)   **5 squares**

REVIEW EXERCISES

C. Find each square root, to the nearest tenth. *(Lessons 110, 111)*

19. $\sqrt{6{,}854}$ **82.8**

20. $\sqrt{3{,}745}$ **61.2**

D. Write these formulas. *(Lessons 100–108)*

21. Surface area of a cube $a_s = 6e^2$
22. Surface area of a rectangular solid $a_s = 2lw + 2wh + 2lh$
23. Surface area of a cylinder $a_s = 2\pi r^2 + 2\pi rh$
24. Surface area of a square pyramid $a_s = 4(\frac{1}{2}b\ell) + b^2$
25. Surface area of a sphere $a_s = 4\pi r^2$
26. Volume of a rectangular solid $v = lwh$
27. Volume of a cube $v = e^3$
28. Volume of a cylinder $v = \pi r^2 h$
29. Volume of a cone $v = \frac{1}{3}\pi r^2 h$
30. Volume of a square pyramid $v = \frac{1}{3}lwh$
31. Volume of a sphere $v = \frac{4}{3}\pi r^3$

E. Find the volume of each sphere. Use 4.18 in your calculations. *(Lesson 108)*

32. $r = 7$ in.   **1,433.74 cu. in.**
33. $r = 25$ m   **65,312.5 m³**

F. Find the surface area of each sphere. *(Lesson 104)*

34. $r = 7$ in.   **616 sq. in.** $\pi = 3\frac{1}{7}$
35. $r = 11$ cm   **1,519.76 cm²** $\pi = 3.14$
36. $r = 25$ ft.   **7,850 sq. ft.** $\pi = 3.14$

G. Solve these multiplication problems. *(Lesson 48)*

37. 5.6 × 1.6 = **8.96**
38. 2.032 × 0.25 = **0.508**

113. Applying the Pythagorean Rule

In Lesson 112 you learned about the Pythagorean rule, which states that in a right triangle the square of the hypotenuse equals the sum of the squares of the two sides. The formula is $c^2 = a^2 + b^2$.

The Pythagorean rule is used in many practical ways, as shown in the following examples.

Example A

If the length of a stairwell is 96 inches and the height of the stairs is 84 inches, what should be the length of the staircase (flight of stairs)? The staircase is the hypotenuse of a right triangle, so the Pythagorean rule can be used.

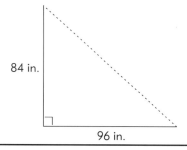

$c^2 = a^2 + b^2$
$c^2 = 96^2 + 84^2$
$c^2 = 9{,}216 + 7{,}056 = 16{,}272$
$\sqrt{c^2} = \sqrt{16{,}272}$
$c = 127.6$ in.

Example B

A house roof is 36 feet wide, and the peak of the roof is 12 feet above the level of the eaves. What is the distance from the peak to the eaves?

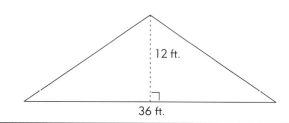

$c^2 = a^2 + b^2$
$c^2 = 18^2 + 12^2$
$c^2 = 324 + 144 = 468$
$\sqrt{c^2} = \sqrt{468}$
$c = 21.6$ ft.

Note that the solution uses 18^2, not 36^2. This is because leg a of the right triangle is only half the total width of the roof.

If you know the length of the hypotenuse and of one leg, you can use the Pythagorean rule to find the length of the other leg. Change the formula to $a^2 = c^2 - b^2$. The letter a represents the unknown leg, c represents the hypotenuse, and b represents the known leg.

Because $c^2 = a^2 + b^2$, therefore $a^2 = c^2 - b^2$.

LESSON 113

Objectives

- To teach several practical applications of the Pythagorean rule.
- To teach *using the Pythagorean rule to find the length of one leg of a right triangle when the length of the other leg and the hypotenuse are known.

Review

1. *Use the method indicated to find each square root.* (Lessons 109–111)

    a. $\sqrt{164}$; table   (12.806)

    b. $\sqrt{7{,}056}$; calculation   (84)

    $$\begin{array}{r} 8\ 4 \\ \sqrt{70{,}56} \\ 64 \\ \end{array}$$

    $(20 \times 8)\ 160\quad 6\ 56$
    $4 \times 164 = \underline{6\ 56}$
    0

2. *Solve these multistep reading problems.* (Lesson 97)

    a. Brother Henry bought 300 bushels of unscreened rye seed for $2.80 per bushel. After cleaning the rye, he found that 5% of the original volume was removed by the cleaner. If he bagged and sold the cleaned rye for $4.50 per bushel, and if his other costs amounted to $80, what was his profit? ($362.50)

    $(300 \times \$2.80) + \$80 = \$920$ costs

    $95\% \times 300 \times \$4.50 = \$1{,}282.50$ income

    $\$1{,}282.50 - \$920.00 = \$362.50$

    b. Jonathan and his family traveled 1,200 miles to visit relatives in a distant state. The journey took a total of $23\frac{1}{2}$ hours, including 3 stops that averaged 25 minutes each. What was their average driving speed, to the nearest whole number? (54 m.p.h.)

    $3 \times 25 = 75$ min. ($1\frac{1}{4}$ hr.)

    $23\frac{1}{2} - 1\frac{1}{4} = 22\frac{1}{4}$ hr. driving

    $1{,}200 \div 22\frac{1}{4} = 53.9$

3. *Find the volume of each cube or rectangular solid.* (Lesson 105)

    a. $l = 23$ in.
    $w = 16$ in.
    $h = 18$ in.   (6,624 cu. in.)

    b. $e = 15$ ft.
    (3,375 cu. ft.)

4. *Divide these decimals.* (Lesson 49)

    a. $16\overline{)0.4}$ (0.025)   b. $5\overline{)0.515}$ (0.103)

Introduction

Review calculating the hypotenuse when the lengths of the two legs are known (Lesson 112).

a. $a = 7$ ft.
    $b = 11$ ft.
    ($c = 13.0$ ft.)

b. $a = 12$ ft.
    $b = 7$ ft.
    ($c = 13.9$ ft.)

Ask the students, "Suppose you were at the edge of a pond. Is there any way you could find the distance across the water without measuring it directly?"

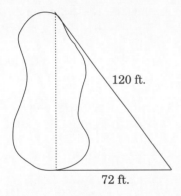

Draw a diagram like this. The students should recognize that the distance across the pond is one leg of a right triangle. We know the length of one leg and the hypotenuse. Can we use the Pythagorean rule to find the length of the other leg?

Think: $c^2 = a^2 + b^2$

Therefore: $a^2 = c^2 - b^2$

$a^2 = 120^2 - 72^2$

$a^2 = 14{,}400 - 5{,}184 = 9{,}216$ (This is a perfect square.)

$\sqrt{a^2} = \sqrt{9{,}216}$

$a = 96$ ft.

Today's lesson teaches this variation of the Pythagorean rule.

Teaching Guide

1. **The Pythagorean rule is used in a number of practical ways.** The previous lesson mentions its use in squaring foundations. Following are two other practical applications.

    a. Finding the length of a staircase when the height and length of the stairwell are known.
        (1) length = 13 ft.
            height = 9 ft.
            (15.8 ft.)
        (2) length = 15 ft.
            height = 10 ft.
            (18.0 ft.)

    b. Finding the distance from the peak to the eaves of a roof when the total width and the height of the peak are known. The total width must be divided by 2 to obtain the distance to the center.
        (1) width of roof = 36 ft.
            height of peak above eaves = 12 ft.
            (21.6 ft.)
        (2) width of roof = 32 ft.
            height of peak above eaves = 12 ft.
            (20 ft.)

2. **The Pythagorean rule can be used to find the length of a leg when the hypotenuse and the other leg are known.** Just as $4 + 3 = 7$ has the related facts of $7 - 3 = 4$ and $7 - 4 = 3$, so the Pythagorean rule $c^2 = a^2 + b^2$ can be restated as $a^2 = c^2 - b^2$ to calculate the unknown length of one leg of a right triangle.

    Find the length of leg a *in each triangle, to the nearest tenth.*

    a. leg $b = 8$ in.
        hypotenuse = 10 in.
        (6 in.)

    b. leg $b = 8$ in.
        hypotenuse = 11 in.
        (7.5 in.)

Example C

Dale has a 6-foot board to nail diagonally across a sagging gate 4 feet wide. If the board reaches exactly from one corner to the other, what is the height of the gate?

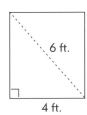

$a^2 = c^2 - b^2$
$a^2 = 6^2 - 4^2$
$a^2 = 36 - 16 = 20$
$\sqrt{a^2} = \sqrt{20}$
$a = 4.5$ ft.

Remember: Always subtract the square of the known leg from the square of the hypotenuse to find the square of the unknown leg.

CLASS PRACTICE

Find the hypotenuse of each right triangle, to the nearest tenth.

a. length: 3 in.   5.8 in.   b. length: 7 ft.   9.9 ft.
   height: 5 in.   height: 7 ft.

Find the length of each staircase, to the nearest tenth.

c. length: 6 ft.   7.2 ft.   d. length: 9 ft.   10.8 ft.
   height: 4 ft.   height: 6 ft.

Find the distance from the peak to the eaves of this roof. Remember to divide the width of the roof in half to find the distance to the center. Answer to the nearest tenth.

e. width of roof: 20 ft.   13.5 ft.
   height of peak above eaves: 9 ft.

Find the length of leg a in each triangle, to the nearest tenth.

f. leg b: 3 in.   5.2 in.   g. leg b: 8 in.   11.5 in.
   hypotenuse: 6 in.   hypotenuse: 14 in.

WRITTEN EXERCISES

A. Find the hypotenuse of each right triangle, to the nearest tenth.

1. length: 1 in.   1.4 in.   2. length: 2 ft.   2.8 ft.
   height: 1 in.   height: 2 ft.

3. length: 9 ft.   11.4 ft.   4. length: 5 ft.   9.4 ft.
   height: 7 ft.   height: 8 ft.

B. Find the length of each staircase, to the nearest tenth.

5. length: 12 ft.   13.9 ft.   6. length: 11 ft.   13.0 ft.
   height: 7 ft.   height: 7 ft.

C. For each roof, find the distance from the peak to the eaves. Remember to divide the width of the roof in half to find the distance to the center. Answer to the nearest tenth.

7. width of roof: 36 ft.   20.1 ft.
   height of peak above eaves: 9 ft.

8. width of roof: 30 ft.   16.2 ft.
   height of peak above eaves: 6 ft.

D. Find the length of leg a in each triangle, to the nearest tenth.

9. leg b: 2 in.   3.5 in.
   hypotenuse: 4 in.

10. leg b: 7 in.   9.7 in.
    hypotenuse: 12 in.

11. leg b: 6 ft.   6.7 ft.
    hypotenuse: 9 ft.

12. leg b: 3 ft.   9.5 ft.
    hypotenuse: 10 ft.

13. leg b: 12 ft.   13.4 ft.
    hypotenuse: 18 ft.

14. leg b: 15 in.   18.7 in.
    hypotenuse: 24 in.

E. Solve these reading problems. Some of them are multistep problems.

15. A pole casts a shadow 6 feet long. If the distance from the top of the pole to the end of the shadow is 10 feet, how tall is the pole?   8 feet

16. A building is being constructed with steel trusses 30 feet high. While workers fasten permanent supports to a truss, it is held in place by cables. One 50-foot cable fastened to the top of a truss is anchored in the ground at a proper distance to position the truss at right angles to the ground. Find the distance from the place where the cable is anchored in the ground, to a point directly beneath the peak of the truss.   40 feet

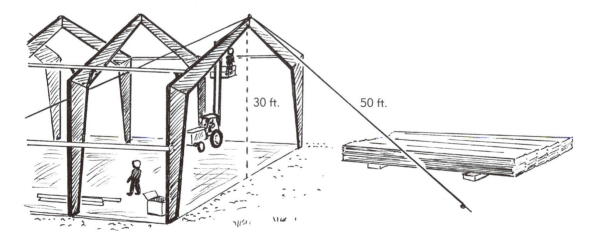

17. The eighth grade class at Meadowview Mennonite School has found that the diagonal measurement of the front classroom wall is about 27 feet. The width of their room is 25 feet. What is the height of the wall, to the nearest foot?   10 feet

18. A field on the Nissley farm forms a right triangle. The diagonal side of the field is 500 feet long, and another side is 420 feet long. How long is the third side, to the nearest foot?   271 feet

Lesson 113   T–384

Solutions for Part E

15. 3-4-5 rule: 6-__-10
16. 3-4-5 rule: 30-__-50
17. $27^2 - 25^2 = 104$; $\sqrt{104}$
18. $500^2 - 420^2 = 73{,}600$; $\sqrt{73{,}600}$

T–385   Chapter 8   Geometric Solids and the Pythagorean Rule

19. 3.14×3^2

20. $13\frac{1}{2} \times 3{,}000 \times 60$

Extractions for Part G

32.
$$
\begin{array}{r}
8.7\ 7 \\
\sqrt{77.00{\scriptstyle\wedge}00} \\
\underline{64}
\end{array}
$$

(20 × 8) 160   13 00
7 × 167 = <u>11 69</u>
(20 × 87) 1740   1 31 00
7 × 1747 = <u>1 22 29</u>
                    8 71

33.
$$
\begin{array}{r}
9\ \ 5.2\ 8 \\
\sqrt{90{\scriptstyle\wedge}80.00{\scriptstyle\wedge}00} \\
\underline{81}
\end{array}
$$

(20 × 9) 180   9 80
5 × 185 = <u>9 25</u>
(20 × 95) 1900   55 00
2 × 1902 = <u>38 04</u>
(20 × 952) 19040   16 96 00
8 × 19048 = <u>15 23 84</u>
                      1 72 16

19. Norman's Garage has a hydraulic cylinder that lifts cars off the floor. What is the surface area of the end of the cylinder if its diameter is 6 inches? 28.26 square inches

20. A paper mill in Wisconsin produces a ribbon of paper $13\frac{1}{2}$ feet wide at the rate of 3,000 feet per minute. How many square feet of paper does it produce in 1 hour? 2,430,000 square feet

REVIEW EXERCISES

F. Write these formulas. *(Lessons 100–108)*

21. Surface area of a cube — $a_s = 6e^2$
22. Surface area of a rectangular solid — $a_s = 2lw + 2wh + 2lh$
23. Surface area of a cylinder — $a_s = 2\pi r^2 + 2\pi rh$
24. Surface area of a square pyramid — $a_s = 4(\frac{1}{2}b\ell) + b^2$
25. Surface area of a sphere — $a_s = 4\pi r^2$
26. Volume of a rectangular solid — $v = lwh$
27. Volume of a cube — $v = e^3$
28. Volume of a cylinder — $v = \pi r^2 h$
29. Volume of a cone — $v = \frac{1}{3}\pi r^2 h$
30. Volume of a square pyramid — $v = \frac{1}{3}lwh$
31. Volume of a sphere — $v = \frac{4}{3}\pi r^3$

G. Calculate each square root, to the nearest tenth. *(Lessons 110, 111)*

32. $\sqrt{77}$ 8.8
33. $\sqrt{9{,}080}$ 95.3

H. Find the volume of each cube or rectangular solid. *(Lesson 105)*

34. $l = 25$ in.
    $w = 18$ in.
    $h = 11$ in.   4,950 cu. in.

35. $e = 29$ ft.   24,389 cu. ft.

I. Solve these division problems. The quotients will come out evenly by the thousandths' place. *(Lesson 49)*

36. $0.15 \overline{)12.48}$ 83.2
37. $22 \overline{)0.308}$ 0.014
38. $1.4 \overline{)5.145}$ 3.675

114. Reading Problems: Using Parallel Problems

Perhaps you have heard it said of someone that he could not see the forest for the trees. Usually that means a person does not have a proper view of a larger situation because he is concentrating too much on the small details.

Details are very important in mathematics. However, focusing on the numbers can sometimes hinder us from seeing the steps needed to solve a problem. This is especially true if the numbers are very large or if they include fractions.

Example A—Original problem

Dwight spent $\frac{3}{5}$ hour doing chores one evening. That was $\frac{3}{4}$ as long as the time he worked on his math lesson. What was the total time that he spent on his chores and his math lesson that evening?

You can simplify this problem by replacing the fractions with simpler numbers. Consider the following parallel problem. Note that only the numbers have been changed.

Parallel problem

Dwight spent 2 hours doing chores one evening. That was $\frac{1}{2}$ as long as the time he worked on his math lesson. What was the total time that he spent on his chores and his math lesson that evening?

The parallel problem makes it easy to see that you must first answer the question "2 is $\frac{1}{2}$ of what number?" To find a number when a fraction of it is known, divide by the fraction: $2 \div \frac{1}{2} = 4$. The second step is to add the two times together: $4 + 2 = 6$. You can solve the original problem by using the same steps.

Solution to parallel problem: 2 is $\frac{1}{2}$ of ___

$2 \div \frac{1}{2} = 4$

$4 + 2 = 6$ hours

Solution to original problem: $\frac{3}{5}$ is $\frac{3}{4}$ of ___

$\frac{3}{5} \div \frac{3}{4} = \frac{3}{5} \times \frac{4}{3} = \frac{4}{5}$

$\frac{4}{5} + \frac{3}{5} = \frac{7}{5} = 1\frac{2}{5}$ hours

When writing a parallel problem, change only the difficult numbers in the original problem. Be sure to use simple numbers. Replace fractions, mixed numbers, and other small numbers with single-digit numbers such as 3, 4, and 5. Replace larger numbers with

LESSON 114

Objective

- To review solving difficult reading problems by using parallel problems with simpler numbers.

Review

1. *Find the length of leg a in each triangle, to the nearest tenth.* (Lesson 113)

   a. leg b = 24 cm
      hypotenuse = 55 cm
      (49.5 cm)

   b. leg b = 14 in.
      hypotenuse = 18 in.
      (11.3 in.)

2. *Find each hypotenuse, to the nearest tenth.* (Lesson 112)

   a. leg a = 12 ft.
      leg b = 12 ft.
      (17.0 ft.)

   b. leg a = 6 in.
      leg b = 8 in.
      (10 in.)

3. *Find the square roots of these perfect squares.* (Lesson 110)

   a. $\sqrt{15{,}376}$ (124)

   $$\sqrt{1{,}53{,}76} \quad \begin{array}{r} 1\ 2\ 4 \end{array}$$

   $\quad\quad\quad\quad$ 1
   $\quad\quad\quad$ 20 $\quad$ 53
   $\quad\quad$ 2 × 22 = 44
   $\quad$ (20 × 12) 240 $\quad$ 9 76
   $\quad\quad$ 4 × 244 = 9 76
   $\quad\quad\quad\quad\quad\quad\quad$ 0

   b. $\sqrt{17{,}956}$ (134)

   $$\sqrt{1{,}79{,}56} \quad \begin{array}{r} 1\ 3\ 4 \end{array}$$

   $\quad\quad\quad\quad$ 1
   $\quad\quad\quad$ 20 $\quad$ 79
   $\quad\quad$ 3 × 23 = 69
   $\quad$ (20 × 13) 260 $\quad$ 10 56
   $\quad\quad$ 4 × 264 = 10 56
   $\quad\quad\quad\quad\quad\quad\quad$ 0

4. *Find the volume of each figure.* (Lesson 106)

   a. Cylinder
      r = 5 ft.
      h = 65 ft.
      v = (5,102.5 cu. ft.)

   b. Cone
      r = 5 in.
      h = 18 in.
      v = (471 cu. in.)

5. *Change each fraction to a decimal. Round to the nearest thousandth.* (Lesson 50)

   a. $\frac{5}{16}$ (0.313)

   b. $\frac{10}{11}$ (0.909)

T–387 Chapter 8 Geometric Solids and the Pythagorean Rule

Introduction

Read the original problem of Example A together in class. Ask the students for suggestions on how to solve the problem. You may get several ideas, including the correct suggestion that first division and then addition are needed. But you may also get the suggestion that the operations needed are multiplication and then addition, or only addition.

Now read the parallel reading problem for Example A. It should become clear, if it was still uncertain, that division and then addition are needed to solve both problems. Replacing the numbers in a reading problem with simpler ones will often clarify the solution.

Teaching Guide

To solve a difficult reading problem, use a parallel problem with simpler numbers. Tell the class to remember the following two points in using this method. Illustrate with lesson examples and *Class Practice* problems.

a. When devising a parallel problem, keep all information parallel. If you do this, you can have confidence that the method of solution will remain parallel. If you change the facts too much, there is the danger of also changing the method of solution.

b. When replacing numbers, be sure to use simpler numbers. Replace fractions, mixed numbers, and other small numbers with single-digit numbers such as 3, 4, and 5. Replace larger numbers with multiples of the smaller numbers you chose. For example, using 4 or 5 and 20 as a set of facts will help make the method of solution much clearer because you hardly need to give any thought to the calculations.

Solutions for CLASS PRACTICE

a. $8 \div 2$

b. $8.12 \div 0.29$

c. $\$9.00 \div 2 = \$4.50$; $\$4.50 = 90\%$ of ___

d. $\$56.10 \div 17 = \$3.30$; $\$3.30 = 88\%$ of ___

Solutions for Part A

1. a. $10 \div 1$ b. $10\frac{2}{3} \div \frac{2}{3}$
2. a. $\frac{2}{3} \times 6$ b. $\frac{1}{4} \times 4\frac{1}{2}$

multiples of the smaller numbers you chose. For example, using 4 or 5 and 20 as a set of facts will help make the steps to a solution much more obvious because you hardly need to give any thought to the calculations.

Example B

Original problem
　　The Lawrences drove 15 miles in $\frac{2}{5}$ hour. What was their average speed?

Parallel problem
　　The Lawrences drove 150 miles in 3 hours. What was their average speed?

Solution to parallel problem
　　distance ÷ time = rate
　　150 mi. ÷ 3 hr. = 50 m.p.h.

Solution to original problem
　　distance ÷ time = rate
　　15 mi. ÷ $\frac{2}{5}$ hr. = $37\frac{1}{2}$ m.p.h.

CLASS PRACTICE

These parallel reading problems are arranged in sets of two (a, b and c, d). In each set, use your solution to the first problem (the simpler one) as a help in solving the second problem (the more difficult one).

a. At a cost of $2.00 each, how many pens can Philip buy for $8.00?   4 pens

b. At a cost of $0.29 each, how many pens can Philip buy for $8.12?   28 pens

c. Ralph sold 2 rabbits at a local auction. He received $9.00 after the auction had deducted a 10% commission from the selling price. What was the selling price of each rabbit? $5.00

d. Ralph sold 17 rabbits at a local auction. He received $56.10 after the auction had deducted a 12% commission from the selling price. What was the selling price of each rabbit? $3.75

WRITTEN EXERCISES

A. Solve these pairs of parallel reading problems. Use your solution to problem *a* (the simpler one) as a help in solving problem *b* (the more difficult one).

1. a. How many potatoes are in a package weighing 10 pounds if the potatoes average 1 pound each?   10 potatoes

   b. How many potatoes are in a package weighing $10\frac{2}{3}$ pounds if the potatoes average $\frac{2}{3}$ pound each?   16 potatoes

2. a. Father planted 6 acres of corn 3 weeks ago, but $\frac{1}{3}$ of the field must be replanted. How many acres do *not* need to be replanted?   4 acres

   b. Father planted $4\frac{1}{2}$ acres of corn 3 weeks ago, but $\frac{3}{4}$ of the field must be replanted. How many acres do *not* need to be replanted? $1\frac{1}{8}$ acres

3. a. Brother Marvin received a 10% commission for selling a lawn mower at Nolt's Garden Supply. His commission was $20. What was the selling price of the mower?
   $200
   b. Brother Marvin received a 4.5% commission for selling a lawn tractor at Nolt's Garden Supply. His commission was $67.05. What was the selling price of the lawn tractor?
   $1,490
4. a. Mark used 10 gallons of gasoline to refill the fuel tank of the family car. The car had been driven 5 hours at an average speed of 50 miles per hour since the tank was last filled. What was the average number of miles per gallon that the car had traveled?
   25 mi. per gal.
   b. Mark used 12.8 gallons of gasoline to refill the fuel tank of the family car. The car had been driven 6.5 hours at an average speed of 48 miles per hour since the tank was last filled. To nearest tenth, what was the average number of miles per gallon that the car had traveled?
   24.4 mi. per gal.
5. a. Father paid $20 for a roll of plastic measuring 4 yards by 20 yards. What was the cost per square yard?
   $0.25
   b. Father paid $68.50 for a roll of plastic measuring $24\frac{1}{2}$ feet by 130 feet. What was the cost per square foot, to the nearest cent?
   $0.02
6. a. Jason can split 1 cord of firewood in 8 hours. His father can split firewood 2 times as fast as he can. How many cords of wood can Father split in the same amount of time?
   2 cords
   b. Jason can split $\frac{1}{2}$ cord of firewood in $3\frac{1}{2}$ hours. His older brother Arnold can split firewood $1\frac{1}{2}$ times as fast as he can. How many cords of wood can Arnold split in the same amount of time?
   $\frac{3}{4}$ cord
7. a. Father bought 5 cases of cherries yesterday for $15.00 a case, and he sold them today for a total of $100.00. What was his profit per case?
   $5.00
   b. Father bought 13 cases of cherries yesterday for $16.85 per case, and he sold them today for a total of $308.75. What was his profit per case?
   $6.90
8. a. Each bench at Hillcrest Mennonite Church is 20 feet long and can seat 10 adults. How much bench space does that allow per person?
   2 feet
   b. The largest bench in the world is in Japan. It measures 1,321.33 feet long and seats 1,282 people. To the nearest whole number, how much space does that allow per person?
   1 foot

REVIEW EXERCISES

B. Write these formulas. *(Lessons 100–108)*

9. Surface area of a cube $\quad a_s = 6e^2$
10. Surface area of a rectangular solid $\quad a_s = 2lw + 2wh + 2lh$
11. Surface area of a cylinder $\quad a_s = 2\pi r^2 + 2\pi rh$
12. Surface area of a square pyramid $\quad a_s = 4(\frac{1}{2}b\ell) + b^2$

Lesson 114   T–388

3. a. $20 = 10% of ___   b. $67.05 = 0.045 of ___
4. a. (5 × 50) ÷ 10   b. (6.5 × 48) ÷ 12.8
5. a. $20 ÷ (4 × 20)   b. $68.50 ÷ (24½ × 130)
6. a. 2 × 1   b. $1\frac{1}{2} \times \frac{1}{2}$
7. a. ($100 ÷ 5) – $15   b. ($308.75 ÷ 13) – $16.85
8. a. 20 ÷ 10   b. 1,321.33 ÷ 1,282

T–389   Chapter 8   Geometric Solids and the Pythagorean Rule

Extractions for Part D

24.
$$\begin{array}{r}2\ 6\ 3\\ \sqrt{6{\scriptstyle\wedge}91{\scriptstyle\wedge}69}\\ 4\end{array}$$

(20 × 2) 40   2 91
6 × 46 = 2 76
(20 × 26) 520   15 69
3 × 523 = 15 69
0

25.
$$\begin{array}{r}9\ 8\ 7\\ \sqrt{97{\scriptstyle\wedge}41{\scriptstyle\wedge}69}\\ 81\end{array}$$

(20 × 9) 180   16 41
8 × 188 = 15 04
(20 × 98) 1960   1 37 69
7 × 1967 = 1 37 69
0

13. Surface area of a sphere $a_s = 4\pi r^2$

14. Volume of a rectangular solid $v = lwh$

15. Volume of a cube $v = e^3$

16. Volume of a cylinder $v = \pi r^2 h$

17. Volume of a cone $v = \frac{1}{3}\pi r^2 h$

18. Volume of a square pyramid $v = \frac{1}{3}lwh$

19. Volume of a sphere $v = \frac{4}{3}\pi r^3$

C. Find the length of the hypotenuse or the missing leg, to the nearest tenth. *(Lessons 112, 113)*

20. leg *a*: 16 ft.   21.3 ft.   21. leg *a*: 9 in.   11.4 in.
    leg *b*: 14 ft.   leg *b*: 7 in.

22. leg *b*: 14 cm   42.8 cm   23. leg *b*: 12 in.   21.9 in.
    hypotenuse: 45 cm   hypotenuse: 25 in.

D. Extract the square roots of these perfect squares. *(Lesson 110)*

24. $\sqrt{69{,}169}$ 263

25. $\sqrt{974{,}169}$ 987

E. Find the volume of each geometric solid. Use 3.14 for pi. *(Lesson 106)*

26. Cylinder   753.6 cu. in.   27. Cone   11,304 mm³
    r = 4 in.   *r* = 12 mm
    h = 15 in.   *h* = 75 mm

F. Change each fraction to a decimal. Round to the nearest thousandth. *(Lesson 50)*

28. $\frac{5}{18}$ 0.278   29. $\frac{8}{15}$ 0.533   30. $\frac{6}{13}$ 0.462

*If the iron be blunt,
and he do not whet the edge,
then must he put to more strength.
Ecclesiastes 10:10*

115. Chapter 8 Review

A. Write these formulas. *(Lessons 100–108)*

1. Surface area of a cube — $a_s = 6e^2$
2. Surface area of a rectangular solid — $a_s = 2lw + 2wh + 2lh$
3. Surface area of a cylinder — $a_s = 2\pi r^2 + 2\pi rh$
4. Surface area of a square pyramid — $a_s = 4(\frac{1}{2}b\ell) + b^2$
5. Surface area of a sphere — $a_s = 4\pi r^2$
6. Volume of a rectangular solid — $v = lwh$
7. Volume of a cube — $v = e^3$
8. Volume of a cylinder — $v = \pi r^2 h$
9. Volume of a cone — $v = \frac{1}{3}\pi r^2 h$
10. Volume of a square pyramid — $v = \frac{1}{3}lwh$
11. Volume of a sphere — $v = \frac{4}{3}\pi r^3$

B. Find the surface areas of these geometric solids. *(Lessons 100–104)*

Cubes

12. $e = 6$ in.   216 sq. in.
13. $e = 12$ ft.   864 sq. ft.
14. $e = 24$ cm   3,456 cm²

Rectangular solids

15. $l = 12$ cm
    $w = 11$ cm
    $h = 10$ cm
    724 cm²

16. $l = 16$ in.
    $w = 16$ in.
    $h = 12$ in.
    1,280 sq. in.

Cylinders

17. $r = 5$ in.
    $h = 7$ in.
    $\pi = 3\frac{1}{7}$
    $377\frac{1}{7}$ sq. in.

18. $r = 13$ ft.
    $h = 23$ ft.
    $\pi = 3.14$
    2,939.04 sq. ft.

Square pyramids

19. $b = 9$ in.
    $\ell = 11$ in.
    279 sq. in.

20. $b = 14$ cm
    $\ell = 8$ cm
    420 cm²

Spheres

21. $r = 6$ mm
    $\pi = 3.14$
    452.16 mm²

22. $r = 9$ cm
    $\pi = 3.14$
    1,017.36 cm²

23. $r = 5\frac{1}{4}$ in.
    $\pi = 3\frac{1}{7}$
    $346\frac{1}{2}$ sq. in.

24. $r = 3\frac{1}{2}$ in.
    $\pi = 3\frac{1}{7}$
    154 sq. in.

LESSON 115

Objective

- To review the material taught in Chapter 8 (Lessons 100–114).

Teaching Guide

1. Give Lesson 115 Quiz (The Pythagorean Rule).
2. Lesson 115 reviews the material taught in Lessons 100–114. For pointers on using review lessons, see *Teaching Guide* for Lesson 98.
3. Be sure to review the following new concepts taught in this chapter.

Lesson number and new concept	Exercises in Lesson 115
101—Finding the lateral area of a rectangular solid.	None
102—Finding the lateral area of a cylinder.	None
103—Finding the surface area of a square pyramid and memorizing the related formula.	4, 19, 20
103—Finding the lateral area of a square pyramid.	None
104—Finding the surface area of a sphere and memorizing the related formula.	5, 21–24
107—Finding the volume of a square pyramid and memorizing the related formula.	10, 33, 34
108—Finding the volume of a sphere and memorizing the related formula.	11, 35, 36
111—Calculating square roots to the nearest tenth.	41–44
112—Using the Pythagorean rule to find the hypotenuse.	45, 46
113—Using the Pythagorean rule to find the length of one leg of a right triangle.	47, 48

T-391   Chapter 8   Geometric Solids and the Pythagorean Rule

Square Root Extractions for Part F

41.
$$\begin{array}{r}4\ \ 4\\ \sqrt{19{\scriptstyle\wedge}36}\\ 16\end{array}$$

(20 × 4) 80   3 36
   4 × 84 = 3 36
                0

42.
$$\begin{array}{r}5\ \ 5\\ \sqrt{30{\scriptstyle\wedge}25}\\ 25\end{array}$$

(20 × 5) 100   5 25
   5 × 105 = 5 25
                0

43.   4 0 . 6   8   (Rounded to 40.7)
           √16ˬ55.00ˬ00
            16
(20 × 4) 80   0 55
   0 × 80 =   0
(20 × 40) 800   55 00
   6 × 806 = 48 36
(20 × 406) 8120   6 64 00
   8 × 8128 = 6 50 24
                13 76

44.   4 4 . 7   2   (Rounded to 44.7)
           √20ˬ00.00ˬ00
            16
(20 × 4) 80   4 00
   4 × 84 = 3 36
(20 × 44) 880   64 00
   7 × 887 = 62 09
(20 × 447) 8940   1 91 00
   2 × 8942 = 1 78 84
                12 16

C. **Find the volumes of these geometric solids, to the nearest hundredth.**
   (Lessons 105–108)

   Cubes
   25. e = 8 in.   512 cu. in.   26. e = 23 cm   12,167 cm³

   Rectangular solids
   27. l = 15 cm   2,970 cm³   28. l = 18 in.   4,032 cu. in.
       w = 11 cm w = 14 in.
       h = 18 cm h = 16 in.

   Cylinders
   29. r = 5 in.   628 cu. in.   30. r = 20 ft.   37,680 cu. ft.
       h = 8 in. h = 30 ft.
       π = 3.14 π = 3.14

   Cones
   31. r = 5 in.   235.5 cu. in.   32. r = 5 ft. $183\frac{1}{3}$ cu. ft.
       h = 9 in. h = 7 ft.
       π = 3.14 $\pi = 3\frac{1}{7}$

   Square pyramids
   33. s = 6 in.   96 cu. in.   34. s = 14 ft. $522\frac{2}{3}$ cu. ft.
       h = 8 in. h = 8 ft.

   Spheres (Use 4.18.)
   35. r = 9 mm   3,047.22 mm³   36. r = 8 in.   2,140.16 cu. in.

D. **Find the square root of each number by using the chart.** *(Lesson 109)*

   37. $\sqrt{161}$ 12.689   38. $\sqrt{162}$ 12.728

E. **Use estimation to find the square roots of these perfect squares.** *(Lesson 109)*

   39. $\sqrt{5{,}776}$ 76   40. $\sqrt{7{,}569}$ 87

F. **Calculate each square root to the nearest tenth (unless it is a whole number). Show your work.** *(Lessons 110, 111)* (See facing page for extractions.)

   41. $\sqrt{1{,}936}$ 44   42. $\sqrt{3{,}025}$ 55   43. $\sqrt{1{,}655}$ 40.7   44. $\sqrt{2{,}000}$ 44.7

G. **Find the length of the hypotenuse or the missing leg, to the nearest tenth.**
   (Lessons 112, 113)

   45. leg a: 3 ft.   4.2 ft.   46. leg a: 2 cm   5.4 cm
       leg b: 3 ft.   leg b: 5 cm

   47. leg b: 5 in.   7.5 in.   48. leg b: 6 in.   11.5 in.
       hypotenuse: 9 in.   hypotenuse: 13 in.

H. Solve these parallel reading problems. Use your solution to problem *a* as a help in solving problem *b*. (Lesson 114)

49. a. At a book sale, Marian bought 2 books for a total of $10.00. If she had bought 5 of them, she would have paid a total of $20.00. How much less would she have paid per book if she had bought 5 of them? $1.00

    b. At a book sale, Marian bought 3 books for a total of $19.50. The store offered 6 such books for $28.50. How much less would she have paid per book if she had bought 6 of them? $1.75

50. a. After the Sensenig reunion, the 2 sisters in the family divided the 4 remaining pies equally among themselves. How much did each of them receive? 2 pies

    b. After the Sensenig reunion, the 4 sisters in the family divided the $2\frac{1}{2}$ remaining pies equally among themselves. How much did each of them receive? $\frac{5}{8}$ pie

51. a. It took Joshua 2 times as long to finish his chores as it took his younger brother Matthew to finish his. Matthew completed his work in 50 minutes. How long did it take Joshua? 100 minutes (or $1\frac{2}{3}$ hours)

    b. It took $\frac{7}{8}$ as much time for Joshua to finish his chores as it took his younger brother Matthew to finish his. Matthew completed his work in $\frac{2}{3}$ hour. How long did it take Joshua? $\frac{7}{12}$ hour (or 35 minutes)

52. a. It took Samuel 2 hours to walk 4 miles. What was Samuel's average speed? 2 m.p.h.

    b. It took Samuel $\frac{3}{5}$ hour to walk $1\frac{4}{5}$ miles. What was Samuel's average speed? 3 m.p.h.

I. Solve these reading problems.

53. The Woodvale Mennonite School purchased 3 new songbooks for $8.95 each, plus 5% sales tax. What was the total cost of the songbooks? $28.19

54. The real estate tax rate in Mountain Township is 18 mills or 0.018. What is the tax on the Mohler property, which is assessed for $125,000? $2,250

116. Chapter 8 Test

LESSON 116

Objective
- To test the students' mastery of the concepts in Chapter 8.

Teaching Guide
1. Correct Lesson 115.
2. Review any areas of special difficulty.
3. Administer the test. For suggestions on giving tests, see *Teaching Guide* for Lesson 99.

Solutions for Part H
49. a. ($10 ÷ 2) − ($20 ÷ 5)   b. ($19.50 ÷ 3) − ($28.50 ÷ 6)
50. a. 4 ÷ 2   b. $2\frac{1}{2} \div 4$
51. a. 50 × 2   b. $\frac{2}{3} \times \frac{7}{8}$
52. a. 4 ÷ 2   b. $1\frac{4}{5} \div \frac{3}{5}$

Solutions for Part I
53. 3 × $8.95 × 1.05
54. 0.018 × $125,000

The Christian who would succeed in managing finances must also be serious about his stewardship. Financial gain can be a deadly snare: "They that will be rich fall into temptation and a snare, and into many foolish and hurtful lusts, which drown men in destruction and perdition. For the love of money is the root of all evil: which while some coveted after, they have erred from the faith, and pierced themselves through with many sorrows" (1 Timothy 6:9, 10).

Chapter 9
Mathematics and Finances

One who would succeed in managing finances must be a careful record keeper. Develop the following habits to spare yourself many financial frustrations.

Neatness. Carelessly formed numerals easily cause errors. ("Is that a six or a zero?")

Promptness. Timely attention to the checkbook balance is essential. Failing to keep the balance current only makes the job harder and increases the frustration of discrepancies.

Faithfulness. Records that are consistently maintained fulfill their purpose as useful tools.

He that is faithful

in that which is least . . .

is faithful also in much.
Luke 16:10

394　Chapter 9　Mathematics and Finances

117. Checking Accounts: Using Deposit Tickets

A **checking account** is one of the most commonly used services that a bank offers. This kind of account operates on a very simple basis. A person opens an account by depositing money in the bank. When he wants to use some of the money, he writes a check. The check tells the bank to transfer money from the depositor's account to the person who received the check.

A checking account is a prudent method of handling money. It is more convenient to carry a checkbook than to carry large amounts of cash. Checks are also safer than cash. A check for $1,000 can be used only by the person whose name is written on the check, whereas $1,000 in cash can be spent by anyone. Furthermore, a checking account provides a record of transactions. This record is useful in preparing financial statements and tax returns.

To increase the funds in his checking account, a person uses a **deposit ticket** to make a deposit. The bank usually offers preprinted forms with the account holder's name, address, and account number. The depositor needs only to fill in the date and the amount of money he is depositing. Following is a sample deposit ticket.

Dale Smith made the following deposit on January 15 to his checking account. He received $40.00 cash and deposited the rest.

This example is given as shown to illustrate a deposit ticket completely filled out. An actual deposit would probably not include both $78.00 in currency and $40.00 in cash received.

Currency $78.00
Coins 7.58
Checks
　58-717 . . . . 75.00
　62-111 . . . . 48.37
　711-14 . . . . 73.00

DEPOSIT TICKET			
Dale N. Smith　　　　56-334/745	CURRENCY	78	00
125 Dartmouth Street	COIN	7	58
Bryan, OH 43506	CHECKS		
	58-717	75	00
	62-111	48	37
Account # 36-5551123	711-14	73	00
	SUBTOTAL	281	95
DATE　　*Jan. 15*　　20 ___	LESS CASH RECEIVED	40	00
Bryan National Bank	TOTAL DEPOSIT	241	95

LESSON 117

Objectives

- To teach *the principles of using a checking account.
- To teach *how to complete deposit tickets and record them in the check register.

Review

1. *Find the length of leg a in each right triangle.* (Lesson 113)

    a. leg b = 8 cm
       hypotenuse = 14 cm
       leg a = (11.489 cm)

    b. leg b = 9 ft.
       hypotenuse = 15 ft.
       leg a = (12 ft.)

2. *Find the square roots by using a table.* (Lesson 109)

    a. $\sqrt{77}$ (8.775)

    b. $\sqrt{54}$ (7.348)

3. *Find the surface area of each rectangular solid.* (Lesson 101)

    a. l = 14 in.
       w = 11 in.
       h = 10 in.
       a_s = (808 sq. in.)

    b. l = 26 ft.
       w = 12 ft.
       h = $8\frac{1}{2}$ ft.
       a_s = (1,270 sq. ft.)

4. *Write these expressions, using symbols.* (Lesson 85)

    a. line segment RS is parallel to line segment TU ($\overline{RS} \parallel \overline{TU}$)

    b. line HI is perpendicular to line JK ($\overleftrightarrow{HI} \perp \overleftrightarrow{JK}$)

5. *Divide by primes to find the prime factors of each number. Use exponents for repeating factors.* (Lesson 34)

    a. 96 (96 = $2^5 \times 3$)

    b. 189 (189 = $3^3 \times 7$)

T–395 *Chapter 9 Mathematics and Finances*

Introduction

Why do people use checks and checking accounts? Why don't they simply pay cash for all business transactions?

(1) *Convenience.* Large amounts of cash are bulky to carry. A checkbook takes up much less room.

(2) *Safety.* If cash is lost or stolen, it is difficult to retrieve or identify. If checks are lost or stolen, it is simple to stop payment with little loss other than the bank fees involved. It is much safer to send checks through the mail than to send cash.

(3) *Transaction record.* After each check is processed by the bank, either the check itself or a copy is returned to the account holder. This establishes positive proof of payment, and it also provides a record that is useful in preparing financial statements and tax returns.

Teaching Guide

1. **With a checking account, funds are deposited in a bank to be withdrawn by the use of checks.** The checking account is one of the most commonly used services that banks offer.

2. **Deposit tickets are used to deposit money into a checking account.** Discuss the three points relating to the sample deposit ticket in the lesson.

    Complete several deposit tickets in class, using the information below and in *Class Practice*. Stress that neatness and accuracy are critical in filling out business forms.

    Nevin Weaver made the following deposit on September 9 to his checking account. He received $130.00 cash and deposited the rest.

Currency	$125.00
Coins	8.96
Check 17-741	235.23
Check 61-963	315.16
Check 49-741	299.95

    (Subtotal: $984.30; total deposit: $854.30)

    Note: In this lesson and several others, some exercises involve forms related to checking and savings accounts. These forms are included with the quiz sheets for this chapter. You may make extra copies of the forms as needed.

Solutions for CLASS PRACTICE b *and* c

b. $182.00 + $312.41 + $202.01 = $696.42 subtotal; $696.42 − $611.42

c. $2,034.82 + $125.00 = $2,159.82 subtotal; $2,159.82 − ($965.72 + $154.35 + $439.78)

Notice the following things about the deposit ticket filled in above.

1. Each check is listed by its bank number. For example, a check from the Bryan National Bank would have the number 56-334/745, the same as the deposit ticket above. For such a check, you would put 56-334 on a deposit ticket. (Only the part before the slash is written.)

2. The deposit ticket is subtotaled. The subtotal shows the total amount of money taken to the bank.

3. If cash is received from the deposit, that amount is written in the proper place and subtracted from the Subtotal in order to arrive at the actual deposit. Both the subtotal and total should be written even if no cash is received.

An identification number is assigned to each bank. The number for Bryan National Bank is 56-334/745. This number is printed on all checks issued by this bank, and it may also be printed on deposit tickets. The 56 is assigned to all banks in the area of Bryan, Ohio. The 334 is assigned specifically to Bryan National Bank. The 745 is a routing symbol to help route the check to the area and bank that holds that account.

CLASS PRACTICE (Forms provided in quiz booklet.)

Complete a deposit ticket for the following deposit. Be sure your calculations are correct. Use the current date.

a. Currency $85.00
    Coins 7.15
    Checks: 35-814 .... 128.50
             556-21 .... 87.00
             70-958 .... 96.31
    Cash received 45.00

CURRENCY	85	00
COIN	7	15
CHECKS		
35-814	128	50
556-21	87	00
70-958	96	31
SUBTOTAL	403	96
LESS CASH RECEIVED	45	00
TOTAL DEPOSIT	358	96

Do these exercises relating to deposit tickets.

b. A deposit ticket showed checks for $182.00, $312.41, and $202.01, and a total deposit of $611.42. Find the amount of cash received. $85.00

c. A deposit ticket showed four checks, $125.00 in cash received, and a total deposit of $2,034.82. Three of the checks were for $965.72, $154.35, and $439.78. Find the amount of the fourth check. $599.97

WRITTEN EXERCISES

A. *Complete deposit tickets for the following deposits. Be sure your calculations are correct. Use the current date.*   (See page T–396)

1. Currency $95.00
    Coins 8.50
    Checks: 58-717 .... 68.00
             62-111 .... 79.00
             731-14 .... 92.00

2. Currency $96.00
    Coins 14.62
    Checks: 133-51 .... 95.00
             276-93 .... 85.00
             41-559 .... 69.00

3. Checks: 62-021 .... 358.00
496-56 .... 85.00
38-553 .... 232.39
Cash received 100.00

4. Coins 26.43
Checks: 98-963 .... 395.00
77-841 .... 272.89
694-89 .... 414.98
Cash received 175.00

B. Solve these reading problems.

5. On Thursday, Fred entered these amounts on his deposit ticket: $82.00 in currency; $0.73 in coins; and $24.52, $15.30, $13.89, and $27.00 in checks. Find the total deposit. $163.44

6. A deposit ticket showed checks for $39.07, $416.59, and $551.85. Cash received was $75.00. Find the total deposit. $932.51

7. The total on a deposit ticket was $1,640.56. Find the amount of cash received if the checks were for $53.75, $387.00, and $1,289.81. $90.00

8. The total amount on a deposit ticket was $2,654.81, cash received was $128.43, and two of three checks were for $1,214.12 and $763.21. Find the amount of the third check. $805.91

9. A window frame has an opening 29 inches wide. How far will a 36-inch stick hold the window open if it is propped diagonally? (Answer to the nearest tenth.)   21.3 inches

10. How high on a wall will a 24-foot ladder reach if its base is 6 feet from the wall? (Answer to the nearest tenth.)   23.2 feet

REVIEW EXERCISES

C. Find the length of leg a of each right triangle, to the nearest tenth. (Lesson 113)

11. leg b = 11 cm   13.0 cm
hypotenuse = 17 cm

12. leg b = 9 ft.   10.7 ft.
hypotenuse = 14 ft.

D. Find these square roots by using the table. (Lesson 109)

13. $\sqrt{86}$ 9.274

14. $\sqrt{69}$ 8.307

E. Find the surface area of each rectangular solid. (Lesson 101)

15. l = 18 in.   744 sq. in.
w = 6 in.
h = 11 in.

16. l = 22 ft.   1,300 sq. ft.
w = 14 ft.
h = $9\frac{1}{2}$ ft.

F. Write each expression, using symbols. (Lesson 85)

17. line segment OP $\overline{OP}$

18. line CD is parallel to line EF $\overleftrightarrow{CD} \parallel \overleftrightarrow{EF}$

G. Divide by primes to find the prime factors of each number. Use exponents for repeating factors. (Lesson 34)

19. 98 $98 = 2 \times 7^2$

20. 135 $135 = 3^3 \times 5$

Lesson 117 T–396

An Ounce of Prevention

1. Stress neatness and accuracy throughout this chapter. Carelessly completed records cause many problems.

2. The deposit ticket form may help to prevent confusion in the reading problems. Using the extra forms on the sheet in the quiz booklet or simply setting up the numbers in form arrangement can help to clarify when to use addition or subtraction.

Answers for Part A

1.

CURRENCY	95	00
COIN	8	50
CHECKS		
58-717	68	00
62-111	79	00
731-14	92	00
SUBTOTAL	342	50
LESS CASH RECEIVED		
TOTAL DEPOSIT	342	50

2.

CURRENCY	96	00
COIN	14	62
CHECKS		
133-51	95	00
276-93	85	00
41-559	69	00
SUBTOTAL	359	62
LESS CASH RECEIVED		
TOTAL DEPOSIT	359	62

3.

CURRENCY		
COIN		
CHECKS		
62-021	358	00
496-56	85	00
38-553	232	39
SUBTOTAL	675	39
LESS CASH RECEIVED	100	00
TOTAL DEPOSIT	575	39

4.

```
             DEPOSIT TICKET
  [Your name]                    56-334/809
  [Your street address]
  [Your city and state]

  Account # 36-5551123

  DATE ___[current date]___ 20 ___

            Bryan National Bank
```

CURRENCY		
COIN	26	43
CHECKS		
98-963	395	00
77-841	272	89
694-89	414	98
SUBTOTAL	1109	30
LESS CASH RECEIVED	175	00
TOTAL DEPOSIT	934	30

Solutions for Part B

5. $82.00 + $0.73 + $24.52 + $15.30 + $13.89 + $27.00

6. $39.07 + $416.59 + $551.85 − $75.00

7. ($53.75 + $387.00 + $1,289.81) − $1,640.56

8. $2,654.81 + $128.43 = $2,783.24 subtotal; $2,783.24 − ($1,214.12 + $763.21)

9. $36^2 - 29^2 = 455$; $\sqrt{455}$

10. $24^2 - 6^2 = 540$; $\sqrt{540}$

T-397 Chapter 9 Mathematics and Finances

LESSON 118

Objective

- To teach *how to write checks.

Review

1. Give Lesson 118 Quiz (Surface Area and Volume).

2. Use problem a as a guide in how to solve problem b. (Lesson 114)

   a. Nathan was plowing a 24-acre field. By noon he had $\frac{3}{4}$ of the field finished. How many acres were plowed by noon? (18 acres)

   b. Gary planted $1\frac{3}{4}$ acre of sweet corn. He estimated that the heifers ruined $\frac{1}{3}$ of the patch when they ran through. What part of an acre did the heifers damage? ($\frac{7}{12}$ acre)

3. Extract the square roots of these perfect squares. (Lesson 110)

   a. $\overset{(95)}{\sqrt{9{,}025}}$ $\quad$ $\sqrt{\overset{9\ \ 5}{90{,}25}}$
   $$\underline{81}$$
   $$(20 \times 9)\ 180 \quad 9\ 25$$
   $$5 \times 185 = \underline{9\ 25}$$
   $$0$$

   b. $\overset{(113)}{\sqrt{12{,}769}}$ $\quad$ $\sqrt{\overset{1\ \ 1\ \ 3}{1{,}27{,}69}}$
   $$\underline{1}$$
   $$20 \quad 27$$
   $$1 \times 21 = \underline{21}$$
   $$(20 \times 11)\ 220 \quad 6\ 69$$
   $$3 \times 223 = \underline{6\ 69}$$
   $$0$$

4. Find the surface area of each cylinder. (Lesson 102)

   a. $r = 2$ in.
      $h = 3\frac{1}{2}$ in.
      $a_s = $ (69.08 sq. in.)

   b. $r = 25$ ft.
      $h = 32$ ft.
      $a_s = $ (8,949 sq. ft.)

5. Identify each angle as acute, obtuse, right, or straight. (Lesson 86)

   a. 84° (acute)
   b. 36° (acute)
   c. 110° (obtuse)
   d. 180° (straight)

6. Find the missing parts of these inverse proportions. (Lesson 54)

   a. $\dfrac{\text{workers (1)}}{\text{workers (2)}}\ \ \dfrac{3}{7} = \dfrac{n}{10}\ \ \dfrac{\text{hours (2)}}{\text{hours (1)}}$
   $$(n = 4\tfrac{2}{7}\text{ hr.})$$

   b. $\dfrac{\text{m. p. h. (1)}}{\text{m. p. h. (2)}}\ \ \dfrac{85}{60} = \dfrac{8\tfrac{1}{2}}{n}\ \ \dfrac{\text{hours (2)}}{\text{hours (1)}}$
   $$(n = 6\text{ hr.})$$

118. Checking Accounts: Writing Checks

A check is a written order directing a bank to transfer money from a depositor's account to another person or business. The person writing the check is the **payer** of the check. The person receiving the check is the **payee.**

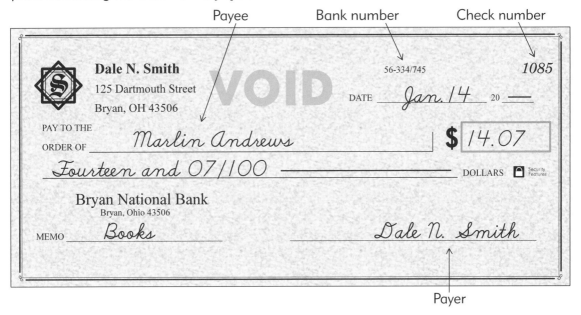

Following are some guidelines for writing a check.

1. Write the date, using either the name of the month or its number. The date above could be written as 1/14.

2. Write the payee's name on the line after the words "Pay to the order of."

3. Write the amount of the check in digits after the dollar sign. Begin close to the dollar sign as a safeguard to prevent someone from inserting digits before the number.

4. Write the amount of the check in words on the line below the payee's name. Use a fraction to write the cents. If there are no cents, write the fraction as 00/100, no/100, or xx/100. Make sure this amount agrees with the amount written in digits. Draw a line above any unused part of the blank to show that no words are to be written there.

5. The memo line is optional. It may be used to write the number of the invoice being paid or to briefly explain the reason for writing the check.

6. The check is worthless until it has been signed by the authorized person. Sign your name in cursive writing, always using the same form (such as always including your middle initial) and always using the same style of handwriting. This helps to reduce the danger of forgery.

Because it is so easy to write checks, dishonest people sometimes use this as an opportunity for fraud. Therefore, banks have developed safeguards such as refusing to accept old or altered checks and issuing preprinted checks that are difficult to photocopy. The individual himself can also do things to help prevent fraud. Some of these safeguards are listed below.

1. Store your checkbook securely, just as you would money.
2. Never sign a check until the rest of it is filled in.
3. Always write with ink.
4. If you make a mistake on a check, void the check by writing *VOID* in large letters over it. The checks in this lesson are voided.
5. Use your checks in numerical order so that you can immediately tell if a check is missing.

CLASS PRACTICE

(Forms are provided in quiz booklet.)

Write the checks indicated. Use the current year in the date, and sign your own name as the payer.

(See page T–399 for answers.)

	Check number	Date	Payee	Amount	Memo
a.	657	Jan. 28	Edwin Steiner	$185.42	Invoice #51
b.	683	Apr. 15	Locust Point Automotive	$54.25	Car repair

WRITTEN EXERCISES

A. Identify the errors on these checks.

1.

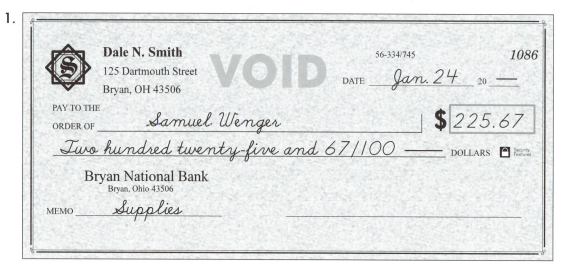

signature missing

Lesson 118 T–398

Introduction

Bring a check to class and have the students identify it. What is the value of the check itself? The intrinsic value of a check is only that of the paper and ink that it is made of. (The same is true of paper currency, and even most United States coins have value only for what they represent and not for the metal they contain.)

What makes a check valuable? A check has value only if sufficient funds are in the bank; otherwise, it is worthless. Knowingly writing a worthless check is dishonest. Christians need to be careful that the checks they write are backed by sufficient funds.

Teaching Guide

1. **A check is a written order directing a bank to transfer money from a depositor's account to another person or business.** The person writing the check is the *payer*. The person receiving the check is the *payee*.

2. **Be sure to follow the proper method when writing a check.** Discuss the guidelines in the lesson text, and illustrate them with the checks described below and in *Class Practice*.

   a. *Check number*: 377
      Date: May 7
      Payee: Elvin Trostle
      Amount: $62.77
      Memo: Repairs

   b. *Check number*: 378
      Date: May 15
      Payee: Lamar Davis
      Amount: $48.35
      Memo: Copier paper

   You may want to explain the method to use when the amount is less than one dollar. If it is 79 cents, the amount in words should be written as "Only seventy-nine cents."

3. **An individual can do certain things that will help to prevent fraud in relation to his checking account.** Discuss the safeguards listed in the lesson text.

T–399   Chapter 9   Mathematics and Finances

Answers for CLASS PRACTICE

a.

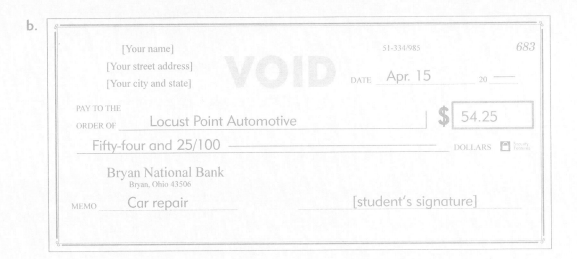

b.

Answers for Part B

5.

2. differing amounts (dollars)

Check #1087, dated Jan. 25, pay to Douglas Lowell, $135.08, written "One hundred five and 08/100", memo: Gas, signed Dale N. Smith.

3. differing amounts (cents)

Check #1088, dated Jan. 25, pay to Mark Dugas, $28.75, written "Twenty-eight and 67/100", memo: Repairs, signed Dale N. Smith.

4. payee missing

Check #1089, dated Jan. 26, pay to the order of _____, $245.00, written "Two hundred forty-five and 00/100", memo: Office expense, signed Dale N. Smith.

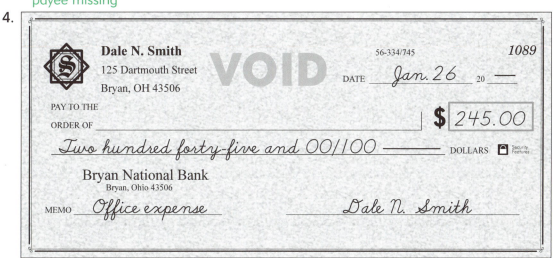

400 Chapter 9 Mathematics and Finances

B. Write the checks indicated. Use the current year in the date, and sign your own name as the payer. (Forms are included with quiz sheets.)

	Check number	Date	Payee	Amount	Memo
5.	151	Jan. 13	Marcus Zendt	$23.54	Invoice #342
6.	152	Feb. 26	Zook Hardware	$41.60	Electric drill
7.	153	Mar. 7	Sherwin Bristol	$250.00	Truck repair
8.	154	Apr. 9	Ronald Smith	$384.64	Electric motor repair

C. Solve these reading problems. Use proportions for numbers 9 and 10. Numbers 13 and 14 are parallel problems.

9. A motor has a 2-inch pulley that spins at 1,725 revolutions per minute. It drives a 7.5-inch pulley on a barn fan. What is the speed of the driven pulley? $n = 460$ r.p.m.

10. If the pulley on this fan motor (problem 9) were changed to 2.5 inches, what would be the speed of the driven pulley? $n = 575$ r.p.m.

11. Find the surface area of a flashlight battery that is $1\frac{3}{4}$ inches long and has a radius of $\frac{1}{2}$ inch. (Use $3\frac{1}{7}$ for pi. and disregard the knob at the positive end.) $7\frac{1}{14}$ square inches

12. A metal can has a diameter of 3 inches and is 7 inches tall. What is its surface area? (Use $3\frac{1}{7}$ for pi.) $80\frac{1}{7}$ square inches

13. Pencils that are 0.8 centimeter wide are packaged in a box 4 centimeters wide. How many pencils can fit side by side in the box?   5 pencils

14. The cone-shaped light receptors on the retina of the eye have a diameter of 0.006 millimeter. The lead of some mechanical pencils is 0.5 millimeter in diameter. How many cones, side by side, would it take to form a line across the diameter of this lead? (Write the remainder as a fraction.) $83\frac{1}{3}$ cones

REVIEW EXERCISES

D. Extract the square roots of these perfect squares. *(Lesson 110)*

15. $\sqrt{4{,}096}$ = 64

16. $\sqrt{69{,}696}$ = 264   (See page T–401 for extractions.)

E. Find the surface area of each cylinder, to the nearest hundredth. *(Lesson 102)*

17. $r = 3.2$ m $h = 12$ m   305.46 m²

18. $r = 31$ ft. $h = 42$ ft.   14,211.64 sq. ft.

19. $r = 5$ in. $h = 8.5$ in.   423.9 sq. in.

20. $r = 11$ ft. $h = 26$ ft.   2,555.96 sq. ft.

F. Identify each angle as acute, obtuse, right, **or** straight. *(Lesson 86)*

21.   obtuse

22.   acute

G. Find the missing parts of these inverse proportions. *(Lesson 54)*

23. $\dfrac{\text{diameter (1)}}{\text{diameter (2)}} = \dfrac{5}{7} = \dfrac{n}{70}$ r.p.m. (2) 50   r.p.m. (1)

24. $\dfrac{\text{m.p.h. (1)}}{\text{m.p.h. (2)}} = \dfrac{65}{45} = \dfrac{13}{n}$ hours (2)   hours (1) 9

Lesson 118 T–400

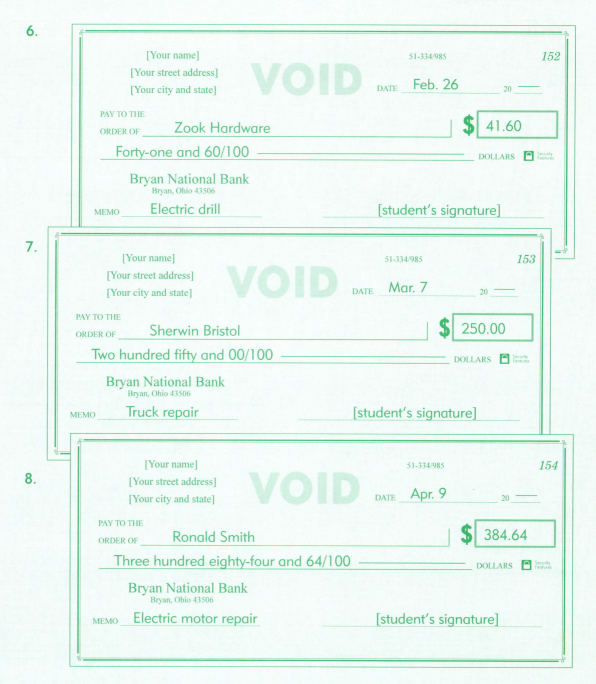

Solutions for Part C

9. $\dfrac{\text{diameter (1)}}{\text{diameter (2)}} \quad \dfrac{2}{7.5} = \dfrac{n}{1{,}725} \quad \dfrac{\text{r.p.m. (2)}}{\text{r.p.m. (1)}}$

10. $\dfrac{\text{diameter (1)}}{\text{diameter (2)}} \quad \dfrac{2.5}{7.5} = \dfrac{n}{1{,}725} \quad \dfrac{\text{r.p.m. (2)}}{\text{r.p.m. (1)}}$

11. $2(3\tfrac{1}{7} \times \tfrac{1}{2} \times \tfrac{1}{2}) + 1\tfrac{3}{4} \times 3\tfrac{1}{7} \times 1$

12. $2(3\tfrac{1}{7} \times 1\tfrac{1}{2} \times 1\tfrac{1}{2}) + 7 \times 3\tfrac{1}{7} \times 3$

13. $4 \div 0.8$

14. $0.5 \div 0.006$

T-401 Chapter 9 *Mathematics and Finances*

LESSON 119

Objective

- To teach *using a check register to record checks and deposits and to maintain the checking account balance.

Review

1. Calculate each square root to the nearest whole number. (Lesson 111)

   a. $\sqrt{15{,}065}$ (123)

   $$\begin{array}{r} 1\ 2\ 2\,.\,7 \\ \sqrt{1{,}50{,}65.00} \\ \underline{1} \\ 20\ 50 \\ 2\times 22 = \underline{44} \\ (20\times 12)\ 240\ 6\ 65 \\ 2\times 242 = \underline{4\ 84} \\ (20\times 122)\ 2440\ 1\ 81\ 00 \\ 7\times 2447 = \underline{1\ 71\ 29} \\ 9\ 71 \end{array}$$

   b. $\sqrt{141{,}032}$ (376)

   $$\begin{array}{r} 3\ 7\ 5\,.\,5 \\ \sqrt{14{,}10{,}32.00} \\ \underline{9} \\ (20\times 3)\ 60\ 5\ 10 \\ 7\times 67 = \underline{4\ 69} \\ (20\times 37)\ 740\ 41\ 32 \\ 5\times 745 = \underline{37\ 25} \\ (20\times 375)\ 7500\ 4\ 07\ 00 \\ 5\times 7505 = \underline{3\ 75\ 25} \\ 31\ 75 \end{array}$$

2. Find the surface area of each square pyramid. (Lesson 103)

   a. $b = 15$ ft.

   $\ell = 20$ ft.

   $a_s = $ (825 sq. ft.)

   b. $b = 78$ m

   $\ell = 95.3$ m

   $a_s = $ (20,950.8 m²)

3. Use direct proportions to solve these reading problems. (Lesson 55)

   a. The primary coil of a transformer has 1,800 turns of wire with an input of 120 volts. How many turns are required for the secondary coil to give an output of 12 volts? (180 turns)

   $$\frac{\text{turns of wire}}{\text{volts}} \quad \frac{1{,}800}{120} = \frac{n}{12} \quad \frac{\text{turns of wire}}{\text{volts}}$$

   b. The secondary coil of a transformer has 1,250 turns of wire and an output of 300 volts. How many turns does the primary coil have if the input is 120 volts? (500 turns)

   $$\frac{\text{turns of wire}}{\text{volts}} \quad \frac{1{,}250}{300} = \frac{n}{120} \quad \frac{\text{turns of wire}}{\text{volts}}$$

Extractions for Lesson 118, Part D

15. $\sqrt{40{,}96}$ 6 4

$$\begin{array}{r} 6\ 4 \\ \sqrt{40{,}96} \\ \underline{36} \\ (20\times 6)\ 120\ 4\ 96 \\ 4\times 124 = \underline{4\ 96} \\ 0 \end{array}$$

16. $\sqrt{6{,}96{,}96}$ 2 6 4

$$\begin{array}{r} 2\ 6\ 4 \\ \sqrt{6{,}96{,}96} \\ \underline{4} \\ (20\times 2)\ 40\ 2\ 96 \\ 6\times 46 = \underline{2\ 76} \\ (20\times 26)\ 520\ 20\ 96 \\ 4\times 524 = \underline{20\ 96} \\ 0 \end{array}$$

119. Checking Accounts: Maintaining Account Balances

A person who has a checking account is responsible to keep an accurate record of the account. A **check register** such as the one shown below provides a place to record every **transaction** (action that affects the balance in the account). Each deposit is added to the balance, and each check or charge is subtracted from the balance. Maintaining an accurate balance helps to keep the person from overdrawing his account.

NUMBER	DATE	DESCRIPTION	✓	FEE	CHECK AMOUNT	DEPOSIT AMOUNT	BALANCE
		Balance carried forward					$335 14
D	4/18	Deposit—Wages				498 56	833 70
401	4/25	Hillview Grocery			72 36		761 34
402	4/26	Springfield Electric			91 91		669 43
403	4/26	Valley Telephone			52 26		617 17
404	4/27	Williams Heating			127 21		489 96
D	4/28	Deposit—Wages				497 31	987 27
C	4/29	April service charge		6.50			980 77

The columns in the register above are typical of check registers and are described below.

1. The Number column shows the check number. If the line is used for a deposit, write D on the line. If a bank charge is entered, write C on the line.

2. The Date column shows the date the check was written or other transaction was done.

3. The Description column shows the name of the payee or the source of a deposit.

4. The checkoff column (✓) is used in account reconciliation, which is discussed in the next lesson.

5. The Fee column shows bank fees such as monthly service charges.

6. The Check Amount column shows the amount of each check that is written.

7. The Deposit Amount column shows the amount of each deposit.

8. The Balance column shows the new balance after each transaction. Add deposits to the balance, and subtract checks and bank charges.

Check registers vary somewhat. Some registers have a column for showing transaction types, including electronic deposits or withdrawals. Some have a column labeled T for marking items that are tax-deductible.

402   Chapter 9   Mathematics and Finances

It is very important to maintain an accurate, up-to-date checkbook balance. One useful habit is to fill in the register before writing a check, to avoid forgetting what the amount was. Keeping an accurate balance will help to prevent overdrawing the checking account, which is a poor testimony for people professing godliness.

CLASS PRACTICE   (Forms are provided in quiz booklet.)

Use this information to fill in a check register.   *Running balance*

a. Balance carried forward on 2/10: $857.32 $857.32
b. Check number 125 on 2/11 to Countryside Fuel for $132.65   724.67
c. Check number 126 on 2/11 to Maple Farms for $50.00   674.67
d. Deposit on 2/14 from sales: $305.26   979.93
e. February service charge on 2/15: $7.50   972.43

(See facing page for completed form.)

WRITTEN EXERCISES

A. Calculate the missing balances in each check register.

1.

NUMBER	DATE	DESCRIPTION	✓	FEE	CHECK AMOUNT	DEPOSIT AMOUNT	BALANCE
		Balance carried forward					$141 | 27
D	2/12	Deposit—Sales				823 | 45	964 | 72
239	2/13	Witmer's Lumber			477 | 22		487 | 50
240	2/14	Valley Power and Light			215 | 61		271 | 89

2.

NUMBER	DATE	DESCRIPTION	✓	FEE	CHECK AMOUNT	DEPOSIT AMOUNT	BALANCE
		Balance carried forward					$322 | 67
314	3/16	Woodville Telephone			199 | 74		122 | 93
D	3/17	Deposit—Sales				141 | 25	264 | 18
C	3/29	March service charge		7.50			256 | 68

3.

NUMBER	DATE	DESCRIPTION	✓	FEE	CHECK AMOUNT	DEPOSIT AMOUNT	BALANCE
		Balance carried forward					$514 | 74
416	4/17	Woodville Bank—Loan			301 | 21		213 | 53
D	4/18	Deposit—Sales				432 | 27	645 | 80

Lesson 119 T–402

Introduction

What will probably happen to a person who takes a stroll in a strange forest without a well-marked trail or a knowledgeable guide? He is likely to wander about aimlessly and get lost.

Something similar can happen with a checking account. The user may think he can freely write checks because he has deposited plenty of money. Unfortunately, such a person will drain his bank account sooner than he realizes.

Most businesses charge a fee of $20 or more when a check is returned for non-sufficient funds (NSF). Writing such checks also harms the testimony of Christian people. Keeping an accurate check register will help us to avoid this mistake.

Teaching Guide

1. **A check register provides a place to record every transaction that affects a checking account.** Although check registers vary in certain details, they are all basically similar to the sample shown in this lesson.

2. **A check register has columns for recording the different kinds of transactions.** Discuss the purpose and function of each column in the sample register. Have students fill in check registers, using the information below and in *Class Practice*.

   a. Balance carried forward on 4/14: $478.79

   b. Check number 821 on 4/15 to Millis Hardware for $15.98
   (Balance: $462.81)

   c. April service charge on 4/15: $6.75
   (Balance: $456.06)

   d. Check number 822 on 4/17 to Hollistan Tractor for $395.26
   (Balance: $60.80)

   e. Deposit on 4/21 from sale of cull cow for $375.00
   (Balance: $435.80)

3. **It is very important to maintain an accurate, up-to-date checkbook balance.** One useful habit is to fill in the register before writing a check, to avoid forgetting what the amount was. Keeping an accurate balance will help to prevent overdrawing the checking account, which is a poor testimony for people professing godliness.

Answers for CLASS PRACTICE

	NUMBER	DATE	DESCRIPTION	✓	FEE	CHECK AMOUNT	DEPOSIT AMOUNT	BALANCE
a.		2/10	Balance carried forward					$857 32
b.	125	2/11	Countryside Fuel			132 65		724 67
c.	126	2/11	Maple Farms			50 00		674 67
d.	D	2/14	Deposit—Sales				305 26	979 93
e.	C	2/15	Feb. service charge		7.50			972 43

T–403 Chapter 9 Mathematics and Finances

An Ounce of Prevention

1. The exercises in this lesson are numbered to give 1 point per form in part A and 1 point per item in part B. A suggestion for part B is to count $\frac{1}{2}$ point for accuracy of the data, and $\frac{1}{2}$ point for the balance.
2. Neatness and accuracy can hardly be overstressed during these lessons. Likewise, keeping the checkbook balance up to date is hardly an overworked item. Far too many people are well familiar with the guilty feeling that comes with failing to keep a current checkbook balance.

   The use of computer accounting does not reduce the importance of watching the checkbook balance. Too many businessmen have blithely assumed that their records are correct because they are kept by computers. Computer accounting does avoid calculation errors, but otherwise it simply expedites either accurate or errant accounting.

Further Study

There are several methods for keeping record of the transactions in a checking account. The following paragraphs describe the two main methods; all others are variations of these two.

The check register shown in the lesson is the simplest and most common method. It has the advantage of bringing all the check records into one location. This is especially effective when a duplicate check system is used, in which a carbon copy is made of each check as it is written. The disadvantage, especially when the duplicate system is not used, is that one must go to a different place to record checks as they are written.

The other method is a system with check stubs at the side or top of the checks. This has the advantage of making it very simple to record each check as it is written. The disadvantage is that the check records are scattered through the check stubs.

Solutions for Part C

12. $834.57 − $32.63 + $235.93

13. $462.33 − ($73.88 + $0.15)

14. $2,657.44 − $49.75 + $3.88

15. $3,652.78 − ($100.72 + $339.00 + $41.02 + $28.98)

16. $\dfrac{\text{oatmeal}}{\text{hamburger}} \; \dfrac{\frac{3}{4}}{1\frac{1}{3}} = \dfrac{n}{10} \; \dfrac{\text{oatmeal}}{\text{hamburger}}$

17. $\dfrac{\text{salt}}{\text{hamburger}} \; \dfrac{1}{1\frac{1}{3}} = \dfrac{n}{10} \; \dfrac{\text{salt}}{\text{hamburger}}$

4.

NUMBER	DATE	DESCRIPTION	✓	FEE	CHECK AMOUNT		DEPOSIT AMOUNT		BALANCE	
		Balance carried forward							$523	35
566	5/19	Blake Hill Supply			171	38			351	97
567	5/20	Davis Office Supply			132	16			219	81

B. Use this information to fill in check registers. (See page T–404.)

5–7.   *Running balance*

    Balance carried forward: $657.32 $657.32

    Check number 357 on 1/19 to Milford Supply for $77.65   579.67

    Check number 358 on 1/20 to Plaza Electric for $95.26   484.41

    Deposit from sales on 1/21 for $275.35   759.76

8–11.

    Balance carried forward: $379.71 $379.71

    Check number 534 on 2/23 to Davis Gasoline for $29.29   350.42

    Check number 535 on 2/24 to Merle Grocery for $92.23   258.19

    Deposit of wages on 2/24 for $466.96   725.15

    February service charge on 2/28 for $5.95   719.20

C. Solve these reading problems. Use proportions for numbers 16 and 17.

12. On April 5, Lester Stover began the day with a balance of $834.57. He wrote a check for $32.63 and made a deposit of $235.93. What was his new balance? $1,037.87

13. Marlene had a balance of $462.33 before she wrote a check for $73.88. The bank charges a 15-cent fee for each check she writes. What was her balance with the check and fee deducted? $388.30

14. Brother Donald has an interest-bearing checking account. On March 25 he wrote a check for $49.75, and the bank deposited $3.88 interest into his account. His balance at the beginning of the day was $2,657.44. What was his new balance? $2,611.57

15. Brother Merle wrote checks for the following amounts: $100.72, $339.00, $41.02, and $28.98. He began with a balance of $3,652.78. What is his new balance? $3,143.06

16. Carol is making meatballs. Her recipe calls for $\frac{3}{4}$ cup oatmeal to be added to $1\frac{1}{3}$ pounds of hamburger. How much oatmeal should she add to 10 pounds of hamburger? $5\frac{5}{8}$ cups

17. Carol's recipe calls for 1 teaspoon of salt to be added to $1\frac{1}{3}$ pounds of hamburger. How much salt should she add to 10 pounds of hamburger? $7\frac{1}{2}$ teaspoons

404   Chapter 9   Mathematics and Finances

REVIEW EXERCISES

D. Calculate each square root to the nearest whole number. *(Lesson 111)*

18. $\sqrt{14,654}$ 121

19. $\sqrt{281,352}$ 530

E. Find the surface area of each square pyramid. *(Lesson 103)*

20. b = 9 in.   297 sq. in.
    ℓ = 12 in.

21. b = 64 cm   13,696 cm²
    ℓ = 75 cm

F. Identify each triangle as right, obtuse, **or** acute. *(Lesson 87)*

22.
    obtuse

23.
    right

24.
    acute

Moreover it is required in stewards, that a man be found faithful.

1 Corinthians 4:2

Lesson 119   T–404

Answers for Part B

5–7.

NUMBER	DATE	DESCRIPTION	✓	FEE	CHECK AMOUNT	DEPOSIT AMOUNT	BALANCE
		Balance carried forward					$657 32
357	1/19	Milford Supply			77 65		579 67
358	1/20	Plaza Electric			95 26		484 41
D	1/21	Deposit—Sales				275 35	759 76

8–11.

NUMBER	DATE	DESCRIPTION	✓	FEE	CHECK AMOUNT	DEPOSIT AMOUNT	BALANCE
		Balance carried forward					$379 71
534	2/23	Davis Gasoline			29 29		350 42
535	2/24	Merle Grocery			92 23		258 19
D	2/24	Deposit—Wages				466 96	725 15
C	2/28	Feb. service charge		5.95			719 20

Extractions for Part D

18.
$$\begin{array}{r} 1\ 2\ 1 . 0 \\ \sqrt{1_\wedge 46_\wedge 54.00} \\ \underline{1} \\ 20 46 \\ 2 \times 22 = \underline{44} \\ (20 \times 12)\ 240 2\ 54 \\ 1 \times 241 = \underline{2\ 41} \\ (20 \times 121)\ 2420 13\ 00 \\ \underline{ 0} \\ 13\ 00 \end{array}$$

19.
$$\begin{array}{r} 5\ 3\ 0 . 4 \\ \sqrt{28_\wedge 13_\wedge 52.00} \\ \underline{25} \\ (20 \times 5)\ 100 3\ 13 \\ 3 \times 103 = \underline{3\ 09} \\ (20 \times 53)\ 1060 4\ 52 \\ \underline{ 0} \\ (20 \times 530)\ 10600 4\ 52\ 00 \\ 4 \times 10604 = \underline{4\ 24\ 16} \\ 27\ 84 \end{array}$$

T-405   Chapter 9   *Mathematics and Finances*

LESSON 120

Objective

- To teach *how to reconcile a checking account.

Review

1. *Find the hypotenuse of each right triangle, to the nearest tenth.* (Lesson 112)

   a. leg a = 11 in.
       leg b = 9 in.
       hypotenuse = (14.2 in.)

   b. leg a = 6 ft.
       leg b = 7 ft.
       hypotenuse = (9.2 ft.)

2. *Find the surface area of each sphere.* (Lesson 104)

   a. r = 14 in.
       a_s = (2,464 sq. in.)

   b. r = 38 ft.
       a_s = (18,136.64 sq. ft.)

3. *Construct these triangles.* (Lesson 88)

   a. 1 in., $1\frac{1}{2}$ in., 1 in.

   b. 5 cm, 3 cm, 3 cm

4. *Find the actual distance represented by each measurement.* (Lesson 56)

   a. Scale: 1 in. = 28 mi.; measurement on map: $2\frac{1}{4}$ in. (63 mi.)

   b. Scale: 1 in. = $4\frac{1}{2}$ mi.; measurement on map: $6\frac{1}{2}$ in. ($29\frac{1}{4}$ mi.)

120. Checking Accounts: Reconciling the Account

In Lesson 119 you read about the check register, which is used to record each transaction and to keep record of the checking account balance. The bank also records each check and deposit and adjusts the account balance accordingly.

Each month the bank sends a **bank statement** to the account holder. The statement shows the balance at the beginning of the month, each check or deposit that has been entered, and the ending balance according to the bank's records. The individual needs to compare his check register with the bank statement to make sure they agree. This is called **reconciling the account.**

Compare the following check register with the items shown on the statement below it. Notice in the register that a check mark is placed after each item shown on the bank statement.

Darwin or Louise Shank, Check Register

NUMBER	DATE	DESCRIPTION	✓	FEE	CHECK AMOUNT	DEPOSIT AMOUNT	BALANCE
		Balance carried forward					$758.98
D	11/20	Deposit—Sales	✓			1,038.41	1,797.39
433	11/26	Handy Hardware	✓		19.87		1,777.52
434	11/27	Wellsville Grocery	✓		92.68		1,684.84
435	11/28	Duville Plumbing Supply			41.56		1,643.28
436	11/29	Lagrange Fuel			456.87		1,186.41
437	11/29	Dartmouth Repair	✓		546.01		640.40
D	11/30	Deposit—Sales				255.00	895.40
C	11/30	Nov. service charge	✓	5.75			889.65

Darwin or Louise Shank, Bank Statement

Date	Description	Amount	Balance
11/20	Deposit	1,038.41	1,797.39
11/26	Check 433	19.87	1,777.52
11/27	Check 434	92.68	1,684.84
11/29	Check 437	546.01	1,138.83
11/30	Service charge	5.75	1,133.08
11/30	Ending balance		1,133.08

Items listed in the register but not on the statement are said to be outstanding. The **outstanding checks** (numbers 435 and 436) have not yet reached the bank for payment. The **outstanding deposit** made on 11/30 was not recorded on this month's statement.

406   Chapter 9   Mathematics and Finances

To reconcile the account, first of all enter any bank charges or fees in the register, and calculate the new balance. Then begin with the ending balance on the statement. Add the total amount of outstanding deposits, and subtract the total amount of outstanding checks. The adjusted balance should agree with the balance in the check register.

Most bank statements include a reconciliation form similar to the one below, as a help in reconciling the account. The following form reconciles the register and the statement above.

Ending bank balance	1,133.08
Add outstanding deposits	255.00
Total	1,388.08
Subtract outstanding checks	498.43
Adjusted balance	889.65

The adjusted balance, $889.65, agrees with the register balance.

CLASS PRACTICE   (Forms are provided in quiz booklet.)

Use reconciliation forms to reconcile the following accounts. Write yes *or* no *to tell whether the balances agree.*

Ending bank balance		Deposits outstanding		Checks outstanding		Ending register balance	
a. $264.84	+	$286.31	−	$121.53	=	$439.62	no
b. $564.85	+	$915.36	−	$573.41	=	$1,336.42	no
c. $385.39	+	$362.75	−	$382.51	=	$365.63	yes
d. $928.42	+	$637.22	−	$742.32	=	$824.32	no

WRITTEN EXERCISES

A. *Use reconciliation forms to reconcile the following accounts. Write* yes *or* no *to tell whether the balances agree.*

Ending bank balance		Deposits outstanding		Checks outstanding		Ending register balance	
1. $695.34	+	$185.62	−	$181.11	=	$699.85	yes
2. $585.22	+	$962.50	−	$661.36	=	$886.35	no
3. $493.39	+	$315.49	−	$333.39	=	$475.59	no
4. $979.13	+	$227.95	−	$268.63	=	$938.45	yes
5. $395.36	+	$868.43	−	$1,049.23	=	$215.46	no
6. $256.94	+	$339.58	−	$436.76	=	$159.76	yes

Lesson 120 T–406

Introduction

Is there any reason that the checkbook balance should be different from the bank's records for that account? Is it possible for the bank's balance and the individual's balance to be different without any mistake in either set of records?

The answer is yes. Can the students tell why?

(1) Some checks might be outstanding, not yet having come in to be cleared.

(2) The bank may have assessed service charges that the individual has not entered in his register.

(3) The bank may have paid interest that the individual has not entered in his register.

(4) The bank may not yet have recorded a deposit, especially if the individual sent it by mail or used the night depository.

Teaching Guide

1. **Each month the bank sends a bank statement to the account holder.** The statement shows the balance at the beginning of the month, each check or deposit that has been entered, and the ending balance according to the bank's records. The individual needs to compare his check register with the bank statement to make sure they agree. This is called *reconciling the account*.

2. **Outstanding checks and deposits are transactions that have been entered in the check register but are not yet on the bank statement.**

3. **Reconciling a checking account involves several steps.** Begin with the ending balance on the statement. Add the total amount of outstanding deposits, and subtract the total amount of outstanding checks. The adjusted balance should agree with the balance in the check register.

Calculations for CLASS PRACTICE

a. Total: 551.15; Adjusted balance: 429.62

b. Total: 1,480.21; Adjusted balance: 906.80

c. Total: 748.14; Adjusted balance: 365.63

d. Total: 1,565.64; Adjusted balance: 823.32

Calculations for Part A

1. Total: 880.96; Adjusted balance: 699.85

2. Total: 1,547.72; Adjusted balance: 886.36

3. Total: 808.88; Adjusted balance: 475.49

4. Total: 1,207.08; Adjusted balance: 938.45

5. Total: 1,263.79; Adjusted balance: 214.56

6. Total: 596.52; Adjusted balance: 159.76

T–407 *Chapter 9 Mathematics and Finances*

Use the exercises below and in *Class Practice* to teach these principles.

a. Ending bank balance: $493.39
    Deposit outstanding: $315.49
        (Total: $808.88)
    Checks outstanding:
        Number 679 $161.13
        Number 681   22.34
        (Total checks: $183.47)
    Ending register balance: $625.41
        (Adjusted balance: $625.41; balances agree)

b. Ending bank balance: $395.36
    Deposits outstanding: $682.81
                          578.34
        (Total: $1,656.51)

Checks outstanding:
    Number 89 $171.79
    Number 593   862.49
    Number 595 $ 14.95
    (Total checks: $1,049.23)
Ending register balance: $617.28
    (Adjusted balance: $607.28; balances do not agree)

4. **If the statement and the checkbook cannot be reconciled, review all the steps you took.** Recheck the calculations done in the check register; review all transactions cleared; and look for any missing, duplicate, or improper entries. If you cannot find the error, ask the bank for assistance.

Solutions for Part B

7. $648.22 + $236.44 − ($28.35 + $106.89 + $65.04 + $34.76)

8. $517.39 + $214.86 + $71.51 − $267.13

9. $2,882.68 + $83.62 − ($77.33 + $17.92 + $49.00)

10. $36^2 + 26^2 = 1{,}972$; $\sqrt{1{,}972}$

11. $4 \times 3.14 \times 10^2$

12. $4 \times 3\frac{1}{7} \times 4\frac{3}{4} \times 4\frac{3}{4}$

Model triangles for Part E

17.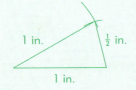

18.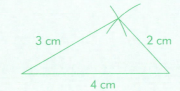

B. Solve these reading problems. Be careful, for some of them contain extra information.

7. Justin's bank statement shows an ending balance of $648.22. A deposit of $236.44 is outstanding. Also outstanding are the following checks: $28.35, $106.89, $65.04, and $34.76. What should be the balance in his check register? $649.62

8. The ending balance on Randall's bank statement is $517.39. Check number 489 for $267.13 is outstanding. Deposits of $214.86 and $71.51 are also outstanding. What should be the balance in his check register? $536.63

9. The ending balance on Marilyn's bank statement is $2,882.68. Outstanding are checks for $77.33, $17.92, and 49.00, and a deposit of $83.62. Find what the balance in her check register should be. $2,822.05

10. A set of shelves is supported by a number of **X** braces. The shelves are 36 inches wide, and the top end of each brace is 26 inches above the bottom end. How long is each arm of one **X** brace, to the nearest whole inch?   44 inches

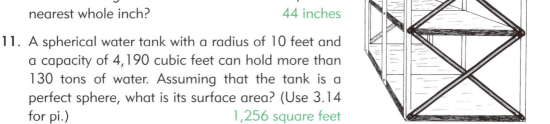

11. A spherical water tank with a radius of 10 feet and a capacity of 4,190 cubic feet can hold more than 130 tons of water. Assuming that the tank is a perfect sphere, what is its surface area? (Use 3.14 for pi.)   1,256 square feet

12. A basketball has a circumference of $2\frac{1}{2}$ feet, a diameter of about $9\frac{1}{2}$ inches, and a weight of about 21 ounces. Find its surface area, using $3\frac{1}{7}$ for pi. $283\frac{9}{14}$ square inches

REVIEW EXERCISES

C. Find the hypotenuse of each right triangle, to the nearest tenth. *(Lesson 112)*

13. leg a = 10 in.   12.2 in.
    leg b = 7 in.

14. leg a = 9 ft.   12.7 ft.
    leg b = 9 ft.

D. Find the surface area of each sphere. Use 3.14 for pi. *(Lesson 104)*

15. r = 11 in.   1,519.76 sq. in.

16. r = 27 ft.   9,156.24 sq. ft.

E. Use each set of facts to construct a triangle. Label it with the facts given. *(Lesson 88)*   (See facing page for model triangles.)

17. 1 in., $\frac{1}{2}$ in., 1 in.

18. 4 cm, 3 cm, 2 cm

F. Find the actual distance represented by each measurement. *(Lesson 56)*

19. Scale: 1 in. = 16 mi.; measurement on map: $1\frac{3}{8}$ in.   22 mi.

20. Scale: 1 in. = $5\frac{1}{2}$ mi.; measurement on map: $4\frac{3}{4}$ in. $26\frac{1}{8}$ mi.

121. Savings Accounts and Simple Interest

Living within our means is an important part of our responsibility as stewards. It is wise to save some of the money God gives us.

A **savings account** yields interest on the money deposited in it. Most savings accounts do not allow check-writing privileges, and most checking accounts pay no interest. However, some checking accounts today do yield interest, and some savings accounts have limited check-writing privileges.

Passbook savings, statement savings, and certificates of deposit are some of the common types of savings accounts. The structure and function of these accounts vary slightly.

Deposit tickets for savings accounts are usually similar to deposit tickets for checking accounts. Notice how the following deposit is prepared. On this ticket, the coins and currency are combined as "Cash."

SAVINGS DEPOSIT			
DATE 2/4 20 —	CASH	290	25
	CHECKS		
Daryl Martin	44-313	75	25
Account name	31-216	80	00
	62-333	125	00
21-2134125			
Account number	SUBTOTAL	570	50
	LESS CASH RECEIVED	50	00
The Longville Bank	TOTAL DEPOSIT	520	50

When a person deposits money into a savings account, he is letting the bank use his money. The money earns interest because the bank can invest the money or lend it to another person at a higher rate of interest. Following are the three main terms used in working with interest.

Principal—Amount of money borrowed.
Rate—Percent used to calculate interest, usually for one year.
Time—Length of the loan, usually in units of years.

This lesson works with **simple interest,** which is kept separate from the principal. To find the amount of interest, multiply the principal, the rate of interest, and the time.

> **Memorize this formula for finding interest.**
> interest = principal × rate × time, or $i = prt$

LESSON 121

Objectives

- To teach *how to complete deposit tickets for savings accounts.
- To practice calculating simple interest, using annual rates and *daily rates.

Review

1. *Reconcile this account balance.* (Lesson 120)

   Statement balance: $634.82
   Deposits outstanding: 362.85
       (Total: $997.67)
   Checks outstanding: 430.71
       (Adjusted balance: $566.96)
   Check register balance: 566.96
       (The two balances agree.)

2. *Find the length of leg* a *in each triangle, to the nearest tenth.* (Lesson 113)

   a. leg b = 12 in.
      hypotenuse = 16 in.
      leg a = (10.6 in.)

   b. leg b = 54 ft.
      hypotenuse = 75 ft.
      leg a = (52.0 ft.)

3. *Find the volume of each cube or rectangular solid.* (Lesson 105)

   a. e = 35 in.
      v = (42,875 cu. in.)

   b. e = 80 ft.
      v = (512,000 cu. ft.)

   c. l = 15 ft.
      w = 8 ft.
      h = 25 ft.
      v = (3,000 cu. ft.)

   d. l = 63 m
      w = 55 m
      h = 11 m
      v = (38,115 m³)

4. *Do these exercises on reading blueprints.* (Lesson 57)

   Scale: 1 in. = 8 ft.

   a. Scale distance: $4\frac{3}{8}$ in.; actual distance: ___ (35 ft.)

   b. Scale distance: $2\frac{3}{4}$ in.; actual distance: ___ (22 ft.)

   Scale: $\frac{1}{4}$ in. = 1 ft.

   c. Actual distance: $5\frac{1}{2}$ ft.; scale distance: ___ ($1\frac{3}{8}$ in.)

   d. Actual distance: 9 ft.; scale distance: ___ ($2\frac{1}{4}$ in.)

Introduction

Ask the students how a savings account differs from a checking account. The general answer is that a savings account bears interest and usually does not include check-writing privileges. A checking account usually does not bear interest.

Banking regulations have been changed in the last several decades to allow interest on some checking accounts and a limited amount of check-writing privileges on savings accounts. Thus the line of demarcation that once existed between checking accounts and savings accounts is not as sharp as it was formerly.

T–409 Chapter 9 Mathematics and Finances

Teaching Guide

1. **A savings account yields interest on the money deposited in it.** Passbook savings, statement savings, and certificates of deposit are a few of the most common types of savings accounts.

2. **Deposit tickets for savings accounts are usually similar to deposit tickets for checking accounts.** On the deposit ticket shown in the lesson, coins and currency are combined as "Cash."

   Use the exercises below and in *Class Practice* to illustrate these principles.

   a. Name: Gary McFarland
      Date: 9/25
      Account number: 31-42536475
      Items deposited
      Coins $ 25.41
      Currency   237.00
      Checks
         13-468   313.91
         46-793   136.54
         79-651   246.82
      Cash Received: 100.00
         (Total deposit: $859.68)

   b. Name: Sanford Myer
      Date: 10/15
      Account number: 97-54651232
      Items deposited
      Coins $131.46
      Currency   343.00
      Checks
         87-542   213.65
         65-231   545.24
         48-658   1,947.41
      Cash Received   200.00
         (Total deposit: $2,980.76)

Answers for CLASS PRACTICE a and b

a.

SAVINGS DEPOSIT		CASH	122	35
DATE [current date] 20 ——		CHECKS		
		15-151	25	56
[student's name]		42-422	69	38
Account name				
09-87654321		SUBTOTAL	217	29
Account number		LESS CASH RECEIVED		
The Longville Bank		TOTAL DEPOSIT	217	29

b.

SAVINGS DEPOSIT		CASH	114	42
DATE [current date] 20 ——		CHECKS		
		85-225	218	54
[student's name]		65-357	185	34
Account name				
09-87654321		SUBTOTAL	518	30
Account number		LESS CASH RECEIVED		
The Longville Bank		TOTAL DEPOSIT	518	30

> **Example A**
> Find the interest on $3,000 at 5% for 3 years.
> $i = prt$
> $i = \$3{,}000 \times 0.05 \times 3$
> $i = \$450.00$

> **Example B**
> Find the interest on $4,500 at $4\frac{1}{2}$% for 4 years.
> $i = prt$
> $i = \$4{,}500 \times .045 \times 4$
> $i = \$810.00$

The interest rate is usually based on one year. For example, the interest on $1,000 at 10% for 1 year is $100. However, shorter time periods such as the month or day are sometimes used. Businesses often add a service charge of 1.5% per month (18% per year) to overdue bills. Interest charged on an unpaid credit card balance may be 0.0493% per day (also about 18% per year).

Using a time period shorter than a year makes the interest rate look small. But the interest adds up faster than a person may think. The "small" daily rate of 0.0685% in Example C is actually 25% per year!

When the rate is based on a period less than a year, use the same interest formula as usual ($i = prt$). Be sure to move the decimal point in the rate before you multiply. (See Example C.) If the answer includes a fraction of a cent, round to the nearest cent.

> **Example C**
> Find the interest on $5,000 at a daily rate of 0.0685% for 45 days.
> $i = prt$
> $i = \$5{,}000 \times 0.000685 \times 45$
> $i = \$154.13$ (rounded)

CLASS PRACTICE

Complete deposit tickets for the following deposits. Use your own name and today's date. (Forms are provided in quiz booklet. See facing page for answers.)

a. Account number: 09-87654321
Items deposited
  Coins $18.35
  Currency 104.00
  Checks
    15-151 .... 25.56
    42-422 .... 69.38

b. Account number: 09-87654321
Items deposited
  Coins $36.42
  Currency 78.00
  Checks
    85-225 .... 218.54
    65-357 .... 185.34

Find the interest.

c. $p = \$2{,}100$ $126.00
   $r = 3\%$
   $t = 2$ yr.

d. $p = \$1{,}500$ $90.00
   $r = 6\%$
   $t = 1$ yr.

e. $p = \$525$ $63.00
   $r = 4\%$
   $t = 3$ yr.

f. $p = \$12{,}000$ $34.32
   $r = 0.0286\%$ per day
   $t = 10$ days

410 Chapter 9 Mathematics and Finances

WRITTEN EXERCISES

A. Complete deposit tickets for the following deposits. Use your own name and today's date.
(See facing page for answers.)

1. Account number: 09-87654321
   Items deposited:
   Coins $25.57
   Currency 125.00
   Checks
       11-111 .... 32.26
       22-222 .... 41.18

2. Account number: 09-87654321
   Items deposited:
   Coins $41.32
   Currency 68.00
   Checks
       12-345 .... 115.00
       67-890 .... 122.29

3. Account number: 09-87654321
   Items deposited:
   Cash $140.00
   Checks
       23-456 .... 16.95
       78-901 .... 94.32
       34-567 .... 67.05

4. Account number: 09-87654321
   Items deposited:
   Cash $75.00
   Checks
       12-345 .... 67.96
       23-456 .... 43.39
       45-678 .... 77.75

B. Find the interest. Numbers 13–16 show daily rates.

5. $p = \$1,000$ $80.00
   $r = 4\%$
   $t = 2$ yr.

6. $p = \$1,500$ $225.00
   $r = 5\%$
   $t = 3$ yr.

7. $p = \$275$ $16.50
   $r = 6\%$
   $t = 1$ yr.

8. $p = \$765$ $76.50
   $r = 2\%$
   $t = 5$ yr.

9. $p = \$925$ $101.75
   $r = 5\frac{1}{2}\%$
   $t = 2$ yr.

10. $p = \$1,500$ $97.50
    $r = 6\frac{1}{2}\%$
    $t = 1$ yr.

11. $p = \$2,500$ $500.00
    $r = 5\%$
    $t = 4$ yr.

12. $p = \$3,600$ $324.00
    $r = 3\%$
    $t = 3$ yr.

13. $p = \$1,000$ $2.47
    $r = 0.0247\%$
    $t = 10$ days

14. $p = \$2,000$ $8.76
    $r = 0.0219\%$
    $t = 20$ days

15. $p = \$10,000$ $3.29
    $r = 0.0329\%$
    $t = 1$ day

16. $p = \$12,000$ $49.80
    $r = 0.0415\%$
    $t = 10$ days

C. Solve these reading problems. Be careful; some problems ask for the ending balance and some for the interest only.

17. Duane opened a savings account with $650. The bank pays $3\frac{1}{2}\%$ simple interest. After 2 years, what will the balance be if he adds his interest to the account? $695.50

18. Vernon had an average balance of $2,650 for the month of May. Find his interest for 28 days at a daily rate of 0.0264%. $19.59

19. Melody deposited $1,160 into a savings account that pays 2.7% simple interest. How much interest will she earn in 4 years? $125.28

3. **Money in a savings account earns interest because the bank can invest the money or lend it to another person at a higher rate of interest.** (The depositor is actually lending his money to the bank.) Following are the three main terms used in working with interest.

   Principal—Amount of money that is borrowed.
   Rate—Percent used to calculate interest, usually for one year.
   Time—Length of the loan.

4. **Simple interest is based only on the principal and not added to the principal.** To find the amount of interest, use this formula: interest = principal × rate × time, or $i = prt$. If the answer includes a fraction of a cent, round to the nearest cent.

   a. $p = \$2,500$
   $r = 4\%$
   $t = 5$ yr.
   $i = (\$500.00)$

   b. $p = \$2,000$
   $r = 7\%$
   $t = 3\frac{1}{2}$ yr.
   $i = (\$490.00)$

Answers for Part A

1.

SAVINGS DEPOSIT

DATE [current date] 20 ——

[student's name]
Account name

09-87654321
Account number

The Longville Bank

CASH	150	57
CHECKS		
11-111	32	26
22-222	41	18
SUBTOTAL	224	01
LESS CASH RECEIVED		
TOTAL DEPOSIT	224	01

2.

SAVINGS DEPOSIT

DATE [current date] 20 ——

[student's name]
Account name

09-87654321
Account number

The Longville Bank

CASH	109	32
CHECKS		
12-345	115	00
67-890	122	29
SUBTOTAL	346	61
LESS CASH RECEIVED		
TOTAL DEPOSIT	346	61

3.

CASH	140	00
CHECKS		
23-456	16	95
78-901	94	32
34-567	67	05
SUBTOTAL	318	32
LESS CASH RECEIVED		
TOTAL DEPOSIT	318	32

4.

CASH	75	00
CHECKS		
12-345	67	96
23-456	43	39
45-678	77	75
SUBTOTAL	264	10
LESS CASH RECEIVED		
TOTAL DEPOSIT	264	10

T-411 Chapter 9 Mathematics and Finances

5. **Although the time of a loan is usually stated in years, it may be expressed as a monthly or daily rate.** A common rate on charge accounts is 18% per year, which is 1.5% per month or 0.0493% per day. (Whenever a monthly or daily rate is given, the law requires that the equivalent annual rate also be stated.)

   a. $p = \$550$
      $r = 0.0247\%$ per day
      $t = 12$ days
      $i = (\$1.63)$
      (0.0247% per day = about 9% per year)

   b. $p = \$3,456$
      $r = 0.0219\%$ per day
      $t = 23$ days
      $i = (\$17.41)$
      (0.0219% per day = about 8% per year)

Further Study

1. Applying the associative law to interest problems can usually simplify the calculations. Consider the following example. Mentally multiply the rate and time; then multiply by the principal.

   $p = \$2,800; r = 5\%; t = 3$ yr.
   $\$2,800(0.05 \times 3) = \$2,800 \times 0.15 = \$420$

2. The Bible speaks about interest ("usury"), condemning it in some situations and seeming to approve it in others. Exodus 22:25, 26 forbids charging a poor person interest, especially one who struggles to have the bare necessities of life. However, when a person borrows for business purposes and makes a profit by using another person's money, it is only right that he pays interest to the lender. This is the case in the parable of the talents, which speaks favorably of receiving interest on an investment (Matthew 25:27).

Solutions for Part C

17. $\$650 \times 0.035 \times 2 = \$45.50$; $\$650 + \$45.50$

18. $\$2,650 \times 0.000264 \times 28$

19. $\$1,160 \times 0.027 \times 4$

20. $\$4,380 \times 0.0575 \times 2 = \$503.70$; $\$4,380 + \$503.70$

21. $40^2 - 18^2 = 1,276$; $\sqrt{1,276}$

22. $5 \times 2.3 \times 1.8$

20. Jacob Sullivan deposited $4,380 into a savings account. The bank will pay $5\frac{3}{4}$% interest if he leaves all the money in the account for 2 years. What will his balance be at the end of 2 years, including interest? $4,883.70

21. A 40-foot elevator is positioned so that its upper end, 18 feet above the ground, is even with the wall of a corn crib. To the nearest foot, how far is the bottom end of the elevator from the corn crib?   36 feet

22. Find the volume of a chest freezer whose interior is 5 feet long, 2.3 feet deep, and 1.8 feet wide.   20.7 cubic feet

REVIEW EXERCISES

D. Find the length of leg a in each right triangle, to the nearest tenth. (Lesson 113)

23. leg b = 10 in.   9.8 in.
    hypotenuse = 14 in.

24. leg b = 45 ft.   46.9 ft.
    hypotenuse = 65 ft.

E. Find the volume of each cube or rectangular solid. (Lesson 105)

25. e = 28 in.   21,952 cu. in.

26. e = 75 ft.   421,875 cu. ft.

27. l = 14 ft.   2,646 cu. ft.
    w = 9 ft.
    h = 21 ft.

28. l = 58 m   34,104 m³
    w = 49 m
    h = 12 m

F. Do these exercises on reading blueprints. (Lesson 57)

29. Scale: 1 in. = 4 ft.; scale distance: $5\frac{3}{8}$ in.; actual distance: $21\frac{1}{2}$ ft.

30. Scale: $\frac{1}{2}$ in. = 1 ft.; actual distance: $6\frac{1}{2}$ ft.; scale distance: $3\frac{1}{4}$ in.

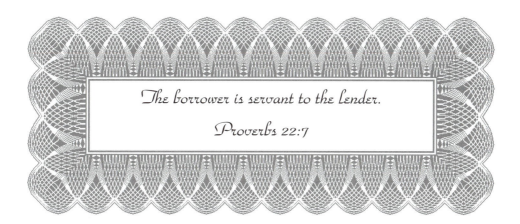

The borrower is servant to the lender.

Proverbs 22:7

122. Calculating Part-year Interest

To fill his greenhouses with plants one spring, Brother Clinton borrowed $6,000 at 4% interest for 9 months. How much interest did he need to pay? The rate for 1 year is 4%, so the interest for 1 year would be $240. Since 9 months is $\frac{3}{4}$ of a year, he needed to pay only $\frac{3}{4}$ of $240, or $180.

To calculate interest for part of a year, use the formula *interest = principal × rate × time* ($i = prt$). The number representing the time will be a fraction.

To simplify calculations, businessmen often consider a year as 12 months of 30 days each, which is a total of 360 business days. Two months is $\frac{1}{6}$ of a year, and 45 days is $\frac{45}{360}$ or $\frac{1}{8}$ of a year.

Example A
Find the interest on $3,200 at 7% for 15 months.
Think: 15 mo. = $1\frac{1}{4}$ yr.
$i = prt$
$i = 3,200 \times 0.07 \times 1\frac{1}{4}$
$i = 224 \times 1\frac{1}{4}$
$i = \$280.00$

Example B
Find the interest on $1,600 at $6\frac{1}{2}$% for 120 days.
Think: 120 days = $\frac{120}{360}$ yr. = $\frac{1}{3}$ yr.
$i = prt$
$i = 1,600 \times 0.065 \times \frac{1}{3}$
$i = 104 \times \frac{1}{3}$
$i = \$34.66\frac{2}{3}$ = \$34.67 (rounded)

CLASS PRACTICE

Find the interest.

	p	r	t			p	r	t	
a.	$600	7%	4 mo.	$14.00	b.	$1,000	9%	11 mo.	$82.50
c.	$1,500	$5\frac{1}{2}$%	7 mo.	$48.13	d.	$900	6%	1 yr. 2 mo.	$63.00
e.	$650	9%	175 days	$28.44	f.	$1,100	9%	110 days	$30.25
g.	$2,600	7%	260 days	$131.44	h.	$900	11%	275 days	$75.63

WRITTEN EXERCISES

A. Find the interest.

	p	r	t			p	r	t	
1.	$500	8%	5 mo.	$16.67	2.	$800	7%	8 mo.	$37.33
3.	$900	9%	7 mo.	$47.25	4.	$600	10%	4 mo.	$20.00
5.	$1,100	8%	3 mo.	$22.00	6.	$1,400	9%	6 mo.	$63.00
7.	$2,200	5%	7 mo.	$64.17	8.	$2,600	6%	11 mo.	$143.00

LESSON 122

Objective

- To review calculating interest for part of a year by using the number of months or days.

Review

1. Give Lesson 122 Speed Test (Horizontal Addition).

2. *Find the interest.* (Lesson 121)

   a. $p = \$7,500; r = 6\%; t = 5$ yr.
      $i = (\$2,250)$

   b. $p = \$3,400; r = 4\%; t = \frac{1}{4}$ yr.
      $i = (\$34.00)$

3. *Solve these parallel problems.* (Lesson 114)

   a. The load limit of a trailer truck is 52,250 pounds. How many 55-pound bags of powdered milk can the truck haul? (950 bags)

   b. The load limit of a certain truck is 2,400 pounds. How many books can the truck haul if each book weighs $\frac{3}{4}$ pound? (3,200 books)

4. *Find the volume of each figure.* (Lesson 106)

   a. Cylinder: $r = 12$ in.
      $h = 50$ in.
      $v = (22,608$ cu. in.)

   b. Cone: $r = 1\frac{1}{2}$ ft.
      $h = 4$ ft.
      $v = (9.42$ cu. ft.)

5. *Find the perimeter of each square or rectangle.* (Lesson 90)

   a. $l = 34$ in.
      $w = 28$ in.
      $p = (124$ in.)

   b. $s = 7.5$ m
      $p = (30$ m)

Introduction

Review calculating interest for a full year.

$p = \$1,000; r = 8\%; t = 1$ yr.

$i = \$1,000 \times 0.08 \times 1 = \$80$

Now calculate interest for a quarter year.

$p = \$2,000; r = 6\%; t = \frac{1}{4}$ yr.

$i = \$2,000 \times 0.06 \times \frac{1}{4} = \$30$

Ask the students, "How would interest be calculated for 10 months?" Interest for part of a year is calculated by stating the time as a fraction of an entire year. Ten months equals $\frac{10}{12}$ year, which reduces to $\frac{5}{6}$ year.

$p = \$3,000; r = 8\%; t = 10$ mo. $= \frac{10}{12}$ yr. $= \frac{5}{6}$ yr.

$i = \$3,000 \times 0.08 \times \frac{5}{6} = \$200$

Teaching Guide

1. **To calculate interest for part of a year, state the time as a fraction of a year and use the basic interest formula:** $i = prt$.

   a. $p = \$1,300$; $r = 7\%$; $t = 5$ mo.
   $i = (\$37.92)$

   b. $p = \$2,300$; $r = 6\%$; $t = 9$ mo.
   $i = (\$103.50)$

   c. $p = \$1,300$; $r = 5\%$; $t = 1$ yr. 5 mo.
   $i = (\$92.08)$

   d. $p = \$1,500$; $r = 4\frac{1}{2}\%$; $t = 1$ yr. 6 mo.
   $i = (\$101.25)$

2. **Businessmen often consider a year as 12 months of 30 days each, or a total of 360 business days.** To write a given number of days as a fractional part of a year, write the number over the denominator of 360 and reduce the fraction.

   a. $p = \$1,200$; $r = 6\%$; $t = 180$ days
   ($t = \frac{1}{2}$ yr.; $i = \$36.00$)

   b. $p = \$2,400$; $r = 7\%$; $t = 200$ days
   ($t = \frac{5}{9}$ yr.; $i = \$93.33$)

   c. $p = \$1,400$; $r = 8\%$; $t = 210$ days
   ($t = \frac{7}{12}$ yr.; $i = \$65.33$)

   d. $p = \$1,600$; $r = 6\frac{1}{2}\%$; $t = 250$ days
   ($t = \frac{25}{36}$ yr.; $i = \$72.22$)

Further Study

A variation of the double-and-divide shortcut can sometimes simplify calculation.

Principal: $1,500
Rate: 6%
Time: $\frac{1}{3}$ year
$\$1,500 \times 6\% = \$15 \times 6 = \$90$; $\frac{1}{3} \times \$90 = \$30$

Another shortcut is to mentally multiply the rate by the time.

$\$1,500 \times 6\% \times \frac{1}{3}$ yr. $= \$1,500 \left(\frac{1}{3} \times 6\%\right) =$
$\$1,500 \times 2\% = \$30$

Solutions for Part C

19. $(\$8,500 \times 0.054 \times 1\frac{1}{2}) + \$8,500$

20. $\$1,800 \times 0.03 \times \frac{1}{6}$

21. $\$4,000 \times 0.07 \times \frac{1}{4}$

22. $(\$2,500 \times 0.12 \times \frac{1}{8}) + \$2,500$

23. 15×6

24. $28 \times \frac{1}{9}$

	p	r	t			p	r	t	
9.	$1,600	4%	1 yr. 3 mo.	$80.00	10.	$1,500	8%	1 yr. 4 mo.	$160.00
11.	$1,600	7%	1 yr. 6 mo.	$168.00	12.	$1,800	8%	1 yr. 8 mo.	$240.00

B. Find the interest by using the 360-day year.

	p	r	t			p	r	t	
13.	$400	7%	180 days	$14.00	14.	$700	8%	90 days	$14.00
15.	$800	8%	75 days	$13.33	16.	$500	11%	225 days	$34.38
17.	$1,200	8%	300 days	$80.00	18.	$1,300	10%	200 days	$72.22

C. Solve these reading problems. Numbers 23 and 24 are parallel problems.

19. Mr. Metzler deposited $8,500 in a savings account that pays 5.4% interest. If the interest is added to the principal after 18 months, what will the balance be? $9,188.50

20. Sister Shirley receives 3% interest on her checking account. How much interest does she earn in 2 months if her average balance is $1,800? $9.00

21. Brother Jacob borrowed $4,000 at 7% interest for 90 days. How much interest did he need to pay? $70.00

22. Wilmer Shirk borrowed $2,500 at 12% interest. What is the total principal and interest that he owes at the end of 45 days? $2,537.50

23. The shipmen in Acts 27:28 measured the depth of the water and found it to be 15 fathoms. One fathom is about 6 feet. How deep was the water in feet? 90 feet

24. According to John 6:19, the disciples had rowed about 28 furlongs when they saw Jesus walking toward them. A furlong is about $\frac{1}{9}$ mile. How many miles had they rowed? $3\frac{1}{9}$ miles

REVIEW EXERCISES

D. Find the interest. *(Lesson 121)*

25. p = $6,500 $1,300.00
    r = 5%
    t = 4 yr.

26. p = $2,800 $49.00
    r = 7%
    t = $\frac{1}{4}$ yr.

E. Find the volume of each figure. *(Lesson 106)*

Cylinders

27. r = 16 in.; h = 60 in.   48,230.4 cu. in.
    π = 3.14

28. r = 11 ft.; h = 35 ft.   13,310 cu. ft.
    π = $3\frac{1}{7}$

Cones

29. r = $2\frac{1}{2}$ ft.; h = 7 ft. $45\frac{5}{6}$ cu. ft.
    π = $3\frac{1}{7}$

30. r = 30 cm; h = 85 cm   80,070 cm³
    π = 3.14

F. Find the perimeter of each square or rectangle. *(Lesson 90)*

31. l = 45 in.   166 in.
    w = 38 in.

32. s = 11.5 m   46 m

123. Calculating Compound Interest

Father deposited $2,000 into a savings account for 1 year at $6\frac{1}{2}\%$ interest. How much interest did he earn? Example A works out the answer with simple interest.

> **Example A**
> Find simple interest on $2,000 at $6\frac{1}{2}\%$ for 1 year.
> $i = prt$
> $i = \$2,000 \times 0.065 \times 1$
> $i = \$130$

Banks commonly pay interest more frequently than once a year. At regular intervals throughout the year, interest is calculated for a fraction of the year and that amount is paid. It may be semiannually (twice a year), quarterly (four times a year), or monthly. Each time interest is paid, it is added to the principal and earns interest at the same rate as the original principal. This is called **compound interest**. Example B works out the answer to the problem above using compound interest.

> **Example B**
> Find the interest on $2,000 at $6\frac{1}{2}\%$ compounded semiannually for 1 year.
>
> First half of year
> $i = prt$
> $i = \$2,000 \times 0.065 \times 0.5$
> $i = \$65$
> $\$65 + \$2,000 = \$2,065$
> The new principal is $2,065.
>
> Second half of year
> $i = prt$
> $i = \$2,065 \times 0.065 \times 0.5$
> $i = \$67.11$ (rounded)
> Total interest = $65.00 + $67.11
> Total interest = $132.11

Compare Examples A and B. Compound interest yields more than simple interest because the principal increases for each new interest period.

To calculate compound interest, use the interest formula ($i = prt$). The time in the formula is the length of the interest period, not the total time of the loan. The length of the interest period for compounding semiannually is $\frac{1}{2}$ year, and for compounding quarterly it is $\frac{1}{4}$ year.

Remember to round to the nearest cent after each calculation. Also remember to add the interest to the principal for each new interest period.

LESSON 123

Objective

- To review calculating compound interest.

Review

1. *Find the interest.* (Lesson 122)

   a. $p = \$1,500$; $r = 10\%$; $t = 7$ mo.
      $i = (\$87.50)$

   b. $p = \$2,000$; $r = 4\frac{1}{2}\%$; $t = 5$ mo.
      $i = (\$37.50)$

2. *Find the volume of each square pyramid.* (Lesson 107)

   a. $b = 15$ in.
      $h = 26$ in.
      $v = (1,950$ cu. in.$)$

   b. $b = 47$ ft.
      $h = 98$ ft.
      $v = (72,160.67$ cu. ft.$)$

3. *Find the circumference of each circle.* (Lesson 91)

   a. $r = 19$ cm
      $c = (119.32$ cm$)$

   b. $r = 5\frac{1}{2}$ ft.
      $c = (34\frac{4}{7}$ ft.$)$

Introduction

Suppose someone borrowed $1,000 at 4% interest for 4 years. At the end of that time, the simple interest formula would be used to determine the amount of the repayment.

$i = prt$, or $i = \$1,000 \times 0.04 \times 4 = \$160$
Total repaid = $\$1,000 + \$160 = \$1,160$

Now suppose someone puts $1,000 into a savings account at 4% interest for 4 years. The bank adds the interest to the account balance at the end of each year. Would the interest be $160 in this case too? No, because the principal increases each year, and this will cause the interest also to increase.

First year
   $\$1,000.00 \times 4\% = \$40.00$;
   new balance = $\$1,040.00$
Second year
   $\$1,040.00 \times 4\% = \$41.60$;
   new balance = $\$1,081.60$
Third year
   $\$1,081.60 \times 4\% = \$43.26$;
   new balance = $\$1,124.86$
Fourth year
   $\$1,124.86 \times 4\% = \$44.99$;
   new balance = $\$1,169.85$

The interest comes to $169.85 rather than $160.00. It is called compound interest when it is calculated this way. Compound interest is generally used when interest is added to the balance in an account rather than paying it out.

T-415   Chapter 9   Mathematics and Finances

Teaching Guide

1. A bank pays compound interest when it adds interest to the balance in a savings account, and this interest becomes part of the principal to earn interest in the next period.

2. To calculate compound interest, use the interest formula ($i = prt$). For compound interest, the time is the length of the interest period rather than the total time of the loan.

   a. Calculate the interest and ending balance for $500 in an account paying 5% interest compounded annually for 4 years. Point out that the last factor in the multiplication is 1 because interest is calculated once a year.
   First year
    $500.00 × 0.05 × 1 = $25.00;
    new balance = $525.00
   Second year
    $525.00 × 0.05 × 1 = $26.25;
    new balance = $551.25
   Third year
    $551.25 × 0.05 × 1 = $27.56;
    new balance = $578.81
   Fourth year
    $578.81 × 0.05 × 1 = $28.94;
    new balance = ($607.75 ending balance)
   (Interest = $25.00 + $26.25 + $27.56 + $28.94 = $107.75)

   b. Calculate the interest and ending balance for $825 in an account paying 6% interest compounded semiannually for 2 years. Point out that the last factor in the multiplication is $\frac{1}{2}$ (0.5) because interest is calculated every half year. Also point out that a step could be saved by multiplying by 0.03, which is equal to 0.06 × 0.5.
   First half of year
    $825.00 × 0.06 × 0.5 = $24.75;
    new balance = $849.75
   Second half of year
    $849.75 × 0.06 × 0.5 = $25.49;
    new balance = $875.24
   Third half of year
    $875.24 × 0.06 × 0.5 = $26.26;
    new balance = $901.50
   Fourth half of year
    $901.50 × 0.06 × 0.5 = $27.05;
    new balance = ($928.55 ending balance)
   (Interest = $24.75 + $25.49 + $26.26 + $27.05 = $103.55)

   c. Calculate the interest and ending balance for $1,500 in an account paying 7% interest compounded quarterly for 1 year.
   First quarter
    $1,500.00 × 0.07 × 0.25 = $26.25;
    new balance = $1,526.25
   Second quarter
    $1,526.25 × 0.07 × 0.25 = $26.71;
    new balance = $1,552.96
   Third quarter
    $1,552.96 × 0.07 × 0.25 = $27.18;
    new balance = $1,580.14
   Fourth quarter
    $1,580.14 × 0.07 × 0.25 = $27.65;
    new balance = ($1,607.79 ending balance)
   (Interest = $26.25 + $26.71 + $27.18 + $27.65 = $107.79)

> **Example C**
> Find the interest on $850 at 7% compounded quarterly for 1 year.
>
> First quarter
> $i = prt$
> $i = \$850 \times 0.07 \times 0.25$
> $i = \$14.88$ (rounded)
>
> Second quarter
> $i = prt$
> $i = \$864.88 \times 0.07 \times 0.25$
> $i = \$15.14$ (rounded)
>
> Third quarter
> $i = prt$
> $i = \$880.02 \times 0.07 \times 0.25$
> $i = \$15.40$ (rounded)
>
> Fourth quarter
> $i = prt$
> $i = \$895.42 \times 0.07 \times 0.25$
> $i = \$15.67$ (rounded)
>
> Total interest = $14.88 + $15.14 + $15.40 + $15.67 = $61.09

CLASS PRACTICE

Find the total interest if it is compounded annually.

	p	r	t			p	r	t	
a.	$2,000	7%	2 yr.	$289.80	b.	$1,500	6%	3 yr.	$286.52

Find the total interest if it is compounded semiannually.

	p	r	t			p	r	t	
c.	$2,000	8%	1 yr.	$163.20	d.	$6,000	4%	2 yr.	$494.60

Find the total interest if it is compounded quarterly.

	p	r	t			p	r	t	
e.	$3,000	5%	1 yr.	$152.83	f.	$1,000	9%	1 yr.	$93.08

WRITTEN EXERCISES

A. *Find the total interest if it is compounded annually.*

	p	r	t			p	r	t	
1.	$1,000	8%	2 yr.	$166.40	2.	$2,000	6%	2 yr.	$247.20
3.	$1,200	4%	4 yr.	$203.83	4.	$3,600	9%	4 yr.	$1,481.69

B. *Find the total interest if it is compounded semiannually.*

	p	r	t			p	r	t	
5.	$2,000	7%	1 yr.	$142.45	6.	$3,000	7%	1 yr.	$213.68

C. *Find the total interest if it is compounded quarterly.*

	p	r	t			p	r	t	
7.	$4,500	9%	1 yr.	$418.88	8.	$5,500	7%	1 yr.	$395.22

416 Chapter 9 Mathematics and Finances

D. Solve these reading problems. When calculating compound interest, round to the nearest cent after each calculation.

9. Ann Weaver deposited $350 into a new savings account that paid 6% interest compounded annually. Find the interest earned in 2 years. $43.26

10. Nevin Witmer deposited $750 into a new savings account that paid 5% interest compounded quarterly. Find his balance at the end of $\frac{1}{2}$ year. $768.87

11. Regina opened a savings account with $350 at a bank that paid 5% interest compounded quarterly. How much interest will she earn in 1 year? $17.84

12. Aaron deposited $4,000 into a new savings account paying 4% interest compounded semiannually. What will his balance be at the end of 1 year? $4,161.60

13. The main mirror of a large telescope on one of the Canary Islands is 165 inches in diameter. Find the circumference of the mirror in feet. 43.175 feet

14. The cornea (bulge on the front) of an adult's eye has a radius of 7.7 millimeters. What is the circumference of the cornea? 48.356 millimeters

REVIEW EXERCISES

E. Find the simple interest. *(Lesson 122)*

	p	r	t			p	r	t	
15.	$1,700	5%	5 mo.	$35.42	16.	$1,000	12%	90 days	$30.00

F. Find the volume of each square pyramid. *(Lesson 107)*

17. s = 14 in.   1,764 cu. in.
    h = 27 in.

18. s = 64 ft.   143,360 cu. ft.
    h = 105 ft.

G. Find the circumference of each circle. *(Lesson 91)*

19. r = 21 cm   132 cm
    π = $3\frac{1}{7}$

20. r = 87 ft.   546.36 ft.
    π = 3.14

Lesson 123   T–416

An Ounce of Prevention

1. Beware of overloading the students with their homework assignment. Just a few exercise numbers on compound interest can mean a whole page of calculations.

2. To have accurate answers, it is important that students use the rounded amount for each new interest period. It might seem that there is greater accuracy in carrying the exact decimal to the final calculation; but in practical use, each interest period ends with a deposit to the principal. That deposit amount is rounded to the nearest whole cent, and the new principal (recorded as dollars and cents) is the basis for the next interest period.

Solutions for Part D

9. First year: $350 × 0.06 = $21
   Second year: $371 × 0.06 = $22.26
   $21.00 + $22.26

10. First quarter: $750 × 0.05 × 0.25 = $9.38
    Second quarter: $759.38 × 0.05 × 0.25 = $9.49
    $750.00 + $9.38 + $9.49

11. First quarter: $350 × 0.05 × 0.25 = $4.38
    Second quarter: $354.38 × 0.05 × 0.25 = $4.43
    Third quarter: $358.81 × 0.05 × 0.25 = $4.49
    Fourth quarter: $363.30 × 0.05 × 0.25 = $4.54
    $4.38 + $4.43 + $4.49 + $4.54

12. First half: $4,000 × 0.04 × 0.5 = $80.00
    Second half: $4,080 × 0.04 × 0.5 = $81.60
    $4,000 + $80.00 + $81.60

13. 165 ÷ 12 × 3.14

14. 3.14 × 15.4

T-417 Chapter 9 Mathematics and Finances

LESSON 124

Objective

- To teach *calculating compound interest by using the compound interest formula:
  $a = p(1 + r)^n$.

Review

1. *Find the total interest paid if interest is compounded semiannually.* (Lesson 123)
   a. $3,000 at 6% for 1 yr. ($182.70)
   b. $4,500 at 9% for 1 yr. ($414.11)
2. *Find the simple interest.* (Lesson 122)
   a. $4,000 at 5% for 7 mo. ($116.67)
   b. $6,000 at 7% for 5 mo. ($175.00)
3. *Find the volume of each sphere.* (Lesson 108)
   a. $r = 7$ in.
      $v = (1,433.74$ cu. in.$)$
   b. $r = 12$ m
      $v = (7,223.04$ m^3)
4. *Find the area of each square or rectangle.* (Lesson 92)
   a. $s = 24$ ft.
      $a = (576$ sq. ft.$)$
   b. $l = 14$ in.
      $w = 10\frac{1}{2}$ in.
      $a = (147$ sq. in.$)$
   c. $l = 92$ ft.
      $w = 68$ ft.
      $a = (6,256$ sq. ft.$)$
5. *Change each of these to a percent.* (Lesson 60)
   a. $\frac{17}{25}$ (68%)
   b. 0.03 (3%)
   c. 17:20 (85%)
   d. 0.8 (80%)

124. Using the Compound Interest Formula

Brother Roy deposited $10,000 into a savings account at 4% interest compounded annually. What was the balance in the account at the end of 2 years?

In Lesson 123, you used the basic interest formula ($i = prt$) to find compound interest. There is another formula that condenses the calculation of compound interest. With this method, you first calculate the total rate of increase, and then multiply the original principal by that rate. This yields the total amount of principal plus interest.

Compound interest formula: $a = p(1 + r)^n$

Example A shows this formula applied to the problem above.

Example A

Find the total amount of principal plus interest if $10,000 is deposited for 2 years at 4% interest compounded annually.

$a = p(1 + r)^n$
$a = \$10,000(1 + 0.04)^2$
$a = \$10,000(1.04 \times 1.04)$
$a = \$10,000 \times 1.0816$
$a = \$10,816$

a = total amount of principal plus interest
p = original principal
r = interest rate per interest period
n = number of interest periods

If interest is compounded semiannually, there are two interest periods per year, and the r for *rate* in the formula should be one-half of the annual rate. (See Example B.) If interest is compounded quarterly, there are four interest periods, and r for *rate* should be one-fourth of the annual rate. (See Example C.)

Example B

Find the total amount of principal plus interest if $2,100 is deposited for 2 years at 6% compounded semiannually.

6% × ½ = 3% per interest period
2 yr. × 2 = 4 interest periods
$a = p(1 + r)^n$
$a = \$2,100(1 + 0.03)^4$
$a = \$2,100(1.03)^4$
$a = \$2,100(1.03 \times 1.03 \times 1.03 \times 1.03)$
$a = \$2,100 \times 1.1255$ (rounded)
$a = \$2,363.55$

Example C

Find the total amount of principal plus interest if $3,400 is deposited for 1 year at 6% compounded quarterly.

6% × ¼ = 1.5% per interest period
1 yr. × 4 = 4 interest periods
$a = p(1 + r)^n$
$a = \$3,400(1 + 0.015)^n$
$a = \$3,400(1.015)^4$
$a = \$3,400(1.015 \times 1.015 \times 1.015 \times 1.015)$
$a = \$3,400 \times 1.0614$ (rounded)
$a = \$3,608.76$

418 Chapter 9 Mathematics and Finances

Notice the rounding that is done before the final multiplications. This simplifies calculation, but it also means that the result will not be as accurate as if exact decimals were used.

Because the many calculations in finding compound interest can be cumbersome, tables have been worked out to simplify the process. One such table is shown below. To use it, look at the appropriate part of the table (semiannual or quarterly compounding), find the correct interest rate, and find the corresponding decimal. Multiply the original principal by this decimal, and you will obtain the total amount of principal plus interest at the end of the stated period.

Compound Interest Table

Interest compounded semiannually

Rate	2 yr.	$2\frac{1}{2}$ yr.	3 yr.
3%	1.0614	1.0773	1.0934
4%	1.0824	1.1041	1.1262
5%	1.1038	1.1314	1.1597
6%	1.1255	1.1593	1.1941
7%	1.1475	1.1877	1.2293
8%	1.1699	1.2167	1.2653
9%	1.1925	1.2462	1.3023
10%	1.2155	1.2763	1.3401

Interest compounded quarterly

Rate	2 yr.	$2\frac{1}{2}$ yr.	3 yr.
3%	1.0616	1.0776	1.0939
4%	1.0829	1.1046	1.1268
5%	1.1045	1.1323	1.1608
6%	1.1265	1.1605	1.1956
7%	1.1489	1.1894	1.2314
8%	1.1717	1.2190	1.2682
9%	1.1948	1.2492	1.3060
10%	1.2184	1.2801	1.3449

The following example shows how to find the total amount of principal plus interest by using the table. The numbers are the same as in Example B above.

Example D

Suppose $2,100 is deposited at 6% interest compounded semiannually for 2 years. Find the combined principal plus interest.

The decimal on the table is 1.1255.
$a = 1.1255 \times \$2,100 = \$2,363.55$

The table can also be used to find the compound interest alone. The 1 in each decimal is there so that the original principal will be included in the product. Subtract the whole number 1 from the multiplier, and the product will be the interest only. Study the following example.

Lesson 124   T–418

Introduction

Use the method taught in Lesson 123 to find the ending balance if $10,000 earns interest at 4% compounded annually for 2 years. Compare the calculations and the answer with Example A in the text.

    First year
        $10,000 × 4% = $400;
        new principal = $10,400
    Second year
        $10,400 × 4% = $416;
        new principal = $10,816

Teaching Guide

1. **Calculating compound interest can be simplified by using the formula $a = p(1 + r)^n$.** Discuss the meaning of the terms in the formula.

   a = total amount of principal plus interest
   p = original principal
   r = interest rate per interest period
   n = number of interest periods

   Interest compounded annually

   a. $p = \$1,000; r = 5\%; t = 2$ yr.
   $1.05^2 = 1.1025$
   $a = 1.1025 \times \$1,000 = (\$1,102.50)$

   b. $p = \$2,000; r = 6\%; t = 2$ yr.
   $1.06^2 = 1.1236$
   $a = 1.1236 \times \$2,000 = (\$2,247.20)$

2. **If interest is compounded semiannually, first divide the rate by 2 since there are 2 interest periods per year. If interest is compounded quarterly, first divide the interest rate by 4 since there are 4 interest periods per year.** Note that the final result will not be completely accurate if rounding is done.

   a. *Interest compounded semiannually*
   $p = \$1,500; r = 6\%; t = 2$ yr.
   $1.03^4 = 1.1255$ (rounded)
   $a = 1.1255 \times \$1,500 = (\$1,688.25)$

   b. *Interest compounded quarterly*
   $p = \$3,200; r = 4\%; t = 1$ yr.
   $1.01^4 = 1.0406$ (rounded)
   $a = 1.0406 \times \$3,200 = (\$3,329.92)$

T-419 *Chapter 9 Mathematics and Finances*

3. **Because the many calculations in finding compound interest can be cumbersome, tables have been worked out to simplify the process.** Discuss the table in the lesson, noting that it can be used to find both the ending balance and the compound interest alone.

   a. *Interest compounded semiannually*
      $p = \$3,700$; $r = 8\%$; $t = 2$ yr.
      decimal = 1.1699
      $a = 1.1699 \times \$3,700 = (\$4,328.63)$
      $i = 0.1699 \times \$3,700 = (\$628.63)$

   b. *Interest compounded quarterly*
      $p = \$2,600$; $r = 10\%$; $t = 2\frac{1}{2}$ yr.
      decimal = 1.2801
      $a = 1.2801 \times \$2,600 = (\$3,328.26)$
      $i = 0.2801 \times \$2,600 = (\$728.26)$

Solutions for CLASS PRACTICE e–k

e. $a = \$3,000 \times 1.04^3 = \$3,000 \times 1.1249$ (rounded)

f. $a = \$4,200 \times 1.04^4 = \$4,200 \times 1.1699$ (rounded)

g. $a = \$4,000 \times 1.0175^3 = \$4,000 \times 1.0534$ (rounded)

h. $a = 1.1925 \times \$5,500$

i. $a = 1.1268 \times \$6,300$

j. $i = 0.1593 \times \$2,200$

k. $i = 0.2682 \times \$3,700$

Example E
Find the interest on $3,400 at 6% compounded quarterly for $2\frac{1}{2}$ years.
 The decimal on the table is 1.1605.
 1.1605 − 1 = 0.1605
 i = 0.1605 × $3,400.00 = $545.70

The decimals on the table are rounded to the nearest ten-thousandth. For this reason, answers will be slightly different from what they would be if calculated with exact decimals.

CLASS PRACTICE

Find the corresponding rates per period for each annual rate.

	Annual	Semiannual	Quarterly
a.	8%	4%	2%
b.	12%	6%	3%
c.	5%	2.5%	1.25%
d.	11%	5.5%	2.75%

Use the compound interest formula to find the principal plus interest. Round the final decimal to the nearest ten-thousandth before multiplying.

e. $3,000 at 4% compounded annually for 3 yr. $3,374.70
f. $4,200 at 8% compounded semiannually for 2 yr. $4,913.58
g. $4,000 at 7% compounded quarterly for $\frac{3}{4}$ yr. $4,213.60

Use the compound interest table to find the principal plus interest.

h. $5,500 at 9% compounded semiannually for 2 yr. $6,558.75
i. $6,300 at 4% compounded quarterly for 3 yr. $7,098.84

Use the compound interest table to find the amount of interest.

j. $2,200 at 6% compounded semiannually for $2\frac{1}{2}$ yr. $350.46
k. $3,700 at 8% compounded quarterly for 3 yr. $992.34

WRITTEN EXERCISES

A. *Find the corresponding rates per period for each annual rate.*

	Annual	Semiannual	Quarterly
1.	4%	2%	1%
2.	10%	5%	2.5%
3.	7%	3.5%	1.75%
4.	9%	4.5%	2.25%

420   Chapter 9   *Mathematics and Finances*

B. Use the compound interest formula to find the principal plus interest. Round the final decimal to the nearest ten-thousandth before multiplying the principal.

Interest compounded annually

5. $2,000 at 5% for 2 yr. $2,205.00
6. $3,000 at 6% for 2 yr. $3,370.80

Interest compounded semiannually

7. $3,200 at 8% for $1\frac{1}{2}$ yr. $3,599.68
8. $4,400 at 6% for 2 yr. $4,952.20

Interest compounded quarterly

9. $8,200 at 7% for $\frac{1}{2}$ yr. $8,489.46
10. $7,500 at 9% for $\frac{3}{4}$ yr. $8,017.50

C. Use the compound interest table to find the principal plus interest.

Interest compounded semiannually

11. $6,650 at 10% for $2\frac{1}{2}$ yr. $8,487.40
12. $9,750 at 9% for 3 yr. $12,697.43

Interest compounded quarterly

13. $7,300 at 7% for 3 yr. $8,989.22
14. $6,700 at 8% for $2\frac{1}{2}$ yr. $8,167.30

D. Use the compound interest table to find the amount of interest.

Interest compounded semiannually

15. $1,200 at 4% for $2\frac{1}{2}$ yr. $124.92
16. $4,500 at 5% for 2 yr. $467.10

Interest compounded quarterly

17. $7,250 at 7% for 2 yr. $1,079.53
18. $6,750 at 8% for 3 yr. $1,810.35

E. Solve these reading problems. Use the compound interest table when interest is compounded semiannually or quarterly.

19. Jolene put $2,250 in a new savings account that pays 6% interest compounded quarterly. What will her balance be at the end of 2 years? $2,534.63
20. Matthew deposited $6,700 in a new savings account that pays 7% interest compounded semiannually. How much interest will he earn in $2\frac{1}{2}$ years? $1,257.59
21. Luke opened a savings account with $750. The bank pays 5% interest compounded semiannually. How much interest will his deposit earn in $2\frac{1}{2}$ years? $98.55
22. If a person deposits $1,500 into a savings account at 8% interest compounded quarterly, what will his balance be in 3 years? $1,902.30
23. What is the volume of a marble with a radius of 11 millimeters? (Use 4.18 and round the answer to the nearest cubic millimeter.)   5,564 cubic millimeters
24. King Solomon's house included a porch that measured 50 cubits by 30 cubits. Find the area of the porch in square feet by first expressing the dimensions in feet.   3,375 square feet

Lesson 124 T–420

Solutions for Part B

5. $a = \$2{,}000 \times 1.05^2 = \$2{,}000 \times 1.1025$
6. $a = \$3{,}000 \times 1.06^2 = \$3{,}000 \times 1.1236$
7. $a = \$3{,}200 \times 1.04^3 = \$3{,}200 \times 1.1249$ (rounded)
8. $a = \$4{,}400 \times 1.03^4 = \$4{,}400 \times 1.1255$ (rounded)
9. $a = \$8{,}200 \times 1.0175^2 = \$8{,}200 \times 1.0353$ (rounded)
10. $a = \$7{,}500 \times 1.0225^3 = \$7{,}500 \times 1.0690$ (rounded)

Solutions for Part C

11. $a = 1.2763 \times \$6{,}650$
12. $a = 1.3023 \times \$9{,}750$
13. $a = 1.2314 \times \$7{,}300$
14. $a = 1.2191 \times \$6{,}700$

Solutions for Part D

15. $i = 0.1041 \times \$1{,}200$
16. $i = 0.1038 \times \$4{,}500$
17. $i = 0.1489 \times \$7{,}250$
18. $i = 0.2682 \times \$6{,}750$

Solutions for Part E

19. $\$2{,}250 \times 1.1265$ (from compound interest table)
20. $\$6{,}700 \times 0.1877$
21. $\$750 \times 0.1314$
22. $\$1{,}500 \times 1.2682$
23. 4.18×11^3
24. $(50 \times 1\tfrac{1}{2}) \times (30 \times 1\tfrac{1}{2})$

T-421 *Chapter 9 Mathematics and Finances*

REVIEW EXERCISES

F. Find the total interest if it is compounded annually. *(Lesson 123)*

	p	r	t			p	r	t	
25.	$4,300	7%	2 yr.	$623.07	26.	$5,800	6%	2 yr.	$716.88

G. Find the simple interest. *(Lesson 122)*

	p	r	t			p	r	t	
27.	$3,000	4%	11 mo.	$110.00	28.	$6,800	7%	300 days	$396.67

H. Find the volume of each sphere. *(Lesson 108)*

29. $r = 3$ in.   112.86 cu. in.   30. $r = 6$ m   902.88 m³

I. Find the area of each square or rectangle. *(Lesson 92)*

31. $s = 18$ ft.   324 sq. ft.   32. $l = 108$ ft.   5,724 sq. ft.
                                      $w = 53$ ft.

J. Change each ratio to a percent. *(Lesson 60)*

33. $\frac{13}{25}$ 52%   34. 11:20   55%

Take four careful steps to use a Compound Interest Table successfully.

1. Correct Section — Semiannual or Quarterly
2. Correct Line — Rate
3. Correct Column — Time
4. Correct Value — With 1 for total amount, without 1 for interest only

125. Calculating Sales Tax and Property Tax

Laundry det.	7.95 T
Milk	2.29
Light bulbs	1.59 T
Subtotal	11.83
6.5% Tax	0.62
Total	12.45
Cash Tend.	15.00
Change	2.55

02/14/00

Thank you for shopping at Taft's Grocery

"Let every soul be subject unto the higher powers. For there is no power but of God: the powers that be are ordained of God. . . . For this cause pay ye tribute also: for they are God's ministers, attending continually upon this very thing" (Romans 13:1, 6).

The civil government not only keeps order but also provides services such as road building and fire protection for its citizens. Governments obtain most of their operating funds by various kinds of taxes. Two of those kinds are sales tax and property tax.

Sales tax is a tax paid by the buyer at the time of a purchase. Most states and some cities collect a sales tax. Some states exempt items such as food, clothing, and farm supplies; others tax almost every item sold. On the sales receipt at the left, the letter T marks each taxable item.

Sales tax is calculated as a percent of the total taxable sale. On the sales receipt at the left, the total taxable amount is $9.54. This amount was multiplied by the tax rate of 6.5% to obtain the sales tax ($0.62). The steps for calculating sales tax are given and illustrated below.

1. Express the tax rate as a decimal.
2. Multiply the taxable amount by the rate, and round the answer to the nearest cent.

To find the total amount due, add the sales tax to the total amount of the sale.

> **Example A**
> Find the total amount due on a taxable item selling for $23.49 if the sales tax rate is 6%.
> 23.49 × 0.06 = 1.4094 = $1.41 (rounded)
> $23.49 + $1.41 = $24.90

Property tax is a tax on land and buildings and is usually collected by a local government such as the county, city, or township. (Some localities collect a personal property tax on vehicles.) The government assesses the value of the property and then levies a tax according to that value. Property tax is generally paid annually, semiannually, or quarterly.

The rate of a property tax is often expressed as a millage (thousandths) rather than a percent. For example, a rate of 15 mills is 0.015 (equal to 1.5%), and 20 mills is 0.020 (equal to 2%).

LESSON 125

Objectives

- To practice calculating sales tax and *property tax.
- To teach that *a mill is one-thousandth.
- To practice *calculating with millage rates.

Review

1. *Use the compound interest formula to find the principal plus interest if interest is compounded semiannually.* (Lesson 124)

   a. $3,000 at 4% for 1 yr. ($3,121.20)

   b. $9,000 at 6% for 1 yr. ($9,548.10)

2. *Use the compound interest table to find the principal plus interest if interest is compounded semiannually.* (Lesson 124)

   a. $3,000 at 7% for 2 yr.
      (1.1475 × $3,000 = $3,442.50)

   b. $4,000 at 8% for $2\frac{1}{2}$ yr.
      (1.2167 × $4,000 = $4,866.80)

3. *Find each square root by using a table.* (Lesson 109)

   a. $\sqrt{158}$ (12.570)

   b. $\sqrt{102}$ (10.100)

4. *Find the area of each figure.* (Lesson 93)

   a. Parallelogram
      $b = 15$ in.
      $h = 13$ in.
      $a = $ (195 sq. in.)

   b. Triangle
      $b = 34$ ft.
      $h = 35$ ft.
      $a = $ (595 sq. ft.)

5. *Solve these percent problems.* (Lesson 61)

   a. 46% of 56 (25.76)

   b. 9% of 18 (1.62)

T–423 Chapter 9 Mathematics and Finances

Introduction

Ask the students to name as many different kinds of taxes as they can. Here is a partial list.

- sales tax
- property (real estate) tax
- use tax
- income tax
- capital gains tax
- excise tax
- estate tax
- personal property tax
- per capita tax
- real estate transfer tax
- gift tax
- luxury tax

These taxes apply in various specific circumstances. The students will probably never pay all of them.

Briefly mention the Scriptural teaching with regard to taxes (Romans 13:6, 7). You may want to use Romans 13 as the text for a classroom devotional.

Teaching Guide

1. **Sales tax is paid by the buyer at the time of a purchase.** Most states and some cities collect a sales tax.

2. **Sales tax is calculated as a percent of the total taxable sale.** To find the sales tax, multiply the amount of the taxable sale by the rate. To find the total amount due, add the sales tax to the total amount of the sale.

	Sale	Tax rate	Tax	Total due
a.	$125	4%	($5.00)	($130.00)
b.	$245	6%	($14.70)	($259.70)
c.	$12.95	3%	($0.39)	($13.34)
d.	$66.55	$4\frac{1}{2}$%	($2.99)	($69.54)

3. **If a sale includes both taxable and nontaxable items, total the taxable sales separately to calculate the sales tax.** Tax-exempt items may include food, clothing, and farm supplies.

   a.
taxable:	$13.76, $12.87
nontaxable:	$15.77, $13.98
tax rate:	5%
subtotal:	($56.38)
taxable sale:	($26.63)
sales tax:	($1.33)
total due:	($57.71)

   b.
taxable:	$25.71, $19.29
nontaxable:	$13.95, $37.93
tax rate:	4%
subtotal:	($96.88)
taxable sale:	($45.00)
sales tax:	($1.80)
total due:	($98.68)

To calculate property tax, follow these steps.

1. Move the decimal point in the millage rate three places to the left.
2. Multiply the assessed value by the decimal found in step 1.
3. Round the product to the nearest cent.

The answer is the annual amount of tax due. Divide the answer by 2 to find the amount of a semiannual payment, and divide it by 4 to find the amount of a quarterly payment.

> **Example B**
> Find the annual property tax on this property.
> Assessed value: $85,600
> Tax rate: 14.3 mills
> $85,600 × 0.0143 = $1,224.08

> **Example C**
> Find the quarterly property tax on this property.
> Assessed value: $125,560
> Tax rate: 38.7 mills
> $125,560 × 0.0387 = $4,859.17   (rounded)
> $4,859.17 ÷ 4 = $1,214.79

CLASS PRACTICE

Find the sales tax on each sale.

	Sale	Tax rate			Sale	Tax rate	
a.	$26.25	5%	$1.31	b.	$64.28	$4\frac{1}{4}$%	$2.73

Find the total amount due, including sales tax.

	Sale	Tax rate			Sale	Tax rate	
c.	$36.94	7%	$39.53	d.	$45.91	$4\frac{1}{2}$%	$47.98

Find the annual property tax.

	Assessment	Tax rate			Assessment	Tax rate	
e.	$95,000	16 mills	$1,520.00	f.	$136,000	18.6 mills	$2,529.60

WRITTEN EXERCISES

A. *Find the sales tax on each sale.*

	Sale	Tax rate			Sale	Tax rate	
1.	$22.00	4%	$0.88	2.	$65.00	5%	$3.25
3.	$48.26	$6\frac{1}{2}$%	$3.14	4.	$95.92	$6\frac{3}{4}$%	$6.47

424   Chapter 9   Mathematics and Finances

B. Find the total amount due, including sales tax.

	Sale	Tax rate			Sale	Tax rate	
5.	$14.25	4%	$14.82	6.	$29.95	6%	$31.75
7.	$32.95	3½%	$34.10	8.	$77.95	4¾%	$81.65

C. Following are two sales receipts with both taxable and nontaxable items. Find the sales tax and the total amount due.

9.
```
              15.95 T
              17.99 T
              18.43
              15.99
Subtotal      68.36        (taxable:
                            $33.94)
6% Tax        $2.04

Total         $70.40
```

10.
```
              16.95
              18.68 T
              22.38
              34.95 T
Subtotal      92.96        (taxable:
                            $53.63)
7% Tax        $3.75

Total         $96.71
```

D. Find the annual property tax.

	Assessment	Tax rate			Assessment	Tax rate	
11.	$50,000	18 mills	$900.00	12.	$60,000	22 mills	$1,320.00
13.	$125,300	16 mills	$2,004.80	14.	$131,800	18 mills	$2,372.40
15.	$310,000	17.8 mills	$5,518.00	16.	$445,000	16.9 mills	$7,520.50

E. Solve these reading problems.

17. Mr. Hawkins paid $62.33 for a breaker panel for his house and $179 each for two fans for his barn. Calculate his total bill if the panel was subject to a 5% sales tax but the fans were tax exempt. $423.45

18. Mark Ramsey's property is assessed at $225,000. If the tax rate is 19 mills and the tax is paid quarterly, what is the amount of each payment? $1,068.75

19. Brother Daniel owns a property assessed at $140,000. The property tax rate is 21 mills. Find the amount of each payment if the tax is paid semiannually. $1,470.00

20. Starner's Auto Repair estimated that Linford's truck would require the following parts: hood for $135.70, bumper for $116.00, and grille for $95.45. The estimated labor charge would be $390 (also taxed). Find the total estimated bill, including 7% sales tax. $788.75

Lesson 125 T–424

4. **Property tax is paid by the owners of land and buildings, and is usually collected by a local government.** The government assesses the value of the property and then levies a tax according to that value. Property tax is generally paid annually, semiannually, or quarterly.

5. **The rate of a property tax is often expressed as a millage (thousandths) rather than a percent.** To calculate property tax, move the decimal point in the millage rate three places to the left, multiply the assessed value by this decimal, and round to the nearest cent.

	Assessment	Tax rate	Tax
a.	$70,000	19 mills	($1,330.00)
b.	$95,000	16 mills	($1,520.00)
c.	$115,300	17 mills	($1,960.10; $980.05 per half year if paid semiannually)
d.	$365,000	18.8 mills	($6,862.00; $1,715.50 per quarter if paid quarterly)

Solutions for Part E

17. $62.33 + (2 × $179) + (0.05 × $62.33)
18. $225,000 × 0.019 ÷ 4
19. $140,000 × 0.021 ÷ 2
20. $135.70 + $116.00 + $95.45 + $390.00 = $737.15; 1.07 × $737.15

T-425 Chapter 9 Mathematics and Finances

Further Study

The students have been using the compound interest formula to find the amount of principal plus interest. The decimal that is obtained with this formula represents exactly the same thing as the decimals on the compound interest table. To find the interest alone, subtract the 1 and multiply by the fractional part of the decimal. The *Challenge Exercises* may be done this way, or by using the formula to find the amount of principal plus interest and then subtracting the original principal.

21. $0.65 \times \$109.60$
22. $0.60 \times \$639.40$

Solutions for Part F

23. $a = \$5,200 \times 1.08^2 = \$5,200 \times 1.1664$
24. $a = \$8,100 \times 1.09^2 = \$8,100 \times 1.1881$

Solutions for Part G

25. $a = 1.2462 \times \$1,800$
26. $a = 1.2293 \times \$3,400$

Solutions for Part K

33. $a = \$10,000 \times 1.06^4$; $1.06^4 = 1.2625$ (rounded);
    $i = 0.2625 \times \$10,000$
34. $a = \$10,000 \times 1.03^8$; $1.03^8 = 1.2668$ (rounded);
    $i = 0.2668 \times \$10,000$
35. $a = \$10,000 \times 1.06^8$; $1.06^8 = 1.5938$ (rounded);
    $i = 0.5938 \times \$10,000$

21. The list price on a ½-horsepower electric motor is $109.60. Green Electric Company can buy the motor at a 35% discount. How much does the company pay for the motor?
$71.24

22. Allen's Hardware sold $639.40 worth of screws and bolts to another dealer at a 40% discount. Find the discount price of the hardware.
$383.64

REVIEW EXERCISES

F. Use the compound interest formula to find the combined principal and interest if interest is compounded semiannually. *(Lesson 124)*

	p	r	t		p	r	t
23.	$5,200	16%	1 yr. $6,065.28	**24.**	$8,100	18%	1 yr. $9,623.61

G. Use the compound interest table on page 418 to find the combined principal and interest if interest is compounded semiannually. *(Lesson 124)*

25. $1,800   9%   2½ yr. $2,243.16   **26.** $3,400   7%   3 yr. $4,179.62

H. Find the square roots from the table on page 584. *(Lesson 109)*

27. $\sqrt{84}$ 9.165

28. $\sqrt{116}$ 10.770

I. Find the area of each figure. *(Lesson 93)*

29. Triangle   609 sq. ft.
b = 29 ft.; h = 42 ft.

30. Parallelogram   546 sq. in.
b = 26 in.; h = 21 in.

J. Find these percentages. *(Lesson 61)*

31. 68% of 21   14.28

32. 27% of 140   37.8

CHALLENGE EXERCISES

K. Use the compound interest formula to find the total interest for these accounts.

33. $10,000 at 12% for 2 years compounded semiannually $2,625.00

34. $10,000 at 12% for 2 years compounded quarterly $2,668.00

35. $10,000 at 12% for 4 years compounded semiannually $5,938.00

126. Calculating Profit

"Charge them that are rich in this world, that they be not highminded, nor trust in uncertain riches, but in the living God, who giveth us richly all things to enjoy" (1 Timothy 6:17).

In the system called **capitalism**, a person operates a business to earn profits for himself. Some people think profits are evil and are obtained by taking unfair advantage of others. The Bible approves making honest profits; however, God does command us to use the profits in ways that are pleasing to Him.

A business earns a profit when its **income** (money received) is greater than its **expenses** (money paid out). Although there are thousands of different businesses in the world, this basic principle applies to all of them.

Businesses earn profits by performing services, selling products, or manufacturing products. This means that businesses can be placed in three main classes: service businesses, retail businesses, and manufacturing businesses. A service business simply provides a service, such as cleaning carpets. Its only expenses are operating expenses that may include heat, electricity, telephone, and vehicle costs. For a service business, profit equals income minus operating expenses.

$$\text{profit} = \text{income} - \text{expenses}$$

A retail business, such as a grocery store, buys and sells products in order to make a profit. Such a business has two kinds of expenses. One is operating expenses, which are called **overhead**; and the other is the **cost of goods** purchased for resale. The **gross profit** of a retail business is the income minus the cost of goods sold.

$$\text{gross profit} = \text{income} - \text{cost of goods}$$

A business can have a gross profit and yet be losing money if its overhead is too high. Therefore, the **net profit** of a business is more important than its gross profit. Net profit equals income minus the total cost of goods plus overhead.

$$\text{net profit} = \text{income} - (\text{cost of goods} + \text{overhead})$$

A manufacturing business makes products to sell. For example, a furniture shop uses wood and other materials to make items such as chairs and tables. A manufacturer, like a retailer, has both costs of goods and overhead. His gross profit and net profit are calculated in the same way as for a retail business.

LESSON 126

Objectives

- To review that income minus expenses equals profit.
- To review that a loss occurs when expenses exceed income.
- To teach *the three basic types of businesses:
  - service businesses
  - retail businesses
  - manufacturing businesses
- To teach *finding gross profit by subtracting the cost of goods from the sales.
- To teach that *overhead is the expenses not directly related to purchasing or manufacturing a product.

Review

1. Give Lesson 126 Quiz (Calculating Interest).

2. *Calculate each sales tax or property tax.* (Lesson 125)

	Sale	Tax Rate	Tax
a.	$164.25	5%	($8.21)
b.	$95.48	$5\frac{1}{2}\%$	($5.25)

	Assessment	Tax Rate	Tax
c.	$105,000	15.8 mills	($1,659)
d.	$245,000	19 mills	($4,655)

3. *Use the compound interest table to find the principal plus interest if interest is compounded quarterly.* (Lesson 124)
   a. $4,400 at 7% for 2 yr.
      (1.1489 × $4,400 = $5,055.16)
   b. $8,000 at 8% for 3 yr.
      (1.2682 × $8,000 = $10,145.60)

4. *Find the simple interest.* (Lesson 122)
   a. $7,500 at 6% for $\frac{1}{4}$ yr. ($112.50)
   b. $11,000 at 0.0308% per day for 15 days ($50.82)

5. *Extract the square roots of these perfect squares.* (Lesson 110)

   a. $\sqrt{2{,}304}$ (48)

   $$\begin{array}{r} 4\ \ 8 \\ \sqrt{23{,}04} \\ \underline{16}\ \ \ \ \\ (20 \times 4)\ 80\quad 7\ 04 \\ 8 \times 88 = \underline{7\ 04} \\ 0 \end{array}$$

   b. $\sqrt{99{,}225}$ (315)

   $$\begin{array}{r} 3\ \ 1\ \ 5 \\ \sqrt{9{,}92{,}25} \\ \underline{9}\ \ \ \ \ \\ (20 \times 3)\ 60\quad 92\ \ \\ 1 \times 61 = \underline{61}\ \ \\ (20 \times 31)\ 620\quad 31\ 25 \\ 5 \times 625 = \underline{31\ 25} \\ 0 \end{array}$$

6. *Find the area of each trapezoid.* (Lesson 94)
   a. $h = 11$ in.
      $b_1 = 16$ in.
      $b_2 = 25$ in.
      $a = (225.5$ sq. in.$)$
   b. $h = 40$ ft.
      $b_1 = 45$ ft.
      $b_2 = 38$ ft.
      $a = (1{,}660$ sq. ft.$)$

7. *Find both the amount of increase or decrease and the new price.* (Lesson 63)
   a. $45.50 decreased by 65%
      (decrease: $29.58
      new price: $15.92)
   b. $455 increased by 15%
      (increase: $68.25
      new price: $523.25)

T–427   Chapter 9   Mathematics and Finances

Introduction

Ask the students, "Suppose you purchased a machine for $250 and spent $75 to pick it up and advertise it. For how much would you need to sell the machine to make a profit of $25?"

This problem is solved by adding the cost of $250, the expenses of $75, and the desired profit of $25. You would need to sell the machine for $350 in order to make a profit of $25.

A profit is made when the income is greater than the expenses.

Teaching Guide

1. **A business earns a profit when its income is greater than its expenses.** Capitalism is the economic system in which people operate businesses to earn profits for themselves.

2. **Businesses can be placed in three main classes: service businesses, retail businesses, and manufacturing businesses.**

3. **The general rule for finding profit is *income – expenses = profit*.**

	Income	Expense	Profit
a.	$375	$105	($270)
b.	$862	$98	($764)

4. **Gross profit equals income minus the cost of goods.** This rule is *gross profit = income – cost of goods*.

	Income	Cost of Goods	Gross Profit
a.	$425	$295	($130)
b.	$497	$374	($123)

5. **Net profit equals income minus the total cost of goods and overhead.** (Overhead refers to operating expenses.) This rule is *net profit = income – (cost of goods + overhead)*.

	Income	Cost of Goods	Overhead	Net Profit
a.	$337	$213	$76	($48)
b.	$377	$211	$91	($75)

6. **A business suffers a loss if expenses exceed income.** To calculate a loss, add the cost of goods and the overhead together, and subtract the income from it. This rule is *loss = (cost of goods + overhead) – income*.

	Income	Cost of Goods	Overhead	Loss
a.	$498	$315	$238	($55)
b.	$738	$618	$210	($90)

> **Example A**
> Find the gross profit if the income is $16,500 and the cost of goods is $11,300.
>
> gross profit = income − cost of goods
> gross profit = $16,500 − $11,300
> gross profit = $5,200

> **Example B**
> Find the net profit if income is $2,300, cost of goods is $1,500, and overhead is $400.
>
> net profit = income − (cost of goods + overhead)
> net profit = $2,300 − ($1,500 + $400)
> net profit = $400

A business will not earn a profit if expenses exceed income. This is called a loss. To calculate a loss, add the cost of goods and the overhead together, and subtract the income from it.

loss = (cost of goods + overhead) − income

> **Example C**
> Find the loss if income is $2,300, cost of goods is $1,700, and overhead is $700.
>
> loss = (cost of goods + overhead) − income
> loss = ($1,700 + $700) − $2,300
> loss = $100

CLASS PRACTICE

Find the profit in each exercise.

	Income	Expenses			Income	Expenses	
a.	$695	$124	$571	b.	$514	$218	$296

Find the gross profit in each exercise.

	Income	Cost of goods			Income	Cost of goods	
c.	$406	$264	$142	d.	$875	$426	$449

Find the net profit or loss in each exercise, and label it profit *or* loss.

	Income	Cost of goods	Overhead	
e.	$216	$112	$25	$79 profit
f.	$624	$465	$160	$1 loss
g.	$281	$211	$84	$14 loss
h.	$342	$178	$51	$113 profit

428 Chapter 9 Mathematics and Finances

WRITTEN EXERCISES

A. Find the profit in each exercise.

	Income	Expenses			Income	Expenses	
1.	$550	$115	$435	2.	$625	$95	$530
3.	$750	$235	$515	4.	$695	$68	$627

B. Find the gross profit in each exercise.

	Income	Cost of goods			Income	Cost of goods	
5.	$278	$143	$135	6.	$389	$196	$193
7.	$1,743	$1,357	$386	8.	$1,643	$980	$663

C. Find the net profit in each exercise.

	Income	Cost of goods	Overhead	
9.	$195	$115	$48	$32
10.	$278	$172	$65	$41
11.	$360	$195	$78	$87
12.	$475	$246	$125	$104

D. Find the loss in each exercise.

	Income	Cost of goods	Overhead	
13.	$415	$360	$75	$20
14.	$556	$369	$218	$31
15.	$621	$430	$295	$104
16.	$875	$882	$195	$202

E. Solve these reading problems.

17. Pine Lane Fabrics buys a certain kind of dress material for $1.62 per yard and sells it for $2.18 per yard. What is the gross profit on a 15-yard bolt of this fabric? $8.40

18. Brother Jason paid $840 for books to put in his bookstore. He sold the books for $1,139, and his overhead was $126. What was his net profit? $173

19. Gabelein Diesel pays $2,200 for a kit to rebuild engines. If they sell the kit for $3,000 and overhead is 15% of the selling price, what is their net profit? $350

20. Sharp's Grocery pays $0.29 per pound for chicken. Overhead is $0.11 per pound. One week the store had chicken on sale at a 15% loss; that is, 15% less than the cost and overhead together. If 1,400 pounds of chicken was sold, what was the amount of the loss? $84

Further Study

Christians have sometimes accepted the mistaken idea that profit is evil. Profit honestly earned is right. Although it is true that some capitalists have taken advantage of other people, the basic premises of the capitalist system are supported by the Scriptures.

Solutions for Part E

17. 15 × ($2.18 − $1.62)

18. $1,139 − ($840 + $126)

19. $3,000 − [$2,200 + (0.15 × $3,000)]

20. 0.15 × ($0.29 + $0.11) × 1,400

T–429   Chapter 9   Mathematics and Finances

21. $160 \times \$0.035 = \$5.60$; $1.055 \times \$5.60$

22. $0.70 \times \$163.50$

Solutions for Part G

27. $a = 1.1605 \times \$6,200$

28. $a = 1.0829 \times \$6,500$

Extractions for Part I

31.
```
                    7 6
              √57̭76
                49
 (20 × 7) 140    8 76
   6 × 146 =     8 76
                    0
```

32.
```
                   1 9 4
              √3̭76̭36
                 1
          20     2 76
  9 × 29 =       2 61
(20 × 19) 380   15 36
  4 × 384 =     15 36
                    0
```

21. Sister Alice made 160 photocopies at 3.5¢ per copy. Find her total bill, including $5\frac{1}{2}$% sales tax. $5.91

22. Midwest Printing offers a 30% discount on the first order of a new customer. The regular price for 2,000 mailing labels is $163.50. What is the discount price? $114.45

REVIEW EXERCISES

F. Calculate each sales tax or property tax. *(Lesson 125)*

23. Sale: $281.95; tax rate: 7% $19.74
24. Sale: $264.53; tax rate: 6% $15.87
25. Assessment: $110,000; tax rate: 17.1 mills $1,881
26. Assessment: $175,000; tax rate: 21 mills $3,675

G. Use the compound interest table on page 418 to find the combined principal and interest if interest is compounded quarterly. *(Lesson 124)*

	p	r	t			p	r	t	
27.	$6,200	6%	$2\frac{1}{2}$ yr.	$7,195.10	28.	$6,500	4%	2 yr.	$7,038.85

H. Find the simple interest. *(Lesson 122)*

	p	r	t			p	r	t	
29.	$5,000	5%	$\frac{1}{4}$ yr.	$62.50	30.	$12,000	6%	7 mo.	$420.00

I. Extract the square roots of these perfect squares. *(Lesson 110)*

31. $\sqrt{5,776}$ 76
32. $\sqrt{37,636}$ 194

J. Find the area of each trapezoid. *(Lesson 94)*

33. h = 36 in.   1,728 sq. in.
    b_1 = 42 in.
    b_2 = 54 in.

34. h = 15.2 m   479.56 m²
    b_1 = 34 m
    b_2 = 29.1 m

K. Find both the amount of increase or decrease and the new price. *(Lesson 63)*

35. $642 decreased by 58% $372.36; $269.64
36. $675 increased by 81% $546.75; $1,221.75

127. Calculating Profit as a Percent of Sales

Two income statements are shown below. According to total sales, Brook View is a much larger business than Spring Valley. It had a gross profit $20,000 higher than Spring Valley's, but the net profit was $1,000 lower than Spring Valley's. Which business was more profitable? Since the businesses are so different in size, it is hard to answer this question by simply comparing the net profits.

A better picture is obtained by expressing the net profit as a percent of sales. This enables the owner to compare this year's profitability with that of other years or of other businesses similar to his. Spring Valley's net profit was 23% of sales, whereas Brook View's net profit was only 6% of sales. This shows that the Spring Valley business was much more profitable than Brook View's, even though Brook View has a much larger business.

Spring Valley Produce
Income Statement
For the Year Ending
December 31, 20—

	Amount	Percent of Sales
Total sales	$35,000	100%
Cost of goods	21,000	60%
Gross profit	14,000	40%
Overhead	6,000	17%
Net profit	8,000	23%

Brook View Produce
Income Statement
For the Year Ending
December 31, 20—

	Amount	Percent of Sales
Total sales	$125,000	100%
Cost of goods	91,000	73%
Gross profit	34,000	27%
Overhead	27,000	22%
Net profit	7,000	6%

The owner of Brook View Produce should analyze his business to see why his profit percentage is so low. Why is his cost of goods such a high percent of sales? Is he buying from an expensive supplier? Does too much produce go to waste? Could he use his labor more efficiently? Analyzing the income statement may help him to find areas in which he can improve his business.

To express profits and expenses as percents of total sales, follow the steps below. The calculations for Spring Valley Produce are shown after the steps.

1. Divide each category on the income statement by the total sales. If the division does not work out evenly, round the quotient to the nearest hundredth.

2. Express the answer as a percent.

3. Check your work by addition. The percents for cost of goods and gross profit should total 100%. The percents for cost of goods, overhead, and net profit should also equal 100%.

LESSON 127

Objective

- To review calculating profit and loss as a percent of sales.

Review

1. Calculate the net profit. (Lesson 126)

	Income	Cost of Goods	Overhead	Net Profit
a.	$655	$482	$126	($47)
b.	$716	$418	$128	($170)

2. Calculate each sales tax or property tax. (Lesson 125)

	Sale	Tax Rate	Tax
a.	$26.95	6%	($1.62)
b.	$84.21	$3\frac{1}{2}\%$	($2.95)

	Assessment	Tax Rate	Tax
c.	$75,000	17 mills	($1,275)
d.	$147,000	19.2 mills	($2,822.40)

3. Use the compound interest formula to find the principal plus interest if interest is compounded semiannually. (Lesson 124)

   a. $4,800 at 8% for 1 yr. ($5,191.68)

   b. $15,000 at 6% for 1 yr.

   ($15,913.50)

4. Extract each square root, to the nearest tenth. (Lesson 111)

   a. $\sqrt{1{,}842}$ (42.9)

   $$\begin{array}{r} 4\;2\,.\,9\;\;1 \\ \sqrt{18{,}42.00{,}00} \\ 16 \end{array}$$

   (20 × 4) 80   2 42
   2 × 82 = 1 64
   (20 × 42) 840   78 00
   9 × 849 = 76 41
   (20 × 429) 8580   1 59 00
   1 × 8581 = 85 81
   73 19

   b. $\sqrt{84{,}075}$ (290.0)

   $$\sqrt{8{,}40{,}75.00{,}00}$$
   4

   (20 × 2) 40   4 40
   8 × 48 = 3 84
   (20 × 28) 560   56 75
   9 × 569 = 51 21
   (20 × 289) 5780   5 54 00
   9 × 5789 = 5 21 01
   (20 × 2899) 57980   32 99 00
   5 × 57985 = 28 99 25
   3 99 75

5. Find the area of each circle. (Lesson 95)

   a. $r = 5\frac{1}{4}$ in.
   $a = (86\frac{5}{8}$ sq. in.)

   b. $r = 36$ cm
   $a = (4{,}069.44$ cm^2)

   c. $r = 45$ ft.
   $a = (6{,}358.5$ sq. ft.)

6. Solve these percent problems. (Lesson 62)

   a. 131% of $45.25 ($59.28)

   b. 214% of 1,008 (2,157.12)

   c. $\frac{5}{8}$% of 21 (0.13125)

   d. $\frac{3}{4}$% of $251.20 ($1.88)

T–431 *Chapter 9 Mathematics and Finances*

Introduction

Compare the two income statements shown at the beginning of the pupil's lesson. Which produce business had the greater profit? Which had the greater *percent* of profit?

What part of his business should the owner of Brook View Produce analyze to determine why his percent of profits is so low? He should analyze the cost of his goods. Why is the cost of his produce such a high percent of the sales? Does he have a good supplier?

There may be a logical explanation for it. Perhaps the owner of Spring Valley Produce is raising more of his own produce and therefore does not need to buy as much.

Percent of sales are used to analyze businesses and to determine weak points that need to be improved.

Teaching Guide

1. **Businesses often evaluate their profitability by expressing profits as a percent of sales.**

2. **To express profits and expenses as percents of total sales, follow these steps.**

   (1) Divide each category on the income statement by the total sales. If the division does not work out evenly, round the quotient to the nearest hundredth.

   (2) Express the answer as a percent.

   (3) Check your work by addition. The percents for cost of goods and gross profit should total 100%. The percents for cost of goods, overhead, and net profit should also equal 100%.

Dry River Garage
Income Statement
For the Year Ending
December 31, 20—

	Amount	*Percent of Sales*
Total sales	$135,000	100%
Cost of goods	69,000	51%
Gross profit	66,000	49%
Overhead	45,000	33%
Net profit	21,000	16%

Spring Valley Produce
Total sales: 35,000 ÷ 35,000 = 1.0 = 100%
Cost of goods: 21,000 ÷ 35,000 = 0.6 = 60%
Gross profit: 14,000 ÷ 35,000 = 0.4 = 40%
Overhead: 6,000 ÷ 35,000 = 0.17 = 17%
Net profit: 8,000 ÷ 35,000 = 0.23 = 23%

Check: 60% + 40% = 100%
60% + 17% + 23% = 100%

CLASS PRACTICE

Calculate the missing percents on each income statement.

West Corner Grocery
Income Statement
For the Year Ending
December 31, 20—

	Amount	Percent of Sales
Total sales	$54,000	100%
Cost of goods	30,000	a. 56%
Gross profit	24,000	b. 44%
Overhead	11,000	c. 20%
Net profit	13,000	d. 24%

Wharves Office Equipment
Income Statement
For the Year Ending
December 31, 20—

	Amount	Percent of Sales
Total sales	$114,000	100%
Cost of goods	94,000	e. 82%
Gross profit	20,000	f. 18%
Overhead	12,000	g. 11%
Net profit	8,000	h. 7%

WRITTEN EXERCISES

A. *Calculate the missing percents on each income statement.*

Route 151 Hardware
Income Statement
For the Year Ending
December 31, 20—

	Amount	Percent of Sales
Total sales	$42,000	100%
Cost of goods	26,000	1. 62%
Gross profit	16,000	2. 38%
Overhead	7,000	3. 17%
Net profit	9,000	4. 21%

Oak Hill Nursery
Income Statement
For the Year Ending
December 31, 20—

	Amount	Percent of Sales
Total sales	$102,000	100%
Cost of goods	62,000	5. 61%
Gross profit	40,000	6. 39%
Overhead	13,000	7. 13%
Net profit	27,000	8. 26%

432 Chapter 9 Mathematics and Finances

Sandy Bluff Lumber
Income Statement
For the Year Ending
December 31, 20—

	Amount	Percent of Sales
Total sales	$275,000	100%
Cost of goods	188,000	9. 68%
Gross profit	87,000	10. 32%
Overhead	62,000	11. 23%
Net profit	25,000	12. 9%

Maple Grove Grocery
Income Statement
For the Year Ending
December 31, 20—

	Amount	Percent of Sales
Total sales	$166,000	100%
Cost of goods	125,000	13. 75%
Gross profit	41,000	14. 25%
Overhead	36,000	15. 22%
Net profit	5,000	16. 3%

B. Solve these reading problems. Round percents to the nearest whole percent. (For problems 17–20, it is helpful to set up the numbers in the form of an income statement as illustrated in number 18.)

17. Witmer Equipment paid $262 for an electrical component. If the part was sold for $349, what percent of sales was the gross profit?
25%

18. During November, Valley Hardware paid $288 for nuts and bolts. Overhead on hardware was $86, and sales of nuts and bolts for the month were $640. What percent of sales was the net profit?
42%

Total sales:	$640
Cost of goods:	288
Gross profit:	(352)
Overhead:	86
Net profit:	(266)

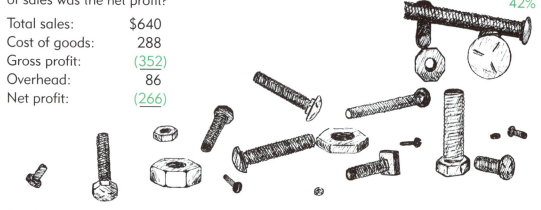

19. For the month of January, West Shore Tire reported $38,400 in total sales, $26,850 in the cost of goods, and $7,296 in overhead. What percent was the net profit of the sales?
11%

20. Faley's Plumbing reported $28,640 in sales for January through March. The cost of goods was $16,611, and overhead was $6,300. The cost of goods was what percent of the sales?
58%

21. A flat disc has a radius of 12 inches. What is its area? (Use 3.14 for pi.)
452.16 square inches

22. Zacchaeus offered to pay back 400% of what he had taken unjustly. If his unjust gains amounted to the equivalent of $125, how much restitution would he have paid?
$500

Lesson 127   T-432

Further Study

Financial ratios serve many different purposes in business. Following are a few of the most commonly used ratios.

Name	How Calculated	Use of Ratio
a. Current ratio	$\dfrac{\text{Current assets}}{\text{Current liabilities}}$	Helps to determine liquidity. A good ratio is 2:1 or better.
b. Acid test ratio	$\dfrac{\text{Cash + Accounts receivable}}{\text{Current liabilities}}$	Helps to determine quick liquidity. A good ratio is 2:1 or better.
c. Accounts receivable turnover in days	$\dfrac{\text{Accounts receivable}}{\text{Sales}} \times 365$	Helps to determine whether Accounts Receivable is too high.
d. Inventory turnover in days	$\dfrac{\text{Inventory}}{\text{Cost of goods}} \times 365$	Helps to determine whether Inventory is too high.

There are no perfect ratios for these and other purposes. However, by making comparisons from year to year and with industry standards, these ratios help to assess the financial health of a business.

Solutions for Part B

17. Total sales: 349
    Cost of goods: 262
    Gross profit: 87   87 ÷ 349

18. 266 ÷ 640

19. Total sales: 38,400
    Cost of goods: 26,850
    Gross profit: 11,550
    Overhead: 7,296
    Net profit: 4,254   4,254 ÷ 38,400

20. Total sales: 28,640
    Cost of goods: 16,611   16,611 ÷ 28,640

21. 3.14×12^2

22. 4 × $125

T-433   Chapter 9   Mathematics and Finances

Solutions for Part E

29. $a = \$5{,}100 \times 1.06^2 = \$5{,}100 \times 1.1236$

30. $a = \$11{,}500 \times 1.07^2 = \$11{,}500 \times 1.1449$

Extractions for Part F

```
                     4 5 . 3
31.             √20ˬ58.00
                     16
     (20 × 4) 80    4 58
       5 × 85 =    4 25
   (20 × 45) 900      33 00
       3 × 903 =     27 09
                       5 91
```

```
                         2 7 4 . 2
32.                √7ˬ51ˬ91.00
                         4
     (20 × 2) 40        3 51
       7 × 47 =        3 29
   (20 × 27) 540          22 91
       4 × 544 =         21 76
 (20 × 274) 5480             1 15 00
      2 × 5482 =           1 09 64
                                5 36
```

REVIEW EXERCISES

C. Find the net profit in each exercise. *(Lesson 126)*

	Income	Cost of goods	Overhead	
23.	$824	$415	$211	$198
24.	$382	$229	$98	$55

D. Calculate each sales tax or property tax. *(Lesson 125)*

25. Sale: $61.82; tax rate: 5% $3.09
26. Sale: $92.95; tax rate: 6% $5.58
27. Assessment: $64,000; tax rate: 21 mills $1,344
28. Assessment: $110,000; tax rate: 18.6 mills $2,046

E. Use the compound interest formula to find the combined principal and interest if interest is compounded semiannually. *(Lesson 124)*

	p	r	t			p	r	t	
29.	$5,100	12%	1 yr.	$5,730.36	30.	$11,500	14%	1 yr.	$13,166.35

F. Calculate each square root to the nearest whole number. *(Lesson 111)*

31. $\sqrt{2,058}$ 45
32. $\sqrt{75,191}$ 274

G. Find the area of each circle. Use $3\frac{1}{7}$ for pi. *(Lesson 95)*

33. $r = 11\frac{2}{3}$ ft. $427\frac{7}{9}$ sq. ft.
34. $r = 28$ cm   2,464 cm²

H. Find these percentages. *(Lesson 62)*

35. 184% of $96.84 $178.19
36. $\frac{1}{4}$% of $351.06 $0.88

434 Chapter 9 Mathematics and Finances

128. Reading Problems Relating to Finances

"If therefore ye have not been faithful in the unrighteous mammon [material wealth], who will commit to your trust the true riches?" (Luke 16:11). This verse shows how important it is to be faithful stewards of the money that God allows us to handle.

The finance studies in this chapter have worked with checking and savings accounts, with simple and compound interest, with sales and property taxes, and with profit and loss. Today's lesson contains reading problems that give further practice with the financial calculations you have learned.

CLASS PRACTICE

a. Rachel deposited some cash and checks into her checking account. The amounts were $45.00 in cash and $45.76, $38.76, and $18.59 in checks. What was the total deposit? $148.11

b. Father put $3,000 into an account that paid 5% interest compounded quarterly. Use the compound interest table on page 418 to find the combined interest and principal in the account at the end of $2\frac{1}{2}$ years. $3,396.90

WRITTEN EXERCISES

A. Solve these reading problems relating to finances.

1. Mother put a day's sales from the family's small bookstore into the checking account. In addition to $45.76 in cash, she deposited checks for $43.57, $21.83, and $23.91. She received $35.00 in cash. What was the amount of the deposit? $100.07

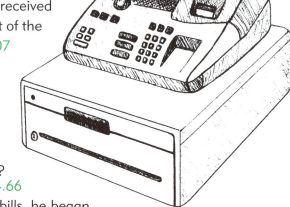

2. The Masts operate a turkey farm. One day they made a deposit that included checks for $315.00 and $285.43, and cash amounting to $54.23. They received a 50-dollar bill in cash from the deposit. What was the total deposit? $604.66

3. One evening when Father was paying bills, he began with a balance of $436.78 in the checking account. He wrote one check to pay the electric bill of $95.67, and one to pay the telephone bill of $76.34. What was the account balance then? $264.77

LESSON 128

Objective
- To give practice with reading problems involving the financial calculations studied in this chapter.

Review

1. Calculate the missing percents on this income statement. (Lesson 127)

   Twin Valley Produce
   Income Statement
   For the Month Ending
   June 30, 20—

	Amount	Percent of Sales
Total sales	$11,000	100%
Cost of goods	6,000	(55%)
Gross profit	5,000	(45%)
Overhead	2,000	(18%)
Net profit	3,000	(27%)

2. Find the net profit. (Lesson 126)

	Income	Cost of Goods	Overhead	Net Profit
a.	$1,582	$795	$431	($356)
b.	$863	$584	$215	($64)

3. Use the compound interest table to find the total interest if it is compounded quarterly. (Lesson 124)
   a. $5,800 at 6% for 2 yr.
      (0.1265 × $5,800 = $733.70)
   b. $14,500 at 9% for 3 yr.
      (0.3060 × $14,500 = $4,437.00))

4. Find the hypotenuse of each right triangle, to the nearest tenth. (Lesson 112)
   a. leg a = 8 in.
      leg b = 3 in.
      hypotenuse = (8.5 in.)
   b. leg a = 6 in.
      leg b = 5 in.
      hypotenuse = (7.8 in.)

5. Solve these problems by first adding or subtracting the percent. (Lesson 64)
   a. $96.15 increased by 9% ($104.80)
   b. $36.95 decreased by 15% ($31.41)

Solutions for CLASS PRACTICE

a. $45.00 + $45.76 + $38.76 + $18.59

b. $3,000 × 1.1323

Solutions for Part A

1. $45.76 + $43.57 + $21.83 + $23.91 − $35.00
2. $315.00 + $285.43 + $54.23 − $50.00
3. $436.78 − ($95.67 + $76.34)

T-435 Chapter 9 *Mathematics and Finances*

Teaching Guide

Solve the *Class Practice* problems in class. Also review any financial calculations from previous lessons that you know have caused difficulty for your class.

---

4. $347.54 + $238.98 − ($215.78 + $22.50 + $45.64)
5. $675.43 + $75.00 − $145.74
6. $798.76 − ($78.34 + $93.45)
7. $2,500 × 0.05 × $\frac{2}{3}$
8. $4,000 × 0.06 × $\frac{1}{3}$
9. $500 × 0.0824
10. $600 × 0.1593
11. $2,500 × 1.1265
12. $3,500 × 1.1608
13. 5 × $2.50 × 0.07
14. $4,500 × 1.06
15. $425,000 × 0.016
16. $125,000 × 0.015
17. $11,340 − ($4,500 + $4,689)
18. $5,527 − ($3,256 + $1,178)

4. Uncle Daryl had $347.54 in his checking account. After making a deposit of $238.98, he wrote checks for $215.78 (car repair), $22.50 (gasoline), and $45.64 (groceries). What was his checkbook balance after these transactions? $302.60

5. When Daniel reconciled his checking account, the ending balance on the statement was $675.43. There was one outstanding check for $145.74 and one outstanding deposit of $75.00. What should his checkbook balance be? $604.69

6. Mark's bank statement shows a balance of $798.76. There are outstanding checks of $78.34 and $93.45, and a service charge of $7.95. What should his checkbook balance be? (Hint: The service charge has already been deducted from the bank statement.) $626.97

7. Father borrowed $2,500 for 8 months at 5% simple interest. How much interest did he owe at the end of the period? $83.33

8. Wayne borrowed $4,000 for 120 days at 6% simple interest. How much interest did he owe at the end of that period? (Use the 360-day year for calculation.) $80.00

9. Vincent deposited $500 in a savings account that drew 4% interest compounded semiannually. Use the compound interest table on page 418 to find the interest for 2 years. $41.20

10. Samuel deposited $600 in an account that paid 6% interest compounded semiannually. What was the total interest at the end of $2\frac{1}{2}$ years? (Use the compound interest table on page 418.) $95.58

11. Uncle Alvin deposited $2,500 in a savings account paying 6% interest compounded quarterly. What was the balance in the account after 2 years? (Use the compound interest table on page 418.) $2,816.25

12. Lewis deposited $3,500 in a savings account earning 5% interest compounded quarterly. Use the compound interest table on page 418 to find the balance in the account at the end of 3 years. $4,062.80

13. Mother bought 5 yards of fabric at $2.50 per yard. How much sales tax did she pay if the rate was 7%? $0.88

14. Father bought a truck for $4,500. Find the total amount that he paid for the truck, including 6% sales tax. $4,770

15. The Martins' farm is assessed at $425,000, and the assessment rate is 16 mills. What is their property tax for one year? $6,800

16. The Seibels live in an area where the assessment rate is 15 mills, and their residence is assessed at $125,000. What is their property tax for one year? $1,875

17. Father paid $4,500 for a number of cattle. He paid a total of $4,689 for feed and other expenses, and he later sold the cattle for $11,340. What was his net profit? $2,151

18. Mr. Phipps paid $3,256 for merchandise that he later sold for $5,527. He calculated that his overhead was $1,178. What was his net profit? $1,093

436   Chapter 9   Mathematics and Finances

REVIEW EXERCISES

B. Calculate the missing percents on this income statement. *(Lesson 127)*

Sun Valley Produce
Income Statement
For the Month Ending
June 30, 20—

	Amount	Percent of Sales
Total sales	$13,000	100%
Cost of goods	8,000	19. 62%
Gross profit	5,000	20. 38%
Overhead	3,000	21. 23%
Net profit	2,000	22. 15%

C. Find the net profit in each exercise. *(Lesson 126)*

	Income	Cost of goods	Overhead	
23.	$1,184	$763	$354	$67
24.	$914	$445	$175	$294

D. Use the compound interest table on page 418 to find the total interest if it is compounded quarterly. *(Lesson 124)*

	p	r	t			p	r	t	
25.	$6,100	7%	2 yr.	$908.29	26.	$10,400	8%	$2\frac{1}{2}$ yr.	$2,277.60

E. Find the hypotenuse of each right triangle, to the nearest tenth of an inch. *(Lesson 112)*

27. leg a = 7 in.   8.1 in.   28. leg a = 6 in.   8.5 in.
    leg b = 4 in.   leg b = 6 in.

F. Solve these problems by first adding or subtracting the percent. *(Lesson 64)*

29. $64.73 increased by 58% $102.27   30. $61.87 decreased by 87% $8.04

Lesson 128 T–436

Solutions for Part D
25. $i = 0.1489 \times \$6{,}100$
26. $i = 0.2190 \times \$10{,}400$

T–437   Chapter 9   Mathematics and Finances

LESSON 129

Objective

- To review the skills taught previously for mental addition, subtraction, multiplication, and division.
  Addition—adding two- and three-digit numbers
  Subtraction—subtracting two- and three-digit numbers
  Multiplication—using the double-and-divide shortcut
     —multiplying by multiples of 10 and 100
  Division—using the divide-and-divide shortcut

Review

1. Give Lesson 129 Quiz (Sales Tax, Property Tax, and Profit).

2. *Calculate the missing percents on this income statement.* (Lesson 127)

   Manchester Furniture
   Income Statement
   For the Month Ending
   July 31, 20—

	Amount	Percent of Sales
Total sales	$54,000	100%
Cost of goods	24,000	(44%)
Gross profit	30,000	(56%)
Overhead	14,000	(26%)
Net profit	16,000	(30%)

3. *Calculate each sales tax or property tax.* (Lesson 125)

Sale	Tax Rate	Tax
a. $64.58	4%	($2.58)

Assessment	Tax Rate	Tax
b. $88,000	18 mills	($1,584)

4. *Find the length of leg a in each triangle, to the nearest tenth. Use the table of square roots.* (Lesson 113)

   a. hypotenuse = 13 in.
      leg b = 9 in.
      leg a = (9.4 in.)

   b. hypotenuse = 15 ft.
      leg b = 11 ft.
      leg a = (10.2 ft.)

5. *Solve these multistep reading problems.* (Lesson 97)

   a. Each day Galen feeds 25 pounds of grain mix per head to his 72 cows. Ground ear corn is 28% of the grain mix. How much ear corn does he feed per week?   (3,528 lb.)

      $72 \times 25 \times 0.28 \times 7 = 3{,}528$

   b. In his pig barn, Brother Merle has 175-watt heat lamps that each draw $1\frac{1}{2}$ amps. How many lamps can he put on a 30-amp circuit without loading it to more than 75% of its rating?   (15 lamps)

      75% of 30 = $22\frac{1}{2}$ amps
      $22\frac{1}{2} \div 1\frac{1}{2} = 15$

6. *Find each percent to the nearest whole number.* (Lesson 65)

   a. 11 is ___ of 85   (13%)
   b. 87 is ___ of 32   (272%)

7. *Read these large numbers.* (Lesson 1)

   a. 2,286,000,000,000   (Two trillion, two hundred eighty-six billion)
   b. 3,082,000,000,000,000   (Three quadrillion, eighty-two trillion)

129. Mental Math: The Four Basic Operations

Mental problem solving is a skill that requires practice both to learn and to maintain. This lesson reviews mental problem solving in the four basic operations.

Adding Mentally

To add mentally, begin with the first addend and add the second to it, starting with the place farthest to the left. It is easier to do mental work from left to right because this is the normal way we read numbers.

Example A	**Example B**
135 + 78 = ___	215 + 187 = ___
Think: 135 + 70 = 205	Think: 215 + 100 = 315
+ 8 = 213	+ 80 = 395
	+ 7 = 402

Subtracting Mentally

To subtract mentally, begin with the place farthest to the left in the subtrahend. Continue by subtracting one place at a time from left to right.

Example C	**Example D**
218 − 46 = ___	361 − 189 = ___
Think: 218 − 40 = 178	Think: 361 − 100 = 261
− 6 = 172	− 80 = 181
	− 9 = 172

Multiplying Mentally

To multiply mentally by a one-digit multiplier, use a left-to-right method similar to that used for mental addition and subtraction.

Example E		**Example F**	
8 × 328 = ___		9 × 276 = ___	
Think: 8 × 300 = 2,400	2,400	Think: 9 × 200 = 1,800	1,800
8 × 20 = 160	+ 160	9 × 70 = 630	+ 630
	2,560		2,430
8 × 8 = 64	+ 64	9 × 6 = 54	+ 54
	2,624		2,484

438 *Chapter 9 Mathematics and Finances*

It is fairly simple to multiply a two-digit number by a multiple of 10 or 100. Multiply the nonzero digits, and annex as many zeroes as there are in the factors.

Example G	**Example H**
60 × 12 = ___	300 × 38 = ___
Think: 6 × 12 = 72	Think: 3 × 30 = 90
Annex 1 zero.	3 × 8 = 24
60 × 12 = 720	90 + 24 = 114
	Annex 2 zeroes.
	300 × 38 = 11,400

Another help for mental multiplication is the double-and-divide method, by which one factor is doubled and the other is divided by 2. This method works well when dividing one factor by 2 produces a single-digit factor. It is also helpful when doubling the other factor produces a more rounded number (such as 70 or 100 from doubling 35 or 50).

Example I	**Example J**
14 × 18 = ___	32 × 50 = ___
Think: 14 × 18 = 7 × 36	Think: 32 × 50 = 16 × 100
7 × 30 = 210	16 × 100 = 1,600
7 × 6 = 42	
210 + 42 = 252	

Dividing Mentally

The divide-and-divide method can simplify some division problems. It works because dividing the dividend and the divisor by the same number does not change the quotient. One way to use this method is to remove the same number of zeroes from the dividend and the divisor (900 ÷ 300 = 9 ÷ 3). This has the effect of dividing both numbers in the problem by the same power of 10.

Some division problems can be simplified by doubling the dividend and the divisor. This works best when doubling a divisor such as 50 or 500 produces a divisor such as 100 or 1,000.

Example K	**Example L**
144 ÷ 18 = ___	3,400 ÷ 50 = ___
Think: Divide both numbers by 2.	Think: Double both numbers.
144 ÷ 18 = 72 ÷ 9	3,400 ÷ 50 = 6,800 ÷ 100
72 ÷ 9 = 8	6,800 ÷ 100 = 68

Introduction

Point out that calculators can become crutches rather than tools if we do not exercise ourselves in mental math. This lesson is designed to sharpen the mental math skills of the students.

Teaching Guide

1. **To add mentally, begin with the first addend and add the second to it, starting with the place farthest to the left.** It is easier to add from left to right because this is the normal way we read numbers.

   a. 228 + 47   (275)
   b. 176 + 86   (262)
   c. 167 + 147   (314)
   d. 193 + 268   (461)

2. **To subtract mentally, begin with the place farthest to the left in the subtrahend. Continue by subtracting one place at a time from left to right.**

   a. 231 − 57   (174)
   b. 145 − 78   (67)
   c. 371 − 189   (182)
   d. 362 − 197   (165)

3. **To multiply mentally by a one-digit multiplier, use a left-to-right method similar to that used for mental addition and subtraction.**

   a. 7 × 278   (1,946)
   b. 9 × 193   (1,737)
   c. 8 × 329   (2,632)
   d. 6 × 426   (2,556)

4. **To multiply a two-digit number by a multiple of 10 or 100, multiply the nonzero digits and annex as many zeroes as there are in the factors.**

   a. 40 × 21   (840)
   b. 30 × 37   (1,110)
   c. 400 × 63   (25,200)
   d. 600 × 54   (32,400)

T–439   Chapter 9   *Mathematics and Finances*

5. **To multiply by the double-and-divide method, double one factor and divide the other by 2.** This method works well when dividing one factor by 2 produces a single-digit factor. It is also helpful when doubling the other factor produces a more rounded number.

   a. 16 × 16   (8 × 32 = 256)
   b. 45 × 18   (90 × 9 = 810)
   c   38 × 50   (19 × 100 = 1,900)
   d. 64 × 50   (32 × 100 = 3,200)

6. **To divide by the divide-and-divide method, divide the dividend and the divisor by the same number.** One way to use this method is to remove the same number of zeroes from the dividend and the divisor. This has the effect of dividing both numbers in the problem by the same power of 10.

   a. 126 ÷ 14   (63 ÷ 7 = 9)
   b. 112 ÷ 16   (56 ÷ 8 = 7)
   c. 154 ÷ 14   (77 ÷ 7 = 11)
   d. 176 ÷ 22   (88 ÷ 11 = 8)

7. **To divide by the double-and-double method, multiply the dividend and the divisor by 2.** This works best when doubling a divisor such as 50 or 500 produces a divisor such as 100 or 1,000. Point out that zeroes could also be dropped in this shortcut, but the problems become so simple that this is unnecessary.

   a. 1,400 ÷ 50   (2,800 ÷ 100 = 28)
   b. 1,900 ÷ 50   (3,800 ÷ 100 = 38)

Solutions for Part E

25. 25 × 2,000 ÷ (6½ × 1,020)

26. 68 × 25 × 365 ÷ 2,000

CLASS PRACTICE

Add mentally.

a. 45 + 81   126
b. 98 + 23   121
c. 184 + 39   223
d. 162 + 78   240
e. 167 + 147   314
f. 193 + 268   461

Subtract mentally.

g. 96 − 58   38
h. 75 − 19   56
i. 112 − 64   48
j. 104 − 37   67
k. 371 − 189   182
l. 362 − 197   165

Multiply mentally.

m. 6 × 25   150
n. 7 × 64   448
o. 9 × 43   387
p. 50 × 38   1,900
q. 400 × 63   25,200
r. 300 × 56   16,800

Divide mentally.

s. 144 ÷ 16   9
t. 112 ÷ 14   8
u. 1,900 ÷ 50   38
v. 4,800 ÷ 120   40
w. 7,500 ÷ 500   15
x. 9,500 ÷ 500   19

WRITTEN EXERCISES

A. *Add mentally.*

1. 78 + 65   143
2. 137 + 67   204
3. 153 + 38   191
4. 178 + 108   286
5. 148 + 169   317
6. 172 + 168   340

B. *Subtract mentally.*

7. 72 − 38   34
8. 115 − 67   48
9. 138 − 69   69
10. 178 − 91   87
11. 215 − 161   54
12. 238 − 149   89

C. *Multiply mentally.*

13. 7 × 46   322
14. 9 × 75   675
15. 7 × 63   441
16. 20 × 61   1,220
17. 40 × 67   2,680
18. 50 × 74   3,700

D. *Divide mentally.*

19. 132 ÷ 22   6
20. 264 ÷ 24   11
21. 1,600 ÷ 50   32
22. 3,100 ÷ 50   62
23. 168 ÷ 14   12
24. 2,700 ÷ 50   54

E. *Solve these multistep reading problems.*

25. When mixing cow feed, Wendell uses 1,020 pounds of shelled corn to make a ton of feed. He uses $6\frac{1}{2}$ tons of feed per week. A 25-ton trailer load of shelled corn will last how many weeks? (Drop any remainder.)   7 weeks

26. Each of Brother Robert's 68 dairy cows eats about 25 pounds of haylage per day. To the nearest whole number, how many tons of haylage does he need in one year?   310 tons

440   Chapter 9   Mathematics and Finances

27. The main fluid area of an adult's eye is a sphere with a radius of about 10 millimeters. What is the volume of this fluid area, to the nearest whole number? (Use 4.18.)
4,180 cubic millimeters

28. On the retina of the eye are about 130 million rods (light-sensing cells) that each have a length of 0.06 millimeter. If all the rods were laid end to end, how many kilometers would they reach? (There are 1,000,000 millimeters in a kilometer.)   7.8 kilometers

29. A high-speed laser printer can print 13,000 lines per minute and 10,000 sheets of paper per hour. How many lines are on each sheet of paper?   78 lines

30. A 100-watt light bulb draws 0.83 amp. How many of these bulbs can be put on a 20-amp circuit without loading it to more than 90% of its rating?   21 bulbs

REVIEW EXERCISES

F. Calculate the missing percents on this income statement. *(Lesson 127)*

Wyman Furniture
Income Statement
For the Month Ending
April 30, 20—

	Amount	Percent of Sales
Total sales	$37,000	100%
Cost of goods	19,000	31. 51%
Gross profit	18,000	32. 49%
Overhead	12,000	33. 32%
Net profit	6,000	34. 16%

G. Calculate each sales tax or property tax. *(Lesson 125)*

35. Sale: $85.14; tax rate: $5\frac{1}{2}$% $4.68
36. Sale: $78.24; tax rate: $7\frac{1}{4}$% $5.67
37. Assessment: $92,000; tax rate: 19 mills $1,748
38. Assessment: $205,000; tax rate: 18.2 mills $3,731

H. Find the length of leg a in each right triangle, to the nearest tenth. Use the table of square roots. *(Lesson 113)*

39. hypotenuse = 12 in.   9.7 in.
    leg b = 7 in.
40. hypotenuse = 14 ft.   9.8 ft.
    leg b = 10 ft.

I. Find each percent to the nearest whole number. *(Lesson 65)*

41. 16 is 21% of 77
42. 74 is 180% of 41

J. Write these numbers, using words. *(Lesson 1)*

43. 1,854,000,000,000
    One trillion, eight hundred fifty-four billion
44. 8,000,000,000,000,000
    Eight quadrillion

27. 4.18×10^3

28. 130×0.06 kilometer (If one rod is 0.06 millimeter, one million rods is 0.06 kilometer.)

29. $13,000 \times 60 \div 10,000$

30. $0.90 \times 20 \div 0.83$

LESSON 130

Objective

- To review the material taught in Chapter 9 (Lessons 117–129).

Teaching Guide

1. Lesson 130 reviews the material taught in Lessons 117–129. For pointers on using review lessons, see *Teaching Guide* for Lesson 98.

2. Be sure to review the following new concepts taught in this chapter.

Lesson number and new concept	Exercises in Lesson 130
117—Using checking account deposit tickets.	1, 2
118—Writing checks.	3, 4
119—Keeping a check register.	5–12
120—Reconciling a checking account.	13, 14
121—Using savings account deposit tickets.	15, 16
121—Calculating simple interest, using daily rates.	21, 22
124—Using the compound interest formula, $a = p(1 + r)^n$.	25, 26
124—Using a compound interest table.	27, 28
125—Calculating property tax.	31, 32
126—Working with gross profit, overhead, and net profit.	33–36

T–441 Chapter 9 Mathematics and Finances

Answers for Part A

1.

DEPOSIT TICKET

[Your name] 56-334/809
[Your street address]
[Your city and state]

Account # 36-5551123

DATE [current date] 20 ____

Bryan National Bank

CURRENCY	37	00
COIN	7	57
CHECKS		
58-717	69	75
62-111	58	94
71-114	91	29
SUBTOTAL	264	55
LESS CASH RECEIVED		
TOTAL DEPOSIT	264	55

2.

DEPOSIT TICKET

[Your name] 56-334/809
[Your street address]
[Your city and state]

Account # 36-5551123

DATE [current date] 20 ____

Bryan National Bank

CURRENCY		
COIN	12	49
CHECKS		
13-351	34	98
27-699	87	61
41-559	74	46
SUBTOTAL	209	54
LESS CASH RECEIVED	90	00
TOTAL DEPOSIT	119	54

Answers for Part B

3.

[Your name] 51-334/985   173
[Your street address]
[Your city and state]   DATE Jan. 23   20 ____

PAY TO THE ORDER OF Lewis Avery $ 79.34

Seventy-nine and 34/100 ———————————— DOLLARS

Bryan National Bank
Bryan, Ohio 43506

MEMO Trucking   [student's signature]

4.

[Your name] 51-334/985   174
[Your street address]
[Your city and state]   DATE Jan. 24   20 ____

PAY TO THE ORDER OF Frank Dudley $ 51.25

Fifty-one and 25/100 ———————————— DOLLARS

Bryan National Bank
Bryan, Ohio 43506

MEMO Office supplies   [student's signature]

Lesson 130

130. Chapter 9 Review

(Forms are provided in quiz booklet.)

A. Complete deposit tickets for these deposits. Use the current date. *(Lesson 117)*

1. Currency $37.00
   Coins 7.57
   Checks:
       58-717 .... 69.75
       62-111 .... 58.94
       71-114 .... 91.29

2. Coins 12.49
   Checks:
       13-351 .... 34.98
       27-699 .... 87.61
       41-559 .... 74.46
   Cash received 90.00

B. Write the checks indicated. Use the current year in the date, and sign your own name as the payer. *(Lesson 118)*

	Check number	Date	Payee	Amount	Memo
3.	173	Jan. 23	Lewis Avery	$79.34	Trucking
4.	174	Jan. 24	Frank Dudley	$51.25	Office supplies

C. Use this information to fill in check registers. *(Lesson 119)*

5–8.   (See page T–442 for completed forms.)   *Running balance*

Balance carried forward: $478.35	$478.35
Check number 427 on 1/29 to Linwood Hardware for $82.62	395.73
Check number 428 on 1/30 to Matthew's Grocery for $68.17	327.56
Deposit from sales on 1/31 for $195.46	523.02

9–12.

Balance carried forward: $419.93	$419.93
Check number 634 on 3/23 to Morris Lumber for $319.69	100.24
Check number 635 on 3/24 to Nolt Supply for $71.14	29.10
Deposit of wages on 3/24 for $419.79	448.89

D. Use reconciliation forms to reconcile the following accounts. Write *yes* or *no* to tell whether the balances agree. *(Lesson 120)*

	Ending bank balance		Deposits outstanding		Checks outstanding		Ending register balance	
13.	$604.50	+	$426.60	–	$565.72	=	$465.37	no
14.	$990.24	+	$339.06	–	$491.96	=	$837.34	yes

335

442   Chapter 9   Mathematics and Finances

E. Complete deposit tickets for these savings deposits. Use your own name and today's date. *(Lesson 121)*

15. Account number: 12-34567890
    Items deposited
        Cash $272.68
        Checks
            11-111 .... 43.37
            22-222 .... 52.29

16. Account number: 12-34567890
    Items deposited
        Cash $121.43
        Checks
            12-345 .... 226.00
            67-890 .... 233.40

F. Calculate simple interest. *(Lessons 121, 122)*

17. $p = \$2{,}000$
    $r = 5\%$ $\$300.00$
    $t = 3$ yr.

18. $p = \$375$
    $r = 7\%$ $\$52.50$
    $t = 2$ yr.

19. $p = \$850$
    $r = 5\%$ $\$258.54$
    $t = 6$ yr. 1 mo.

20. $p = \$3{,}400$
    $r = 6\%$ $\$527.00$
    $t = 2$ yr. 7 mo.

G. Calculate simple interest, using the 360-day year. *(Lesson 122)*

	p	r	t			p	r	t	
21.	$700	5%	210 days	$20.42	22.	$800	9%	105 days	$21.00

H. Find the total interest if it is compounded annually. *(Lesson 123)*

	p	r	t			p	r	t	
23.	$1,000	8%	2 yr.	$166.40	24.	$2,600	7%	2 yr.	$376.74

I. Use the compound interest formula to find the combined principal and interest if interest is compounded semiannually. *(Lesson 124)*

	p	r	t			p	r	t	
25.	$3,000	14%	1 yr.	$3,434.70	26.	$4,000	12%	1 yr.	$4,494.40

J. Use the compound interest table on page 418 to find the combined principal and interest if interest is compounded quarterly. *(Lesson 124)*

	p	r	t			p	r	t	
27.	$5,000	7%	2 yr.	$5,744.50	28.	$6,000	9%	$2\frac{1}{2}$ yr.	$7,495.20

K. Following are two sales receipts with both taxable and nontaxable items. Find the sales tax and the total amount due. *(Lesson 125)*

29.
    27.39 T
    18.63 T
    29.55
    16.36   (taxable: $46.02)
Subtotal   91.93
6% Tax $2.76
Total $94.69

30.
    48.63 T
    69.85
    16.14 T
    28.75   (taxable: $64.77)
Subtotal   163.37
7% Tax $4.53
Total $167.90

Lesson 130 T–442

Answers for Part C

5–8.

NUMBER	DATE	DESCRIPTION	✓	FEE	CHECK AMOUNT	DEPOSIT AMOUNT	BALANCE
		Balance carried forward					$478 35
427	1/29	Linwood Hardware			82 62		395 73
428	1/30	Matthew's Grocery			68 17		327 56
D	1/31	Deposit—Sales				195 46	523 02

9–12.

NUMBER	DATE	DESCRIPTION	✓	FEE	CHECK AMOUNT	DEPOSIT AMOUNT	BALANCE
		Balance carried forward					$419 93
634	3/23	Morris Lumber			319 69		100 24
635	3/24	Nolt Supply			71 14		29 10
D	3/24	Deposit—Wages				419 79	448 89

Calculations for Part D

13. Total: 1,031.10; Adjusted balance: 465.38
14. Total: 1,329.30; Adjusted balance: 837.34

Answers for Part E

15.

SAVINGS DEPOSIT			
DATE [current date] 20 ——	CASH	272	68
	CHECKS		
[student's name]	11-111	43	37
Account name	22-222	52	29
12-34567890			
Account number	SUBTOTAL	368	34
The Longville Bank	LESS CASH RECEIVED		
	TOTAL DEPOSIT	368	34

T–443 Chapter 9 Mathematics and Finances

16.

SAVINGS DEPOSIT		CASH	121	43
DATE [current date] 20 —		CHECKS		
		12-345	226	00
[student's name]		67-890	233	40
Account name				
12-34567890		SUBTOTAL	580	83
Account number		LESS CASH RECEIVED		
The Longville Bank		**TOTAL DEPOSIT**	580	83

Solutions for Part H

23. First year: $1,000 × 0.08 = $80.00
Second year: $1,080 × 0.08 = $86.40
$80.00 + $86.40

24. First year: $2,600 × 0.07 = $182.00
Second year: $2,782 × 0.07 = $194.74
$182.00 + $194.74

Solutions for Part I

25. $a = \$3{,}000 \times 1.07^2 = \$3{,}000 \times 1.1449$
26. $a = \$4{,}000 \times 1.06^2 = \$4{,}000 \times 1.1236$

Solutions for Part J

27. $a = 1.1489 \times \$5{,}000$
28. $a = 1.2492 \times \$6{,}000$

L. Calculate these property taxes. *(Lesson 125)*

31. Assessment: $60,000; tax rate: 19 mills $1,140
32. Assessment: $70,000; tax rate: 21 mills $1,470

M. Calculate the profit in each exercise. *(Lesson 126)*

	Income	Expenses			Income	Expenses	
33.	$3,574	$2,741	$833	34.	$4,489	$2,576	$1,913

N. Calculate the net profit in each exercise. *(Lesson 126)*

	Income	Cost of goods	Overhead	
35.	$379	$216	$38	$125
36.	$483	$297	$77	$109

O. Calculate the missing percents on this income statement. *(Lesson 127)*

Oak Glen Landscaping
Income Statement
For the Year Ending
December 31, 20—

	Amount	Percent of Sales
Total sales	$36,000	100%
Cost of goods	23,000	**37.** 64%
Gross profit	13,000	**38.** 36%
Overhead	7,000	**39.** 19%
Net profit	6,000	**40.** 17%

P. Solve these problems mentally. *(Lesson 129)*

41. 79 + 66   145   42. 138 + 68   206
43. 73 − 39   34   44. 84 − 49   35
45. 8 × 47   376   46. 9 × 59   531
47. 28 × 50   1,400   48. 46 × 50   2,300
49. 162 ÷ 18   9   50. 216 ÷ 24   9
51. 2,600 ÷ 50 52   52. 3,800 ÷ 50 76

Q. Solve these reading problems. *(Lesson 128)*

53. Using the 360-day year, find the interest on $1,000 at 6% for 120 days. (This problem is easy to solve mentally when you find what fraction 120 is of 360.) $20

444   Chapter 9   Mathematics and Finances

54. Using the 360-day year, find the interest on $2,000 at 6% for 60 days. (This problem is easy to solve mentally when you find what fraction 60 is of 360.) $20

55. A residence is assessed at $95,000 in a town with a tax rate of 19 mills per year. One-half of the annual tax is due by May 1. What is the amount of the semiannual tax? $902.50

56. In the same town, a commercial building is assessed at $135,000 at a tax rate of 21 mills. What is the amount of the semiannual tax that is due by May 1? $1,417.50

57. A business with sales of $78,200 has a net profit of $14,600. What percent was the net profit of the sales, to the nearest whole percent?   19%

58. The same business had a gross profit of $28,000 on its annual sales of $78,200. What percent was the gross profit of the sales, to the nearest whole percent?   36%

Solutions for Part Q

53. $1,000 × 0.06 × $\frac{1}{3}$

54. $2,000 × 0.06 × $\frac{1}{6}$

55. $95,000 × 0.019 × $\frac{1}{2}$

56. $135,000 × 0.021 × $\frac{1}{2}$

57. 14,600 ÷ 78,200

58. 28,000 ÷ 78,200

131. Chapter 9 Test

LESSON 131

Objectives

- To test the students' mastery of the concepts in Chapter 9.

Teaching Guide

1. Correct Lesson 130.
2. Review any areas of special difficulty.
3. Administer the test. For pointers on giving tests, see *Teaching Guide* for Lesson 99.

The principles of algebra could be considered laws of nature. They are statements of truth that are just as consistent as elementary arithmetic facts. Abstract mathematical truths have divine origin, just as truly as the natural law that physical fruits always grow from their own specific seed.

Chapter 10
Introduction to Algebra

Algebra is a study of how numbers work in relation to each other. Algebraic principles can be stated without using numbers. For instance, if you know that an onion has the same weight as a certain potato, and that a tomato also has the same weight as that potato, you know that the tomato weighs the same as the onion. Algebra uses letters (literal numbers) to represent numerical values. All the basic mathematical operations can be performed with literal numbers.

The word *algebra* comes from two Arabic words that mean "the reuniting of broken parts." With algebra, you can take facts apart and put them back together in ways that yield answers to problems that might otherwise seem impossible to solve.

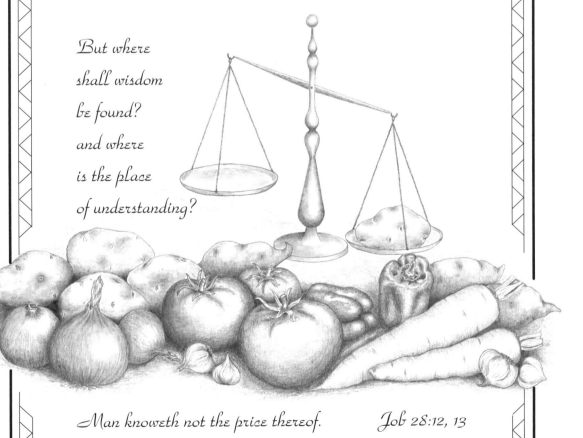

But where shall wisdom be found? and where is the place of understanding?

Man knoweth not the price thereof.   *Job 28:12, 13*

132. Identifying Algebraic Expressions

Many people recognize several symbols for the number three. The Arabic numeral 3, the Roman numeral III, and three fingers are commonly used. You could use a symbol such as ♣ to represent three as long as others know what it means. The branch of mathematics known as **algebra** often uses letters to represent numbers.

Literal Numbers

Letters that represent unknown values are called **literal numbers**. A given letter can represent any numerical value. For example, in $4 + x = 7$, the value of x is 3. In $4 + 3 = x$, the value of x is 7. Within a given expression such as $2 + x + x = 10$, the value of a literal number is always the same. (In this case, both x's equal 4.) If different letters are used, each letter has a different value.

Symbols of Operation

In algebra, addition and subtraction are indicated by the regular symbols + and −. But since the times sign (×) could easily be confused with the literal number x, multiplication is normally shown by placing the factors side by side as in $2lw$. If this does not clearly indicate multiplication, one or both numbers are placed within parentheses. For example, 35 means "thirty-five," but 3(5) means "three times five." Multiplication is also indicated with a raised dot (·). Thus, 5 times 6 can be written as $5 \cdot 6$, 5(6), (5)6, or (5)(6). Division is usually indicated by using the fraction form, as in the formula $r = \frac{d}{t}$.

> **Correct forms in algebra**
> Addition: $4 + a$ $b + c$ Subtraction: $n - 9$ $7 - m$
> Multiplication (arithmetic numbers): 6(7), (6)7, (6)(7), $6 \cdot 7$
> Multiplication (literal numbers): xyz
> Multiplication (arithmetic and literal numbers): $6xyz$
> Division: $\frac{h}{2}$ $\frac{6}{g}$

Algebraic Terms and Expressions

The chalkboard at the front of a classroom is divided into 4 sections. Each section is n feet wide. The total width of the chalkboard can be written as $4n$. This is sometimes called a product because it shows multiplication ($4 \times n$).

A product that contains a literal number, such as $4n$, is an **algebraic term** (ăl′ jə brā′ ĭk). Other examples of terms are n and $6xy$ (meaning $6 \times x \times y$). The literal number n is an algebraic term because it means the same as $1 \times n$, though the 1 is not usually written.

In an algebraic term, each number or letter is called a **factor** or a **coefficient**. (*Coefficient* means "working together.") In the term $6n$, 6 is the **numerical coefficient** of n, and n is the **literal coefficient** of 6. Notice that the numerical coefficient is always placed

LESSON 132

Objectives

- To review that literal numbers are letters that represent numbers.
- To teach *new algebraic terms: factor, coefficient, literal coefficient, numerical coefficient, polynomial, monomial, binomial, trinomial.
- To review the algebraic methods for indicating the four basic mathematical operations.

Review

1. Give Lesson 132 Quiz (Calculating Interest).

2. *Use the compound interest formula to find the total interest if it is compounded quarterly.* (Lesson 124)
   a. $7,500 at 8% for 6 mo. ($303.00)
   b. $8,400 at 12% for 6 mo. ($511.56)

3. *Find the surface area of each cube.* (Lesson 100)
   a. $e = 8.6$ cm
      $a_s = (443.76$ cm$^2)$
   b. $e = 41$ ft.
      $a_s = (10,086$ sq. ft.)

4. *Use the commission formula to solve these problems.* (Lesson 68)
   a. $s = \$195.00; r = 4\frac{1}{2}\%; c = (\$8.78)$
   b. $c = \$25.84; r = 5\%; s = (\$516.80)$

5. *Tell whether each problem illustrates the commutative or associative law of addition.* (Lesson 4)
   a. $4 + 9 = 9 + 4$ (commutative)
   b. $w + (x + y) = (w + x) + y$
      (associative)

T–447   Chapter 10   Introduction to Algebra

Introduction

Ask the students what the word *literal* means. They will probably offer a definition such as "actual; not figurative."

In algebra, the word *literal* means "expressed with a letter." A literal number is a number represented by a letter. Think of the letters in the formulas for area and volume in geometry. Each letter represents a number.

Algebra is a branch of mathematics that uses literal numbers (letter numbers) as well as numerals to calculate.

Teaching Guide

1. **Algebra is the branch of mathematics in which numerals as well as letters are used to represent quantities.** Any letter can be used to represent a number.

2. **Letters that represent unknown values are called literal numbers.** A given letter may represent any numerical value. (This is why literal numbers are also called variables.) Although the numerical value of a given letter will vary in different expressions, the value of one literal number will always be the same within a given expression.

3. **Algebra uses the following symbols to indicate mathematical operations.**

    a. Addition—the symbol +
       five plus $a = 5 + a$.

    b. Subtraction—the symbol −
       a minus $6 = a - 6$.

    c. Multiplication—a raised dot (·) or placing the factors side by side. If placing factors side by side does not clearly indicate multiplication, one or both numbers are placed within parentheses.
       6 times $8 = 6 \cdot 8$ or $6(8)$ or $(6)8$ or $(6)(8)$
       8 times a times $b = 8ab$

    d. Division—the fraction form
       18 divided by $6 = \frac{18}{6}$.

first. For example, 6 times e is always stated as $6e$ and never as $e6$. In a term such as cd, the numerical coefficient is 1 although it is not written.

An **algebraic expression** consists of one or more terms. An expression with only one term is a **monomial** (mŏ nō′ mē əl). If an expression has more than one term, the terms are separated by plus (+) signs or minus (−) signs. An expression with more than one term is a **polynomial** (pŏl′ ē nō′ mē əl). A polynomial with two terms is a **binomial**, and one with three terms is a **trinomial**.

Monomial	Polynomials				
	Binomial		Trinomial		
$4\pi r^2$ — term	$2\pi r^2$ — term	+ $2\pi rh$ — term	$3ab$ — term	+ $2b^2$ — term	− c^2 — term

CLASS PRACTICE

Identify the one or two operations indicated by each expression.

a. $11 - t$ — subtraction b. $r + s$ — addition c. $4 \cdot 8$ — multiplication

d. $\frac{k}{2} + 7$ — division, addition e. $3(6v)$ — multiplication f. $\frac{9-3}{z}$ — subtraction, division

Identify each algebraic expression as a monomial, polynomial, binomial, or trinomial. More than one term will apply for some.

g. $7m$ — monomial h. $5rs + 2st + v$ — polynomial, trinomial i. $fgh - 9n$ — polynomial, binomial

j. $ab + \frac{a}{b} + \frac{6b}{9c} - \frac{8}{5}$ — polynomial k. $3vw^2x^3y^4z^5$ — monomial l. $3n^3 - 2n^2 + 4n$ — polynomial, trinomial

Identify the numerical and literal coefficients in these monomials.

m. $8vw$ — numerical: 8, literal: v, w n. xyz — numerical: 1, literal: x, y, z o. $9ab$ — numerical: 9, literal: a, b

WRITTEN EXERCISES

A. *Identify the one or two operations indicated by each expression.*

1. $9 + x$ — addition
2. $\frac{2a}{b}$ — multiplication, division
3. $4 \cdot 23$ — multiplication
4. $\frac{15}{y} - 2$ — division, subtraction
5. $5m + n$ — multiplication, addition
6. $(6)\frac{18}{k}$ — multiplication, division

B. *Identify each algebraic expression as a monomial or polynomial.*

7. $5b$ — monomial
8. $8n - 3n$ — polynomial
9. $6m^3 + 3m^2 - 4m$ — polynomial
10. $\frac{7k}{3m}$ — monomial

C. *Write the number of terms in each polynomial.*

11. $7m^3 + 7m + 7$ — 3
12. $6n^3 - 7xy$ — 2
13. $7n^3 - n^2 + n + 8$ — 4
14. $10a^3 - a^2 + b^3$ — 3

448 Chapter 10 *Introduction to Algebra*

D. Copy these terms. Underline the numerical coefficients and circle the literal coefficients.

15. 6ab   6<u>ab</u> (ab circled)
16. 7mn   <u>7</u>(mn)
17. 10xyz   <u>10</u>(xyz)
18. 15xy   <u>15</u>(xy)

E. Solve these reading problems.

19. Calculate the interest on $5,000 put into a certificate of deposit bearing 6% interest compounded semiannually for one year. $304.50

20. Calculate the interest on $6,000 put into a certificate of deposit bearing 8% interest compounded semiannually for one year. $489.60

21. Jewel's younger sister is coloring a cardboard box that is a 6-inch cube. Calculate the surface area of the cube.   216 square inches

22. Dale built a rabbit cage in the form of a 3-foot cube. What is the outside surface area of the cage?   54 square feet

23. Uncle Dwight receives a 5% sales commission for his wages. If he sells $350.00 in merchandise, what is his commission? $17.50

24. At a 5% commission, what must be the amount of Uncle Dwight's sales in order to earn $45.00? $900.00

REVIEW EXERCISES

F. Use the compound interest formula to find the total principal plus interest if interest is compounded semiannually. *(Lesson 124)*

	p	r	t			p	r	t	
25.	$6,000	12%	1 yr.	$6,741.60	26.	$3,200	16%	1 yr.	$3,732.48

G. Find the surface area of each cube. *(Lesson 100)*

27. $e = 7.4$ cm   328.56 cm²
28. $e = 34$ ft.   6,936 sq. ft.

H. Use the commission formula to solve these problems. *(Lesson 68)*

29. $s = \$3{,}620$; $c = \$126.70$; $r = \underline{3\frac{1}{2}}\%$
30. $c = \$253.50$; $r = 7\%$; $s = \underline{\$3{,}621.43}$

I. Write *commutative law* or *associative law* to tell what each problem illustrates. *(Lesson 4)*

31. $b + c = c + b$ commutative
32. $(2 + 4) + (m + 6) = 2 + (4 + m) + 6$ associative

4. **In an algebraic expression that indicates multiplication, each number or letter is called a factor or a coefficient.** A number is a numerical coefficient, and a letter is a literal coefficient. The numerical coefficient always comes first in such an expression. If an expression contains only literal numbers, the numerical coefficient is 1.

n times $8 = 8n$, not $n8$
x times $1 = x$

Have students pick out the coefficients in the following algebraic expressions.

  a. $6d$ $(6, d)$
  b. $9ab$ $(9, a, b)$
  c. $6abc$ $(6, a, b, c)$

5. **When an algebraic expression consists of two or more parts joined by a plus or minus sign, each part is called a term.** An expression with only one term is a monomial, and one with two or more terms is a polynomial. A polynomial with two terms is a binomial, and one with three terms is a trinomial.

Identify each expression as a monomial or a polynomial.

  a. $asg + 2wqf$ (polynomial)
  b. $3klm$ (monomial)
  c. $4klm - nop + pqr$ (polynomial)
  d. $s + t + u - v + \frac{x}{w}$ (polynomial)

Identify each expression as a monomial, a binomial, or a trinomial.

  a. $4abc + 2def$ (binomial)
  b. $3mno$ (monomial)
  c. $3abc - def + ghi$ (trinomial)
  d. $\frac{14fg}{abcd}$ (monomial)

Solutions for Part E

19. $\$5{,}000 \times 0.06 \times \frac{1}{2} = \$150$
    $\$5{,}150 \times 0.06 \times \frac{1}{2} = \$154.50$
    $\$150.00 + \$154.50$

20. $\$6{,}000 \times 0.08 \times \frac{1}{2} = \$240$
    $\$6{,}240 \times 0.08 \times \frac{1}{2} = \$249.60$
    $\$240.00 + \$249.60$

21. $6 \times 6 \times 6$

22. $3 \times 3 \times 3$

23. $0.05 \times \$350$

24. $0.05 \times \underline{} = \$45$; $\$45 \div 0.05$

LESSON 133

Objectives

- To review the correct order of operations: solve multiplication and division first and then addition and subtraction.
- To review that calculations within parentheses should be done first.
- To teach the use of *grouping symbols: parentheses, brackets, and the vinculum.

Review

1. *Give the numerical coefficients of these terms.* (Lesson 132)

    a. $8x^2y$ (8)

    b. a^3b^4 (1)

2. *Solve by mental calculation.* (Lessons 129 and 5)

    a. 365 + 191 (556)

    b. 74 + 36 (110)

    c. 891 + 221 (1,112)

    d. 1,250 + 1,175 (2,425)

3. *Calculate these property taxes.* (Lesson 125)

	Assessment	Tax Rate	Tax
a.	$42,500	18 mills	($765.00)
b.	$185,400	19.1 mills	($3,541.14)

4. *Find the surface area of each rectangular solid.* (Lesson 101)

    a. l = 54 in.
    w = 34 in.
    h = 16 in.
    a_s = (6,488 sq. in.)

    b. l = 41.8 m
    w = 19.1 m
    h = 9 m
    a_s = (2,692.96 m^2)

5. *Solve these percent problems mentally.* (Lesson 69)

    a. $33\frac{1}{3}$% of 18 = ___ (6)

    b. 80% of 40 = ___ (32)

133. Order of Operations

Algebraic expressions can be **simplified** by doing the mathematical operations indicated. For example, $2n + 3n$ is simplified as $5n$. But if you simplify $3 + 5 \cdot 7$, do you get 56 or 38? Does $12 - 2 \cdot 5$ equal 50 or 2? Is it possible to simplify an algebraic expression in two different ways? No, there can be only one correct result.

There are two basic rules for the order of operations.
(1) Do any multiplications or divisions in the order they appear.
(2) Do additions and subtractions in the order they appear.

You have used these rules with various formulas, such as $a_s = 2lw + 2wh + 2lh$ for the surface of a rectangular solid. With this formula, you (1) multiply each set of dimensions, and then (2) add the products.

Example A	**Example B**
Simplify $3 + 5 \cdot 7$	Simplify $4 + 2(5) - \frac{10}{2} + 3 \cdot 6 - (2)3(4)$
$= 3 + 35$	
$= 38$	$= 4 + 10 - 5 + 18 - 24$
	$= 14 - 5 + 18 - 24$
	$= 9 + 18 - 24$
	$= 27 - 24$

Sometimes the meaning of an algebraic expression requires that the operations be done in a special order. Grouping symbols must be used when operations are to be done in an order different from the basic rules. These symbols are parentheses ( ), brackets [ ], and the fraction vinculum ―. An expression enclosed within grouping symbols should be simplified before working with any terms outside the symbols.

$(12 - 2)5$ means $10 \cdot 5$, or 50.
$12 - 2 \cdot 5$ means $12 - 10$, or 2.

If grouping symbols enclose polynomials, apply the basic rules for order of operations inside the parentheses or brackets. In $4(11 - 5 \cdot 2) + 6$, the multiplication inside the parentheses must be done before the subtraction.

If grouping symbols enclose another set of grouping symbols, solve the operations inside the inner group first. In $2 + [4(11 - 5 \cdot 2) + 6]2$, start inside the parentheses. When that is simplified, work on the steps inside the brackets, as illustrated below.

$2 + [4(11 - 5 \cdot 2) + 6]2$
$2 + [4(11 - 10) + 6]2$
$2 + [4(1) + 6]2$ (The parentheses are retained to show multiplication of $4 \cdot 1$.)
$2 + [4 + 6]2$
$2 + [10]2$ (The brackets are retained to show multiplication of $10 \cdot 2$.)
$2 + 20$
22

If an expression includes a numerical coefficient with an exponent, simplify this value first of all. The first step in simplifying $2^2 + [4(11 - 5 \cdot 2) + 6]2$ is to change 2^2 to 4.

450   Chapter 10   Introduction to Algebra

The steps below give the basic order for simplifying algebraic expressions.

Memorize these rules for the order of operations.

1. Simplify numerical coefficients with exponents.
2. Simplify expressions within parentheses. Drop the parentheses unless they are needed to separate numerical coefficients.
3. Simplify expressions within brackets. Drop the brackets unless they are needed to separate numerical coefficients.
4. Do multiplications and divisions in the order they appear.
5. Do additions and subtractions in the order they appear.

Example C

Simplify $\frac{14-8}{5-2} + 7$.

Simplify above and below vinculum.
$$\frac{14-8}{5-2} + 7$$
$$= \frac{6}{3} + 7$$

Do multiplication and division in order.
$$= 2 + 7$$

Do addition and subtraction in order.
$$= 9$$

Example D

Simplify $4[3^3(7-3)]$.
Simplify numerical coefficient with exponent.
$$4[3^3(7-3)]$$
$$= 4[27(7-3)]$$
Simplify within parentheses.
$$= 4[27(4)]$$
Simplify within brackets.
$$= 4[108]$$
Do the multiplication.
$$= 432$$

CLASS PRACTICE

Simplify these expressions.

a. $\frac{36}{4} + 18 \cdot 4$ 81   b. $(6-3)(4+7)$ 33   c. $3 + \frac{7+3}{2}$ 8

d. $2[3 + 3(12 - 3^2)]$ 24   e. $3 + 5 \cdot 6 - 4^2$ 17   f. $\frac{12}{8 - 2 \cdot 2} + 5 \cdot 6$ 33

WRITTEN EXERCISES

A. *Copy the part of each expression that you must simplify first.*

1. $3 \cdot 2 + 5$ $3 \cdot 2$ 2. $4(3^2 \times 2)$ 3^2 3. $2 + \frac{10-4}{3} \cdot 2$ $10 - 4$

4. $3[4 + (7-2) + 6]$ $7-2$ 5. $5[\frac{15}{2} - (2+5)]$ $2+5$ 6. $5 \cdot \frac{8+6}{7}$ $8+6$

B. *Choose the correct answer for each of these.*

7. $15 + 7 \cdot 9 = (78 \text{ or } 198)$ 78   8. $48 - 12 \cdot 3 = (12 \text{ or } 108)$ 12

9. $\frac{42}{14}(16 + \frac{60}{4}) = (63 \text{ or } 93)$ 93   10. $(17 + 3^2 \cdot 2) + 4 \cdot 6 = (59 \text{ or } 76)$ 59

Lesson 133 T-450

Introduction

Write the following problem on the board.

16 − 3 · 2 + 5 = (5, 15, 31, 91)

Which of the answers is correct? Could they all be right? According to the rules for order of operations, the only correct answer is 15.

CLASS PRACTICE *Step-by-step*

a. 9 + 18 · 4   (divide)
   9 + 72   (multiply)

b. (3)(11)   (simplify within parentheses)

c. $3 + \frac{10}{2}$ (simplify above vinculum)
   3 + 5   (divide)

d. 2[3 + 3(12 − 9)]   (simplify exponent)
   2[3 + 3(3)]   (simplify within parentheses)
   2[3 + 9]   (within brackets: multiply first)
   2[12]   (simplify within brackets)

e. 3 + 5 · 6 − 16   (simplify exponent)
   3 + 30 − 16   (multiply)
   33 − 16   (add)

f. $\frac{12}{8-4} + 5 \cdot 6$ (below vinculum: multiply first)
   $\frac{12}{4} + 5 \cdot 6$ (simplify below vinculum)
   3 + 5 · 6   (divide)
   3 + 30   (multiply)

Teaching Guide

1. **There are two basic rules for the correct order of operations.**

   a. Do any multiplications and divisions in the order they appear.

   b. Do additions and subtractions in the order they appear.

   3 + 6 × 2 − 9 ÷ 3 × 2 − 2 + 12 ÷ 4 − 2 × 3 − 3

   If this is done as a mixed-computation drill, working every calculation in the order it comes, the result is 3. Applying the first rule for order of operations changes it to the following chain of additions and subtractions.

   3 + 12 − 6 − 2 + 3 − 6 − 3

   When the computations are done in this order, the result is 1. Inclusion of grouping symbols could produce a wide variety of other results.

2. **Grouping symbols are used when operations are to be done in an order different from that stated above.** These symbols are parentheses ( ), brackets [ ], and the fraction vinculum —. Solve the operations within these symbols by doing multiplication and division before addition and subtraction. Then proceed to solve the rest of the problem.

T-451   Chapter 10   Introduction to Algebra

3. **Memorize these rules for the order of operations.**

   (1) Simplify numerical coefficients with exponents.

   (2) Simplify expressions within parentheses. Drop the parentheses unless they are still needed in an expression such as 3(4).

   (3) Simplify expressions within brackets. Drop the brackets unless they are still needed in an expression such as 5[6].

   (4) Do multiplications and divisions in the order they appear.

   (5) Do additions and subtractions in the order they appear.

   a. $5 \cdot 7(6 + 3^2) =$ ___ (525)
       $5 \cdot 7(6 + 9)$
       $5 \cdot 7(15)$

   b. $(8 + 2)4 - 6(3^2 - 5) =$ ___ (16)
       $(8 + 2)4 - 6(9 - 5)$
       $(10)4 - 6(4)$
       $40 - 24$

   c. $4 + 9[8(12 - 6)] =$ ___ (436)
       $4 + 9[8(6)]$
       $4 + 9[48]$
       $4 + 432$

   d. $(10 - 4)[6(\dfrac{5^2 + 7}{24 \div 6})] =$ ___ (288)
       $(10 - 4)[6(\dfrac{25 + 7}{24 \div 6})]$
       $(10 - 4)[6(\dfrac{32}{4})]$
       $6[6(8)]$

An Ounce of Prevention

Be sure students understand the basic rule taught in the lesson: Look for the innermost portion and apply the order of operations there. Work out through each set of grouping symbols, applying the order of operations at each level.

Part C Step-by-step

11. $6 + 48$ (multiply)

12. $12 + 24$ (multiply)

13. $\dfrac{18}{3} \cdot 25 - 20$ (simplify exponent)
    $6 \cdot 25 - 20$ (divide)
    $150 - 20$ (multiply)

14. $24 + 9 \cdot 7$ (simplify exponent)
    $24 + 63$ (multiply)

15. $(24)4$ (simplify within parentheses)

16. $6(5)$ (simplify within parentheses)

17. $5(2 + 9) + 3$ (simplify exponent)
    $5(11) + 3$ (simplify within parentheses)
    $55 + 3$ (multiply)

18. $6[(2)2 + 5]$ (simplify within parentheses)
    $6[4 + 5]$ (within brackets: multiply first)
    $6[9]$ (simplify within brackets)

19. $2[(18-4) - 2(2)]$ (simplify exponent)
    $2[14 - 2(2)]$ (simplify within parentheses)
    $2[14 - 4]$ (within brackets: multiply first)
    $2[10]$ (simplify within brackets)

20. $[10(4) + 2]2$ (simplify within parentheses)
    $[40 + 2]2$ (within brackets: multiply first)
    $[42]2$ (simplify within brackets)

Solutions for Part D

21. $3 \times (540 + 185)$

22. $6 \times (12 + 5)$

23. $11 + 2(21 - 11)$

24. $5 \times 9 + 3$

25. $2(12 \times 8) + 2(12 \times 4) + 2(8 \times 4)$

26. $2(20 \times 7\frac{1}{2})$ sides $+ 1(7 \times 7\frac{1}{2})$ front $+ 1(20 \times 7)$ roof

C. Simplify these expressions.

11. 6 + 12 · 4   54
12. 12 + 4 · 6   36
13. $\frac{18}{3} \cdot 5^2 - 20$ 130
14. $24 + 3^2 \cdot 7$ 87
15. (33 – 9)4   96
16. 6(3.6 + 1.4)   30
17. $5(2 + 3^2) + 3$ 58
18. 6[(4 – 2)2 + 5]   54
19. $2[(18 – 2^2) – 2(2)]$ 20
20. [10(14 – 10) + 2]2   84

D. Solve these reading problems.

21. Marcus is mixing a triple batch of feed for the steers. A single batch contains 540 pounds of silage plus 185 pounds of ground corn. How many pounds of feed will Marcus mix?   2,175 pounds

22. In 6 days one week, the Mussers' average daily sales were 12 dozen ears of yellow corn and 5 dozen ears of white corn. How many dozen ears of sweet corn did they sell in all?
                    102 dozen

23. Sherry wants to bake 2 cookies for each adult and 1 cookie for each child that is invited to a Sunday dinner. Of the 21 people invited, 11 are children. How many cookies does Sherry need to bake?   31 cookies

24. The width of the hallway at Lakeview School is 9 feet. Its length is 3 feet longer than 5 times its width. How long is the hall?   48 feet

25. What is the outside surface area of a cereal box 12 inches high, 8 inches long, and 4 inches wide?   352 square inches

26. A truck has a box 20 feet long, 7 feet wide, and $7\frac{1}{2}$ feet high for carrying cargo. How many square feet of aluminum sheeting are needed to cover the sides, front, and roof of this box? (Do not include the back end or the floor.) $492\frac{1}{2}$ square feet

REVIEW EXERCISES

E. Identify each expression as a *monomial* or *polynomial*. (Lesson 132)

27. $7n^2 + 4n^2$ polynomial
28. $\frac{24}{4}xy$ monomial

F. Solve by mental calculation. (Lessons 129 and 5)

29. 219 + 174   393
30. 157 + 318   475

G. Calculate these property taxes. (Lesson 125)

31. Assessment: $64,100; tax rate: 17 mills $1,089.70
32. Assessment: $175,400; tax rate: 21.2 mills $3,718.48

H. Find the surface area of each rectangular solid. (Lesson 101)

33. *l* = 31 in.; *w* = 28 in.; *h* = 8 in.
                    2,680 sq. in.
34. *l* = 16.8 m; *w* = 11.5 m; *h* = 7 m
                    782.6 m²

I. Solve these percent problems mentally. (Lesson 69)

35. 50% of 46 = 23
36. $16\frac{2}{3}$% of 540 = 90

452 Chapter 10 Introduction to Algebra

134. Evaluating Algebraic Expressions

Evaluating an algebraic expression means replacing the literal numbers with numerical values and then finding the value of the expression. This replacing is known as **substitution**.

You have used substitution when working with various formulas. For example, the formula for finding the perimeter of a square is $p = 4s$. To find the perimeter of a 6-inch square, you replace s with 6 and multiply: $p = 4 \cdot 6 = 24$. The perimeter is 24 inches.

Here are the steps for evaluating algebraic expressions.

1. Rewrite the algebraic expression, replacing each literal number with its given value. For this initial rewriting, place each substitution within parentheses. This will help to avoid confusing separate coefficients, and establishing the habit now will benefit you later in work with signed numbers.

2. Simplify the expression. Follow the mathematical rules for the order of operations as given in Lesson 133.

Example A

Evaluate $\frac{1}{2}bh$ if $b = 8$ and $h = 6$

$\frac{1}{2}bh$
$= \frac{1}{2}(8)(6)$
$= 24$

Example B

Evaluate $\frac{4a(2b - c)}{b^2 + 9}$ if $a = 5$, $b = 4$, and $c = 3$

$\frac{4a(2b - c)}{b^2 + 9}$

$= \frac{4(5)[2(4) - (3)]}{(4)^2 + 9}$

$= \frac{20[8 - 3]}{16 + 9}$

$= \frac{20[5]}{25}$

$= \frac{100}{25}$

$= 4$

CLASS PRACTICE

Evaluate each expression if $a = 6$, $b = 2$, and $c = 8$.

a. $c - a + b$ 4   b. $2ab + 2bc + 2ac$ 152   c. $5c - 5b^2$ 20

Evaluate each expression if $x = 7$, $y = 8$, and $z = 9$.

d. $\frac{y^2 - y}{x}$ 8   e. $4x(2z - y)$ 280   f. $x^2(2y - z - x)$ 0

LESSON 134

Objectives

- To review finding the value of algebraic expressions by substituting the given values for the literal numbers.
- To teach *how to evaluate expressions with exponents.

Review

1. Simplify each expression, using the correct order of operations. (Lesson 133)

    a. $(12 + 2^2)(11 - 8) + 5^2 =$ ___ (73)
    $(12 + 4)(11 - 8) + 25$
    $16(3) + 25$

    b. $3[6 + 4(3 - 1)] - 21 =$ ___ (21)
    $3[6 + 4(2)] - 21$
    $3[6 + 8] - 21$
    $3(14) - 21$

2. Give the number of terms in each polynomial. (Lesson 132)

    a. $2x + yz$ (2)
    b. $5a + b - 2c$ (3)

3. Calculate the net profit in each exercise. (Lesson 126)

	Income	Cost of Goods	Overhead	Net Profit
a.	$215	$104	$48	($63)
b.	$342	$165	$112	($65)

4. Find the surface area of each cylinder. (Lesson 102)

    a. $r = 2$ in.
    $h = 8$ in.
    $a_s = (125.6$ sq. in.$)$

    b. $r = 6.1$ m
    $h = 8.4$ m
    $a_s = (555.466$ m$^2)$

5. Use a sketch to solve each problem. (Lesson 70)

a. Matthew is helping his father put aluminum roofing on a small shed. It takes 12 sheets placed side by side to reach from one end to the other. Matthew uses 9 screws along each edge of a sheet (lengthwise), plus 2 additional screws at each end (crosswise). The edge of one sheet overlaps the next one, thus reducing the number of screws needed. How many screws does Matthew need to secure 12 sheets? (165 screws)

11 sheets like this →
(13 screws)
11×13
$= 143$ screws

1 sheet like this →
(22 screws)
$143 + 22$
$= 165$

CLASS PRACTICE Step-by-step

a. $(8) - (6) + (2)$ (substitute given values)

b. $2(6)(2) + 2(2)(8) + 2(6)(8)$ (substitute given values)
$24 + 32 + 96$ (multiply)

c. $5(8) - 5(2)^2$ (substitute given values)
$5(8) - 5(4)$ (simplify exponent)
$40 - 20$ (multiply)

d. $\dfrac{(8)^2 - (8)}{(7)}$ (substitute given values)

$\dfrac{64 - 8}{7}$ (simplify exponent)

$\dfrac{56}{7}$ (simplify above vinculum)

e. $4(7)[2(9) - (8)]$ (substitute given values)
$4(7)[18 - 8]$ (within brackets: multiply first)
$4(7)[10]$ (simplify within brackets)
$4(70)$ (multiply)

f. $(7)^2[2(8) - (9) - (7)]$ (substitute given values)
$49[2(8) - 9 - 7]$ (simplify exponent)
$49[16 - 9 - 7]$ (within brackets: multiply first)
$49[0]$ (simplify within brackets)

T–453 Chapter 10 Introduction to Algebra

b. Margaret must allow enough space on her market stand to accommodate 20 quart boxes that are $5\frac{1}{2}$ inches square. She sets the boxes $\frac{1}{2}$ inch apart in 4 rows. What is the length and width of the area that is needed? ($29\frac{1}{2}$ in. by $23\frac{1}{2}$ in.)

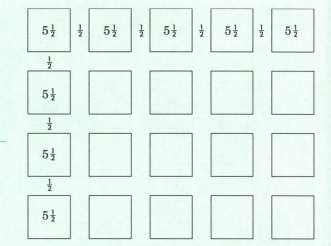

Solutions for Parts A–D
Part A

1. (3)(5) (substitute given values)

2. $(5)^2 - (3)(4)$ (substitute)
   25 – (3)(4) (simplify exponent)
   25 – 12 (multiply)

3. 2(3) + 4(5) (substitute)
   6 + 20 (multiply)

4. (3)[2(4) – (5)] (substitute)
   3[8 – 5] (within brackets: multiply first)
   3[3] (simplify within brackets)

5. $(3)^3 + (5) - (4)^2$ (substitute)
   27 + 5 – 16 (simplify exponents)

6. (3)[(5) + (4)] (substitute)
   3[9] (simplify within brackets)

Part B

7. $(6)^2 - (4)$ (substitute)
   36 – 4 (simplify exponent)

8. $\frac{2(6)}{(6)}$ (substitute)
   $\frac{12}{6}$ (simplify above vinculum)

9. 8(6) – 3(4) (substitute)
   48 – 12 (multiply)

10. $2(4)^2 - 5(6)$ (substitute)
    2(16) – 5(6) (simplify exponent)
    32 – 30 (multiply)

11. $\frac{(4)^2 + 2}{(6)}$ (substitute)
    $\frac{16 + 2}{6}$ (simplify exponent)
    $\frac{18}{6}$ (simplify above vinculum)

12. (6) + 2(4) (substitute)
    6 + 8 (multiply)

Part C

13. (3)(5)(7) (substitute)

14. (3)(5) – (7) (substitute)
    15 – 7 (multiply)

15. (7)[(5) – (3) – 2] (substitute)
    7[0] (simplify within brackets)

16. $\frac{(7)}{(5) - (3)}$ (substitute)
    $\frac{7}{2}$ (simplify below vinculum)

Part D

17. (7) + (7)(8) – (8) (substitute)
    7 + 56 – 8 (multiply)

18. 2[(6) – 1](5) (substitute)
    2[5](5) (simplify within brackets)

19. [(5) + (6)][(6) – (5)] (substitute)
    [11][1] (simplify within brackets)

20. $\frac{(6) + (8)}{(8) - (6)}$ (substitute)
    $\frac{14}{2}$ (simplify above and below vinculum)

Solutions for Part E

21. $t = \frac{d}{r}$; 20 ÷ 50

22. $t = \frac{d}{r}$; 60 ÷ 90

23. $6\frac{1}{2} \times 60 \times 0.21$

24. 33 + 22(7 – 1)

WRITTEN EXERCISES

A. *Evaluate each expression if* $w = 3$, $x = 5$, *and* $y = 4$.

1. wx — 15
2. $x^2 - wy$ — 13
3. $2w + 4x$ — 26
4. $w(2y - x)$ — 9
5. $w^3 + x - y^2$ — 16
6. $w(x + y)$ — 27

B. *Evaluate each expression if* $a = 6$ *and* $b = 4$.

7. $a^2 - b$ — 32
8. $\dfrac{2a}{a}$ — 2
9. $8a - 3b$ — 36
10. $2b^2 - 5a$ — 2
11. $\dfrac{b^2 + 2}{a}$ — 3
12. $a + 2b$ — 14

C. *Evaluate each expression if* $x = 3$, $y = 5$, *and* $z = 7$.

13. xyz — 105
14. $xy - z$ — 8
15. $z(y - x - 2)$ — 0
16. $\dfrac{z}{y - x}$ — $3\frac{1}{2}$

D. *Evaluate each expression by substitution, using the values indicated.*

$a = 7$; $b = 8$; $x = 6$; $y = 5$

17. $a + ab - b$ — 55
18. $2(x - 1)y$ — 50
19. $(y + x)(x - y)$ — 11
20. $\dfrac{x + b}{b - x}$ — 7

E. *Solve these reading problems. Draw sketches for numbers 25 and 26.*

21. How long does it take a car traveling 50 miles per hour to go 20 miles? — $\frac{2}{5}$ hour (or 24 minutes)

22. A train travels at the rate of 90 miles per hour between two cities 60 miles apart. How long does it take to travel this distance? — $\frac{2}{3}$ hour (or 40 minutes)

23. A person breathes approximately 0.21 cubic foot of air per minute. At that rate, how many cubic feet of air do you breathe during $6\frac{1}{2}$ hours at school? — 81.9 cubic feet

24. In 1999, the first-class postage rate in the United States was 33 cents for the first ounce and 22 cents for each additional ounce. What was the first-class postage for a 7-ounce parcel? — $1.65

25. Frank counted 13 telephone poles along the road that runs by his house. The poles were evenly spaced at 190-foot intervals; but where the road curved, one pole stood only 120 feet away from the poles on either side. What was the total distance that the poles stretched from one end of the road to the other? — 2,140 feet
(See page T–455 for sample sketch.)

454 Chapter 10 Introduction to Algebra

26. Maribeth is setting the table for Sunday dinner. The table is 15 feet long. The 10-inch plates are spaced 10 inches apart along each side. She allows a space of 15 inches between the ends of the table and where she places the first and the last plates along the sides. If she places 2 plates at both ends of the table, how many plates are there in all? (See page T–455 for sample sketch.) 20 plates

We give thee thanks, O Lord God Almighty, which art, and wast, and art to come . . . — Revelation 11:17

REVIEW EXERCISES

F. Simplify each expression, using the correct order of operations. *(Lesson 133)*

27. $(14 - 2)(18 - 9) - 8$ 100

28. $4[5 + 3(5 - 4)] - 15$ 17

G. Write the number of terms in each polynomial. *(Lesson 132)*

29. $2b + 7c - 3d$ 3

30. $9xy - w$ 2

H. Calculate the net profit in each exercise. *(Lesson 126)*

31. Income: $218; cost of goods: $99; overhead: $45 $74

32. Income: $352; cost of goods: $172; overhead: $105 $75

I. Find the surface area of each cylinder. *(Lesson 102)*

33. $r = 3$ in.; $h = 9$ in.   226.08 sq. in.

34. $r = 12$ ft.; $h = 35$ ft.   3,541.92 sq. ft.

J. Solve these subtraction problems. *(Lesson 6)*

35. 26,125
    − 6,948
    19,177

36. $19,894.08
    − 6,248.94
    $13,645.14

Lesson 134   T-454

Introduction

Write the formula $p = 4s$ on the board. Ask the students if they can find the value of p. This cannot be done because the value of s has not been stated.

The students should recognize $p = 4s$ as the formula for the perimeter of a square. By assigning a value to s, the value of p can be found for any square. When we assign such a value, we can evaluate (find the value of) an algebraic expression.

Teaching Guide

1. **Evaluating an algebraic expression means replacing the literal numbers with numerical values and then finding the value of the expression.** This replacing is known as substitution.

2. **To evaluate algebraic expressions, use the following steps.**

   (1) Rewrite the algebraic expression.

   (a) Replace each literal number with empty parentheses.
   As the first step in evaluating $x + x(y - z)$ if $x = 8$, $y = 9$, $z = 3$, write $(\ ) + (\ )[(\ ) - (\ )]$.

Note: Parentheses in the original expression become brackets when parentheses are inserted with the substitution. Brackets are not necessarily changed back to parentheses in the simplifying process.

   (b) Fill in each set of parentheses with the corresponding given value. $(8) + (8)[(9) - (3)]$

   (2) Simplify the expression. Follow the mathematical rules for the order of operations as given in Lesson 133. Remember that the fraction vinculum is a grouping symbol as well as a division sign.

Evaluate each expression by using substitution.

a. $ab - a + b$ if $a = 8$ and $b = 7$
$ab - a + b$
$= (\ )(\ ) - (\ ) + (\ )$
$= (7)(8) - (8) + (7)$
$= 56 - 8 + 7$
$= 55$

b. $3(a + 5)y$ if $a = 4$ and $y = 6$
$3(a + 5)y$
$= 3[(\ ) + 5](\ )$
$= 3[4 + 5]6$
$= 3[9]6$
$= 162$

c. $5(a + b)[(b - a)5 + b]$
if $a = 4$ and $b = 5$
$5(a + b)[(b - a)5 + b]$
$5[(\ ) + (\ )]\{[(\ ) - (\ )]5 + (\ )\}$
$= 5[4 + 5]\{[5 - 4]5 + 5\}$
$= 5[9]\{[1]5 + 5\}$
$= 5[9]\{5 + 5\}$
$= 5[9]\{10\}$
$= 450$

Note the use of braces to enclose operations in brackets.

d. $y + (\frac{2y - z}{5} + z)x$
if $x = 5$, $y = 9$, and $z = 3$
$y + (\frac{2y - z}{5} + z)x$
$= (\ ) + [\frac{2(\ ) - (\ )}{5} + (\ )](\ )$
$9 + [\frac{2(9) - 3}{5} + 3]5$
$= 9 + [\frac{18 - 3}{5} + 3]5$
$= 9 + [\frac{15}{5} + 3]5$
$= 9 + [3 + 3]5$
$= 9 + [6]5$
$= 9 + 30$
$= 39$

Solutions for Part F

27. $(12)(9) - 8$ (simplify within parentheses)
    $108 - 8$ (multiply)

28. $4[5 + 3(1)] - 15$ (simplify within parentheses)
    $4[5 + 3] - 15$ (within brackets: multiply first)
    $4[8] - 15$ (simplify within brackets)
    $32 - 15$ (multiply)

T–455 Chapter 10 Introduction to Algebra

LESSON 135

Objectives

- To teach *recognizing like terms and unlike terms.
- To teach *adding and subtracting monomials.

Review

1. *Evaluate these algebraic expressions by using substitution.* (Lesson 134)

   a. $2lw + 2wh + 2lh$
   if $l = 12$, $w = 8$, and $h = 9$ (552)

   b. $3[w + wx(x - w) - x]$
   if $w = 5$ and $x = 7$ (204)

2. *Simplify by using the correct order of operations.* (Lesson 133)

   a. $3 + [7 - (8 - 2^2)] =$ ___   (6)
   $3 + [7 - (8 - 4)]$
   $3 + [7 - 4]$
   $3 + 3$

   b. $12[9 - (5 - 2)] =$ ___   (72)
   $12[9 - 3]$
   $12[6]$

3. *Calculate the missing percents on this income statement.* (Lesson 127)

   Delbert's Welding Shop
   Income Statement
   For the Quarter Ending
   July 31, 20—

	Amount	Percent of Sales
Total sales	$25,000	100%
Cost of goods	12,000	(48%)
Gross profit	13,000	(52%)
Overhead	9,000	(36%)
Net profit	4,000	(16%)

4. *Find the surface area of each square pyramid.* (Lesson 103)

   a. $b = 6$ cm
   $\ell = 6$ cm
   $a_s = (108 \text{ cm}^2)$

   b. $b = 10$ in.
   $\ell = 8$ in.
   $a_s = (260 \text{ sq. in.})$

5. *Do these subtractions mentally.* (Lesson 7)

   a. $91 - 57$ (34)
   b. $238 - 79$ (159)

Sketches for Lesson 134, Part E

25.

26.

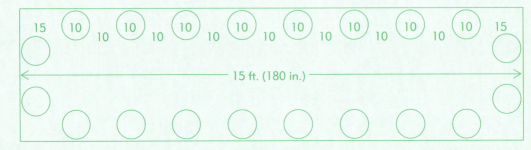

135. Adding and Subtracting Monomials

One Saturday afternoon two men ($2m$) came in one vehicle and three men ($3m$) came in another vehicle to do some work on the church building. How many men came for the work project?

Monomials such as $2m$ and $3m$ are said to be **like terms** because they have the same literal coefficients. Like terms may include literal numbers raised to a certain power; for example, $7x^2y$ and $2x^2y$ are like terms. The numerical coefficients of like terms may be alike or different.

Monomials that have different literal coefficients, or literal coefficients raised to different powers, are said to be **unlike terms**. For example, $8xy$ and $4x$ are unlike terms, and $6a^2$ and $6a^3$ are unlike terms. Only terms that have exactly the same literal coefficients are like terms.

Like terms

$8a$ and $6a$	$3xy$ and xy	$7a^3b^2$ and $12a^3b^2$	$6xy^2z$ and $\frac{1}{4}xy^2z$

Unlike terms

$6a$ and $6b$	$4a^2b$ and $2ab$	$9a^2b^2c^2$ and abc^2	$8x^2y^3$ and $7x^2y$

If polynomials have like terms, they can be simplified by combining those terms. Do this by adding or subtracting the numerical coefficients of the like terms; leave the literal coefficient the same. For example, $2m + 3m$ is $5m$, and $7a^2 - 3a^2$ is $4a^2$. A literal number standing alone has a numerical coefficient of 1.

Example A
Simplify this polynomial.
$3b - b + 9$
Think: $(3b - b) + 9$
$= 2b + 9$

Example B
Simplify this polynomial.
$a + b + 2a + 3b + b + 10$
Think: $(a + 2a) + (b + 3b + b) + 10$
$= 3a + 5b + 10$

Example B uses the commutative law of addition in rearranging addends to put like terms together. The associative law of addition permits the grouping of any addends. Notice that the answer is written with the literal numbers in alphabetical order. A number with no literal coefficient is written last.

CLASS PRACTICE

Tell whether each pair of terms is like or unlike.

a. $5a$ and $3a$ — like
b. $16xy$ and $12ay$ — unlike
c. a^2b^3 and $3a^2b^3$ — like
d. $4xy^3$ and $5x^3y^3$ — unlike
e. $\frac{1}{2}b^3$ and $9b^3$ — like
f. $2xyz$ and $4xyz$ — like

Simplify these polynomials.

g. $11h - 6h$ — $5h$
h. $14n + 9n$ — $23n$
i. $2x + 4y + 6x$ — $8x + 4y$
j. $14c + 13d + d$ — $14c + 14d$

456   Chapter 10   Introduction to Algebra

WRITTEN EXERCISES

A. Write whether each pair of terms is like (L) or unlike (U).

1. $3a^2$ and $4a^2$ — L
2. $2xy$ and xy — L
3. $4ab$ and $4a^2b^2$ — U
4. $4x^2y$ and $6x^2$ — U
5. $8xy^3z$ and $6x^2y^3z$ — U
6. $8a^3b^2c$ and a^3b^2c — L

B. Simplify these polynomials.

7. $16a + 7a$ — $23a$
8. $12n + 7n$ — $19n$
9. $3z + 7z + z$ — $11z$
10. $11a - 4a - a$ — $6a$
11. $18m - 12m$ — $6m$
12. $13n - 5n$ — $8n$
13. $5s + 6s - 2s$ — $9s$
14. $8w - 2w + 5w$ — $11w$
15. $6r + r - 2r$ — $5r$
16. $3m - 2m + 7$ — $m + 7$
17. $6n^2 + 5n^2 - n$ — $11n^2 - n$
18. $4m - 3m + 6$ — $m + 6$
19. $3a + 4 + a + 8 + 6a$ — $10a + 12$
20. $14a + 3a + a$ — $18a$
21. $7n - 2n + m + 6 - 5$ — $m + 5n + 1$
22. $3x - 2x + x$ — $2x$
23. $8x + 5 + 2x + y + 2y$ — $10x + 3y + 5$
24. $7c + 2c + c + d + 4d$ — $10c + 5d$

C. Solve these reading problems.

25. Leroy and his family raise vegetable and flower seedlings to sell in the spring. One evening Leroy calculated that they had 15 flats of pepper seedlings, 30 flats of tomato seedlings, and 20 flats of various other plants. They had a total of 2,340 seedlings. If each flat is the same size, how many seedlings fit in one flat?   36 seedlings

26. At the feed mill where they work, Barry, Marlin, and Richard were unloading a train car of pellet binder onto skids. They unloaded 18 skids in the morning, 12 skids in the afternoon, and 15 skids the next morning. The total number of bags was 3,645. What was the average number of bags on each skid?   81 bags

27. A typical finback whale is 23 feet longer than a sperm whale. If a common length for a sperm whale is 59 feet, what is the combined length of the finback and the sperm whale?   141 feet

28. A typical finback whale weighs 15 tons more than a sperm whale. If a finback whale weighs 65 tons, what is the combined weight of the finback and the sperm whale?   115 tons

29. Joshua found that the length of his schoolhouse was 3 times as great as its width. If the length is 120 feet, what is the perimeter of the building?   320 feet

30. Henry is building a fence around a rectangular chicken yard that is 12 feet longer than it is wide. The chicken house serves as the wall at one end, so he needs to fence only two long sides and one width. If the yard is 30 feet long, what length of fence is needed?   78 feet

Lesson 135   T–456

Introduction

Ask the students, "Does $3a + 4a = 7a$?" Use the following calculations to show that the answer is yes.

Let $a = 5$.
$3a + 4a = 7a$
$3(5) + 4(5) = 7(5)$
$15 + 20 = 35$

This works because of the distributive law, which the pupils studied in Lesson 10.

Teaching Guide

1. **Monomials are said to be like terms if they have the same literal coefficients.** Only terms that have exactly the same literal coefficients are like terms. If monomials have different literal coefficients, or literal coefficients raised to different powers, they are unlike terms.

   Identify the like and unlike terms in each expression.

   a. $6a, 7a, 6b, 8a, 6c$ (like terms: $6a$, $7a$, $8a$; unlike terms: $6b$, $6c$)

   b. $7n, 8m, 6m, n, 3n^2$ (like terms: $7n$, n; also $8m$, $6m$; unlike term: $3n^2$)

2. **If polynomials have like terms, they can be simplified by combining those terms.** Simply add or subtract the numerical coefficients, and leave the literal coefficient the same.

   a. $6a + 8a = (14a)$
   b. $5n - 3n = (2n)$

Solutions for Part C

25. $2{,}340 \div (15 + 30 + 20)$
26. $3{,}645 \div (18 + 12 + 15)$
27. $59 + (59 + 23)$
28. $15 + (65 - 15)$
29. $l = 120$; $w = 120 \div 3 = 40$; $p = 2(l + w)$
30. $l = 30$; $w = 30 - 12 = 18$; fence $= 2 \times 30 + 18$

T–457   *Chapter 10   Introduction to Algebra*

Solutions for Part E

33. $49 - 2[(16 - 9) + 2(8 - 4)]$ (simplify exponents)
  $49 - 2[7 + 2(4)]$ (simplify within parentheses)
  $49 - 2[7 + 8]$ (within brackets: multiply first)
  $49 - 2[15]$ (simplify within brackets)
  $49 - 30$ (multiply)

34. $14 + 6 \cdot 4 - 12$ (divide)
  $14 + 24 - 12$ (multiply)
  $38 - 12$ (add)

REVIEW EXERCISES

D. Evaluate these algebraic expressions by using substitution. *(Lesson 134)*

31. $(a + b)(b - a)$ 52
    $a = 12; b = 14$

32. $ab - c$ 13
    $a = 5; b = 4; c = 7$

E. Simplify by using the correct order of operations. *(Lesson 133)*

33. $49 - 2[(16 - 3^2) + 2(8 - 2^2)]$ 19

34. $\frac{56}{4} + 6 \cdot 4 - 12$ 26

F. Calculate the missing percents on this income statement. *(Lesson 127)*

Byler's Fruit Farm
Income Statement
For the Year Ending
December 31, 20—

	Amount	Percent of Sales
Total sales	$325,000	100%
Cost of goods	197,000	**35.** 61%
Gross profit	128,000	**36.** 39%
Overhead	98,000	**37.** 30%
Net profit	30,000	**38.** 9%

G. Find the surface area of each square pyramid. *(Lesson 103)*

39. $b = 5$ cm   95 cm²
    $\ell = 7$ cm

40. $b = 9$ in.   243 sq. in.
    $\ell = 9$ in.

H. Do these subtractions mentally. *(Lesson 7)*

41. $94 - 58$ 36

42. $4{,}750 - 2{,}825$ 1,925

458   Chapter 10   Introduction to Algebra

136. Using Addition and Subtraction to Solve Equations

An **equation** is a mathematical statement saying that two values are equal. One or more mathematical operations may be included on each side of the equation. Performing the operations does not change the value of either side; it only proves that both sides are equal.

Equations: $6 + 5 = 14 - 3$ $2 + 8 - 4 = 12 - 6$ $17 - 8 - 5 = 2^2$
Proof: $11 = 11$ $6 = 6$ $4 = 4$

An equation often includes at least one literal number, as in $x - 8 = 10$. Such an equation cannot be proven until you know the value of x. Finding the value of the unknown number is called *solving the equation*. The solution will have the unknown number on one side of the equal sign and all other numbers combined on the other side.

If you could add 8 to the expression $x - 8$, you would have the literal number alone. However, that would break the equality of the equation unless you also add 8 on the other side of the equal sign. Equal expressions may change and still be equal to each other, as stated in the addition axiom.

> **Addition Axiom:** The same value can be added to both sides of an equation without changing its equality.

The addition axiom can be applied when there is subtraction in an expression with the literal number. Conversely, if the literal number is part of an expression with addition, it can be isolated by applying the subtraction axiom.

> **Subtraction Axiom:** The same value can be subtracted from both sides of an equation without changing its equality.

Here are the two basic steps for solving equations involving addition or subtraction.

1. Simplify the side of the equation that has only numerals.
2. Isolate the unknown number by using the inverse operation to what appears on the side containing the literal number. Use addition to eliminate subtraction (Examples A and B). Use subtraction to eliminate addition (Examples C and D).

> Add to eliminate subtraction in an equation.

Note in Example A that simplifying $16 - 3$ does not change the value of that side of the equation. However, adding 8 to $n - 8$ does change its value, so the other side of the equation must be changed in the same way. Adding 8 to both sides keeps the two sides equal.

LESSON 136

Objective

- To review solving equations by using addition and subtraction.

Review

1. Give Lesson 136 Quiz (Order of Operations).

2. *Simplify these polynomials.* (Lesson 135)

    a. $2n + 6 + 4n + 3m + n$ $(3m + 7n + 6)$

    b. $5x - 4x + 3y - 6$ $(x + 3y - 6)$

3. *Evaluate these algebraic expressions by using substitution.* (Lesson 134)

    a. $\dfrac{d+e}{d-e}$ if $d = 8$ and $e = 4$ (3)

    b. $\dfrac{bh}{2}$ if $b = 6$ and $h = 8$ (24)

4. *Write the numerical coefficient of each term.* (Lesson 132)

    a. $8hm^2$ (8)

    b. w^4y^3 (1)

5. *Find the surface area of each sphere, to the nearest hundredth.* (Lesson 104)

    a. $r = 5.4$ m
    $a_s =$ (366.25 m²)

    b. $r = 21$ ft.
    $a_s =$ (5,538.96 sq. ft.)

6. *Do these multiplications.* (Lesson 8)

    a. 589
        × 142
        (83,638)

    b. 564
        × 948
        (534,672)

T-459 Chapter 10 Introduction to Algebra

Introduction

Write the following equation on the board: $9 + n = 14$. Call on a student to read the equation, substituting "a certain number" for n. ("Nine plus a certain number equals 14.")

Ask the class, "Is the number that n represents more or less than 14? How do you know?" It is less because the addends that make up a sum are less than the sum.

Now ask, "What operation do we use to solve this equation?" We use subtraction because it is the inverse operation of addition, which is used in the equation.

Teaching Guide

1. **An equation is a mathematical statement saying that two values are equal.** This statement of equality sets equations apart from algebraic expressions. Unless the two values are equal, the statement is not a true equation.

    a. $3 + 5 = 8$ b. $6 + 9 \neq 8$

2. **An equation usually includes at least one literal number, whose value is to be found by solving the equation.** The solution will have the unknown number on one side of the equal sign and all other numbers combined on the other side.

3. **There are the two basic steps for solving equations involving addition or subtraction.**

    (1) Simplify the side of the equation that has only numerals.

    (2) Isolate the unknown number by using the inverse (opposite) operation to what appears on the side containing the literal number.

4. **Use addition to eliminate subtraction in an equation.**

    a. $n - 12 = 2$
        $n - 12 + 12 = 2 + 12$
        $n = 14$
        Check: $14 - 12 = 2$
        $2 = 2$

    b. $z - 13 = 16$
        $z - 13 + 13 = 16 + 13$
        $z = 29$
        Check: $29 - 13 = 16$
        $16 = 16$

CLASS PRACTICE Step-by-step

c. $x - 4 = 14$
    $x - 4 + 4 = 14 + 4$ (add to eliminate subtraction)
    $x = 18$

d. $n + 3 + 4 = 9$
    $n + 7 = 9$
    $n + 7 - 7 = 9 - 7$ (subtract to eliminate addition)
    $n = 2$

e. $b + 8 = 5^2$
    $b + 8 = 25$
    $b + 8 - 8 = 25 - 8$ (subtract to eliminate addition)
    $b = 17$

f. $x - 6 = \dfrac{4^2}{8}$

    $x - 6 = \dfrac{16}{8}$

    $x - 6 = 2$
    $x - 6 + 6 = 2 + 6$ (add to eliminate subtraction)
    $x = 8$

Example A

$$n - 8 = 16 - 3$$
$$n - 8 = 13$$
$$n - 8 + 8 = 13 + 8$$
$$n = 21$$
Check: $(21) - 8 = 13$
$$13 = 13$$

Example B

$$n - \frac{12}{3} = 7$$
$$n - 4 = 7$$
$$n - 4 + 4 = 7 + 4$$
$$n = 11$$
Check: $(11) - \frac{12}{3} = 7$
$$7 = 7$$

In Example B, simplifying $\frac{12}{3}$ to 4 does not change the value of that side of the equation. So the other side does not need a corresponding change.

> Subtract to eliminate addition in an equation.

Example C

$$n + 7 + 4 = \frac{57}{3}$$
$$n + 11 = 19$$
$$n + 11 - 11 = 19 - 11$$
$$n = 8$$
Check: $(8) + 7 + 4 = \frac{57}{3}$
$$19 = 19$$

Example D

$$n + (12 - 5) = 26$$
$$n + 7 = 26$$
$$n + 7 - 7 = 26 - 7$$
$$n = 19$$
Check: $(19) + (12 - 5) = 26$
$$19 + 7 = 26$$
$$26 = 26$$

Note in Example D that simplifying $12 - 5$ does not change the value of that side of the equation.

Remember: Whatever is done to change the value of one side of the equation must also be done to the other side.

CLASS PRACTICE

Tell whether addition or subtraction is needed to solve each equation.

a. $6 + e = 14$ subtraction
b. $c - 8 = 12$ addition

Solve these equations. Show each step on paper.   (See facing page for steps.)

c. $x - 4 = 14$ $x = 18$
d. $n + 3 + 4 = 9$ $n = 2$
e. $b + 8 = 5^2$ $b = 17$
f. $x - 6 = \frac{4^2}{8}$ $x = 8$

Solve each reading problem by first completing the equation.

g. Father bought a new rocking chair at a discounted price of $140. That was $35 less than the regular price (n). What was the regular price?
$n = \$175$
$n - \underline{35} = 140$

h. The sum of three numbers is 48. Two of the numbers are 9 and 15. What is the third number?
$n = 24$
$\underline{n} + \underline{9} + \underline{15} = 48$

WRITTEN EXERCISES

A. Write *addition* or *subtraction* to tell what is needed to solve each equation.

1. $3 + c = 15$ — subtraction
2. $z + 4 = 22$ — subtraction
3. $n - 11 = 67$ — addition
4. $n - 22 = 25$ — addition

B. Solve these equations. Show each step in your work. (See facing page.)

5. $5 + c = 11$
6. $8 + c = 23$
7. $n - 4 = 12$
8. $m - 8 = 15$
9. $x - 9 = 14 \cdot 2$
10. $n - 8 = 3^2 + 4$
11. $a + 5 + 3 = 21$
12. $a - 4 = 2(8 + 5)$
13. $n + (7 - 3) = 12$
14. $n + (8 - 3) = 14$
15. $y - 2^3 = \frac{27}{3}$
16. $n - (12 - 8) = 4^2$

C. Solve these reading problems. For numbers 17–20, first copy and complete the equations.

17. The sum of three numbers is 25. Two of the numbers are 7 and 8. What is n, the third number? $n = 10$ $\underline{n} + \underline{7} + \underline{8} = 25$

18. The difference between a certain number (n) and the sum of 11 and 12 is 13. What is the unknown number? $n = 36$ $\underline{n} - (\underline{11} + \underline{12}) = 13$

19. The life span of men after the Flood was much shorter than before the Flood. Abraham lived 175 years, which was 755 years less than Adam's life span (n). How long did Adam live? $175 = \underline{n} - 755$
$n = 930$ years

20. Methuselah lived 969 years, which was 19 years more than Noah's life span (n). How many years did Noah live? $969 = n + \underline{19}$
$n = 950$ years

21. Father bought a plow for $1,500 at an out-of-state auction and sold it on his equipment lot for $1,800. His gross profit was what percent of the selling price? (Answer to the nearest whole percent.)   17%   ($1,800 − $1,500 = $300 gross profit; $300 ÷ $1,800)

22. Father also bought a garden tractor for $2,100 and sold it for $550 more than his cost. To the nearest whole number, what percent was his gross profit of the selling price?
21%   ($2,100 + $550 = $2,650 selling price; $550 ÷ $2,650)

REVIEW EXERCISES

D. Simplify these polynomials. *(Lesson 135)*

23. $3f - 2f + 4n$ $f + 4n$
24. $8a + 5a + 3b$ $13a + 3b$

E. Evaluate these algebraic expressions by using substitution. *(Lesson 134)*

25. $\frac{4a}{3b} + 5b$ $a = 9$; $b = 6$ 32
26. $2x(y + 5)$ $x = 6$; $y = 3$ 96

F. Write the numerical coefficient of each term. *(Lesson 132)*

27. $6mn^3$ 6
28. x^2y^3 1

G. Find the surface area of each sphere. *(Lesson 104)*

29. $r = 6$ m   452.16 m²
30. $r = 5$ ft.   314 sq. ft.

H. Do these multiplications. *(Lesson 8)*

31. 281 × 109 = 30,629
32. 280 × 654 = 183,120

5. Use subtraction to eliminate addition in an equation.

   a. $h + 14 = 22$
      $h + 14 - 14 = 22 - 14$
      $h = 8$
      Check: $8 + 14 = 22$
               $22 = 22$

   b. $r - 12 = 15$
      $r - 12 + 12 = 15 + 12$
      $r = 27$
      Check: $27 - 12 = 15$
               $15 = 15$

   c. $n + (6 - 3) = 9$
      $n + 3 = 9$
      $n + 3 - 3 = 9 - 3$
      $n = 6$
      Check: $6 + (6 - 3) = 9$
               $6 + 3 = 9$
               $9 = 9$

   d. $n + (10 - 4) = 13$
      $n + 6 = 13$
      $n + 6 - 6 = 13 - 6$
      $n = 7$
      Check: $7 + (10 - 4) = 13$
               $7 + 6 = 13$
               $13 = 13$

6. **In solving equations, it is imperative that whatever is done to change the value of one side must also be done to the other side.** Emphasize this point as you work through the exercises in class. Show the steps as you go. These lessons are designed to teach how to keep equations equal while solving them, not how to omit steps in solving them.

Answers for Part B

5. $5 + c = 11$
   $5 + c - 5 = 11 - 5$
   $c = 6$

6. $8 + c = 23$
   $8 + c - 8 = 23 - 8$
   $c = 15$

7. $n - 4 = 12$
   $n - 4 + 4 = 12 + 4$
   $n = 16$

8. $m - 8 = 15$
   $m - 8 + 8 = 15 + 8$
   $m = 23$

9. $x - 9 = 14 \cdot 2$
   $x - 9 = 28$
   $x - 9 + 9 = 28 + 9$
   $x = 37$

10. $n - 8 = 3^2 + 4$
    $n - 8 = 9 + 4$
    $n - 8 = 13$
    $n - 8 + 8 = 13 + 8$
    $n = 21$

11. $a + 5 + 3 = 21$
    $a + 8 = 21$
    $a + 8 - 8 = 21 - 8$
    $a = 13$

12. $a - 4 = 2(8 + 5)$
    $a - 4 = 2(13)$
    $a - 4 = 26$
    $a - 4 + 4 = 26 + 4$
    $a = 30$

13. $n + (7 - 3) = 12$
    $n + 4 = 12$
    $n + 4 - 4 = 12 - 4$
    $n = 8$

14. $n + (8 - 3) = 14$
    $n + 5 = 14$
    $n + 5 - 5 = 14 - 5$
    $n = 9$

15. $y - 2^3 = \dfrac{27}{3}$
    $y - 2^3 = 9$
    $y - 8 = 9$
    $y - 8 + 8 = 9 + 8$
    $y = 17$

16. $n - (12 - 8) = 4^2$
    $n - (12 - 8) = 16$
    $n - 4 = 16$
    $n - 4 + 4 = 16 + 4$
    $n = 20$

Solutions for Part E

25. $\dfrac{4(9)}{3(6)} + 5(6)$ (substitute)

    $\dfrac{36}{18} + 5(6)$ (simplify above and below vinculum)

    $2 + 5(6)$ (divide)

    $2 + 30$ (multiply)

26. $2(6)[(3) + 5]$ (substitute)
    $2(6)[8]$ (simplify within brackets)
    $2(48)$ (multiply)

Chapter 10 Introduction to Algebra

LESSON 137

Objective

- To review solving equations by using multiplication and division.

Review

1. *Solve these equations by using addition or subtraction.* (Lesson 136)

    a. $m + (7 - 5) = 9$ $(m = 7)$

    b. $19 - x = 3$ $(x = 16)$

2. *Simplify these monomials by combining like terms.* (Lesson 135)

    a. $3n + 2m - n + 4m - 7$
    $(6m + 2n - 7)$

    b. $3x + 7 + 2y - y$ $(3x + y + 7)$

3. *Use the correct order of operations to simplify each expression.* (Lesson 133)

    a. $3^2 - 2 \cdot 4 + 2(7 - 3) =$ ___ (9)

    $9 - 2 \cdot 4 + 2(4)$

    $9 - 8 + 8$

    $1 + 8$

    b. $8 + 2(2 + 2^2 + 2 \cdot \dfrac{6-2}{6-4}) =$ ___ (28)

    $8 + 2(2 + 4 + 2 \cdot \dfrac{4}{2})$

    $8 + 2(2 + 4 + 4)$

    $8 + 2(10)$

    $8 + 20$

4. *Add or subtract mentally.* (Lesson 129)

    a. $164 + 218$ (382)

    b. $381 - 276$ (105)

5. *Calculate the simple interest.* (Lesson 121)

    a. $1,650 at 4% for 2 yr. ($132.00)

    b. $16,500 at $6\frac{1}{2}$% for 3 yr.
    ($3,217.50)

6. *Find the volume of each cube or rectangular solid.* (Lesson 105)

    a. $l = 62$ ft.
    $w = 38$ ft.
    $h = 17$ ft.
    $v = $ (40,052 cu. ft.)

    b. $e = 16$ in.
    $v = $ (4,096 cu. in.)

7. *Calculate the arithmetic mean.* (Lesson 73)

    a. 88, 97, 78, 100, 72, 95, 100 (90)

    b. 215, 301, 257, 264, 248, 263 (258)

8. *Do these multiplications mentally.* (Lesson 9)

    a. 8×281 (2,248)

    b. $10,000 \times 7,215$ (72,150,000)

    c. $6 \times 1,094$ (6,564)

    d. 50×91 (4,550)

137. Using Multiplication and Division to Solve Equations

In Lesson 136 you studied two basic principles of solving equations. First, whatever is done to change the value of one side of the equation must also be done to the other side. Second, to isolate the unknown number, use the inverse operation of the one that is used. These principles also apply to equations requiring multiplication and division.

Multiplication Axiom: Both sides of an equation may be multiplied by the same number without changing the equality.
Division Axiom: Both sides of an equation may be divided by the same number without changing the equality.

Here are the two basic steps for solving equations involving multiplication or division.

1. Simplify the side of the equation that has only numerals.

2. Isolate the unknown number by using the inverse (opposite) operation to what appears on the side containing the literal number. Use multiplication to eliminate division (Examples A and B). Use division to eliminate multiplication (Examples C and D).

Multiply to eliminate division in an equation.

In algebra, division is indicated by using the fraction form: $\frac{8}{4}$ means $8 \div 4$. Multiplying a fraction by its denominator cancels the denominator. For example, $\frac{1}{4} \cdot 4$ (denominator) = 1 (numerator); and $\frac{3}{4} \cdot 4$ (denominator) = 3 (numerator). For this reason, multiplying $\frac{n}{6}$ by 6 cancels the denominator of 6 and isolates n (Example A). In Example B, multiplying by 4 isolates $k + 3$, and then subtraction isolates k.

Example A

$$\frac{n}{6} = 8$$
$$\frac{n}{6} \cdot 6 = 8 \cdot 6$$
$$n = 48$$
Check: $\frac{(48)}{6} = 8$
$$8 = 8$$

Example B

$$\frac{k+3}{4} = 6$$
$$\frac{k+3}{4} \cdot 4 = 6 \cdot 4$$
$$k + 3 = 24$$
$$k + 3 - 3 = 24 - 3$$
$$k = 21$$
Check: $\frac{(21)+3}{4} = 6$
$$6 = 6$$

Divide to eliminate multiplication in an equation.

In Example C, $x + 2x + 3x$ contains like terms that can be added without changing their value. Then x is isolated by dividing both sides of the equation by 6.

Example C
$$x + 2x + 3x = 42$$
$$6x = 42$$
$$\frac{6x}{6} = \frac{42}{6}$$
$$x = 7$$
Check: $(7) + 2(7) + 3(7) = 42$
$$7 + 14 + 21 = 42$$
$$42 = 42$$

Example D
$$n + 3n + 3n - 7 = 14$$
$$7n - 7 = 14$$
$$7n - 7 + 7 = 14 + 7$$
$$7n = 21$$
$$\frac{7n}{7} = \frac{21}{7}$$
$$n = 3$$
Check: $(3) + 3(3) + 3(3) - 7 = 14$
$$3 + 9 + 9 - 7 = 14$$
$$14 = 14$$

More than one operation is often required to solve an equation. These operations should be done in whatever order is needed to isolate the unknown number. As a result, you may sometimes do addition or subtraction before multiplication or division. But whenever you find multiple operations in one expression to be simplified (such as $6 + 3 \cdot 5$), remember to follow the correct order of operations as given in Lesson 133.

CLASS PRACTICE

Solve these equations. Give answers in the form n = __.

a. $\frac{n}{6} = 2$ $n = 12$

b. $\frac{x}{8} = 4$ $x = 32$

c. $\frac{a}{9} = 5$ $a = 45$

d. $3x = 18$ $x = 6$

e. $7n = 42$ $n = 6$

f. $2n + 4n + 2 = \frac{64}{2}$ $n = 5$

g. $\frac{4n}{2} = 3 \cdot 2$ $n = 3$

h. $\frac{12x}{8} = 3$ $x = 2$

i. $x + 2x + 2x = 30 - 5$ $x = 5$

j. $\frac{n}{15} = \frac{20}{5}$ $n = 60$

k. $n + 4n + 3 = 33$ $n = 6$

l. $2n + 1 + n + 2 = 12$ $n = 3$

Write equations to solve these reading problems.

m. At Rockville Mennonite School, there are 7 students in ninth grade. That is 1 less than twice the number (n) of students in eighth grade. How many students are in eighth grade? $7 = 2n - 1$; $n = 4$ students

n. The earth orbits the sun in 365 days. That is 13 more than 4 times the number of days required for Mercury's orbit. What is the length of Mercury's orbit (m) in days? $365 = 4m + 13$; $m = 88$ days

Lesson 137   T–462

Introduction

Write $3x = 18$ on the board. Ask the students, "If addition is shown in an equation, what operation do you use to solve it?" (subtraction) "If subtraction is shown in an equation, what operation do you use to solve it?" (addition)

Now ask, "What operation do you think should be used if multiplication is shown in an equation?" Division is used because division is the inverse operation of multiplication.

The solutions in these lessons show every single step in order to teach the logic with which equations are solved. After the students gain experience, they will be able to do several of the steps mentally and omit them from their written work.

Solutions for CLASS PRACTICE

a. $\dfrac{n}{6} = 2$

$\dfrac{n}{6} \cdot 6 = 2 \cdot 6$ (multiply to eliminate division)
$n = 12$

b. $\dfrac{x}{8} = 4$

$\dfrac{x}{8} \cdot 8 = 4 \cdot 8$ (multiply to eliminate division)
$x = 32$

c. $\dfrac{a}{9} = 5$

$\dfrac{a}{9} \cdot 9 = 5 \cdot 9$ (multiply to eliminate division)
$a = 45$

d. $3x = 18$
$\dfrac{3x}{3} = \dfrac{18}{3}$ (divide to eliminate multiplication)
$x = 6$

e. $7n = 42$
$\dfrac{7n}{7} = \dfrac{42}{7}$ (divide to eliminate multiplication)
$n = 6$

f. $2n + 4n + 2 = \dfrac{64}{2}$

$2n + 4n + 2 = 32$ (simplify numerical side of the equation)
$6n + 2 = 32$ (add like terms)
$6n + 2 - 2 = 32 - 2$ (subtract to eliminate addition)
$6n = 30$
$\dfrac{6n}{6} = \dfrac{30}{6}$ (divide to eliminate multiplication)
$n = 5$

g. $\dfrac{4n}{2} = 3 \cdot 2$

$\dfrac{4n}{2} = 6$ (simplify numerical side of the equation)
$\dfrac{4n}{2} \cdot 2 = 6 \cdot 2$ (multiply to eliminate division)
$4n = 12$
$\dfrac{4n}{4} = \dfrac{12}{4}$ (divide to eliminate multiplication)
$n = 3$

h. $\dfrac{12x}{8} = 3$

$\dfrac{12x}{8} \cdot 8 = 3 \cdot 8$ (multiply to eliminate division)
$12x = 24$
$\dfrac{12x}{12} = \dfrac{24}{12}$ (divide to eliminate multiplication)
$x = 2$

i. $x + 2x + 2x = 30 - 5$
$x + 2x + 2x = 25$ (simplify numerical side of equation)
$5x = 25$
$\dfrac{5x}{5} = \dfrac{25}{5}$ (divide to eliminate multiplication)
$x = 5$

j. $\dfrac{n}{15} = \dfrac{20}{5}$

$\dfrac{n}{15} = 4$ (simplify numerical side of equation)
$\dfrac{n}{15} \cdot 15 = 4 \cdot 15$ (multiply to eliminate division)
$n = 60$

Chapter 10 Introduction to Algebra

Teaching Guide

1. **Use multiplication to eliminate division in an equation.**

   a. $\dfrac{n}{12} = 7$

   $\dfrac{n}{12} \cdot 12 = 7 \cdot 12$

   $n = 84$

   Check: $\dfrac{84}{12} = 7$

   $7 = 7$

   b. $\dfrac{n+3}{4} = 6$

   $\dfrac{n+3}{4} \cdot 4 = 6 \cdot 4$

   $n + 3 = 24$

   $n + 3 - 3 = 24 - 3$

   $n = 21$

   Check: $\dfrac{21+3}{4} = 6$

   $\dfrac{24}{4} = 6$

   $6 = 6$

2. **Use division to eliminate multiplication in an equation.**

   a. $7n = 35$

   $\dfrac{7n}{7} = \dfrac{35}{7}$ Check: $7(5) = 35$

   $n = 5$ $35 = 35$

   b. $(5d + 3d)3 = 48$

   $(8d)3 = 48$

   $\dfrac{(8d)3}{3} = \dfrac{48}{3}$ Check:

   $8d = 16$ $[5(2) + 3(2)]3 = 48$

   $\dfrac{8d}{8} = \dfrac{16}{8}$ $[10 + 6]3 = 48$

   $d = 2$ $[16]3 = 48$

   $48 = 48$

3. **More than one step is often needed to isolate the unknown number.** Simplify any expression that has like terms on the same side of the equation. Do inverse operations in whatever order required by grouping symbols, to isolate the unknown number.

   k. $n + 4n + 3 = 33$

   $5n + 3 = 33$ (add like terms)

   $5n + 3 - 3 = 33 - 3$ (subtract to eliminate addition)

   $5n = 30$

   $\dfrac{5n}{5} = \dfrac{30}{5}$ (divide to eliminate multiplication)

   $n = 6$

   l. $2n + 1 + n + 2 = 12$

   $3n + 3 = 12$ (add like terms)

   $3n + 3 - 3 = 12 - 3$ (subtract to eliminate addition)

   $3n = 9$

   $\dfrac{3n}{3} = \dfrac{9}{3}$ (divide to eliminate multiplication)

   $n = 3$

Solutions for Part A

1. $\dfrac{n}{9} = 3$ 2. $\dfrac{c}{5} = 52$

   $\dfrac{n}{9} \cdot 9 = 3 \cdot 9$ $\dfrac{c}{5} \cdot 5 = 52 \cdot 5$

   $n = 27$ $c = 260$

3. $\dfrac{x}{7} = 6 + 2$ 4. $\dfrac{x}{12} = 6 - 2$

   $\dfrac{x}{7} = 8$ $\dfrac{x}{12} = 4$

   $\dfrac{x}{7} \cdot 7 = 8 \cdot 7$ $\dfrac{x}{12} \cdot 12 = 4 \cdot 12$

   $x = 56$ $x = 48$

5. $6m = 7 \cdot 6$ 6. $3n + 9 = 42$

   $6m = 42$ $3n + 9 - 9 = 42 - 9$

   $\dfrac{6m}{6} = \dfrac{42}{6}$ $3n = 33$

   $m = 7$ $\dfrac{3n}{3} = \dfrac{33}{3}$

   $n = 11$

WRITTEN EXERCISES

A. *Solve these equations. Write answers in the form n = __.*

1. $\frac{n}{9} = 3$ $n = 27$
2. $\frac{c}{5} = 52$ $c = 260$
3. $\frac{x}{7} = 6 + 2$ $x = 56$
4. $\frac{x}{12} = 6 - 2$ $x = 48$
5. $6m = 7 \cdot 6$ $m = 7$
6. $3n + 9 = 42$ $n = 11$
7. $4y - 2 = 10$ $y = 3$
8. $k + 2k - 4 = 17$ $k = 7$
9. $x + 4x + 3 = 32 - 4$ $x = 5$
10. $\frac{n}{10} = 13$ $n = 130$
11. $\frac{d}{8} = 16 - 9$ $d = 56$
12. $2x + 6x = 24$ $x = 3$
13. $a + a + 2 = 5 \cdot 4$ $a = 9$
14. $\frac{n}{12} = \frac{18}{3}$ $n = 72$
15. $6n - 4 = 17 + 3$ $n = 4$
16. $4x = 9 \cdot 8$ $x = 18$
17. $8n - n = 28$ $n = 4$
18. $\frac{5x}{6} = 10$ $x = 12$
19. $2n + 4n + 3n = 27$ $n = 3$
20. $5n - 12 = 48$ $n = 12$

B. *Use the equations given to solve these reading problems. Write answers in the form n = __.*

21. Lorene is 2 more than $\frac{1}{9}$ of Grandfather's age. Lorene is 10. What is Grandfather's age (g)?
$$\frac{g}{9} + 2 = 10 \qquad g = 72 \text{ years}$$

22. Delbert is 3 less than $\frac{1}{3}$ of Uncle Roy's age. Delbert is 12. What is Uncle Roy's age (r)?
$$\frac{r}{3} - 3 = 12 \qquad r = 45 \text{ years}$$

23. The Webers recently moved into a house that is 175 years old. That is 5 less than 15 times the age of the house they just sold. How old is the house (h) that they sold?
$$175 = 15h - 5 \qquad h = 12 \text{ years}$$

24. Last year the Weber children attended a school that had 105 students. That is 3 more than 6 times larger than the school they are now attending. What is the enrollment (e) in their new school?
$$105 = 6e + 3 \qquad e = 17 \text{ students}$$

25. Last week Stanley worked 8 hours less than this week. The hours for the two weeks combined is 10^2. How many hours (h) did he work this week?
$$h - \underline{8} + \underline{h} = 10^2 \qquad h = 54 \text{ hours}$$

26. The difference between a certain number (n) and the sum of 8 and 5 is 13. What is the number?
$$n - (\underline{8} + \underline{5}) = 13 \qquad n = 26$$

REVIEW EXERCISES

C. Solve these equations by using addition or subtraction. *(Lesson 136)*

27. $x - 6 = 12$ $x = 18$ 28. $a + 7 = 15$ $a = 8$

D. Simplify these monomials by combining like terms. *(Lesson 135)*

29. $4m + n + 2m$ $6m + n$ 30. $4x + x - 2x - 15$ $3x - 15$

E. Use the correct order of operations to simplify each expression. *(Lesson 133)*

31. $4^2 - 2(8 - 3)$ 6   32. $12 + 2[3 + \frac{11 + 7}{3^2} - (6 - 3)]$ 16

F. Add or subtract mentally. *(Lesson 129)*

33. $264 + 281$ 545   34. $614 - 484$ 130

G. Calculate the simple interest. *(Lesson 121)*

35. $2,460 at 5% for 2 yr. $246   36. $11,400 at $8\frac{1}{2}$% for 3 yr. $2,907

H. Find the volume of each cube or rectangular solid. *(Lesson 105)*

37. $l = 45$ ft.   15,840 cu. ft.   38. $e = 22$ in.   10,648 cu. in.
    $w = 32$ ft.
    $h = 11$ ft.

I. Calculate the arithmetic mean. *(Lesson 73)*

39. 61, 84, 77, 78, 89, 63, 71, 61   73   40. 948, 954, 958, 935, 940, 959   949

J. Do these multiplications mentally. *(Lesson 9)*

41. 6×342 2,052   42. 40×86 3,440

Lesson 137   T–464

7. $4y - 2 = 10$
$4y - 2 + 2 = 10 + 2$
$4y = 12$
$\frac{4y}{4} = \frac{12}{4}$
$y = 3$

8. $k + 2k - 4 = 17$
$3k - 4 = 17$
$3k - 4 + 4 = 17 + 4$
$3k = 21$
$\frac{3k}{3} = \frac{21}{3}$
$k = 7$

9. $x + 4x + 3 = 32 - 4$
$x + 4x + 3 = 28$
$5x + 3 = 28$
$5x + 3 - 3 = 28 - 3$
$5x = 25$
$\frac{5x}{5} = \frac{25}{5}$
$x = 5$

10. $\frac{n}{10} = 13$
$\frac{n}{10} \cdot 10 = 13 \cdot 10$
$n = 130$

11. $\frac{d}{8} = 16 - 9$
$\frac{d}{8} = 7$
$\frac{d}{8} \cdot 8 = 7 \cdot 8$
$d = 56$

12. $2x + 6x = 24$
$8x = 24$
$\frac{8x}{8} = \frac{24}{8}$
$x = 3$

13. $a + a + 2 = 5 \cdot 4$
$a + a + 2 = 20$
$2a + 2 = 20$
$2a + 2 - 2 = 20 - 2$
$2a = 18$
$\frac{2a}{2} = \frac{18}{2}$
$a = 9$

14. $\frac{n}{12} = \frac{18}{3}$
$\frac{n}{12} = 6$
$\frac{n}{12} \cdot 12 = 6 \cdot 12$
$n = 72$

15. $6n - 4 = 17 + 3$
$6n - 4 = 20$
$6n - 4 + 4 = 20 + 4$
$6n = 24$
$\frac{6n}{6} = \frac{24}{6}$
$n = 4$

16. $4x = 9 \cdot 8$
$4x = 72$
$\frac{4x}{4} = \frac{72}{4}$
$x = 18$

17. $8n - n = 28$
$7n = 28$
$\frac{7n}{7} = \frac{28}{7}$
$n = 4$

18. $\frac{5x}{6} = 10$
$\frac{5x}{6} \cdot 6 = 10 \cdot 6$
$5x = 60$
$\frac{5x}{5} = \frac{60}{5}$
$x = 12$

19. $2n + 4n + 3n = 27$
$9n = 27$
$\frac{9n}{9} = \frac{27}{9}$
$n = 3$

20. $5n - 12 = 48$
$5n - 12 + 12 = 48 + 12$
$5n = 60$
$\frac{5n}{5} = \frac{60}{5}$
$n = 12$

Solutions for Part E

31. $16 - 2(8 - 3)$ (simplify exponent)
    $16 - 2(5)$ (simplify within parentheses)
    $16 - 10$ (multiply)

32. $12 + 2[3 + \frac{11 + 7}{9} - (6 - 3)]$ (simplify exponent)

    $12 + 2[3 + \frac{11 + 7}{9} - 3]$ (simplify within parentheses)

    $12 + 2[3 + \frac{18}{9} - 3]$ (within brackets: simplify above vinculum first)

    $12 + 2[3 + 2 - 3]$ (within brackets: divide second)
    $12 + 2[2]$ (simplify within brackets)
    $12 + 4$ (multiply)

LESSON 138

Objective

- To review using equations to solve reading problems.

Review

1. *Solve these equations involving the four basic operations.* (Lessons 136, 137)

    a. $2n + 3 = 23$ ($n = 10$)

    b. $5a - 5 = 35$ ($a = 8$)

2. *Evaluate these algebraic expressions by using substitution.* (Lesson 134)

    a. $a - b^2 =$ ___ (3)
       $a = 12; b = 3$

    b. $a + b(a + b) =$ ___ (37)
       $a = 2; b = 5$

3. *Calculate the simple interest.* (Lesson 122)

    a. $1,000 at 5% for 10 mo. ($41.67)

    b. $4,000 at 6% for 210 days
       ($140.00)

4. *Find the volume of each figure.* (Lesson 106)

    a. Cylinder
       $r = 6$ in.
       $h = 10$ in.
       $v = $ (1,130.4 cu. in.)

    b. Cone
       $r = 12$ cm
       $h = 45$ cm
       $v = $ (6,782.4 cm^3)

5. *Use the distributive law to find the missing numbers.* (Lesson 10)

    a. $8 \times 16 = (8 \times$ __$) + (8 \times$ __$)$ [10, 6]

    b. $3 \times 94 = (3 \times 90) + ($__ $\times$ __$)$ [3, 4]

138. Reading Problems: Writing Equations

For several years, you have worked with reading problems that use the following phrases to show the four basic operations. This lesson uses these phrases to indicate algebraic expressions. Notice that the order of a subtraction or division expression is critical. The expression $x - y$ will not give the same result as $y - x$ because $5 - 3$ is not the same as $3 - 5$.

Operation	Phrase	Example	Algebraic expression
Addition	added to	five added to n	$n + 5$
	plus	n plus seven	$n + 7$
	more than	r more than six	$6 + r$
	the sum of	the sum of a and b	$a + b$
	combined	m and n combined	$m + n$
	increased	s increased by seven	$s + 7$
Subtraction	less than	nine less than n	$n - 9$
	subtracted from	m subtracted from eight	$8 - m$
	minus	m minus n	$m - n$
	difference	the difference between six and d	$6 - d$
	decreased by	eight decreased by z	$8 - z$
Multiplication	times	eight times n	$8n$
	product of	the product of n and five	$5n$
Division	divided by	n divided by five	$\dfrac{n}{5}$
	quotient of	the quotient of h and three	$\dfrac{h}{3}$
Equals sign	equals	four times x equals twenty	$4x = 20$
	is, are, was, etc.	seven less than b is ten	$b - 7 = 10$
	totals	x plus y added to four totals nine	$x + y + 4 = 9$

The exercises in this lesson are multistep reading problems. In previous lessons you have solved multistep problems piece by piece, one step at a time. Now you will learn to use equations to put all the steps together into one statement. An equation shows how each fact is related to other facts and to the whole.

466 *Chapter 10 Introduction to Algebra*

Use these steps to write equations for reading problems such as the ones below.

1. Choose a literal number to represent the answer to the question. If you must also find other unknown numbers, express them in relation to your literal number. In Example A below, p represents the acres of pasture and $p + 60$ represents the acres of cropland. You could not solve the problem by using p for the pastureland and c for the cropland.

2. Set up the equation. Be alert for words that tell how the facts are related. See the table above.

3. Solve the equation.

4. Check the equation.

Example A

The Shanks' farm has 76 acres of pasture and cropland combined. There are 60 more acres of cropland than pasture. How many acres of pasture are there?

The pasture plus cropland is 76 acres.
Let p represent the acres of pasture. (This is the answer to be found.)
Then $p + 60$ represents the acres of cropland.

$p + (p + 60) = 76$
$2p + 60 = 76$
$2p + 60 - 60 = 76 - 60$
$2p = 16$
$\frac{2p}{2} = \frac{16}{2}$
$p = 8$ acres

(The parentheses in this equation simply clarify that "$p + 60$" is a unit. They are not needed for the correct order of operations and can be dropped when solving the equation.)

Check: $(8) + [(8) + 60] = 76$
$76 = 76$

Example B

The sum of 3 consecutive numbers (numbers following each other) is 78. What is the smallest number?

Three numbers added together equal 78.
Let n represent the smallest number.
Then $n + 1$ represents the next larger number.
And $n + 2$ represents the largest number.

$n + (n + 1) + (n + 2) = 78$
$3n + 3 = 78$
$3n + 3 - 3 = 78 - 3$
$3n = 75$
$\frac{3n}{3} = \frac{75}{3}$
$n = 25$

Check: $(25) + [(25) + 1] + [(25) + 2] = 78$
$25 + 26 + 27 = 78$
$78 = 78$

Lesson 138 T–466

Introduction

Ask the students, "Is there any difference between a formula and an equation?" The students may not have made the connection, but a formula is a standardized equation that applies to many similar circumstances. For example, the formula for finding the area of a circle is the same regardless of the size of a circle.

An equation is a formula applied to a specific situation. It too would fit for any other situation with the same relationships of facts. To write an equation for a reading problem, then, is to write a formula that is true for those facts. An equation is a written plan for solving the problem.

Teaching Guide

1. **In an equation for a multistep reading problem, all the steps are put together into one statement.** An equation shows how each fact is related to other facts and to the whole.

2. **The phrases used in a reading problem indicate the operations to be performed in an equation.** Discuss the phrases shown in the lesson. This is largely review, since the pupils have studied these phrases earlier in learning to solve reading problems.

3. **To write an equation for a reading problem, use the following steps.**

    (1) Choose a literal number to represent the answer to the question. If you must also find other unknown numbers, express them in relation to your literal number.

    (2) Set up the equation. Be alert for words that tell how the facts are related.

    (3) Solve the equation.

    (4) Check the equation.

    Discuss these steps in relation to the problems in the lesson text and in *Class Practice*.

T–467 Chapter 10 Introduction to Algebra

Further Study

Here is another way to find three consecutive numbers whose total is a given sum.

Find three consecutive numbers whose sum is 93.

Since the three numbers are so close together, each one will be about $\frac{1}{3}$ of the sum. So the first step is to divide 93 by 3. The result, 31, is the middle number of the three addends. The other addends are 30 and 32; 30 + 31 + 32 = 93.

Note that only a multiple of 3 can be the sum of three consecutive whole numbers.

This method can be used to find any odd number of consecutive numbers whose total is a given sum. To find five addends whose total is a given sum, dividing the sum by 5 will yield the middle number. Example: 45 ÷ 5 = 9; 7 + 8 + 9 + 10 + 11 = 45.

For an even number of consecutive whole numbers, the middle value is always a mixed number including $\frac{1}{2}$. For example, 12 + 13 + 14 + 15 = 54; and 54 ÷ 4 = $13\frac{1}{2}$. Since this mixed number is always halfway between the two middle addends, the others can easily be found by counting backward and forward from that point. In this example, there are four addends; the two whole numbers before $13\frac{1}{2}$ (12 and 13) and the two whole numbers after $13\frac{1}{2}$ (14 and 15) are the four addends.

Only if a number is 2 more than a multiple of 4 can it be the sum of four consecutive whole numbers.

Example C

Naomi spent a total of 42 minutes working on her English and math assignments. She spent 5 times as many minutes on math as on English. How many minutes did she spend on her English lesson?

English and *math* totals 42 minutes.
Let e represent the minutes spent on English.
Then $5e$ represents the minutes spent on math.

$$e + 5e = 42$$
$$6e = 42$$
$$\frac{6e}{6} = \frac{42}{6}$$
$$e = 7 \text{ min.}$$

Check: $(7) + 5(7) = 42$
$42 = 42$

CLASS PRACTICE

Write these phrases as algebraic expressions.

a. eleven and t combined $\quad 11 + t$ b. the difference between r and s $\quad r - s$

c. k divided by two $\quad \frac{k}{2}$ d. the product of three, a, and b $\quad 3ab$

e. four times x equals six $\quad 4x = 6$ f. nine is the product of c and d $\quad 9 = cd$

Write an equation, and then find the solution to each reading problem.

g. Kenneth is twice as old as his brother Darrel. The sum of their ages is 12. How old is Darrel?

   Think: Darrel's age added to Kenneth's age is 12.

   (1) Let $\underline{d}$ represent Darrel's age.
   (2) Let $\underline{2d}$ represent Kenneth's age.
   (3) Complete the equation and solve. $d + 2d = 12$ $\qquad d = 4$ years

h. One Saturday it took Dwayne 3 times as long to mow the yard as to wash the car. Then he spent 10 minutes organizing the garage. The total time he spent doing these things was 70 minutes. How many minutes (m) did it take to wash the car?

   Think: Time to wash the car, mow the yard, and organize the garage is 70 minutes.

   (1) Let $\underline{m}$ represent time to wash the car.
   (2) Let $\underline{3m}$ represent time to mow the yard.
   (3) Add $\underline{10}$ minutes to organize the garage.
   (4) Complete the equation and solve. $m + 3m + 10 = 70$ $\qquad m = 15$ minutes

i. A broiler chicken house holds 3,000 less than twice as many birds as a layer chicken house. The total number of birds in the two houses is 60,000. How many chickens does the layer house hold? $c + (2c - 3{,}000) = 60{,}000$ $\qquad c = 21{,}000$ chickens

j. The sum of three numbers is 78. The second number is 2 times the first number (n). The third number is 6 more than the first number. What is the first number?
$n + 2n + (n + 6) = 78$ $\qquad n = 18$

WRITTEN EXERCISES

A. Write these phrases as algebraic expressions.

1. eight more than n — $n + 8$
2. n less than fifteen — $15 - n$
3. the product of four and a — $4a$
4. the sum of b and c — $b + c$
5. y decreased by 3 is 8 — $y - 3 = 8$
6. k minus m equals 27 — $k - m = 27$
7. 5 in addition to t — $t + 5$
8. the product of g and h — gh
9. k and m combined — $k + m$
10. the product of r, s, and t — rst

B. Write an equation for each reading problem. Then find the solution. Problems 11–14 give steps to help you write equations.

11. Rosanne is 6 years older than Marlene. The total of their ages is 28. How old is Marlene?
    Think: Marlene's age plus Rosanne's age totals 28.
    (1) Let m represent Marlene's age.
    (2) Let ___ represent Rosanne's age. — $m + 6$
    (3) Complete the equation and solve. — $m + (m + 6) = 28$ — $m = 11$ years

12. At Lundville Christian School, the number of students in the lower grade classroom is 3 less than in the upper grade room. The total enrollment in both rooms is 25. What is the enrollment in the upper grade classroom?
    Think: Upper grade students plus lower grade students is 25.
    (1) Let u represent upper grade students.
    (2) Let ___ represent lower grade students. — $u - 3$
    (3) Complete the equation and solve. — $u + (u - 3) = 25$ — $u = 14$ students

13. One week Henry picked strawberries for twice as many hours as Dwight. Their combined time was 24 hours. How many hours did Dwight spend picking strawberries?
    Think: Dwight's hours plus Henry's hours were 24.
    (1) Let d represent Dwight's hours.
    (2) Let $2d$ represent Henry's hours.
    (3) Complete the equation and solve. — $d + 2d = 24$ — $d = 8$ hours

14. The city of Jerusalem lies between the Mediterranean Sea and the Dead Sea. The distance from Jerusalem to the Dead Sea is 19 miles less than from Jerusalem to the Mediterranean. The sum of these distances is 47 miles. How far is Jerusalem from the Mediterranean?
    Think: Distance from Jerusalem to Mediterranean plus Jerusalem to Dead Sea is 47 miles.
    (1) Let m represent the distance from Jerusalem to the Mediterranean.
    (2) Let ___ represent the distance from Jerusalem to the Dead Sea. — $m - 19$
    (3) Complete the equation and solve. — $m + (m - 19) = 47$ — $m = 33$ miles

T-469   Chapter 10   Introduction to Algebra

15. Carla picked 8 more quarts of blueberries than Susan did. Together they picked 26 quarts. How many quarts of blueberries did Susan pick?
$s + (s + 8) = 26$ $s = 9$ quarts

16. Grandfather is 10 times as old as Yvonne. The sum of their ages is 77. How old is Yvonne?
$10y + y = 77$ $y = 7$ years

17. The sum of 3 consecutive numbers is 99. What is the smallest number?
$n + (n + 1) + (n + 2) = 99$ $n = 32$

18. The sum of 3 consecutive numbers is 63. What is the smallest number?
$n + (n + 1) + (n + 2) = 63$ $n = 20$

19. Last month Hoover's Furniture built 15 times as many chairs as hutches. The total hutches and chairs were 96. How many hutches did they build?
$h + 15h = 96$ $h = 6$ hutches

20. The house that Mark lives in is 2 years more than 3 times as old as Mark. The combined age of the house and Mark is 58 years. How old is Mark?
$m + (3m + 2) = 58$ $m = 14$ years

REVIEW EXERCISES

C. Solve these equations involving the four basic operations. *(Lessons 136, 137)*

21. $4x + 5 = 49$ $x = 11$ 22. $3a - 11 = 82$ $a = 31$

D. Evaluate these algebraic expressions by using substitution. *(Lesson 134)*

23. $a + b^2$ 29
     $a = 4; b = 5$

24. $a - b(a - b)$ 0
     $a = 4; b = 2$

E. Calculate the simple interest. *(Lesson 122)*

25. $2,000 at 7% for 9 mo. $105   26. $3,600 at 10% for 160 days $160

F. Find the volume of each figure. *(Lesson 106)*

27. Cylinder   1,177.5 cu. in.
     $r = 5$ in.; $h = 15$ in.

28. Cone   7,234.56 cm³
     $r = 12$ cm; $h = 48$ cm

G. Use the distributive law to find the missing numbers. *(Lesson 10)*

29. $7 \times 17 = (7 \times 10) + (7 \times \underline{7})$ 30. $6 \times 83 = (6 \times \underline{80}) + (6 \times \underline{3})$

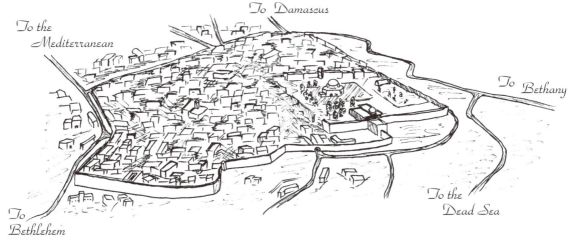

139. Exponents and Literal Numbers

If the same number is used repeatedly as a factor in a multiplication problem, that factor is often expressed with an exponent. This is a small raised number that shows how many times the **base** (the repeated factor) is used. In the number below, the base is 7 and the exponent is 4. That number is read "seven to the fourth power."

Exponential form		**Expanded form**		**Standard form**
7^4	=	$7 \cdot 7 \cdot 7 \cdot 7$	=	2,401

Literal numbers are like numerals except that it is not known what value a literal number represents. Any operation that can be performed with numerals can also be done with literal numbers, including those with exponents. Here are some rules for using exponents when the base is either a numeral or a literal number.

(1) The exponent indicates the number of times the base is a factor.
$a^3 = aaa = a \cdot a \cdot a$ $\qquad$ $2^3 = 2 \cdot 2 \cdot 2$

(2) The exponent applies only to the number that immediately precedes it.
$ac^2 = acc = a \cdot c \cdot c$

(3) When writing coefficients, write the numerical coefficient first and then the literal coefficients in alphabetical order.
$2a^2bc^2$

Example A	**Example B**
Write 3^4 in expanded form.	Write a^4 in expanded form.
$3^4 = 3 \cdot 3 \cdot 3 \cdot 3$	$a^4 = aaaa$ or $a \cdot a \cdot a \cdot a$
Example C	**Example D**
Write $3 \cdot 3 \cdot 3 \cdot 3 \cdot 3 \cdot 3$, using exponents.	Write $(3)(3)(3)(e)(e)(f)(f)$, using exponents.
$3 \cdot 3 \cdot 3 \cdot 3 \cdot 3 \cdot 3 = 3^6$	$(3)(3)(3)(e)(e)(f)(f) = 3^3e^2f^2$

As stated in Lesson 138, monomials containing exponents can be added or subtracted just like other monomials as long as they are like terms. This means that each monomial must have exactly the same literal number or combination of literal numbers, and the exponents must also match. If a numeral is raised to a certain power (such as 3^3), it can be simplified by writing it in the standard form. See Example G.

LESSON 139

Objectives

- To review the meaning of exponents.
- To teach that *exponents can be used with literal numbers just as with numerals.

Review

1. *Solve these equations involving the four basic operations.* (Lessons 136, 137)

   a. $3c - 21 = 3$ $(c = 8)$

   b. $\frac{x}{12} + 3 = 5$ $(x = 24)$

2. *Simplify these polynomials.* (Lesson 135)

   a. $2c + 8c - c$ $(9c)$

   b. $5n + 8m + n - 4$ $(8m + 6n - 4)$

3. *Find the total interest if it is compounded semiannually.* (Lesson 123)

   a. $21,400 at 6% for 1 yr. ($1,303.26)

   b. $1,800 at 4% for 1 yr. ($72.72)

4. *Find the volume of each square pyramid.* (Lesson 107)

   a. $b = 12$ in.
   $h = 16$ in.
   $v =$ (768 cu. in.)

   b. $b = 24$ m
   $h = 31$ m
   $v =$ (5,952 m³)

5. *Solve these division problems.* (Lesson 11)

   a. $32 \overline{)15{,}891}$ (496 R 19)

   b. $268 \overline{)21{,}584}$ (80 R 144)

T–471   *Chapter 10   Introduction to Algebra*

Introduction

Ask how we can write (3)(3)(3)(3) in a shorter way. A shorter way is to write 3^4.

Now see if the students can think of a way to write *eeeee*. If they suggest writing 5*e*, see if that works out. Let *e* equal 3. Does 5(3) equal (3)(3)(3)(3)(3)? No; 5 × 3 = 15, but 3 × 3 × 3 × 3 × 3 equals 243.

The way to write *eeeee*, as the students may have already suggested, is e^5.

Teaching Guide

1. **If the same number is used repeatedly as a factor in a multiplication problem, that factor is often expressed with an exponent.** This is a small raised number that shows how many times the base (the repeated factor) is used.

   a. 4^2 has a base of 4 and an exponent of 2 and is read "four to the second power." (It may also be read "four squared.")

   b. 8^3 has a base of 8 and an exponent of 3 and is read "eight to the third power." (It may also be read "eight cubed.")

2. **Literal numbers are like numerals except that it is not known what value a literal number represents.** Any operation that can be performed with numerals can also be done with literal numbers, including those with exponents. Following are some rules for using exponents when the base is either a numeral or a literal number.

   (1) The exponent indicates the number of times the base is a factor.
   $a^3 = aaa = a \cdot a \cdot a \qquad 2^3 = 2 \cdot 2 \cdot 2$

   (2) The exponent applies only to the number that immediately precedes it.
   $ac^2 = acc = a \cdot c \cdot c$

   (3) When writing coefficients, write the numerical coefficient first and then the literal coefficients in alphabetical order.
   $2a^2bc^2$

   Use the following expressions for practice with expanding numbers.

   a. $a^2b^2 = (a)(a)(b)(b)$
   b. $3a^3bc^2 = (3)(a)(a)(a)(b)(c)(c)$

Example E
Simplify $3a^2b + 4a^2b$ if possible.
$3a^2b + 4a^2b = 7a^2b$

Example F
Simplify $3a^2b + 4ab^2$ if possible.
$3a^2b + 4ab^2$ does not have like terms. It cannot be simplified.

Example G
Simplify.
Combine any numerical coefficients.
$3^3e^2 + 2e^2$
$27e^2 + 2e^2 = 29e^2$

Example H
Simplify.
Combine any numerical coefficients.
$4g^3h^2 + 5^2g^3h^2$
$4g^3h^2 + 25g^3h^2 = 29g^3h^2$

CLASS PRACTICE

(Accept varied forms for multiplication: mmm or $(m)(m)(m)$ or $m \cdot m \cdot m$)

Write these in expanded form.

a. 6^2 $6 \cdot 6$
b. s^5 $s \cdot s \cdot s \cdot s \cdot s$
c. w^3 $w \cdot w \cdot w$
d. m^4 $m \cdot m \cdot m \cdot m$

Use exponents to write these multiplications.

e. $(7)(7)(e)(e)(e)(e)$ 7^2e^4
f. $(2)(2)(2)(f)(f)(f)(f)(f)$ 2^3f^5
g. $(4)(4)(4)(4)(c)(c)(c)(g)(g)$ $4^4c^3g^2$
h. $(3)(3)(s)(t)(t)(u)(v)(v)(v)(x)$ $3^2st^2uv^3x$

Simplify these polynomials. Combine any numerical coefficients.

i. $6b^3 - 4b^3$ $2b^3$
j. $k^2 + 8k^2$ $9k^2$
k. $4^2x^4y^3 - x^4y^3$ $15x^4y^3$
l. $2^2y^2 + 3y^2$ $7y^2$

WRITTEN EXERCISES

A. *Write these in expanded form.*

1. 4^2 $4 \cdot 4$
2. 5^2 $5 \cdot 5$
3. 6^3 $6 \cdot 6 \cdot 6$
4. 8^4 $8 \cdot 8 \cdot 8 \cdot 8$
5. d^4 $d \cdot d \cdot d \cdot d$
6. n^3 $n \cdot n \cdot n$
7. t^5 $t \cdot t \cdot t \cdot t \cdot t$
8. k^5 $k \cdot k \cdot k \cdot k \cdot k$

B. *Use exponents to write these multiplications.*

9. $(5)(5)(5)$ 5^3
10. $(6)(6)(6)(6)$ 6^4
11. $(e)(e)(e)(e)(e)$ e^5
12. $(p)(p)(p)(p)(p)(p)$ p^6
13. $(2)(2)(2)(m)(m)(n)(n)(n)(n)$ $2^3m^2n^4$
14. $(4)(4)(c)(c)(c)(c)(d)(d)$ $4^2c^4d^2$
15. $(x)(x)(x)(x)(x)(x)(y)(z)(z)(z)$ x^6yz^3
16. $(2)(2)(2)(g)(g)(h)(h)(h)(h)$ $2^3g^2h^4$

C. *Simplify these polynomials. Combine any numerical coefficients.*

17. $4a^3 + 5a^3$ $9a^3$
18. $5n^2 + 4n^2$ $9n^2$
19. $8n^3 - n^3$ $7n^3$
20. $4m^3 - 2m^3$ $2m^3$
21. $2^2a^2b + 3a^2b$ $7a^2b$
22. $2q^4 + 3^2q^4$ $11q^4$

472   Chapter 10   Introduction to Algebra

D. Write equations to solve these reading problems.

23. Three less than Grandfather's age is equal to Timothy's age raised to the third power. Timothy is 4. How old is Grandfather?
$g - 3 = 4^3$ $g = 67$ years

24. When Sharlene was 13, she calculated that her age to the third power minus 210 equals the year she was born. In what year was she born?
$y = 13^3 - 210$ $y = 1987$

25. The Red Sea is about 1,200 miles longer than it is wide. Its length and width together are 1,600 miles. What is the width of the Red Sea?
$w + (w + 1,200) = 1,600$ $w = 200$ miles

26. The length of Arnold's math book is 3 inches shorter than twice its width. The total of its length and width is 15 inches. How wide is the book?
$w + (2w - 3) = 15$ $w = 6$ inches

27. The tallest tree on record is an Australian eucalyptus. It was 62 feet higher than the tallest tree currently standing (as measured in 1990) in Redwood National Park, California. The height of both trees together would be 808 feet. How tall was the eucalyptus in Australia?
$h + (h - 62) = 808$ $h = 435$ feet

28. The maximum life span of the Asian elephant is 78 years. That is 8 more than 5 times Samuel's age. How old is Samuel?
$78 = 5s + 8$ $s = 14$ years

REVIEW EXERCISES

E. Solve these equations. *(Lessons 136, 137)*

29. $5k + 13 = 28$ $k = 3$
30. $\frac{a}{8} - 2 = 6$ $a = 64$

F. Simplify these polynomials. *(Lesson 135)*

31. $4n - n + 3n$ $6n$
32. $6b + 6r + b - 2 + 2r$ $7b + 8r - 2$

G. Find the total interest if it is compounded semiannually. *(Lesson 123)*

33. $11,000 at 12% for 1 yr. $1,359.60
34. $3,000 at 18% for 1 yr. $564.30

H. Find the volume of each square pyramid. *(Lesson 107)*

35. $b = 11$ in.   726 cu. in.
    $h = 18$ in.
36. $b = 12$ m   1,104 m³
    $h = 23$ m

I. Solve these division problems. *(Lesson 11)*

37. 28)41,530   1,483 R 6
38. 271)297,054   1,096 R 38

3. **Monomials containing exponents can be added or subtracted just like other monomials as long as they are like terms.** This means that each monomial must have exactly the same literal number or combination of literal numbers, each raised to the same power. If a numeral is raised to a certain power (such as 3^3), it can be simplified by writing it in the standard form.

a. $a^2 + a^2 = 2a^2$

b. $5a^2b^2 - 2a^2b^2 = 3a^2b^2$

c. $2c^3d^2 + 2c^3d^2 = 4c^3d^2$

d. $3e^4f^3 + 2^3e^4f^3 = 11e^4f^3$

e. $3^2g^3h^2 - 6g^3h^2 = 3g^3h^2$

f. $6^2m^2n^3 - 5^2m^2n^3 = 11m^2n^3$

Solutions for Part G

33. First half: $11,000 × 0.06 = $660
    Second half: $11,600 × 0.06 = $699.60
    $660.00 + $699.60

34. First half: $3,000 × 0.09 = $270
    Second half: $3,270 × 0.09 = $294.30
    $270.00 + $294.30

T–473   Chapter 10   Introduction to Algebra

LESSON 140

Objective

- To teach *how to multiply numbers with exponents.

Review

1. *Simplify these expressions.* (Lesson 139)
   a. $(6)(6)(6)(v)(v)(v)(w)(x)(x) = $ ___ $(216v^3wx^2)$
   b. $2s^2 + 6s^2 = $ ___ $(8s^2)$

2. *Write algebraic expressions for these.* (Lesson 132)
   a. b less than eighteen $(18 - b)$
   b. nine in addition to r $(r + 9)$

3. *Use the compound interest formula to find the interest if it is compounded quarterly.* (Lesson 124)
   a. $10,000 at 4% for 6 mo.   ($201.00)
   b. $6,000 at 8% for 6 mo.   ($242.40)

4. *Find the volume of each sphere.* (Lesson 108)
   a. $r = 6$ in.
      $v = $ (902.88 cu. in.)
   b. $r = 13$ ft.
      $v = $ (9,183.46 cu. ft.)

5. *Do these divisions mentally.* (Lesson 12)
   a. $432 \div 48$ (9)
   b. $128 \div 16$ (8)
   c. $648 \div 72$ (9)

140. Exponents in Multiplication

How can you do a multiplication such as $2^3 \cdot 2^4$? You could begin by writing the factors in expanded form: $(2 \cdot 2 \cdot 2) \cdot (2 \cdot 2 \cdot 2 \cdot 2)$.

Because of the associative law of multiplication, you can drop the parentheses and write the expression like this: $2 \cdot 2 \cdot 2 \cdot 2 \cdot 2 \cdot 2 \cdot 2$, or 2^7. The new exponent (7) is the sum of the two exponents in the original factors (3 and 4).

What happens if you multiply $a^2 \cdot a^3$? Think $(a \cdot a) \cdot (a \cdot a \cdot a) = a^5$. Again, the new exponent (5) is the sum of the two exponents in the original factors (2 and 3).

> To multiply matching factors that have exponents, add the exponents in the factors.

This can be done only if both factors have the same base numeral or literal number.

Example A	**Example B**
$n^5 \cdot n^2 = n^{(5+2)} = n^7$	$n^2 \cdot n^3 \cdot n^4 = n^{(2+3+4)} = n^9$

When a factor shows no exponent, that exponent is 1. Thus, $b^2 \cdot b^4 \cdot b = b^{(2+4+1)} = b^7$.

When factors have more than one literal number with exponents, add exponents together for only the literal numbers that match.

$5m^2n \cdot 5mn^3 = 5 \cdot m \cdot m \cdot n \cdot 5 \cdot m \cdot n \cdot n \cdot n = 5 \cdot 5 \cdot m \cdot m \cdot m \cdot n \cdot n \cdot n \cdot n = 25m^3n^4$

Example C	**Example D**
$4x^3y^2z \cdot 2^2x^4y^2z^2 = 16x^7y^4z^3$	$a^5bc^2 \cdot a^6b^2d = a^{11}b^3c^2d$

CLASS PRACTICE

Write each multiplication in expanded form; then write the answer with an exponent.

a. $5^2 \cdot 5^4$
   $(5 \cdot 5) \cdot (5 \cdot 5 \cdot 5 \cdot 5) = 5^6$

b. $d^3 \cdot d^5$
   $(d \cdot d \cdot d) \cdot (d \cdot d \cdot d \cdot d \cdot d) = d^8$

c. $c^4 \cdot c^3$
   $(c \cdot c \cdot c \cdot c) \cdot (c \cdot c \cdot c) = c^7$

Write each answer in exponential form and in standard form.

d. $4^3 \cdot 4^4$ 4^7 16,384

e. $7 \cdot 7^3$ 7^4 2,401

Simplify these multiplications.

f. $b^2 \cdot b^6 \cdot b^5$ b^{13}

g. $2^3mn^3 \cdot m^4n$ $8m^5n^4$

h. $5r^2t^2 \cdot 7s^2t^4$ $35r^2s^2t^6$

474 Chapter 10 Introduction to Algebra

WRITTEN EXERCISES

A. Write each multiplication in expanded form; then write the answer with an exponent.

1. $6^2 \cdot 6^3$ $(6 \cdot 6) \cdot (6 \cdot 6 \cdot 6) = 6^5$ 2. $b^3 \cdot b^4$ $(b \cdot b \cdot b) \cdot (b \cdot b \cdot b \cdot b) = b^7$

3. $n^4 \cdot n^2$ $(n \cdot n \cdot n \cdot n) \cdot (n \cdot n) = n^6$ 4. $x^5 \cdot x^3$ $(x \cdot x \cdot x \cdot x \cdot x) \cdot (x \cdot x \cdot x) = x^8$

B. Write each answer in exponential form and in standard form.

5. $2^2 \cdot 2^2$ 2^4 16   6. $3 \cdot 3^3$ 3^4 81

7. $5^3 \cdot 5^2$ 5^5 3,125   8. $4^3 \cdot 4^3$ 4^6 4,096

C. Simplify these multiplications.

9. $c^2 \cdot c^3$ c^5 10. $2a^5 \cdot a^2$ $2a^7$

11. $b^2 \cdot 4b^3$ $4b^5$ 12. $n^2 \cdot n^5$ n^7

13. $3n^2 \cdot 3n^4 \cdot n^3$ $9n^9$ 14. $z^5 \cdot z^3 \cdot z^2$ z^{10}

15. $3b^3 \cdot 2b^4 \cdot 2b$ $12b^8$ 16. $2k^3 \cdot 7k \cdot k^4$ $14k^8$

17. $a^2b^2 \cdot a^2b^2$ a^4b^4 18. $c^3d^2 \cdot c^2d^4$ c^5d^6

19. $3g^3h^2 \cdot 5g^2h$ $15g^5h^3$ 20. $6m^2n^4 \cdot 4m^3n^3$ $24m^5n^7$

D. Solve these reading problems. Write equations for numbers 21–24.

21. Faye's father is 8 years more than twice as old as Faye. Their combined age is 56 years. How old is Faye?
$$f + (2f + 8) = 56 \qquad f = 16 \text{ years}$$

22. Keith's great-grandmother is 2 years more than 6 times his age. Their combined age is 100 years. How old is Keith?
$$k + (6k + 2) = 100 \qquad k = 14 \text{ years}$$

23. Wilson read about a blizzard that struck Buffalo, New York, in the winter of 1976–77. It left snowdrifts 30 feet high, which is 3 feet less than 6 times Wilson's height. How tall is Wilson?
$$6w - 3 = 30 \qquad w = 5\tfrac{1}{2} \text{ feet}$$

24. Wilson read that the 1993–94 winter also brought much snow to the eastern United States. The state of Kentucky used 140,000 tons of salt on its highways that winter. How many dump trucks would this salt fill if each truck carried 12 tons? (Answer to the nearest whole number.)
$$t = \frac{140,000}{12} \qquad t = 11,667 \text{ trucks}$$

25. Calculate the volume of a steel ball 2 inches in diameter, to the nearest hundredth.
    4.18 cubic inches (4.18×1^3)

26. Gerald blew up a round balloon to a diameter of 14 inches. What was the volume of the air that it contained? (Use $\tfrac{4}{3}$ and $3\tfrac{1}{7}$ in your calculations.)
    $1{,}437\tfrac{1}{3}$ cubic inches $(\tfrac{88}{21} \times 7^3)$

Lesson 140 T-474

Introduction

What happens when we multiply $3^2 \times 3^3$? Find out by using the expanded form.

$(3 \times 3) \times (3 \times 3 \times 3) = 9 \times 27 = 243$

The problem could also be written as $3^2 \times 3^3 = 3^5$; and $3^5 = 243$. Do the students see the relationship between the exponents in the factors and the exponent in the product? The exponent in the product is the sum of the exponents in the factors.

Teaching Guide

1. **When factors that have exponents are multiplied, the exponent of the product is the sum of the exponents in the factors.**

    a. $2^2 \cdot 2^3$ (2^5)

    b. $3^4 \cdot 3^5$ (3^9)

    c. $4^8 \cdot 4^5$ (4^{13})

    d. $5^9 \cdot 5^7$ (5^{16})

2. **When more than one literal number is used in multiplication, the exponents for each one need to be kept separate.** Remember, if a literal number has no numerical coefficient, that coefficient is 1. If no exponent is included, that exponent is also 1.

    a. $ab \cdot ab$ (a^2b^2)

    b. $c^2d^3 \cdot c^3d$ (c^5d^4)

    c. $2m^4n^2 \cdot 6m^2n^2$ ($12m^6n^4$)

    d. $3x^3y^5 \cdot 5x^2y^4$ ($15x^5y^9$)

T–475 Chapter 10 Introduction to Algebra

Solutions for Part G
31. $a = \$12{,}100 \times 1.02^2$
32. $a = \$7{,}400 \times 1.03^2$

REVIEW EXERCISES

E. Simplify these polynomials. *(Lesson 139)*

27. $8^2g^3h - 14g^3h$ $50g^3h$ 28. $4^2n^2 + 2n^2$ $18n^2$

F. Write algebraic expressions for these. *(Lesson 132)*

29. the total of e and f $e + f$ 30. g times h gh

G. Use the compound interest formula to find the amount of principal plus interest if interest is compounded quarterly. *(Lesson 124)*

31. $12,100 at 8% for 6 mo. $12,588.84   32. $7,400 at 12% for 6 mo. $7,850.66

H. Find the volume of each sphere. *(Lesson 108)*

33. $r = 9$ in.   3,047.22 cu. in.   34. $r = 8$ m   2,140.16 m³

I. Do these divisions mentally. *(Lesson 12)*

35. $216 \div 18$ 12   36. $144 \div 24$ 6

A blizzard is defined as a snowstorm with winds of 35 miles per hour or more, temperatures of 10°F or lower, and visibility approaching zero.

476   Chapter 10   Introduction to Algebra

141. Exponents in Division

In Lesson 140, you learned that to multiply identical numerals or literal numbers that have exponents, you can add the exponents in the factors.

$$3^2 \cdot 3^3 = 3^5$$
$$\downarrow \quad \downarrow \quad \downarrow$$
$$9 \cdot 27 = 243$$

$$n^2 \cdot n^3 = n^5$$

Division is the inverse operation of multiplication. If $3^2 \cdot 3^3 = 3^5$, then $3^5 \div 3^3 = 3^2$. This can be demonstrated with the expanded form. The division $3 \cdot 3 \cdot 3 \cdot 3 \cdot 3 \div 3 \cdot 3 \cdot 3$ may be written in fraction form.

$$\frac{\cancel{3} \cdot \cancel{3} \cdot \cancel{3} \cdot 3 \cdot 3}{\cancel{3} \cdot \cancel{3} \cdot \cancel{3}} = \frac{3 \cdot 3}{1} = 3^2$$

Use cancellation to cross out one numerator and one denominator at a time.

To divide matching base numerals or letters that have exponents, subtract the exponents.

Remember that a numeral or letter without an exponent is the same as having the exponent 1. Numerical coefficients may be canceled in division just as with regular fractions. See Example B.

Example A
$$n^5 \div n^2 = n^{(5-2)} = n^3$$
$$\frac{n^5}{n^2} = \frac{\cancel{n} \cdot \cancel{n} \cdot n \cdot n \cdot n}{\cancel{n} \cdot \cancel{n}} = n \cdot n \cdot n = n^3$$

Example B
$$8a^3 \div 2a = 4a^{(3-1)} = 4a^2$$
$$\frac{8a^3}{2a} = \frac{\overset{4}{\cancel{8}} \cdot \cancel{a} \cdot a \cdot a}{\underset{1}{\cancel{2} \cdot \cancel{a}}} = 4a^2$$

Finding square roots is a division-related process that is very simple when terms have even exponents. Consider n^6 in expanded form.

$$n^6 = n \cdot n \cdot n \cdot n \cdot n \cdot n = (n \cdot n \cdot n) \cdot (n \cdot n \cdot n) = n^3 \cdot n^3$$

The associative law of multiplication allows placing the factors in two equal groups, as shown. The product of each group is the same factor, and multiplying those factors yields the original term. Therefore, the square root of a term with an even exponent is the same term with an exponent half as great. The square root of n^6 is n^3.

Example C $\sqrt{5^2} = 5^1 = 5$ **Example D** $\sqrt{d^4} = d^2$

LESSON 141

Objectives

- To teach *how to divide numbers with exponents.

- To teach that *the square root of a number with an exponent can be found by dividing the exponent in half.

Review

1. Give Lesson 141 Quiz (Solving Equations).

2. *Write these products with exponents.* (Lesson 140)

    a. $c^2 \cdot c \cdot c^6$ (c^9)

    b. $d^2e^3 \cdot de^2$ (d^3e^5)

3. *Write equations to solve these reading problems.* (Lessons 136, 137)

    a. Clay and starch are two ingredients used in manufacturing paper. The amount of clay needed for one ton of paper is 742 pounds. If this is 164 pounds more than 2 times the amount of starch, how much starch is needed?

    ($2s + 164 = 742$; $s = 289$ lb.)

    b. It takes 560 kilowatt-hours of electricity to manufacture 5 tons of paper. This is 20 more than 90 times the tons of coal needed. How many tons of coal are required?

    ($90c + 20 = 560$; $c = 6$ tons)

4. *Simplify each expression, using the correct order of operations.* (Lesson 133)

    a. $(14 - 5)2 - 3(4 + 2) = \underline{}$ (0)
    $(9)2 - 3(6)$
    $18 - 18$

    b. $53 - 2^2[1 + 2(8 - 2^2)] = \underline{}$ (17)
    $53 - 4[1 + 2(8 - 4)]$
    $53 - 4[1 + 2(4)]$
    $53 - 4[1 + 8]$
    $53 - 4[9]$
    $53 - 36$

5. *Calculate these property taxes.* (Lesson 125)

	Assessment	Tax Rate	Tax
a.	$125,000	19 mills	($2,375.00)
b.	$205,000	17.1 mills	($3,505.50)

6. *Find square roots by using a table.* (Lesson 109)

    a. 156   (12.490)

    b. 94   (9.695)

7. *Do these divisions mentally.* (Lesson 13)

    a. $135 \div 5$ (27)

    b. $1,600 \div 50$ (32)

    c. $16,500 \div 500$ (33)

Introduction

Write the problem $d^5 \div d^2$ on the board. Expand the problem like this: $d \cdot d \cdot d \cdot d \cdot d \div d \cdot d$. Then ask the students, "How is division usually shown in algebra?" It is usually shown by using the fraction form.

$$\frac{d \cdot d \cdot d \cdot d \cdot d}{d \cdot d}$$

Use cancellation to remove the same number of like factors in the divisor and the dividend.

$$\frac{d \cdot d \cdot d \cdot \cancel{d} \cdot \cancel{d}}{\cancel{d} \cdot \cancel{d}}$$

This demonstrates that $d^5 \div d^2 = d^3$.

Teaching Guide

1. **When factors that have exponents are divided, the exponent of the quotient is the difference between the exponents of the dividend and the divisor.** The resulting exponent is dropped if it is 1.

   a. $f^6 \div f^2 = f^4$
   b. $g^7 \div g^4 = g^3$
   c. $y^8 \div y^2 = y^6$
   d. $z^9 \div z^3 = z^6$
   e. $m^3 \div m^2 = m$
   f. $n^6 \div n^5 = n$

2. **To find the square root of a number with an even exponent, divide the exponent by 2.**

   a. $\sqrt{e^4}$ (e^2)
   b. $\sqrt{t^6}$ (t^3)
   c. $\sqrt{g^8}$ (g^4)
   d. $\sqrt{h^{10}}$ (h^5)

Lesson 141

CLASS PRACTICE

Write these problems and their answers with exponents.

a. $h \cdot h \cdot h \cdot h \cdot h \div h \cdot h \cdot h$ $\dfrac{h^5}{h^3} = h^2$ b. $w \cdot w \cdot w \cdot w \cdot w \cdot w \div w \cdot w \cdot w \cdot w \cdot w$ $\dfrac{w^6}{w^5} = w$

Write each problem in expanded form; then write the answer with an exponent.

c. $r^5 \div r^2$ $\dfrac{r \cdot r \cdot r \cdot r \cdot r}{r \cdot r} = r^3$ d. $e^4 \div e^3$ $\dfrac{e \cdot e \cdot e \cdot e}{e \cdot e \cdot e} = e$ (no exponent needed)

Write the answers with exponents.

e. $r^5 \div r^3$ r^2 f. $24v^{12} \div 6v^9$ $4v^3$

g. $\dfrac{t^8}{t^6}$ t^2 h. $\dfrac{14x^7}{2x}$ $7x^6$

Find the square roots.

i. $\sqrt{y^8}$ y^4 j. $\sqrt{g^{16}}$ g^8 k. $\sqrt{n^6}$ n^3 l. $\sqrt{k^{18}}$ k^9

WRITTEN EXERCISES

A. *Write these problems and their answers with exponents.*

1. $b \cdot b \cdot b \cdot b \cdot b \cdot b \div b \cdot b \cdot b \cdot b$ $\dfrac{b^6}{b^4} = b^2$ 2. $c \cdot c \cdot c \cdot c \cdot c \cdot c \div c \cdot c \cdot c$ $\dfrac{c^6}{c^3} = c^3$

3. $d \cdot d \cdot d \cdot d \cdot d \div d \cdot d \cdot d \cdot d$ $\dfrac{d^5}{d^4} = d$ 4. $f \cdot f \cdot f \cdot f \cdot f \cdot f \cdot f \div f \cdot f$ $\dfrac{f^7}{f^2} = f^5$

B. *Write each problem in expanded form; then write the answer with an exponent.*

5. $n^4 \div n^2$ $\dfrac{n \cdot n \cdot n \cdot n}{n \cdot n} = n^2$ 6. $q^3 \div q$ $\dfrac{q \cdot q \cdot q}{q} = q^2$

7. $y^5 \div y^2$ $\dfrac{y \cdot y \cdot y \cdot y \cdot y}{y \cdot y} = y^3$ 8. $s^3 \div s^2$ $\dfrac{s \cdot s \cdot s}{s \cdot s} = s$ (no exponent needed)

C. *Write the answers with exponents.*

9. $d^4 \div d^2$ d^2 10. $e^5 \div e^2$ e^3

11. $12m^6 \div 4m^3$ $3m^3$ 12. $g^6 \div g^2$ g^4

13. $\dfrac{h^7}{h^3}$ h^4 14. $\dfrac{18y^8}{3y^6}$ $6y^2$

15. $\dfrac{25x^8}{5x^5}$ $5x^3$ 16. $\dfrac{k^6}{k^2}$ k^4

D. *Find the square roots.*

17. $\sqrt{z^6}$ z^3 18. $\sqrt{v^{12}}$ v^6 19. $\sqrt{u^2}$ u 20. $\sqrt{s^8}$ s^4

478 Chapter 10 Introduction to Algebra

E. Write equations and solve these reading problems.

21. Dale is d years old and Melvin is 2 years more than 3 times as old as Dale. The sum of their ages is 58 years. How old is Dale? $d + (3d + 2) = 58$ $d = 14$ years

22. The upper grade classroom at the Belmont Christian School has 19 students. There are 2 less than twice as many boys as girls. How many girls are there?
$g + (2g - 2) = 19$ $g = 7$ girls

23. The sum of 3 consecutive numbers is 66. What is the smallest number?
$n + (n + 1) + (n + 2) = 66$ $n = 21$

24. The sum of 3 consecutive numbers is 48. What is the smallest number?
$n + (n + 1) + (n + 2) = 48$ $n = 15$

25. Two important ingredients in making paper are lime and water. It takes 29,600 pounds of water to produce a ton of paper. This is 200 pounds more than the amount of lime required. How many pounds of lime are required?
$p + 200 = 29,600$ $p = 29,400$ pounds

26. It takes p pounds of dye and pigments to produce a ton of paper. It takes 2 pounds more than twice as much sulfur as dye and pigments. Altogether, 122 pounds of these ingredients are required. How many pounds of dye and pigments are needed?
$p + (2p + 2) = 122$ $p = 40$ pounds

REVIEW EXERCISES

F. Write these products with exponents. *(Lesson 140)*

27. $n^3 \cdot n \cdot n^8$ n^{12} 28. $n^5 r^2 \cdot n^3 r^4$ $n^8 r^6$

G. Solve these equations involving the four basic operations. *(Lessons 136, 137)*

29. $3b - 5 = 55$ $b = 20$ 30. $5a - 2 = 63$ $a = 13$

H. Simplify each expression, using the correct order of operations. *(Lesson 133)*

31. $(16 - 7)3 - 2(5 + 3)$ 11   32. $31 - 2[1 + 2(9 - 3^2) + 3]$ 23

I. Calculate these property taxes. *(Lesson 125)*

33. Assessment: $108,500; tax rate: 14 mills $1,519.00
34. Assessment: $195,400; tax rate: 21.2 mills $4,142.48

J. Find the square roots by using a table. *(Lesson 109)*

35. 198   14.071   36. 82   9.055

K. Do these divisions mentally. *(Lesson 13)*

37. 155 ÷ 5   31   38. 12,000 ÷ 500   24

CHALLENGE EXERCISES

L. See if you can solve these for extra credit.

39. $n^4 \div n^4$ 1 (or n^0)   40. $m^5 \div m^5$ 1 (or m^0)   41. $\sqrt{v^{2n}}$ v^n 42. $\sqrt{u^{6n}}$ u^{3n}

Lesson 141   T–478

Further Study

Exponents are interesting to work with. Consider the following table.

$10^6 =$	1,000,000	$4^6 = 4,096$
$10^5 =$	100,000	$4^5 = 1,024$
$10^4 =$	10,000	$4^4 = 256$
$10^3 =$	1,000	$4^3 = 64$
$10^2 =$	100	$4^2 = 16$
$10^1 =$	10	$4^1 = 4$
$10^0 =$	1	$4^0 = 1$
$0^{-1} =$	0.1	$4^{-1} = \frac{1}{4}$
$10^{-2} =$	0.01	$4^{-2} = \frac{1}{16}$
$10^{-3} =$	0.001	$4^{-3} = \frac{1}{64}$
$10^{-4} =$	0.0001	$4^{-4} = \frac{1}{256}$
$10^{-5} =$	0.00001	$4^{-5} = \frac{1}{1,024}$
$10^{-6} =$	0.000001	$4^{-6} = \frac{1}{4,096}$

Also consider these examples.

$1,000 \times 100 = 100,000$ or $10^3 \times 10^2 = 10^{5\,(3+2)}$

$10,000 \times 100 = 1,000,000$ or $10^4 \times 10^2 = 10^{6\,(4+2)}$

$4,096 \div 64 = 64$ or $4^6 \div 4^3 = 4^{3\,(6-3)}$

$16 \div 4,096 = \frac{1}{256}$ or $4^2 \div 4^6 = 4^{-4\,(2-6)}$

$1,000 \times 0.00001 = 0.01$ or $10^3 \times 10^{-5} = 10^{-2\,(3+-5)}$

$\frac{1}{16} \div \frac{1}{4,096} = 256$ or $4^{-2} \div 4^{-6} = 4^{4\,(-2--6)}$

$100 \times \frac{1}{100} = 1$ or $10^2 \times 10^{-2} = 10^{0\,(2+-2)}$

T-479 Chapter 10 Introduction to Algebra

LESSON 142

Objective

- To review the material taught in Chapter 10 (Lessons 132–141).

Teaching Guide

1. Lesson 142 reviews the material taught in Lessons 132–141. For pointers on using review lessons, see *Teaching Guide* for Lesson 98.
2. Be sure to review the following new concepts taught in this chapter.

Lesson number and new concept	Exercises in Lesson 142
132—Algebraic terms: literal coefficient, numerical coefficient, polynomial, monomial, binomial, trinomial.	1–6
133—Use of brackets and vinculum.	11, 12
134—Evaluating expressions with exponents.	13, 14
135—Recognizing like and unlike terms.	15, 16
135—Adding and subtracting monomials.	35–40
139—Using exponents with literal numbers.	17–20
140—Multiplying numbers with exponents.	25–28
141—Dividing numbers with exponents.	29–32
141—Finding square roots of numbers with exponents.	33, 34

142. Chapter 10 Review

A. Answer these questions. *(Lessons 132, 136, 137)*

1. Is $7c + 4d$ a monomial or a polynomial?   polynomial
2. Is $8k^2m^2n$ a monomial or a polynomial?   monomial
3. Is $9d + 4e - 5y$ a monomial, a binomial, or a trinomial?   trinomial
4. Is $4d^2ef + 4g^2h$ a monomial, a binomial, or a trinomial?   binomial
5. What is the numerical coefficient in $7ab$?   7
6. What are the literal coefficients in $9de$? d, e
7. Which operation should be used to solve $z + 4 = 7$?   subtraction
8. Which two operations should be used to solve $5n - 6 = 54$?   addition, division

B. Simplify each expression. *(Lesson 133)*

9. $3 + 16 \cdot 2$ 35
10. $4 \cdot 4^2 - 3$ 61
11. $4[5 - (8 - 5)]$ 8
12. $2(\frac{20}{2} - 3 + 4)$ 22

C. Do these exercises with exponents. *(Lessons 134, 135, 139)*

13. Evaluate a^2b^2 if $a = 3$ and $b = 5$.   225
14. Evaluate $2ab^2$ if $a = 5$ and $b = 7$.   490
15. Compare $6m^3n$ and $4m^3n$. Are they *like* or *unlike* terms?   like
16. Compare $2xy^2$ and $2x^2y$. Are they *like* or *unlike* terms?   unlike
17. Write b^6 in expanded form. $b \cdot b \cdot b \cdot b \cdot b \cdot b$
18. Write e^5 in expanded form. $e \cdot e \cdot e \cdot e \cdot e$
19. Write $(3)(3)(a)(a)(a)(a)$, using exponents. 3^2a^4
20. Write $(7)(7)(7)(x)(x)(x)(x)(x)(x)$, using exponents. 7^3x^6

D. Write an equation for each statement. *(Lesson 138)*

21. n equals three less than the product of eight and two $n = 8(2) - 3$
22. d equals five more than the product of six and k $d = 6k + 5$
23. p equals four times the sum of five and nine $p = 4(5 + 9)$
24. q equals four more than eight divided by m $q = \frac{8}{m} + 4$

E. Write the answer to each problem, using exponents. *(Lessons 140, 141)*

25. $3a^5 \cdot 7a^2$ $21a^7$
26. $3n^8 \cdot 8n^7$ $24n^{15}$

480   Chapter 10   Introduction to Algebra

27. $g^4k^2 \cdot g^2k$ g^6k^3 **28.** $x^5y^3 \cdot x^3y$ x^8y^4

29. $n^6 \div n^3$ n^3 **30.** $q^5 \div q^2$ q^3

31. $\dfrac{c^5}{c^4}$ c (no exponent needed)   **32.** $\dfrac{e^6}{e^3}$ e^3

33. $\sqrt{m^{12}}$ m^6 **34.** $\sqrt{m^{14}}$ m^7

F. Simplify these polynomials. *(Lessons 135, 139)*

35. $st + 3st + 2w - 4$ $4st + 2w - 4$ **36.** $ab + 2c + 5ab + c$ $6ab + 3c$

37. $4b + 2c + 5 + b + 5c + 4$ $5b + 7c + 9$ **38.** $12c + 7d + 3c - e$ $15c + 7d - e$

39. $6b^2 + 7b^2$ $13b^2$ **40.** $7kn - 4kn$ $3kn$

G. Solve these equations. Show each step in your work. *(Lesson 136)*

(See facing page.)

41. $\dfrac{z}{5} = 4$ **42.** $n - 7 = 11$

43. $k - 8 = 7$ **44.** $4x = 24$

45. $3x - 4 = 20$ **46.** $7 + 4y = 39$

H. Write equations to solve these reading problems. *(Lesson 138)*

47. Doris is two years more than twice as old as Marvin. Their combined age is 32. How old is Marvin? $m + (2m + 2) = 32$ $m = 10$ years

48. The sum of three consecutive numbers is 33. What is the smallest of the numbers?
$n + (n + 1) + (n + 2) = 33$ $n = 10$

49. The sperm whale can stay underwater 18 times longer than a sea otter. The combined total of their underwater limit is 95 minutes. How long can a sea otter hold its breath?
$18m + m = 95$ $m = 5$ minutes

50. A porpoise can hold its breath for 9 minutes less than twice the number of minutes a muskrat can go without breathing. The combined total of their underwater limit is 27 minutes. How long can a muskrat hold its breath?
$m + (2m - 9) = 27$ $m = 12$ minutes

51. The largest known bird eggs were laid by an extinct bird of Madagascar in Africa. The length of the eggs is 4 inches more than their diameter. The length and the diameter together are 23 inches. What is the diameter of the eggs?
$d + (d + 4) = 23$ $d = 9\tfrac{1}{2}$ inches

52. The smallest mature bird eggs are laid by the vervain hummingbird of Jamaica. Adding 0.1 inch to 10 times the length of the eggs equals 4 inches. How long are the eggs?
$10e + 0.1 = 4$ $e = 0.39$ inch

---

143. Chapter 10 Test

LESSON 143

Objective
- To test the students' mastery of the concepts in Chapter 10.

Teaching Guide
1. Correct Lesson 142.
2. Review any areas of special difficulty.
3. Administer the test. For pointers on giving tests, see *Teaching Guide* for Lesson 99.

Solutions for Part G

41. $\frac{z}{5} \cdot 5 = 4 \cdot 5$
    $z = 20$

42. $n - 7 + 7 = 11 + 7$
    $n = 18$

43. $k - 8 + 8 = 7 + 8$
    $k = 15$

44. $\frac{4x}{4} = \frac{24}{4}$
    $x = 6$

45. $3x - 4 + 4 = 20 + 4$
    $3x = 24$
    $\frac{3x}{3} = \frac{24}{3}$
    $x = 8$

46. $7 + 4y - 7 = 39 - 7$
    $4y = 32$
    $\frac{4y}{4} = \frac{32}{4}$
    $y = 8$

A diligent shepherd understands the equations of animal health and life. He knows how to care for his sheep by applying the factors to keep these equations in balance.

Chapter 11
Signed Numbers, Tables, and Graphs

The branch of mathematics called algebra includes working with signed numbers (numbers marked with + or − signs to show whether they are positive or negative). Signed numbers are already familiar in the context of temperatures above or below zero. Now you will learn to use signed numbers in the four basic mathematical operations.

This chapter extends practice in writing and solving equations, and introduces tables and graphs as another way to show relationships between numbers.

Be thou diligent to know the state of thy flocks, and look well to thy herds.
Proverbs 27:23

144. Introduction to Signed Numbers

Mathematics is the study of numbers. Numbers can be grouped in several sets. **Natural numbers**, such as 1, 2, and 3, are the whole numbers that you use for counting. In fact, these are sometimes called **counting numbers**. A second set of numbers is the **integers**. This includes all positive and negative whole numbers; –2, –1, 0, 1, and 2 are all integers. A third set of numbers, the **real numbers**, includes the partial steps between the integers. Whole numbers, fractions, and decimals—both positive and negative—are all real numbers.

Natural numbers: 1, 2, 3, 4, 5, 6, 7, 8, 9, 10, 11, 12 . . .

Integers: . . . –6, –5, –4, –3, –2, –1, 0, 1, 2, 3, 4, 5, 6 . . .

Real numbers (examples): 7, 3, –4, –7, 0, $\frac{1}{4}$, $-\frac{2}{5}$

Usually the number 0 means "nothing." If no pens are in your pocket, you have 0 pens in your pocket. For integers and real numbers, 0 is a reference point. For example, 0° Celsius is not the absence of temperature; it is rather the temperature at which water freezes. On a number line, numbers to the right of 0 or above 0 are marked with the + sign, read "positive." Numbers to the left of 0 or below 0 are marked with the – sign, read "negative." If no sign precedes a number, the number is assumed to be positive. (The number 0 is neither positive nor negative.)

```
–10 –9 –8 –7 –6 –5 –4 –3 –2 –1  0 +1 +2 +3 +4 +5 +6 +7 +8 +9 +10
  |  |  |  |  |  |  |  |  |  |  |  |  |  |  |  |  |  |  |  |  |
```

Positive and negative numbers are known as **signed numbers**. The following examples list some practical uses for signed numbers. Notice that often no sign is shown for a positive number.

Temperature	85 degrees above zero	+85° or 85°
	30 degrees below zero	–30°
Profit and loss	profit of $75	+$75 or $75
	loss of $10	–$10
Elevation	620 feet above sea level	+620 feet or 620 feet
	440 feet below sea level	–440 feet
Change	increase of $7	+$7
	decrease of $0.65	–$0.65
	forward 8 yards	+8 yards
	backward 2 yards	–2 yards

LESSON 144

Objective

- To review positive and negative numbers.
- To review that the absolute value of a signed number is its value without considering the sign.
- To teach *comparing the values and the absolute values of signed numbers.

Review

1. Write the answers to these problems, using exponents. (Lesson 140)

   a. $r^3s^2 \cdot r^2s^2$ (r^5s^4)

   b. $4^2f^2g \cdot 2f^3g$ $(32f^5g^2)$

2. Solve these equations, using the correct order of operations. (Lessons 136, 137)

   a. $6x + 4 = 20 + 2$ $(x = 3)$
   $6x + 4 = 22$
   $6x = 18$

   b. $\dfrac{x}{14} + 14 = 2^2 \cdot 4$ $(x = 28)$

   $\dfrac{x}{14} + 14 = 4 \cdot 4$

   $\dfrac{x}{14} + 14 = 16$

   $\dfrac{x}{14} = 2$

3. Write an equation to solve this reading problem. (Lesson 138)

   The number of pages in a certain book is 20 more than 20 times the number of chapters. If there are 400 pages, how many chapters are in the book?

   $(20n + 20 = 400; n = 19$ chapters$)$

4. Use the Pythagorean rule to find each hypotenuse, to the nearest tenth. (Lesson 112)

   a. leg a = 10 ft.
   leg b = 8 ft.
   hypotenuse = (12.8 ft.)

   b. leg a = 16 in.
   leg b = 12 in.
   hypotenuse = (20 in.)

T–483 Chapter 11 Signed Numbers, Tables, and Graphs

Introduction

Discuss the value of the number 0. How many students are sick today? How much of this lesson have you done already? If 0 is the answer to a question asking "how many" or "how much," then 0 means "nothing." In such cases it is not possible to have less than zero.

Sometimes zero is simply a reference point, and numbers can go below zero. Discuss the difference between 0°F and 0°C. The freezing point of water is at 32°F and 0°C. This does not mean that the freezing temperature varies. Rather, 0° on the Fahrenheit scale refers to a different temperature from 0° on the Celsius scale.

Teaching Guide

1. **Several sets of numbers are used in mathematics.**

   a. Natural numbers are the whole numbers used to count things (1, 2, 3, 4, 5, . . .).

   b. Integers are all the positive and negative whole numbers; that is, natural numbers and their opposites (. . . –5, –4, –3, –2, –1, 0, 1, 2, 3, 4, 5, . . .)

   c. Real numbers include all whole, fractional, and decimal numbers, both positive and negative and both rational and irrational. Examples of real numbers are listed below.

5	0
–3	+8
$\frac{3}{4}$	$\frac{4}{5}$
3.1617	4.333 . . .
–1.2727 . . .	22%
π (3.14159 . . .)	$\sqrt{7}$ (2.64575 . . .)

2. **Positive and negative numbers are known as signed numbers.** Signed numbers have various practical uses in everyday life.

   a. profits and losses

   b. temperatures above and below zero

   c. elevation above and below sea level

   d. debits and credits in double-entry accounting

3. **The value of a signed number is indicated by its relationship to zero.** For positive numbers, the farther from zero a number is, the greater its value. For negative numbers, the closer to zero a number is, the greater its value. Movement to the right on the number line represents increasing value, and

The value of a signed number is known by its relationship to zero. For positive numbers, the farther from zero a number is, the greater is its value. A profit of +$12 is greater than a profit of +$7. For negative numbers, however, the closer to zero a number is, the greater is its value. If the temperature was −4° one morning and −10° the next morning, the temperature was higher on the first morning than on the second.

The value of zero or any positive number is greater than the value of any negative number. (Zero itself is neither positive nor negative.) Read the following number relationships with *greater than* or *less than*.

$$+9 > +6 \qquad -3 > -8 \qquad 11 < 14 \qquad +2 > -7$$
$$0 < +4 \qquad 0 > -4 \qquad -6 < +6 \qquad -12 < -5$$

The **absolute value** of a number is the value of that number without its sign. Absolute value is indicated by placing vertical bars before and after the number. Thus, |−100| is read "the absolute value of negative 100." That value is 100. When absolute values are compared, the distance from zero is considered rather than the actual values of the numbers. In Example A, the absolute value of −24 is greater than the absolute value of +7 because −24 is farther from zero than is +7. Absolute values can be compared as though all the numbers were positive.

Example A
Compare |−24| with |+7|.
The absolute value of −24 is 24.
The absolute value of +7 is 7.
|−24| > |+7|

Example B
Compare |−8| with |−6|.
The absolute value of −8 is 8.
The absolute value of −6 is 6.
|−8| > |−6|

CLASS PRACTICE

Write a signed number for each phrase. Include a label.

a. an electrical surge of 3 volts   +3 volts
b. a price reduction of $15   −$15
c. a loss of $81   −$81
d. 324 feet above sea level   +324 feet

Write the absolute value of each number.

e. |+54| 54
f. |−34| 34
g. |−18| 18
h. |+21| 21

Compare each set of numbers, and write < or > between them.

i. 5 < 6
j. 6 > 4
k. −4 < −2
l. −1 > −5
m. |−4| > |+3|
n. |−4| > |−3|
o. |−3| > |+2|
p. |+3| < |−4|

484   Chapter 11   *Signed Numbers, Tables, and Graphs*

WRITTEN EXERCISES

A. *Write a signed number for each phrase. Include a label if there is one.*

1. an increase of 15   +15
2. a decrease of 73   −73
3. a price reduction of $10   −$10
4. a price markup of $17   +$17
5. a temperature rise of 45°   +45°
6. a temperature drop of 38°   −38°
7. a loss of $76   −$76
8. 282 feet below sea level   −282 feet
9. 1,296 feet below sea level   −1,296 feet
10. 6,288 feet above sea level   +6,288 feet

B. *Write the absolute value of each number.*

11. $|+62|$ 62
12. $|+71|$ 71
13. $|-31|$ 31
14. $|-1|$ 1
15. $|+14|$ 14
16. $|-6|$ 6
17. $|+2|$ 2
18. $|+45|$ 45

C. *Compare each set of numbers, and write < or > between them. Check your answers by using the number line.*

19. 3 _<_ 5
20. 4 _<_ 7
21. −2 _<_ $|-3|$
22. −2 _>_ −5
23. $|-7|$ _>_ $|-5|$
24. $|+6|$ _>_ $|-5|$
25. $|-2|$ _>_ $|+1|$
26. $|+8|$ _>_ $|-5|$

D. *Solve these reading problems. Write equations for numbers 31 and 32.*

27. Mount Nebo, from which Moses viewed the Promised Land, reaches a height of 2,631 feet above sea level. Write this altitude, using a signed number.   +2,631 feet

28. One morning the wind chill was 58°F below zero. Express this temperature as a signed number.   −58°

29. Mount Gerizim and Mount Ebal are twin mountains mentioned in Joshua 8. Mount Gerizim rises to an altitude of 2,890 feet, and Mount Ebal rises to 3,084 feet. Write these altitudes with signed numbers; then place > or < between them.
    +2,890 feet < +3,084 feet

30. The altitude of Lake Merom (Lake Hula) is 270 feet below sea level. The Sea of Galilee is 700 feet below sea level. Write these altitudes with signed numbers; then place > or < between them.   −270 feet > −700 feet

31. Lois's butternut squash plants bore 100 squash, or 5 more than 5 times the number of sugar pumpkins that she harvested. How many sugar pumpkins did she harvest?
    $5p + 5 = 100$ $p = 19$ pumpkins

32. Sound travels through water at a speed of 3,043 miles per hour. That is 43 miles per hour faster than 4 times the speed of sound through air. How fast does sound travel through air? $4s + 43 = 3,043$ $s = 750$ m.p.h.

Lesson 144 T–484

movement to the left represents decreasing value. Be sure to discuss examples such as *a* and *c* below.

a. –6 (<) –2

b. +8 (>) –1

c. 0 (>) –5

d. +8 (<) 15

e. –5 (<) +3

f. –9 (<) 0

4. **The absolute value of a number is the value of that number without its sign.** Absolute value is indicated by placing a vertical bar before and after the number. Thus, |–50| is read "the absolute value of negative 50." That value is 50.

Give the absolute value of the following numbers.

a. |+4| (4)

b. |–3| (3)

c. |–11| (11)

d. |0| (0)

e. |–9| (9)

f. |+11| (11)

5. **When absolute values are compared, consider the numbers as positive numbers.**

a. |–3| < |+6|

b. |+3| > |–2|

c. |–4| > |0|

d. |–3| < |+4|

e. |+9| > |–4|

f. |–8| > |+3|

REVIEW EXERCISES

E. Write the answers to these problems, using exponents. *(Lesson 140)*

33. $x^4y^3 \cdot x^2y^5$ x^6y^8

34. $2^3a^3b \cdot 3a^2b^6$ $24a^5b^7$

F. Solve these equations, using the correct order of operations. *(Lessons 136, 137)*

35. $7x - 4 = 31$ $x = 5$

36. $\frac{x}{4} + 14 = 19$ $x = 20$

G. Use the Pythagorean rule to find each hypotenuse, to the nearest tenth. *(Lesson 112)*

37. leg a = 10 ft.   12.8 ft.
    leg b = 8 ft.

38. leg a = 8 in.   10 in.
    leg b = 6 in.

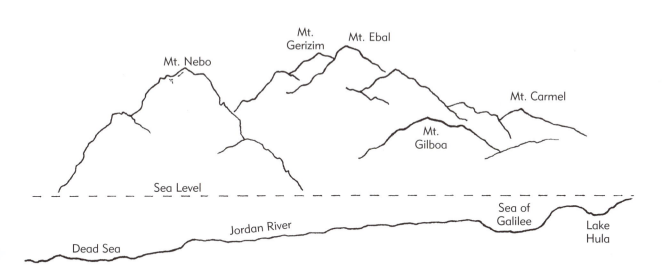

In his hand are the deep places of the earth:

the strength of the hills is his also.

Psalm 95:4

145. Adding Signed Numbers

Addition is combining the values of two or more numbers. For example, 4 plus 5 is 9 (4 + 5 = 9). No positive or negative signs are used, so the numbers are assumed to be positive. In fact, all numbers are positive in normal addition.

Example A shows that adding positive numbers is the same as adding natural numbers. Beginning at positive 2 and moving (adding) 4 in the positive direction puts you at positive 6.

Example A $+2 + (+4) = +6$

−10 −9 −8 −7 −6 −5 −4 −3 −2 −1 0 +1 +2 +3 +4 +5 +6 +7 +8 +9 +10

Example B $-2 + (-4) = -6$

Example B illustrates the exact opposite of Example A. Starting at negative 2 and moving 4 in the negative direction puts you at negative 6. This shows that the signs can be considered as directional symbols. Adding a positive number means moving in the positive direction, and adding a negative number means moving in the negative direction.

Rule 1: To add signed numbers with like signs, add their absolute values. Give the sum the same sign as the addends.

Example C

$$\begin{array}{r} +12 \\ + (+9) \\ \hline +21 \end{array}$$

The numbers have like signs. Add 12 + 9 and use the + sign for the sum.

Example D

$$\begin{array}{r} -8 \\ + (-5) \\ \hline -13 \end{array}$$

The numbers have like signs. Add 8 + 5, and use the − sign for the sum.

LESSON 145

Objective

- To review adding signed numbers with like and unlike signs.

Review

1. Give Lesson 145 Quiz (Multiplying and Dividing With Exponents).

2. *Compare each set of numbers, and write < or > between them.* (Lesson 144)

   a. |−8| (>) |+6|

   b. |−2| (<) |+3|

3. *Find these square roots.* (Lesson 141)

   a. $\sqrt{r^{14}}$ (r^7)

   b. $\sqrt{k^{12}}$ (k^6)

4. *Write equations to solve these reading problems.* (Lesson 137)

   a. In a classroom with 15 students, there are 3 more boys than girls. How many girls are in the classroom?
      $(g + [g + 3] = 15; g = 6$ girls$)$

   b. A recipe calls for twice as many cups of flour as oatmeal. The total of these two ingredients is 6 cups. How many cups of oatmeal does the recipe call for?
      $(2n + n = 6; n = 2$ cups$)$

5. *Solve these problems mentally.* (Lesson 129)

   a. 164 + 147 (311)

   b. 615 − 381 (234)

   c. 7 × 615 (4,305)

   d. 1,400 ÷ 50 (28)

6. *Use the Pythagorean rule to find these answers to the nearest tenth.* (Lesson 113)

   a. Width of two-sided shed roof: 16 ft.
      Height of peak above eaves: 9 ft.
      Distance from peak to eaves: ___
      (12.0 ft.)

   b. Width of rectangular opening: 12 in.
      Height of rectangular opening: ___
      Diagonal measurement: 18 in.
      (13.4 in.)

7. *Find these equivalents.* (Lesson 17)

   a. 3 mi. = ___ ft. (15,840)

   b. 96 in. = ___ ft. (8)

T–487 Chapter 11 Signed Numbers, Tables, and Graphs

Introduction

Write a set of two integers with like signs on the board (such as 6 and 8). Also write a set having the same absolute values as the first set but having opposite signs (such as +6 and –8). Ask the students what the sum is if they combine +6 and +8. The obvious answer is +14.

Do the students know what the sum is if they combine +6 and –8?

Teaching Guide

1. **To add signed numbers with like signs, add their absolute values. Give the sum the same sign as the addends.**

  a. +8
     + (+9)
     (+17)

  b. +8
     + (+6)
     (+14)

  c. +7
     + (+9)
     (+16)

  d. +8
     + (+4)
     (+12)

  e. –3
     + (–8)
     (–11)

  f. –4
     + (–9)
     (–13)

  g. –8
     + (–5)
     (–13)

  h. –7
     + (–8)
     (–15)

2. **To add signed numbers with unlike signs, find the difference between their absolute values. Give the sum the same sign as the addend with the larger absolute value.**

  a. +8
     + (–9)
     (–1)

  b. –8
     + (+6)
     (–2)

  c. +9
     + (–9)
     (0)

  d. –8
     + (+4)
     (–4)

  e. –3
     + (+8)
     (+5)

  f. +4
     + (–9)
     (–5)

  g. +8
     + (–5)
     (+3)

  h. –7
     + (+8)
     (+1)

Solutions for CLASS PRACTICE i and j

i. +6 + (–14)

j. +14,700 + (+800)

Now look at Example E. Beginning at negative 2 and moving (adding) 4 in the positive direction puts you at positive 2.

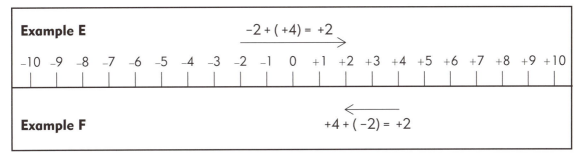

Example F shows that starting at positive 4 and moving (adding) 2 in the negative direction puts you at positive 2.

Rule 2: To add signed numbers with unlike signs, find the difference between their absolute values. Give the sum the same sign as the addend with the greater absolute value.

Example G

$$\begin{array}{r} +6 \\ + (-9) \\ \hline -3 \end{array}$$

The numbers have unlike signs. Subtract 9 – 6, and use the – sign for the sum.

Example H

$$\begin{array}{r} +12 \\ + (-8) \\ \hline +4 \end{array}$$

The numbers have unlike signs. Subtract 12 – 8, and use the + sign for the sum.

CLASS PRACTICE

Add these signed numbers.

a. $\begin{array}{r} +6 \\ + (+5) \\ \hline +11 \end{array}$ b. $\begin{array}{r} -3 \\ + (-7) \\ \hline -10 \end{array}$ c. $\begin{array}{r} -6 \\ + (+4) \\ \hline -2 \end{array}$ d. $\begin{array}{r} -8 \\ + (+8) \\ \hline 0 \end{array}$

e. $\begin{array}{r} +6 \\ + (-2) \\ \hline +4 \end{array}$ f. $\begin{array}{r} -7 \\ + (+8) \\ \hline +1 \end{array}$ g. $\begin{array}{r} +26 \\ + (-15) \\ \hline +11 \end{array}$ h. $\begin{array}{r} -18 \\ + (+11) \\ \hline -7 \end{array}$

Solve these reading problems, and use signed numbers in the answers.

i. At 8:00 one evening the temperature was 6° F. By the next morning the temperature had dropped 14°. What was the temperature in the morning? –8°

j. A plane flying at an altitude of 14,700 feet rose 800 feet. What was its new altitude? +15,500 feet

488 Chapter 11 Signed Numbers, Tables, and Graphs

WRITTEN EXERCISES

A. Add these signed numbers. The addends have like signs.

1. +7
   + (+8)
   +15

2. +9
   + (+3)
   +12

3. −3
   + (−1)
   −4

4. −9
   + (−2)
   −11

5. −9
   + (−5)
   −14

6. −5
   + (−7)
   −12

7. +6
   + (+3)
   +9

8. +7
   + (+14)
   +21

B. Add these signed numbers. The addends have unlike signs.

9. +9
   + (−11)
   −2

10. −3
    + (+7)
    +4

11. −8
    + (+6)
    −2

12. +15
    + (−14)
    +1

13. +14
    + (−11)
    +3

14. −9
    + (+6)
    −3

15. −11
    + (+16)
    +5

16. +13
    + (−18)
    −5

C. Add these signed numbers. Watch the signs.

17. +4
    + (−6)
    −2

18. −2
    + (−8)
    −10

19. +5
    + (+9)
    +14

20. −8
    + (+6)
    −2

21. +8
    + (−2)
    +6

22. −9
    + (+5)
    −4

23. +11
    + (−13)
    −2

24. +3
    + (+16)
    +19

D. Solve these reading problems. Use signed numbers in the answers for numbers 25–28.

25. When the Jordan River leaves the Sea of Galilee, its surface is 700 feet below sea level. The water drops 610 feet to reach the Dead Sea. What is its altitude at the Dead Sea?
    −1,310 feet

26. One day the temperature at Browning, Montana, was 44°F at noon. Over the next 24 hours the temperature dropped 100°F. What was the temperature the next day?
    −56°F

27. On December 24, 1924, Fairfield, Montana, recorded 63°F at noon. By midnight the temperature had dropped 84°F. What was the temperature at midnight?
    −21°F

28. On January 19, 1892, chinook winds swept across a Montana town and raised the temperature 42°F in 15 minutes. The beginning temperature was −5°F. What was the ending temperature?
    +37°F

Lesson 145   T–488

Solutions for Part D

25. −700 + (−610)
26. +44 + (−100)
27. +63 + (−84)
28. −5 + (+42)

29. Father is 3 years more than 3 times Wilmer's age. The sum of their ages is 59. How old is Wilmer? (Write an equation and solve.) $w + (3w + 3) = 59$ $w = 14$ years

30. The Wilsons planted 142 acres of corn and wheat. The corn acreage was 22 more than 4 times the acres of wheat. How many acres of wheat did they plant? (Write an equation and solve.) $w + (4w + 22) = 142$ $w = 24$ acres

31. A certain triangle has 15 more degrees in the second angle than in the first one, and 30 more degrees in the third angle than in the first one. Write an equation and solve to find the size of the first angle. (Hint: The sum of the angles in a triangle always equals 180 degrees.) $a + (a + 15) + (a + 30) = 180$ $a = 45$ degrees

REVIEW EXERCISES

E. Compare each set of numbers, and write < or > between them. *(Lesson 144)*

32. $|-9|$ __>__ $|+8|$

33. 0 __>__ -5

F. Find these square roots. *(Lesson 141)*

34. $\sqrt{a^{16}}$ a^8

35. $\sqrt{d^{10}}$ d^5

G. Solve these problems mentally. *(Lesson 129)*

36. $251 + 284$ 535

37. $2{,}000 \div 500$ 4

H. Use the Pythagorean rule to find the missing dimensions, to the nearest tenth. *(Lesson 113)*

38.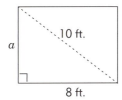

39.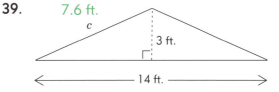

7.6 ft.

I. Find these equivalents. *(Lesson 17)*

40. 12 ft. = __144__ in.

41. 6 mi. = ____ yd.
10,560

146. Subtracting Signed Numbers

One day the Witmer family traveled to their grandparents' house. When they left home, the odometer read 62,040 miles. When they reached their destination, the odometer showed 62,185 miles. How far did the Witmers travel? Subtracting 62,185 – 62,040 gives the answer of 145 miles.

Subtraction finds the difference between two numbers or finds out how far apart two numbers are. For natural numbers, the minuend is always larger than the subtrahend. For example, 7 – 5 = 2.

Subtracting signed numbers is different because the subtrahend may be either larger or smaller than the minuend. See Examples A and B. The answers show the direction of change as well as the difference between the two numbers.

To subtract signed numbers, follow these steps.
1. Change the sign of the subtrahend.
2. Add according to the rules for adding signed numbers.

Example A

How much does the temperature change when it falls from +4° to –9°?

$$\begin{array}{cc} -9° & -9° \\ -\,(+4°) \quad \text{Change sign} & +\,(-4°) \\ \text{and add.} & -13° \end{array}$$

The answer indicates a negative change (a decrease) of 13°.

Example B

How much does the temperature change when it rises from –5° to +4°?

$$\begin{array}{cc} +4° & +4° \\ -\,(-5°) \quad \text{Change sign} & +\,(+5°) \\ \text{and add.} & +9° \end{array}$$

The answer indicates a positive change (an increase) of 9°.

Reading problems involving subtraction of signed numbers can be a special challenge because the minuend is not always the larger number. (See Example A.) To determine which number is which, follow the rule *destination minus origin equals difference*.

destination – origin = difference

This rule was followed in solving the problem at the beginning of the lesson. The odometer reading at the Witmers' destination was 62,185. The odometer reading at the origin (beginning) of their trip was 62,040 miles. Subtracting destination (62,185) minus origin (62,040) gives the difference of 145 miles.

LESSON 146

Objectives

- To review subtracting signed numbers with like and unlike signs.
- To teach that in subtracting signed numbers, *destination minus origin equals difference. (The destination is the minuend and the origin the subtrahend.)

Review

1. *Add these signed numbers.* (Lesson 145)

    a. +5
        + (−3)
        (+2)

    b. −3
        + (+4)
        (+1)

    c. +15
        + (−14)
        (+1)

    d. −12
        + (−31)
        (−43)

2. *Compare each set of numbers, and write < or > between them.* (Lesson 144)

    a. |−9| (>) |+6|

    b. −5 (<) 3

3. *Solve these parallel reading problems.* (Lesson 114)

    a. At 60 miles per hour, how many hours does it take to go 120 miles? (2 hr.)

    b. The golden eagle has been known to dive at speeds of 3 miles per minute. At that rate, how long would it take to go $\frac{1}{2}$ mile? ($\frac{1}{6}$ minute or 10 seconds)

4. *Find these equivalents.* (Lesson 18)

    a. 9 tbsp. = ___ tsp. (27)

    b. 16 pk. = ___ bu. (4)

T-491   Chapter 11   *Signed Numbers, Tables, and Graphs*

Introduction

Write the subtraction problem 8 – 3 = 5 on the board. How does the answer (5) compare 8 and 3? The number 5 tells how far apart 8 and 3 are.

Subtraction, then, is the calculation of how far apart two numbers are. That is why the answer is called the difference.

Now write the problem +6 – (–3) on the board. How far apart are +6 and –3? Referring to a number line reveals that they are 9 apart.

Teaching Guide

1. **To subtract signed numbers, change the sign of the subtrahend and add according to the rules for adding signed numbers.**

   a.   +6   Change to: +6
       – (–5)   + (+5)
       ─────   ──────
                            +11

   b.   –8   Change to: –8
       – (+5)   + (–5)
       ─────   ──────
                            –13

   c.   –8
       – (–9)
       ─────
       (+1)

   d.   –6
       – (+8)
       ─────
       (–14)

   e.   +7
       – (–12)
       ─────
       (+19)

2. **In subtraction of signed numbers, destination minus origin equals difference.** The destination is the minuend, and the origin is the subtrahend. This rule is especially helpful in reading problems to determine which signed number to subtract from which. Use the example below and the one in *Class Practice* to illustrate this point.

   One morning the temperature was –8° at sunrise and +19° by noon. What was the temperature change during that period?

   The destination (ending temperature) is +19°, so the minuend is +19. The origin (beginning temperature) is –8°, so the subtrahend is –8.

   +19°
   – (–8°)
   ─────
   +27°

> **Example C**
> One winter morning the temperature was −14°. By noon it had risen to −4°. What was the change in temperature?
> The destination (ending temperature) is −4°, so the minuend is −4°.
> The origin (beginning temperature) is −14°, so the subtrahend is −14.
>
> ```
>      −4°                        −4°
>   − (−14°)   Change sign    + (+14°)
>              and add.          +10°
> ```
> The answer indicates a positive change (an increase) of 10°.

CLASS PRACTICE

Subtract these signed numbers.

a. +6   b. −6   c. −8   d. +15
  − (−8)   − (+6)   − (+6)   − (−17)
    +14   −12   −14   +32

e. +15   f. −5   g. −16   h. +31
  − (+16)   − (−9)   − (−14)   − (+16)
    −1   +4   −2   +15

Identify the destination and the origin; then find the solution.

i. The temperature rose from −25° at sunrise one morning to −9° an hour later. What was the change in temperature?

destination (minuend)   −9°
origin (subtrahend)   − (−25°)
difference   +16°

WRITTEN EXERCISES

A. *Subtract these signed numbers.*

1.   +6   2.   +8   3.   −4   4.   −8
   − (+9)   − (−1)   − (−2)   − (+1)
     −3   +9   −2   −9

5.   +8   6.   −2   7. −17   8. +14
   − (−9)   − (+6)   − (+5)   − (−14)
    +17   −8   −22   +28

9. +13   10. −6   11. −12   12. +12
   − (+12)   − (−6)   − (−15)   − (+17)
     +1   0   +3   −5

13. +5   14. −3   15. +4   16. −7
   − (−7)   − (−7)   − (+8)   − (+5)
    +12   +4   −4   −12

17. +15   18. −18   19. +12   20. +4
   − (−13)   − (+6)   − (+14)   − (+14)
    +28   −24   −2   −10

492 Chapter 11 Signed Numbers, Tables, and Graphs

B. Solve these reading problems. For numbers 21–24, calculate with the signed numbers that belong in the blanks.

21. On the morning of January 22, 1943, chinook winds swept across Spearfish, South Dakota. The temperature rose from −4°F to 45°F in 2 minutes. What was the temperature change?

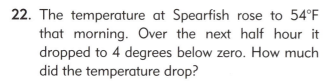

destination (minuend)	+45°
origin (subtrahend)	− (−4°)
difference	+49°

22. The temperature at Spearfish rose to 54°F that morning. Over the next half hour it dropped to 4 degrees below zero. How much did the temperature drop?

destination (minuend)	−4°
origin (subtrahend)	− (+54°)
difference	−58°

23. On the morning of January 10, 1911, the temperature at Rapid City, South Dakota, dropped from 55°F to 8°F in 15 minutes. What was the temperature change during that time?

destination (minuend)	+8°
origin (subtrahend)	− (+55°)
difference	−47°

24. On January 12, 1911, Rapid City recorded 49°F at 6:00 A.M. and −13°F at 8:00 A.M. What was the temperature change?

destination (minuend)	−13°
origin (subtrahend)	− (+49°)
difference	−62°

25. Last summer, Mother bought peaches at $8 per half bushel. Find the price per quart, to the nearest cent. $0.50

26. Wendell took $7\frac{1}{4}$ tons of hay to an auction and sold it for $190 per ton. How much was that per pound? (Answer to the nearest tenth of a cent.) $0.095 (9.5¢)

REVIEW EXERCISES

C. Add these signed numbers. *(Lesson 145)*

27. +6
    + (−4)
    ───
    +2

28. +16
    + (−18)
    ────
    −2

D. Compare each set of numbers, and write < or > between them. *(Lesson 144)*

29. |−8| > |+6|

30. −4 < 5

E. Find these equivalents. *(Lesson 18)*

31. 5 pk. = 40 qt.

32. 18 cups = 9 pt.

Solutions for Exercises 25 and 26
25. $8 ÷ 16
26. $190 ÷ 2,000

T-493 Chapter 11 Signed Numbers, Tables, and Graphs

LESSON 147

Objectives

- To review multiplying and dividing signed numbers.
- To teach that *when more than one factor is multiplied, the product is positive if there is an even number of negative factors, and the product is negative if there is an odd number of negative factors.

Review

1. Give Lesson 147 Quiz (Adding and Subtracting Signed Numbers).

2. *Subtract these signed numbers.* (Lesson 146)

    a. +5
        −(−6)
        (+11)

    b. −8
        −(+8)
        (−16)

    c. +18
        −(−21)
        (+39)

    d. −52
        −(−61)
        (+9)

3. *Add these signed numbers.* (Lesson 145)

    a. +2
        +(−8)
        (−6)

    b. −6
        +(+9)
        (+3)

    c. +35
        +(−46)
        (−11)

    d. −14
        +(−11)
        (−25)

4. *Add these monomials.* (Lesson 139)

    a. $3a^2 + 2a^2 − 3$ $(5a^2 − 3)$

    b. $3m + 3m − n − 6$ $(6m − n − 6)$

5. *Find these equivalents.* (Lesson 19)

    a. 576 sq. in. = ___ sq. ft. (4)

    b. $5\frac{1}{2}$ sq. mi. = ___ a. (3,520)

147. Multiplying and Dividing Signed Numbers

Multiplication is the combining of a given number a multiple of times. Thus $4 \cdot 8$ is equal to $8 + 8 + 8 + 8 = 32$. Likewise, $4 \cdot -8$ is equal to $-8 + (-8) + (-8) + (-8) = -32$. A negative value taken a positive number of times yields a negative product.

Can a negative value be taken a negative number of times? Suppose Sanford removes three scoops of rabbit pellets from the bag each day. The daily effect on the contents of the bag is −3 scoops. In five days the effect on the contents is 5 times −3. Positive 5 times negative 3 equals −15.

Now suppose Sanford empties the bag of rabbit pellets today, and he wants to express the contents of the bag four days back in time (−4). This could be stated as −4 times −3. Simple logic tells you that the bag had a positive amount of 12 scoops four days ago (−4) if the daily effect is −3 scoops. A negative value taken a negative number of times yields a positive product.

Use these simple rules for multiplying signed numbers.

> If two numbers with like signs are multiplied, the product is positive.
> If two numbers with unlike signs are multiplied, the product is negative.

Example A

The average high temperature for the day increased by 3° (+3) per day for 5 days (+5). What was the overall change in the average high temperature?

$$\begin{array}{r} +3° \\ \times \ (+5) \\ \hline +15° \end{array}$$

Example B

The temperature change during a hailstorm was a drop of 3° per minute (−3°). At that rate, how many degrees different was the temperature 5 minutes ago (−5)?

$$\begin{array}{r} -3° \\ \times \ (-5) \\ \hline +15° \end{array}$$

Example C

One night the temperature dropped 3° (−3) per hour for 5 hours (+5). What was the change in temperature?

$$\begin{array}{r} -3° \\ \times \ (+5) \\ \hline -15° \end{array}$$

Example D

The temperature rose 3° (+3) per hour until noon. How many degrees different was the temperature 5 hours earlier (−5)?

$$\begin{array}{r} +3° \\ \times \ (-5) \\ \hline -15° \end{array}$$

If there are more than two factors in a problem, they can be multiplied in any order. Remember to consider the signs with each step. Also remember that multiplication is indicated if numbers in parentheses are directly side by side.

494   Chapter 11   Signed Numbers, Tables, and Graphs

Example E	Example F	Example G
$+3(-4)(-2)(+2)$	$-2(-3)(-5)(+2)$	$-2(-3)(-4)(0)(-2)$
$= -12(-4)$	$= +6(-10)$	$= +6(0)(-2)$
$= +48$	$= -60$	$= +6(0)$
		$= 0$

Notice the pattern in the examples above. Example E has an even number of negative factors, so the final product is positive. Example F has an odd number of negative factors, so the final product is negative. In Example G, one factor is 0, so the final product is 0 because any number times 0 is 0. Use these tips to check your answers.

Because division is the inverse operation of multiplication, the same rules apply to division of signed numbers as to multiplication.

> If two numbers with like signs are divided, the quotient is positive.
> If two numbers with unlike signs are divided, the quotient is negative.

The following examples show division of signed numbers in equation form and in fraction form. Notice the application of the rules above, in the signs of the quotients.

Example H	Example I	Example J	Example K
$+10 \div (+5) = +2$	$-10 \div (-5) = +2$	$\frac{-10}{+5} = -2$	$\frac{+10}{-5} = -2$

CLASS PRACTICE

Solve these multiplication problems.

a. $+5$
     $\times (+6)$
     $+30$

b. $+6$
     $\times (-5)$
     -30

c. -4
     $\times (-8)$
     $+32$

d. -7
     $\times (+6)$
     -42

e. $+2(-3)(-2)(+3)$ $+36$

f. $-4(-1)(+4)(-2)(-3)$ $+96$

Solve these division problems.

g. $+16 \div (-4)$ -4 h. $-9 \div (-3)$ $+3$ i. $+8 \div (-4)$ -2 j. $-8 \div (-1)$ $+8$

k. $\frac{-18}{-9}$ $+2$ l. $\frac{-36}{+18}$ -2 m. $\frac{+42}{-6}$ -7 n. $\frac{-48}{+16}$ -3

Solve these reading problems, using equations with signed numbers.

o. If the milk price drops $0.20 per hundredweight each month for the next 7 months, what will the price change be in that time? $-\$0.20(+7) = -\$1.40$

p. If the average daily temperature is dropping 3° this week, how many degrees different was the average temperature 3 days ago? $-3°(-3) = +9°$

Introduction

Ask the students to write a negative number that indicates a loss of $40. (–$40)

Now ask, "What is the loss if it is 5 times –$40?" (Five times a loss of $40 is a loss of $200. Therefore, 5 × –$40 = –$200.)

Teaching Guide

1. **If two numbers with like signs are multiplied, the product is positive.**

   a. $+7$
   $\underline{\times (+3)}$
   $(+21)$

   b. -4
   $\underline{\times (-3)}$
   $(+12)$

   c. -2
   $\underline{\times (-2)}$
   $(+4)$

   d. -5
   $\underline{\times (-3)}$
   (-15)

2. **If two numbers with unlike signs are multiplied, the product is negative.**

   a. -7
   $\underline{\times (+2)}$
   (-14)

   b. $+5$
   $\underline{\times (-3)}$
   (-15)

   c. -2
   $\underline{\times (+4)}$
   (-8)

   d. $+6$
   $\underline{\times (-2)}$
   (-12)

3. **If there are more than two factors in a problem, they can be multiplied in any order.** Use the following points to check your answer. If there is an even number of negative factors, the product is positive. If there is an odd number of negative factors, the product is negative.

   a. +2(–3)(–4)   (+24)

   b. +2(–2)(–3)(–4)   (–48)

   c. –2(–2)(–3)(–3)   (+36)

   d. –2(–2)(+3)(+3)(–4)   (–144)

4. **If two numbers with like signs are divided, the result is positive. If two numbers with unlike signs are divided, the result is negative.**

   a. +12 ÷ (–4) = ___   (–3)

   b. –8 ÷ (–2) = ___   (+4)

   c. $\dfrac{+24}{+12} = $ ___   (2)

   d. $\dfrac{-18}{+3} = $ ___   (–6)

T–495 Chapter 11 Signed Numbers, Tables, and Graphs

Solutions for Part C

25. $+5(+5°)$

26. $-9(-2°)$

27. $+6(-3°)$

28. $-4(+5°)$

29. $8 \times \frac{3}{4} \times 1$

30. 3.14×200^2

WRITTEN EXERCISES

A. Solve these multiplication problems.

1. +5
    × (+8)
    +40

2. +7
    × (−2)
    −14

3. −3
    × (−4)
    +12

4. −5
    × (+2)
    −10

5. +6
    × 0
    0

6. −4
    × (−8)
    +32

7. +5
    × (+9)
    +45

8. −8
    × (+4)
    −32

9. +3(−4)(−5)   +60

10. −2(−3)(−4)   −24

11. +2(−3)(−4)(0)(+5)   0

12. −3(−2)(+3)(−4)(−1)   +72

B. Solve these division problems.

13. +12 ÷ (−4)   −3
14. −6 ÷ (−2)   +3
15. +8 ÷ (−2)   −4
16. −7 ÷ (−1)   +7

17. −16 ÷ (−4)   +4
18. +9 ÷ (+3)   +3
19. −6 ÷ (−3)   +2
20. −20 ÷ (−5)   +4

21. $\frac{-12}{+6}$ −2
22. $\frac{-15}{-3}$ +5
23. $\frac{+26}{+13}$ +2
24. $\frac{+32}{-8}$ −4

C. Solve these reading problems. Use signed numbers for problems 25–28.

25. One morning the temperature increased 5° (+5) per hour for a total of 5 (+5) hours. What was the change of temperature during that period?   +25°

26. One morning David saw that the temperature had dropped 2° (−2) per hour during the previous night. At that rate, how many degrees different was the temperature 9 hours earlier (−9)?   +18°

27. If the temperature drops 3 degrees per hour for the next 6 hours, what will the temperature change be in that time?   −18°

28. If the temperature is rising 5 degrees per hour, how many degrees different was the temperature 4 hours ago?   −20°

29. Carol used eight 9-inch by 12-inch pieces of construction paper to form a large rectangle on the bulletin board. How many square feet does the rectangle cover?   6 square feet

30. An air traffic controller can maintain radio contact with all the planes in a 200-mile radius. How many square miles are in this circle?   125,600 square miles

Chapter 11 Signed Numbers, Tables, and Graphs

REVIEW EXERCISES

D. Subtract these signed numbers. (Lesson 146)

31. +6
    − (−7)
    ―――
    +13

32. +25
    − (−19)
    ―――
    +44

E. Add these signed numbers. (Lesson 145)

33. +6
    + (−9)
    ―――
    −3

34. +51
    + (−58)
    ―――
    −7

F. Add these monomials. (Lesson 139)

35. $6^2tu^2 + 3tu^2$ $39tu^2$

36. $3r^2s^5 + 3r^2s^5$ $6r^2s^5$

G. Find these equivalents. (Lesson 19)

37. 5 sq. ft. = __720__ sq. in.

38. 162 sq. ft. = __18__ sq. yd.

CHALLENGE EXERCISES

H. Write equations to solve these reading problems.

39. The sum of four consecutive numbers is 66. What is the smallest number?
    $n + (n + 1) + (n + 2) + (n + 3) = 66$ $n = 15$

40. The sum of seven consecutive numbers is 161. What is the largest number?
    $n + (n − 1) + (n − 2) + (n − 3) + (n − 4) + (n − 5) + (n − 6) = 161$ $n = 26$

41. The sum of three consecutive odd numbers is 261. What is the smallest number?
    $n + (n + 2) + (n + 4) = 261$ $n = 85$

42. The sum of three consecutive even numbers is 138. What is the middle number?
    $n + (n + 2) + (n − 2) = 138$ $n = 46$

43. The sum of three consecutive signed numbers is −12. What is the lowest number?
    $n + (n + 1) + (n + 2) = −12$ $n = −5$

44. The sum of three consecutive signed numbers is −27. What is the highest number?
    $n + (n − 1) + (n − 2) = −27$ $n = −8$

Lesson 147   T–496

Solutions for Part H

Note that these exercises ask for the smallest, middle, or largest number of the series. All may be solved by using this type of equation: $n + (n + 1) \ldots$ (or $n + [n + 2] \ldots$ for consecutive odd or even numbers), which will yield the smallest number. To then determine the middle or largest number, identify the portion of the equation that represents that number, and solve accordingly.

The solutions given here are set up so that n represents the answer to be found.

39. $n + (n + 1) + (n + 2) + (n + 3) = 66$
$4n + 6 = 66$
$4n = 60$
$n = 15$

40. $n + (n - 1) + (n - 2) + (n - 3) + (n - 4) + (n - 5) + (n - 6) = 161$
$7n - 21 = 161$
$7n = 182$
$7n$
$n = 26$

41. $n + (n + 2) + (n + 4) = 261$
$3n + 6 = 261$
$3n = 255$
$n = 85$

42. $n + (n + 2) + (n - 2) = 138$
$3n = 138$
$n = 46$

43. $n + (n + 1) + (n + 2) = -12$
$3n + 3 = -12$
$3n = -15$
$n = -5$

44. $n + (n - 1) + (n - 2) = -27$
$3n - 3 = -27$
$3n = -24$
$n = -8$

LESSON 148

Objective

- To review finding the value of algebraic expressions by substituting the literal numbers with given values, *including negative values. (This lesson is an extension of Lesson 134.)

Review

1. *Multiply or divide these signed numbers.* (Lesson 147)

   a. $\begin{array}{r} +9 \\ \times(-1) \\ \hline (-9) \end{array}$ b. $\begin{array}{r} +9 \\ \times(+8) \\ \hline (+72) \end{array}$

   c. $\dfrac{-36}{+9}$ (−4)

   d. $\dfrac{+48}{+12}$ (+4)

2. *Subtract these signed numbers.* (Lesson 146)

   a. $\begin{array}{r} +8 \\ -(-5) \\ \hline (+13) \end{array}$ b. $\begin{array}{r} +24 \\ -(-38) \\ \hline (+62) \end{array}$

3. *Compare each set of numbers, and write < or > between them.* (Lesson 144)

   a. |−8| (>) |+5|

   b. +5 (<) +6

4. *Multiply these monomials.* (Lesson 140)

   a. $4^2 d^3 \cdot 3 d^4$ ($48 d^7$)

   b. $3^3 w^4 x^2 \cdot 2^2 w^3 x^3$ ($108 w^7 x^5$)

5. *Write these phrases as algebraic expressions.* (Lesson 132)

   a. The sum of s and t ($s + t$)

   b. The product of 4, b, and c ($4bc$)

6. *Find these equivalents.* (Lesson 20)

   a. 300 min. = ___ hr. (5)

   b. 6 millennia = ___ decades (600)

148. Evaluating Expressions With Signed Numbers

In Lesson 134 you used the following steps to evaluate algebraic expressions. You will evaluate algebraic expressions in this lesson by using the same steps, but some of the numbers are negative.

1. Rewrite the algebraic expression, replacing each literal number with its given value.
2. Simplify the expression. Follow the mathematical rules for the order of operations as given in Lesson 133.

To calculate with signed numbers, use the rules given in the last three lessons. Those rules are listed below.

Addition

To add signed numbers with like signs, add their absolute values. Give the sum the same sign as the addends.

To add signed numbers with unlike signs, find the difference between their absolute values. Give the sum the same sign as the addend with the larger absolute value.

Subtraction

To subtract signed numbers, change the sign of the subtrahend and follow the rules for adding signed numbers.

Multiplication and division

If numbers with like signs are multiplied or divided, the answer is positive.
If numbers with unlike signs are multiplied or divided, the answer is negative.

Example A	**Example B**
Evaluate $3x + 2y$	Evaluate $\dfrac{2x^2}{4y}$
$x = 4$ and $y = -3$	$x = -6$ and $y = -2$
$3x + 2y$	$\dfrac{2x^2}{4y}$
$= 3(4) + 2(-3)$	$= \dfrac{2(-6)(-6)}{4(-2)}$
$= 12 + (-6)$	$= \dfrac{2(+36)}{-8}$
$= 6$	$= \dfrac{72}{-8}$
	$= -9$

498 Chapter 11 Signed Numbers, Tables, and Graphs

CLASS PRACTICE

Evaluate each expression, using the values indicated.

a. $b - d$ +3
$b = -3;\ d = -6$

b. $w + y$ −16
$w = -7;\ y = -9$

c. $\dfrac{2gh}{4k}$ +5
$g = -4;\ h = -5;\ k = +2$

d. $\dfrac{2rs}{t}$ −9
$r = -9;\ s = -6;\ t = -12$

WRITTEN EXERCISES

A. *Evaluate each expression, using the values indicated.*

Exercises 1–4: $m = +3;\ n = -4$

1. $m - n$ +7
2. $m + n$ −1
3. $2m - 3n$ +18
4. $\dfrac{3n}{2m}$ −2

Exercises 5–8: $g = +2;\ h = -3$.

5. $3g + 2h$ 0
6. $7 - 3g + 2h$ −5
7. $-2 - 3g$ −8
8. $7 - 4h$ +19

Exercises 9–10: $p = -3;\ q = -6$

9. $\dfrac{pq}{3p}$ −2
10. $\dfrac{p + 2q}{p - q}$ −5

Exercises 11–12: $x = -2;\ y = +4;\ z = -5$

11. x^2y^3z −1,280
12. $\dfrac{yz^2}{x + y}$ +50

B. *Solve these reading problems.*

13. A large puffball mushroom may be packed with as many as 7,000,000 spores. Find the volume of a spherical puffball that is 6 centimeters in diameter, to the nearest tenth. (Use 3.14 and 1.33.)   112.8 cubic centimeters

14. Tuna fish rarely stop swimming; they usually maintain a steady motion of about 9 miles per hour. At that rate, how far will a tuna swim in 1 year (365 days)?   78,840 miles

15. The inside dimensions of an oven are 3 feet by $2\tfrac{1}{2}$ feet by 2 feet. How many square feet of stainless steel will it take to line the oven? (Subtract $2\tfrac{1}{2}$ square feet for the door opening.) $34\tfrac{1}{2}$ square feet

16. What was the perimeter of Noah's ark if it had the shape of a rectangle 450 feet long and 75 feet wide?   1,050 feet

17. Although most birds seldom fly higher than 3,000 feet, some fly very high. Once a condor collided with an airplane at 20,000 feet. Find the difference from the lower elevation to the higher elevation, using signed numbers.   +17,000 feet

18. Plants grow on the ocean floor to a depth of around 400 feet below the surface. Starfish have been found as deep as 32,800 feet. Find the difference from the higher elevation to the lower elevation, using signed numbers.   −32,400 feet

Lesson 148 T–498

Introduction

Evaluate the following algebraic expression by substitution.

$3s - 2t$ if $s = 7$ and $t = 5$
$3(7) - 2(5)$
$= 21 - 10$
$= 11$

Teaching Guide

1. **To evaluate algebraic expressions, remember the following steps from Lesson 134.** In today's lesson, some of the given values (step 1) are negative numbers.

   (1) Rewrite the algebraic expression, replacing each literal number with its given value.

   (2) Simplify the expression. Follow the mathematical rules for the order of operations as given in Lesson 133.

2. **To calculate with signed numbers, use the rules in Lessons 145–147.** Review the rules as shown in the lesson text; then give some practice with evaluating algebraic expressions.

   a. $2b + 2c$ $b = -2; c = -3$ (-10)
   b. $3d + 2e$ $d = -3; e = -5$ (-19)
   c. $\dfrac{2bc}{d}$ $b = -2; c = 3; d = -4$ $(+3)$
   d. $\dfrac{4fg}{gh}$ $f = -2; g = -3; h = -4$ $(+2)$

Solutions for Part B

13. $1.33 \times 3.14 \times 3^3$
14. $365 \times 24 \times 9$
15. $2(3 \times 2\frac{1}{2}) + 2(3 \times 2) + 2(2\frac{1}{2} \times 2) - 2\frac{1}{2}$
16. $(2 \times 450) + (2 \times 75)$
17. $+20,000 - (+3,000)$
18. $-32,800 - (-400)$

Lesson 148

REVIEW EXERCISES

C. Multiply or divide these signed numbers. *(Lesson 147)*

19. +8
     × (−6)
     −48

20. +8
     × (+7)
     +56

21. $\frac{-54}{+9}$ −6

22. $\frac{+108}{+12}$ +9

D. Subtract these signed numbers. *(Lesson 146)*

23. +9
     − (−6)
     +15

24. +34
     − (−28)
     +62

E. Compare each set of numbers, and write < or > between them. *(Lesson 144)*

25. |−7| > |+6|

26. +5 < +7

F. Multiply these monomials. *(Lesson 140)*

27. $3^2 t^2 \cdot 4t^3 u$ $36t^5 u$

28. $2^3 m^2 \cdot 3m^4 n$ $24m^6 n$

G. Write the number of terms in each polynomial. *(Lesson 132)*

29. $4g + 5gh$ 2

30. $7 - 3m + 2mn - 4p$ 4

H. Find these equivalents. *(Lesson 20)*

31. 300 yr. = __3__ centuries

32. 65 days = __$9\frac{2}{7}$__ wk.

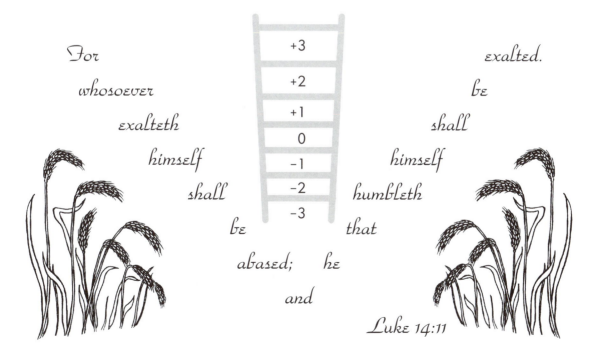

For whosoever exalteth himself shall be abased; and he that humbleth himself shall be exalted.

Luke 14:11

149. Constructing Tables From Formulas

The tables in this lesson show the answers obtained when the literal numbers in various formulas are replaced with certain numerals. For example, the following table shows how travel time decreases as the rate of speed increases. To use the table, find the desired rate in the row labeled r, and look at the corresponding number in the row labeled t. That number tells how many hours are required to travel 550 miles at the given rate.

Formula: $t = \frac{550}{r}$	**Time required to travel 550 miles**					
r (m.p.h.)	10	20	30	40	50	60
t (hr.)	55	27.5	18.33	13.75	11	9.17

To prepare a table like the one above, evaluate the given formula by replacing one literal number with a series of values as shown in the first row. Write each solution in the correct space in the second row.

The following table is based on the formula for the area of a circle. The radius and the area are related exponentially because the radius must be squared to find the area.

Formula: $a = \pi r^2$	**Areas of circles** (to nearest whole number)						
r (ft.)	1	2	3	4	5	6	7
a (sq. ft.)	3	13	28	50	79	113	154

In the next table, the y value increases at a uniform rate according to the x value.

Formula: $y = 2x + 6$							
x	0	1	2	3	4	5	6	7
y	6	8	10	12	14	16	18	20

CLASS PRACTICE

Copy and complete this table.

Formula: $a = 3b + 2$							
b	1	2	3	4	5	6	7	8
a	a. 5	b. 8	c. 11	d. 14	e. 17	f. 20	g. 23	h. 26

LESSON 149

Objective

- To review making tables to show information based on formulas.

Review

1. *Evaluate each expression.* (Lesson 148)

   a. $2g + 4h$ if $g = 3$ and $h = -4$ (-10)

   b. $9 - 2m + 3n$ if $m = -5$ and $n = -9$ (-8)

2. *Multiply or divide these signed numbers.* (Lesson 147)

   a. $\quad +7$ $\quad\quad$ b. $\quad +8$
   $\quad\quad \times (-6)$ $\quad\quad\quad\quad \times (+4)$
   $\quad\quad\;\; (-42)$ $\quad\quad\quad\quad\;\; (+32)$

   c. $\dfrac{-72}{+8}$ (-9) $\quad$ d. $\dfrac{-96}{-12}$ ($+8$)

3. *Add these signed numbers.* (Lesson 145)

   a. $\quad +9$ $\quad\quad$ b. $\quad +49$
   $\quad\quad + (-6)$ $\quad\quad\quad\quad + (-51)$
   $\quad\quad\;\; (+3)$ $\quad\quad\quad\quad\;\; (-2)$

4. *Find these square roots.* (Lesson 141)

   a. $\sqrt{k^8}$ (k^4) $\quad$ b. $\sqrt{d^{22}}$ (d^{11})

5. *Simplify each expression, using the correct order of operations.* (Lesson 133)

   a. $3[3(40 - 6^2) - 6] = $ ___ (18)
   $3[3(40 - 36) - 6]$
   $3[3(4) - 6]$
   $3[12 - 6]$
   $3[6]$

   b. $3[4 - (\dfrac{38 - 6}{9 - 1})] = $ ___ (0)
   $3[4 - \dfrac{32}{8}]$
   $3[4 - 4]$
   $3[0]$

T-501   Chapter 11 *Signed Numbers, Tables, and Graphs*

Introduction

Prepare the following chart for the students, having them help with completing the second row.

Formula: $d = 55t$ Distance traveled at 55 m.p.h.

t (hr.)	1	2	3	4	5	6
d (mi.)	55	110	165	220	275	330

This chart tabulates the values of the formula $d = rt$ when the rate is 55 miles per hour.

Teaching Guide

1. **A table can show the answers obtained when the literal numbers in a formula are replaced with certain numerals.**

2. **To complete a table based on a formula, evaluate the formula by replacing the literal number with each value given in the first row. Write each answer in the correct space in the second row.** Use the tables in the lesson text and in *Class Practice* to illustrate these points.

Solutions for Part B

23. $3\frac{1}{7} \times 1\frac{3}{4} \times 1\frac{3}{4}$

24. $2 \times 15 \times 2{,}900 \div 43{,}560$

25. $-\$1.92 \div (+4)$

26. $+5(-73)$

27. $\$84.00 + \$211.54 + \$96.31 + \$103.09 + \$75.50$

28. $\$23.27 + \$39.74 + \$753.26 + \$91.62 - \$85.00$

29. $a + (a + 12) + (a + 48) = 180$
$3a + 60 = 180$
$3a = 120$
$a = 40$

Lesson 149

WRITTEN EXERCISES

A. Copy and complete each table.

Formula: $a = 3l$			Areas of rectangles 3 inches wide					
l (in.)	1	2	3	4	5	6	7	8
a (sq. in.)	1. 3	2. 6	3. 9	4. 12	5. 15	6. 18	7. 21	8. 24

Formula: $a = 2b + 5$								
b	1	2	3	4	5	6	7	8
a	9. 7	10. 9	11. 11	12. 13	13. 15	14. 17	15. 19	16. 21

Formula: $p = 2(7 + w)$			Perimeters of rectangles 7 feet long				
w (ft.)	1	2	3	4	5	6	
p (ft.)	17. 16	18. 18	19. 20	20. 22	21. 24	22. 26	

B. Solve these reading problems.

23. Father cut a $3\frac{1}{2}$-foot circle from a sheet of plywood. Find the area of this circle, using $3\frac{1}{7}$ for pi. $9\frac{5}{8}$ square feet

24. A combine cuts a swath 15 feet wide through a field 2,900 feet long. How many acres will it harvest in 2 passes through the field? (Round to the nearest acre.)   2 acres

25. Brandon's father reported that over a 4-month period (+4) the milk price dropped $1.92 per hundredweight. Use a signed number to show the average monthly change.   −$0.48

26. The sailfish, one of the ocean's fastest fish, can dive at a speed of 73 feet per second. Use a signed number to show the total change of altitude during a dive lasting 5 seconds (+5)   −365 feet

27. Brother Chester filled out a deposit ticket with currency of $84.00 and checks of $211.54, $96.31, $103.09, and $75.50. What was the total deposit? $570.44

28. Clayton completed a deposit ticket with coins of $23.27 and checks of $39.74, $753.26, and $91.62. He received $85.00 in cash. What was the total deposit? $822.89

The fishes of the sea shall declare unto thee . . . that the hand of the LORD hath wrought this.

Job 12:8, 9

29. The smallest angle in a certain triangle is 12 degrees smaller than the second angle, and 48 degrees smaller than the third one. Write an equation, and find the size of the smallest angle. (Think: What is the sum of the three angles in a triangle?) $a = 40°$
$$a + (a + 12) + (a + 48) = 180°$$

REVIEW EXERCISES

C. Evaluate each expression if $y = 2$ and $z = -8$. (Lesson 148)

30. $2y + 4z$ -28

31. $6 - 2z + 4y$ 30

D. Multiply or divide these signed numbers. (Lesson 147)

32. $+8$
     $\times (-5)$
     $\overline{-40}$

33. $+9$
     $\times (+1)$
     $\overline{+9}$

34. $\dfrac{-56}{+8}$ -7

35. $\dfrac{+88}{+11}$ $+8$

E. Add these signed numbers. (Lesson 145)

36. $+8$
     $+ (-6)$
     $\overline{+2}$

37. $+48$
     $+ (-33)$
     $\overline{+15}$

F. Find these square roots. (Lesson 141)

38. $\sqrt{s^4}$ s^2

39. $\sqrt{y^{32}}$ y^{16}

G. Simplify each expression, using the correct order of operations. (Lesson 133)

40. $3[14 - (36 - 3^3)]$ 15

41. $3(30 - \dfrac{73 - 1}{8 - 2^2})$ 36

H. Draw these figures. (Lesson 85)

42. decagon

(or)

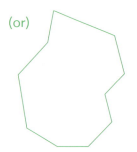

43. parallelogram

LESSON 150

Objective

- To teach making line graphs to show information based on formulas, *including formulas with exponents to the third power. (Previous experience: Exponents to the second power.)

Review

1. *Find the missing numbers for this table, to the nearest whole number.* (Lesson 149)

Formula: $a_s = 4\pi r^2$	Surface area of spheres				
r	1	2	3	4	5
a_s	(13)	(50)	(113)	(201)	(314)

2. *Evaluate these algebraic expressions by substitution.* (Lessons 148, 134)

   a. $-4 - 6e$ if $e = -4$ (20)

   b. $8 - 3v$ if $v = -5$ (23)

3. *Subtract these signed numbers.* (Lesson 146)

   a. $+7$
       $-(-6)$
       $(+13)$

   b. $+35$
       $-(-49)$
       $(+84)$

4. *Identify these angles.* (Lesson 86)

   a. 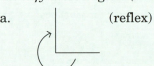 (reflex)

   b.  (obtuse)

   c.  (acute)

   d.  (right)

5. *Find these equivalents.* (Lesson 21)

   a. 12 hm = ___ m   (1,200)

   b. 125 mm = ___ dkm   (0.0125)

150. Constructing Graphs From Formulas

Formulas and other equations are sometimes graphed to show how the result changes as one of the variables in the equation changes. Such a graph also makes a handy tool for quickly finding areas, volumes, or other facts.

When a graph represents a formula, the data is taken from a table like one of those in Lesson 149. Use the following steps to construct such a graph.

1. Decide on an appropriate interval for the vertical scale. The graph should have between 10 and 20 horizontal lines, with the scale extending a little below the lowest value and a little above the highest value.

2. Mark off the vertical scale, labeling the lines according to the intervals chosen in step 1. Write a label for the vertical units.

3. Mark off and label the lines of the horizontal scale. Space the labels evenly across the bottom of the graph. Write a label for the horizontal units.

4. Plot a line on the graph by placing dots at the correct positions and drawing a line to join them.

5. Title the graph with the formula that the graph represents.

The following formula is said to be **directly proportionate** because the length of the sides and the perimeter both increase at a constant rate. For example, if the sides of a 3-foot square are doubled, the perimeter also doubles. A formula that is directly proportionate produces a straight line on the graph.

Formula: $p = 4s$			**Perimeters of squares**					
s	0	1	2	3	4	5	6	7
p	0	4	8	12	16	20	24	28

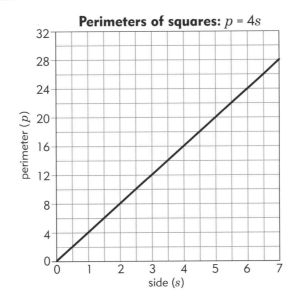

504   Chapter 11 Signed Numbers, Tables, and Graphs

The volume of a cube does not increase at a constant rate according to the length of its edges. Rather, the volume increases according to the edge cubed. This means that if the edges are multiplied by 2, the volume is multiplied by 2^3, or 8. For example, increasing an edge twofold (as from 3 to 6) will increase the volume eightfold (as from 27 to 216). So the formula for the volume of a cube is said to be **exponentially proportionate.** A formula that is exponentially proportionate produces a curved line on the graph.

Formula: $v = e^3$	**Volumes of cubes**						
e	0	1	2	3	4	5	6	7
v	0	1	8	27	64	125	216	343

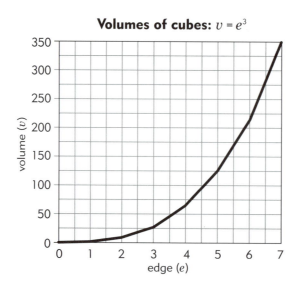

Volumes of cubes: $v = e^3$

CLASS PRACTICE

Prepare graphs from these tables.   (See facing page.)

Formula: $i = 1{,}000(0.08t)$	**Simple interest on $1,000 at 8%**						
t (yr.)	0	1	2	3	4	5	6	7
i	0	$80	$160	$240	$320	$400	$480	$560

Formula: $a = \pi r^2$	**Areas of circles** (to nearest whole number)						
r	0	1	2	3	4	5	6	7
a	0	3	13	28	50	79	113	154

Introduction

Sketch a graph to show simple interest at 5% by plotting two points, one at each end of the graph. Demonstrate that by connecting those two points with a straight line, in effect you are calculating the interest for all points in between. You could illustrate that extending the graph shows the interest for longer periods of time as well.

Also point out that a graph is only as accurate as it is consistent.

Simple interest on $1,000 at 5%:
$i = 1,000(0.05t)$

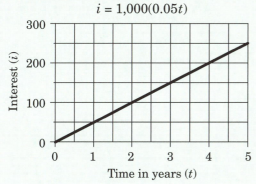

Teaching Guide

1. **Formulas and other equations are sometimes graphed to show how the result changes as one of the variables in the equation changes.** Such a graph also makes a handy tool for quickly finding areas, volumes, or other facts.

2. **When a graph represents a formula, the data is taken from a table like one of those in Lesson 149.** Discuss the steps for constructing a graph based on a table.

3. **A formula that is directly proportionate produces a straight line on a graph. A formula that is exponentially proportionate produces a curved line on a graph.** (A line representing an exponentially proportionate formula is a parabolic curve.)

Graphs for CLASS PRACTICE

Simple interest on $1,000 at 8%
$i = 1,000(0.08t)$

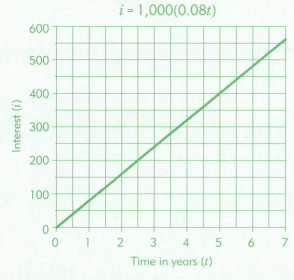

Areas of circles
$a = \pi r^2$

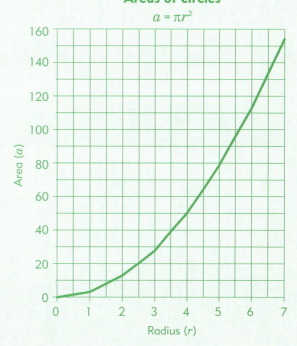

Graphs for WRITTEN EXERCISES

1–8.
Circumference of circles: $c = \pi d$

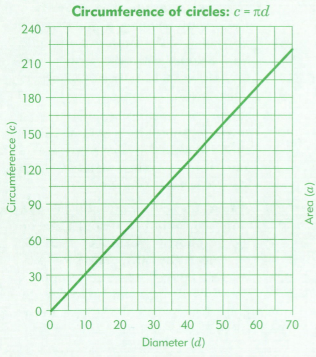

9–16.
Areas of squares: $a = s^2$

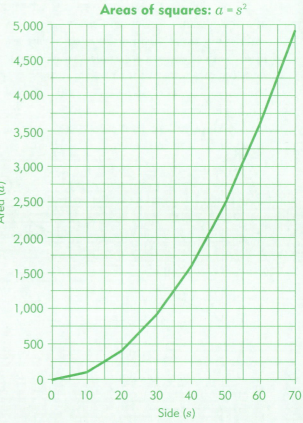

17–24.
Volumes of spheres: $v = \frac{4}{3}\pi r^3$

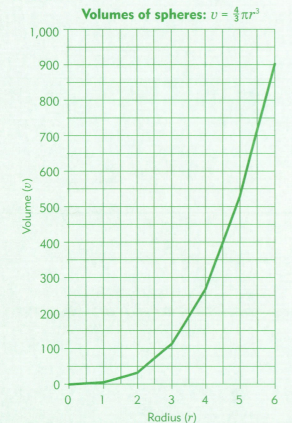

Solutions for Part B

25. $\frac{1}{2} \times 14 \times 7$
26. $4 \times 3.14 \times 4.5^2$
27. -395 destination $- (+732)$ origin
28. -59 destination $- (-126.9)$ origin
29. $3 \times 3 \times 2.4$
30. 68.5×40

Lesson 150

WRITTEN EXERCISES

A. Prepare graphs from these tables.   (See facing page.)

1–8.

Formula: $c = \pi d$	**Circumferences of circles** (to nearest whole number)						
d	0	10	20	30	40	50	60	70
c	0	31	63	94	126	157	188	220

9–16.

Formula: $a = s^2$	**Areas of squares**						
s	0	10	20	30	40	50	60	70
a	0	100	400	900	1,600	2,500	3,600	4,900

17–24.

Formula: $v = \frac{4}{3}\pi r^3$	**Volumes of spheres** (to nearest whole number)					
r	0	1	2	3	4	5	6
v	0	4	33	113	268	523	903

B. Solve these reading problems.

25. Twila used some triangular patches to make the top of a comforter. What is the area of each patch if it has a 14-inch base and a 7-inch height?   49 square inches

26. The sphere on a paper weight has a radius of 4.5 centimeters. What is the surface area of the sphere? (Use 3.14 for pi.)   254.34 square centimeters

27. Jerusalem lies 732 meters above sea level. If David's flight from Absalom took him across the Jordan River just north of the Dead Sea, he would have descended to 395 meters below sea level. Use a signed number to express this change in altitude.
   −1,127 meters

28. Over a 7-year period, the average temperature near the South Pole was 59°F below zero. Antarctica also holds the world record for the lowest temperature: −126.9°F. What was the difference between the lowest recorded temperature (origin) and the average (destination)? (Use a signed number in your answer.)   +67.9°

29. A large envelope has 3 postage stamps on it. Each stamp measures 3 centimeters by 2.4 centimeters. What is the total area covered by the stamps?   21.6 square centimeters

30. The state of Rhode Island is as large as a rectangle 68.5 kilometers long and 40 kilometers wide. What is the area of Rhode Island?   2,740 square kilometers

Chapter 11 Signed Numbers, Tables, and Graphs

REVIEW EXERCISES

C. Find the missing numbers for this table, to the nearest whole number. *(Lesson 149)*

Formula: $a = \pi r^2$	Areas of circles			
r	11	12	13	14
a	**31.** 380	**32.** 452	**33.** 531	**34.** 615

D. Evaluate these algebraic expressions by substitution. *(Lessons 148, 134)*

35. $-7 - 3g$ if $g = -3$ +2   **36.** $7 - 2x$ if $x = -4$ +15

E. Subtract these signed numbers. *(Lesson 146)*

37.   +9
     − (−5)
     +14

38.   +44
     − (−26)
     +70

F. Identify these angles. *(Lesson 86)*

39.  obtuse

40.  reflex

G. Find these equivalents. *(Lesson 22)*

41. 64 kg = ____ g   64,000   **42.** 850 kg = 0.85 MT

Lesson 150   T–506

T-507 Chapter 11 Signed Numbers, Tables, and Graphs

LESSON 151

Objectives

- To give more practice with simplifying and evaluating algebraic expressions.
- To teach that in algebra, *addition is the only operation used in polynomials; and that + and – are considered as positive and negative signs.

Review

1. Give Lesson 151 Quiz (Calculations With Signed Numbers).

2. *Find the missing numbers for this table.* (Lesson 149)

   Formula: $k = 5h^2 + 5$

h	0	2	4	6	8	10
k	(0)	(25)	(85)	(185)	(325)	(505)

3. *Multiply or divide these signed numbers.* (Lesson 147)

   a. $+6 \times (-5) \quad (-30)$

   b. $+10 \times (+11) \quad (+110)$

   c. $\dfrac{-96}{+8} \quad (-12)$

   d. $\dfrac{+52}{+13} \quad (+4)$

4. *Compare each set of numbers, and write < or > between them.* (Lesson 144)

   a. $|-9| \;(>)\; |+8|$

   b. $-7 \;(<)\; 7$

5. *Simplify these polynomials.* (Lesson 135)

   a. $5m - 2m + 2n - n \quad (3m - n)$

   b. $4g + g + 3h \quad (5g + 3h)$

6. *Use each set of facts to construct a triangle. Label it with the dimensions given.* (Lesson 88)

   a. 1 in., 45°, $1\tfrac{1}{2}$ in.

   b. 3 cm, 4 cm, 3 cm

151. Polynomials and Signed Numbers

A polynomial is an algebraic expression that has more than one term. For example, $3x + 4x - 2x$ is a polynomial. The terms are separated by + or − signs. In Chapter 10, you treated these as signs of operation indicating addition or subtraction, and you considered the numbers in each term as positive.

$3x + 4x - 2x$ Simplified: $3x + 4x = 7x$; $7x - 2x = 5x$

The + or − signs in polynomials can also be understood as signs of direction indicating positive or negative terms. The following illustration shows why this is so.

$+3x$ plus $+4x$ plus $-2x$ Simplified: $+3x$ plus $+4x = +7x$; $+7x$ plus $-2x = +5x$

With either method of simplifying, the result is positive $5x$. This points to an important principle of algebra: **Addition is the only operation used between terms of a polynomial.** That is, a polynomial simply contains two or more terms that are being combined (added). If a minus sign precedes a term, it means that a negative term is being added—which is the same as subtracting a positive.

This principle is important because it allows you to apply the commutative law of addition to polynomials that also have subtraction (negative) signs.

The rule for the order of operations would tell you that you must do steps 1 and 2 before you can subtract $5a$ in the polynomial below.

$$\begin{array}{cccc} 1 & 2 & 3 & \\ 8a & - 6b & + 6c & - 5a \end{array}$$

But if this polynomial is actually an addition problem with four addends, the commutative law of addition permits you to change the order of the addends. Thus you can easily combine like terms $8a$ and $-5a$. The expression is then simplified to $3a - 6b + 6c$.

It is very important to associate the sign before each term with that term, and recognize it as positive or negative. If the first term has no sign, it is understood to be positive. All other terms are clearly positive or negative.

In the following examples, polynomials are simplified by combining like terms and treating + and − as directional signs. The column form is a helpful way to set like terms together.

Example A	Example B
Simplify $3b + 2c - b + 3c$.	Simplify $4a - 2b^2 - 9a + 4b^2$.
$\ +3b\ \ +2c$	$\ +4a\ \ -2b^2$
$\underline{(+)\ -b\ \ +3c}$	$\underline{(+)\ -9a\ \ +4b^2}$
$\ +2b\ \ +5c\ = 2b + 5c$	$\ -5a\ \ +2b^2\ = -5a + 2b^2$

Chapter 11 Signed Numbers, Tables, and Graphs

Note again that no directional sign is used for the first term in a polynomial if that term is positive. If the first term in the answer is negative, the negative sign is used and no space separates it from the term (as in $-5a$). The other $+$ and $-$ signs have spaces before and after them.

In the answers for Examples A and B, note that the literal coefficients are written in alphabetical order.

CLASS PRACTICE

Simplify these polynomials.

a. $6w + 6w - 9w$ $\qquad$ $3w$

b. $s^2 + 6s^2 - 3s^2$ $\qquad$ $4s^2$

c. $3m - 6m - 3n + 4m$ $\qquad$ $m - 3n$

d. $2b^2 + b^2 - 5b^2 - b^2$ $\qquad$ $-3b^2$

e.
$\;+h^3 \quad -6$
$(+)\;-2h^3 \quad +4$
$-h^3 \quad -2 \quad (-h^3 - 2)$

f.
$\;+4a \quad -b^3$
$(+)\;-2a \quad +b^3$
$\;2a$

g.
$\;+5a \quad +3b$
$(+)\;+4a \quad -7b \quad +2$
$\;9a \quad -4b \quad +2 \quad (9a - 4b + 2)$

h.
$\;+v \quad +2w$
$(+)\;-3v \quad +4w \quad -8$
$-2v \quad +6w \quad -8 \quad (-2v + 6w - 8)$

WRITTEN EXERCISES

A. Simplify these polynomials.

1. $3a + 3a + 2a$ $\qquad$ $8a$

2. $c + 2c + 3c$ $\qquad$ $6c$

3. $12a + 2b - 4a$ $\qquad$ $8a + 2b$

4. $15n + 2n - 3p$ $\qquad$ $17n - 3p$

5. $3e + 2e - 4e$ $\qquad$ e

6. $2b + 2b - 6b$ $\qquad$ $-2b$

7. $d + 2d - 4d$ $\qquad$ $-d$

8. $-a^2 - a^2 + 4a^2$ $\qquad$ $+2a^2$

9.
$\;+2e \quad +g$
$(+)\;+5e \quad -3g$
$\;7e \quad -2g \quad (7e - 2g)$

10.
$\;-5c \quad +4d$
$(+)\;+c \quad -d$
$-4c \quad +3d \quad (-4c + 3d)$

11.
$\;+2k \quad +3m \quad -4$
$(+)\;+2k \quad -5m$
$\;4k \quad -2m \quad -4 \quad (4k - 2m - 4)$

12.
$\;-4y^2 \quad +5y$
$(+)\;+6y^2 \quad -4y$
$\;2y^2 \quad +y \quad (2y^2 + y)$

B. Solve these reading problems.

13. The average temperature at Honolulu (sea level) is 76°F. At an altitude of 11,000 feet on Mauna Loa, a nearby volcano, the average is 44°F. What is the average temperature change per thousand feet of altitude, to the nearest whole number? (Use a signed number in your answer.) $\qquad$ $-3°$

14. The duck hawk can dive at a rate of 264 feet per second. At that rate, what would its change in altitude be during a 4-second dive? (Use a signed number in your answer.) $\qquad$ $-1,056$ feet

Lesson 151 T–508

Introduction

Ask the students to combine the following amounts of time: 2 weeks, 1 week, 3 days, 7 days, and 2 months. The answer is 2 months 3 weeks 10 days.

Would it work to simply add all the numbers, use the middle unit (weeks) for the label, and give the answer as 15 weeks? No, only like units of measure can be added.

Likewise, when combining terms in polynomials, only like terms can be combined. Therefore, $2w + w + 3d + 7d + 2m = 10d + 2m + 3w$.

Teaching Guide

1. **In algebra, the only operation used in polynomials is addition.** That is, a polynomial simply contains two or more terms that are being combined (added). If a minus sign precedes a term, it means that a negative term is being added—which is the same as subtracting a positive.

2. **In a polynomial, the sign before a term always goes with that term, showing whether it is positive or negative.** The first term often has no sign, so it is understood to be positive. All the other terms are clearly marked as positive or negative.

   a. $7a + 2a - 3a$
   $= 7a$ plus $+2a$ plus $-3a$
   $= 9a$ plus $-3a$
   $= 6a$

   b. $2n + 3p - 4n$
   $= 2n$ plus $-4n$ plus $+3p$
   $= -2n$ plus $+3p$
   $= -2n + 3p$

Solutions for Part B

13. $[+44 \text{ destination} - (+76) \text{ origin}] \div 11$
14. $4(-264)$

T–509   Chapter 11   Signed Numbers, Tables, and Graphs

Further Study

This lesson states that "the sign before a term always goes with that term, showing whether it is positive or negative." This fact becomes important later in algebra when equations are solved by transposition. In this method, a term on one side of an equation is transposed to the other side and its sign is changed. Study the following examples.

$n + 8 = 13$
Transpose +8 and change its sign.
$n = 13 - 8$
$n = 5$

$9 = n - 7$
Transpose –7 and change its sign.
$9 + 7 = n$
$n = 16$

15. –80 destination – (–69) origin
16. –22 destination – (–17) origin

15. On a typical day in July (winter), the daytime high temperature in Antarctica may reach −69°F and the nighttime low temperature may be −80°F. What is the change in temperature from daytime to nighttime?   −11°

16. A typical Antarctica day in January (summer) may have a high temperature of −17°F and a low temperature of −22°F. What is the change in temperature from daytime to nighttime?   −5°

17. Clymer's Quick Print took an order for 7,000 sale bills. That was 1,000 more than twice the number of fliers that the company printed for a local supermarket. How many fliers were printed for the supermarket? (Write an equation and solve.)
$2f + 1{,}000 = 7{,}000$ $f = 3{,}000$ flyers

18. Clymer's Quick Print gives a discount on orders of 10,000 copies or more. One day a local business placed an order for information brochures. If the order had been for 3 times as many copies plus 1,600 more, it would have qualified for the discount. How many brochures were ordered? (Write an equation and solve.)
$3b + 1{,}600 = 10{,}000$ $b = 2{,}800$ brochures

REVIEW EXERCISES

C. Find the missing numbers for this table. *(Lesson 149)*

Formula: $a = s^2$	Areas of squares			
s	5	10	15	20
a	**19.** 25	**20.** 100	**21.** 225	**22.** 400

D. Multiply or divide these signed numbers. *(Lesson 147)*

23.   +3
     × (−8)
     ─────
     −24

24.   +78
        +13   +6

E. Compare each set of numbers, and write < or > between them. *(Lesson 144)*

25. $|-8|$ __>__ $|+7|$ **26.** -6 __<__ 4

F. Simplify these polynomials. *(Lesson 135)*

27. $3n + 2p − n − 4$ $2n + 2p − 4$ **28.** $3x + 2y − 6 − x + y$ $2x + 3y − 6$

G. Use each set of facts to construct a triangle. Label it with the facts given. *(Lesson 88)*

29. 55°, 2 in., 1½ in.

30. 4 cm, 3 cm, 5 cm

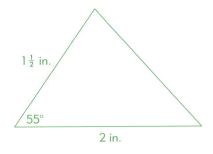
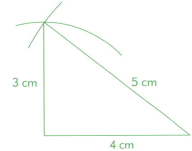

152. Reading Problems: More Practice With Equations

Equations are an effective tool for solving difficult problems. An equation is a mathematical statement showing how the problem will be solved. Writing the correct equation can be half of the work.

This lesson gives further practice with the points taught in Lesson 138. The following steps review writing equations.

1. Choose a literal number to represent the answer to the question. If you must also find another number, show that number in relation to your literal number. For example, in the problem below, s represents Samuel's age and $s - 5$ represents Dale's age. You could not solve the problem by using s for Samuel's age and d for Dale's age.

2. Set up the equation. Be alert for words that tell how the facts are related.

3. Solve the equation.

4. Check the equation.

> **Example**
> Samuel is a certain age, and Dale is 5 years younger. Seven more than their combined ages is 26 years. How old is Samuel?
>
> 1. Let s represent Samuel's age. (This is the answer to be found.)
>   Then $s - 5$ represents Dale's age.
>   And 7 more than their combined ages is 26 years.
>
> 2. Samuel's age + Dale's + 7 = 26
> $s + (s - 5) + 7 = 26$
>
> 3. $s + s - 5 + 7 = 26$
> $2s + 2 = 26$ (Combining -5 and $+7$ yields $+2$.)
> $2s + 2 - 2 = 26 - 2$
> $2s = 24$
> $\frac{2s}{2} = \frac{24}{2}$
> $s - 12$ yr.
>
> 4. Check: $(12) + (12) - 5 + 7 = 26$
> $26 = 26$

LESSON 152

Objective

- To give further practice with writing equations to solve reading problems.

Review

1. *Simplify these polynomials.* (Lesson 151)

   a. $3m + 5 - 4m - 3n + 5n$

   $(-m + 2n + 5)$

   b. $k + 7f - 3f + 7 - 6k$ $(4f - 5k + 7)$

2. *Evaluate these algebraic expressions.* (Lesson 148)

   a. $3y - 6z$ if $y = 5$ and $z = -2$ (27)

   b. $7 - 2a + 4b$ if $a = -4$ and $b = -8$

   (-17)

3. *Compare each set of numbers, and write < or > between them.* (Lesson 144)

   a. $|-9|$ (>) $|+8|$

   b. -5 (<) 3

4. *Draw line segments having these lengths, and use a compass to bisect them.* (Lesson 89)

   a. $1\frac{1}{2}$ in.   b. 2 in.

5. *Find these equivalents.* (Lesson 24)

   a. 16 ha = ___ m²   (160,000)

   b. 900 mm² = ___ m²   (0.0009)

T–511   *Chapter 11 Signed Numbers, Tables, and Graphs*

Introduction

Review using the four operations to solve equations as presented in Lessons 136 and 137.

Addition is used to eliminate subtraction.
$$n - 7 = 21$$
$$n - 7 + 7 = 21 + 7$$
$$n = 28$$

Subtraction is used to eliminate addition.
$$n + 7 = 21$$
$$n + 7 - 7 = 21 - 7$$
$$n = 14$$

Multiplication is used to eliminate division.
$$\frac{n}{7} = 21$$
$$\frac{n}{7} \times 7 = 21 \times 7$$
$$n = 147$$

Division is used to eliminate multiplication.
$$7n = 21$$
$$\frac{7n}{7} = \frac{21}{7}$$
$$n = 3$$

Teaching Guide

Review equation-writing concepts by discussing the steps in the lesson text. Then discuss the sample reading problem and its solution, and work through the *Class Practice* problems together.

CLASS PRACTICE

Write an equation, and then find the solution to each reading problem.

a. Mr. Price has a sheep farm and also raises some peafowl. Twenty times the number of peafowl, plus 100 more, equals the number of sheep that he has. The total number of sheep and peafowl together is 415. How many peafowl does Mr. Price have?
   (1) Let _p_ represent the number of peafowl.
   (2) Let ___ represent the number of sheep. $20p + 100$
   (3) Complete the equation and solve.
   $p + (20p + 100) = 415$
   $p = 15$ peafowl

b. Darla tied three colors of flowers in bunches. The number of white bunches and yellow bunches was the same, and there were 6 more bunches of pink flowers than either of the other two. She had 30 bunches of flowers in all. How many bunches of white flowers did Darla have?
   (1) Let _w_ represent the white bunches.
   (2) Let _w_ represent the yellow bunches.
   (3) Let ___ represent the pink bunches. $w + 6$
   (4) Complete the equation and solve. $w + w + (w + 6) = 30$ $w = 8$ bunches

c. Three times Kevin's age equals his father's age increased by 5. Kevin's father is 40 years old. How old is Kevin? $3k = 40 + 5$ $k = 15$ years

WRITTEN EXERCISES

A. Write equations to solve these reading problems.

1. One year an orchard harvested 100 more than twice as many bushels of Red Delicious as of Golden Delicious apples. The yield of both varieties together was 3,700 bushels. What was the harvest of Golden Delicious apples?
   a. Let _g_ represent the yield of Golden Delicious.
   b. Let ___ represent the yield of Red Delicious. $2g + 100$
   c. Complete the equation and solve. $g + (2g + 100) = 3{,}700$ $g = 1{,}200$ bushels

Chapter 11 Signed Numbers, Tables, and Graphs

2. A bin of Red Delicious and a bin of Golden Delicious apples together weighed 1,900 pounds. The Golden Delicious apples weighed 60 pounds less than the Red Delicious apples. What was the weight of the bin of Red Delicious apples?
   a. Let _r_ represent the pounds of Red Delicious.
   b. Let ___ represent the pounds of Golden Delicious. $r - 60$
   c. Complete the equation and solve. $r + (r - 60) = 1,900$ $r = 980$ pounds

3. Darrell left one copy of the *Star of Hope* at each house along Main Street, Second Street, and Third Street. The number of houses along Second Street was 9 less than the number of houses on Third Street, and the number of houses on Main Street was 3 times the number of houses on Third Street. If Darrell distributed 101 copies in all, how many houses were along Third Street?
   a. Let _t_ represent the houses along Third Street.
   b. Let ___ represent the houses along Second Street. $t - 9$
   c. Let ___ represent the houses along Main Street. $3t$
   d. Complete the equation and solve. $t + (t - 9) + 3t = 101$ $t = 22$ houses

4. Last summer Mother canned twice as many quarts of applesauce as cherries, and 20 more quarts of pears than cherries. Altogether she had 180 quarts. How many quarts of cherries did she can?
   a. Let _c_ represent the quarts of cherries.
   b. Let _2c_ represent the quarts of applesauce.
   c. Let ___ represent the quarts of pears. $c + 20$
   d. Complete the equation and solve. $c + 2c + (c + 20) = 180$ $c = 40$ quarts

5. The Bears' farm has 177 acres of corn and timothy hay together. There are 16 more acres of corn than 6 times the acres of timothy. How many acres of timothy are there?
   $t + (6t + 16) = 177$ $t = 23$ acres

6. One Sunday, the number of pupils in the junior Sunday school class was 4 more than the number in the intermediate class. The total in these two classes was 16. How many pupils were in the intermediate class?
   $i + (i + 4) = 16$ $i = 6$ pupils

7. Of the 38 people in Sunday school, the number of adults was 4 less than twice as many as the number of children. How many children were in Sunday school?
   $c + (2c - 4) = 38$ $c = 14$ children

8. Naomi used 6 quarts of peaches and pears to make a large batch of fruit salad. She used 2 times more peaches than pears. How many quarts of pears did she use?
   $p + 2p = 6$ $p = 2$ quarts

9. One week Earl noticed that the best cow in the herd had averaged 126 pounds of milk per day. That was 10 pounds less than twice the average daily production per cow that week. What was the average daily production per cow?
   $2d - 10 = 126$ $d = 68$ pounds

Model Constructions for Part E

19.

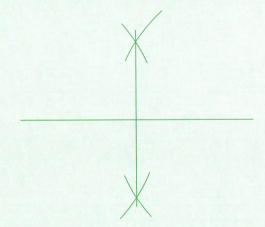

20.

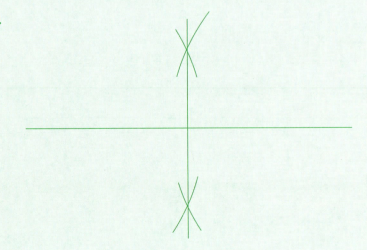

10. At the Verden School, there are 16 students in the seventh and eighth grade classroom. The number of seventh graders is 2 less than twice the number of eighth graders. How many students are in eighth grade?

$e + (2e - 2) = 16$ $e = 6$ students

11. Dorcas noted that the number of exercises in her math lesson was 4 less than that in her English assignment. The two lessons had 80 exercises in all. How many exercises were in her math lesson?

$m + (m + 4) = 80$ $m = 38$ exercises

12. The number of exercises in Steven's math lesson was 3 more than 3 times the number of exercises in his Bible lesson. The number of exercises in the two subjects together was 39. How many exercises were in his Bible lesson?

$b + (3b + 3) = 39$ $b = 9$ exercises

REVIEW EXERCISES

B. Simplify these polynomials. *(Lesson 151)*

13. $5s - 4t - 2s + 3t + 6$ $3s - t + 6$ 14. $-3t + u + 4t - 5u + 7$ $t - 4u + 7$

C. Evaluate these expressions if $a = 6$ **and** $b = -7$. *(Lesson 148)*

15. $4a - 5b$ 59   16. $9 - 3a + 2ab$ -93

D. Compare each set of numbers, and write < or > between them. *(Lesson 144)*

17. $|-5|$ $<$ $|+6|$ 18. -6 $<$ 1

E. Draw line segments having these lengths, and use a compass to bisect them. *(Lesson 89)*   (See facing page for model constructions.)

19. $2\frac{1}{2}$ inches   20. $3\frac{1}{2}$ inches

F. Find these equivalents. *(Lesson 24)*

21. 15,000 m² = _1.5_ ha   22. 7,500 m² = ____ km²

0.0075

153. Chapter 11 Review

A. Write a signed number for each phrase. Include a label. *(Lesson 144)*

1. a loss of $13   −$13
2. a profit of $7   +$7
3. 45° above zero   +45°
4. 799 feet below sea level   −799 feet

B. Write the absolute value of each number. *(Lesson 144)*

5. $|+3|$ 3
6. $|-16|$ 16
7. $|-5|$ 5
8. $|+12|$ 12

C. Compare each set of numbers, and write < or > between them. *(Lesson 144)*

9. −2 _<_ +8
10. −4 _<_ −2
11. $|-2|$ _<_ $|-7|$
12. $|-3|$ _>_ $|+1|$

D. Solve these problems containing signed numbers. *(Lessons 145–148)*

13. −4 + (−6) = −10
14. −3 + (−9) = −12
15. −5 + (+2) = −3
16. +6 + (−11) = −5

17. −12 − (−10) = −2
18. −7 − (+8) = −15
19. +18 − (+18) = 0
20. +21 − (−11) = +32

21. +4 × 0 = 0
22. −3 × (−7) = +21
23. +4 × (−8) = −32
24. −9 × (+5) = −45

25. +3(−2)(−2)(+3)   +36
26. −4(−3)(+2)(−3)(+4)   −288

27. +16 ÷ −4   −4
28. −8 ÷ −2   +4
29. $\frac{-48}{-16}$ +3
30. $\frac{+30}{-10}$ −3

E. Evaluate these algebraic expressions, using the values given. *(Lesson 148)*

$a = 3$ $g = -3$ $t = +9$
$b = -2$ $h = -5$ $u = -14$
$c = 2$

31. $g - h$ +2
32. $t + u$ −5
33. $\frac{4ab}{c}$ −12
34. $\frac{6c}{ab}$ −2

LESSON 153

Objective

- To review the material taught in Chapter 11 (Lessons 144–152).

Teaching Guide

1. Lesson 153 reviews the material taught in Lessons 144–152. For pointers on using review lessons, see *Teaching Guide* for Lesson 98.
2. Be sure to review the following new concepts taught in this chapter.

Lesson number and new concept	Exercises in Lesson 153
144—Comparing the values and absolute values of signed numbers.	9–12
146—In subtracting signed numbers, destination minus origin equals difference.	None
147—Multiplying more than two signed numbers.	25, 26
148—Substituting literal numbers with negative values.	31–34
150—Making graphs based on formulas including exponents.	43–54
151—In algebra, addition is the only operation used in polynomials.	55–58

Graphs for Part G

43–48.

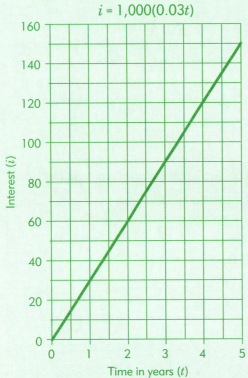

Simple interest on $1,000 at 3%:
$i = 1,000(0.03t)$

49–54.

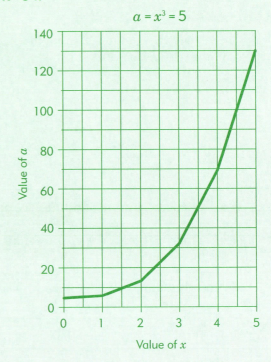

$a = x^3 = 5$

F. Find the missing numbers for each table, to the nearest whole number. *(Lesson 149)*

Formula: $a = \pi r^2$				Areas of circles				
r		11		12		13		14
a	35.	380	36.	452	37.	531	38.	615

Formula: $x = 3y + 4$				Areas of circles				
y		5		6		7		8
x	39.	19	40.	22	41.	25	42.	28

G. Prepare graphs from these tables. *(Lesson 150)*   (See facing page.)

43–48.

Formula: $i = 1{,}000(0.03t)$	Simple interest on $1,000 at 3%					
t (yr.)	0	1	2	3	4	5
i	$0	$30	$60	$90	$120	$150

49–54.

Formula: $a = x^3 + 5$						
x	0	1	2	3	4	5
a	5	6	13	32	69	130

H. Simplify each expression by combining like terms. *(Lesson 151)*

55. $6n - 3n$ $3n$ 56. $23b - 18b$ $5b$

57. $4d + 5e - 6d$ $-2d + 5e$ 58. $2x - 5y - x + 2y$ $x - 3y$

I. Write equations to solve these reading problems. *(Lesson 152)*

59. There are 176 verses in Psalm 119. That is 8 more than 12 times the number of verses in Psalm 19. How many verses does Psalm 19 have?
$12v + 8 = 176$ $v = 14$ verses

60. According to Genesis 7 and 8, 1 year and 10 days passed from the time the Flood began until the earth was dried. (Assume that the total was 375 days.) That was 15 more than 9 times the number of days that rain fell. How many days did it rain?
$9r + 15 = 375$ $r = 40$ days

61. One year Brother Arnold's farm received 74 inches of rainfall. He calculated that that was 6 inches less than two times the average rainfall for his area. What was the average amount of rainfall for his area?
$74 = 2r - 6$ $r = 40$ inches

62. A certain triangle has 10 more degrees in the second angle than in the first one, and 20 more degrees in the third angle than in the first one. What is the size of the first angle? (Think: What is the sum of the three angles in a triangle?)

$f + (f + 10) + (f + 20) = 180$ $f = 50$ degrees

63. Jesse picked 24 more bushels of apples than Harold did. Together they picked 168 bushels of apples. How many bushels did Harold pick?

$h + (h + 24) = 168$ $h = 72$ bushels

64. Galen calculated the daily average milk production of his father's two best cows. The second best cow gives 4 pounds less than the best cow. The daily milk production of these two cows together is 208 pounds. How much does the best cow produce?

$b + (b - 4) = 208$ $b = 106$ pounds

154. Chapter 11 Test

LESSON 154

Objective

- To test the students' mastery of the concepts in Chapter 11.

Teaching Guide

1. Correct Lesson 153.
2. Review any areas of special difficulty.
3. Administer the test. For pointers on giving tests, see *Teaching Guide* for Lesson 99.

In the Old Testament, very large amounts were typically expressed in figurative language. "The stars of heaven," "the sand of the sea," and "grasshoppers for multitude" all represent numbers beyond normal comprehension. Abraham, Isaac, and Jacob were promised offspring like the stars of heaven, and their descendants were considered so (Genesis 22:17; 26:4; Exodus 32:13; Deuteronomy 10:22; 28:62; 1 Chronicles 27:23; Nehemiah 9:23).

The simile of sand was used to describe Israel's number (Isaiah 10:22), as well as the amount of corn Joseph stored (Genesis 41:49), the number of enemy troops and their camels (Joshua 11:4; Judges 7:12), the greatness of Solomon's perception (1 Kings 4:29), Job's expectation of days before his calamities (Job 29:18), the supply of quail in the wilderness (Psalm 78:27), and the thoughts of God toward us (Psalm 139:18).

Chapter 12
Other Numeration Systems and Final Reviews

The decimal system is a sensible and efficient numeration system. But when you need to work with numbers twenty or thirty digits long, it is easy to lose your place. Likewise, extremely small numbers are difficult to read because of the many digits in the denominator (either decimal or common fractions). Scientific notation is a compact system that uses exponents for writing extremely large or small numbers.

A value other than ten can be the base of a place-value system. This concept is similar to carrying and borrowing in working with compound measures. When the "inches' place" is full, you carry 12 (not 10) units to make 1 in the place for feet. When borrowing in subtraction of pounds and ounces, you change 1 pound to 16 (not 10) in the "ounces' place."

Stretching your mind in these different ways of expressing numbers will enhance your overall keenness in mathematics.

She judged him faithful who had promised. Therefore sprang there even of one, and him as good as dead, so

many as the stars of the sky in multitude, and as the sand which is by the sea shore innumerable. Hebrews 11:11, 12

155. Reading Electric Meters

Electric energy is measured in watts as water is measured in gallons. One watt of energy used for one hour is called one watt-hour. One thousand watt-hours is one kilowatt-hour. Ten 100-watt light bulbs shining for one hour use one kilowatt-hour of electricity: 100 watts × 10 = 1,000 watts × 1 hour = 1,000 watt-hours = 1 kilowatt-hour. Electricity is sold by the kilowatt-hour (kwh).

An electric meter is a device that measures the amount of electricity used by a customer. Many electric meters have five dials that show the amount of electricity used. Each dial represents a place value. The dial farthest to the right shows ones, the second dial shows tens, the third dial shows hundreds, and so forth. Thus, the dials on the meter represent digits in the same order that they are written in Arabic numerals.

Hands next to each other turn in opposite directions. For this reason the dials are numbered in alternating directions. The dials for the ones', hundreds', and ten thousands' places are numbered clockwise, whereas the dials for the tens' and thousands' places are numbered counterclockwise.

To read an electric meter, follow these directions.

1. Beginning at the right, write down the digit to which the hand points. If the hand is between two digits, write the lower one. If the hand is between 9 and 0, the lower digit is 9 because the zero then represents 10.

2. Go left to the next dial, and repeat step 1 until every dial is read.

> **Example A**
> Read this electric meter (April reading).
>
>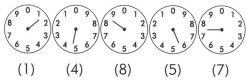
>
> (1)   (4)   (8)   (5)   (7)
>
> Reading: 14,857 kilowatt-hours
>
> **Example B**
> Read this electric meter (May reading).
>
>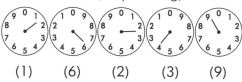
>
> (1)   (6)   (2)   (3)   (9)
>
> Reading: 16,239 kilowatt-hours

LESSON 155

Objective

- To teach *reading numbers shown by dials, as on some electric meters.

Review

1. *Simplify these polynomials.* (Lessons 139, 151)

    a. $5v + 2v^2w - 3w^2 - 8v$
    $(2v^2w - 3v - 3w^2)$

    b. $d^2 - 3d^2 - 5e + 5e + 3d^2$ (d^2)

2. *Multiply these signed numbers.* (Lesson 147)

    a. $\quad +8$
    $\quad \times (-7)$
    $\quad (-56)$

    b. $\quad +6$
    $\quad \times (+4)$
    $\quad (+24)$

3. *Use decimals from the table to find the interest compounded semiannually.* (Lesson 124)

    a. $5,400 at 7% for $1\frac{1}{2}$ yr. ($586.98)
    Decimal: 1.1087

    b. $14,000 at 4% for $1\frac{1}{2}$ yr. ($856.80)
    Decimal: 1.0612

4. *Find the circumference of each circle.* (Lesson 91)

    a. $r = 15$ in. $c = $ (94.2 in.)

    b. $r = 1.4$ m $c = $ (8.792 m)

5. *Find these equivalents, using the tables.* (Lesson 27)

    a. 315 g = ___ oz. (11.025)

    b. 26 m = ___ ft. (85.28)

Introduction

Ask the students what units of measure are used to sell the following.

milk (pint, quart, gallon, hundredweight, liter)

gasoline (gallon, liter)

carpet (square yard, square meter)

land (acre, hectare)

concrete (cubic yard, cubic meter)

electricity (kilowatt-hour)

How is the number of kilowatt-hours determined? Electricity cannot be weighed on a scale or measured with a yardstick.

Electricity is measured by an electric meter. As the electricity passes through a small motor in the meter, electromagnetism causes the meter wheel to turn. The rotating wheel turns the dials. The more electricity that is used, the faster the wheel and dials will turn. If no electricity passes through, the wheel and dials are still.

Teaching Guide

1. **Electric energy is measured in watts.** One watt of energy used for one hour is called one watt-hour. One thousand watt-hours is one kilowatt-hour (kwh). Ten 100-watt light bulbs shining for one hour use one kilowatt-hour of electricity: 100 watts × 10 = 1,000 watts × 1 hour = 1,000 watt-hours = 1 kilowatt-hour.

2. **An electric meter is a device that measures the amount of electricity used by a customer.** The electric supplier (or the customer himself) periodically reads the meter, and the customer is billed at a set rate per kilowatt-hour.

3. **Many electric meters have five dials that show the amount of electricity used.** Reading such a meter is not difficult if a person understands the following facts.

    a. Each dial represents a place value. The dial farthest to the right shows ones, the second dial shows tens, the third dial shows hundreds, and so forth. The dials on the meter represent digits in the same order that they are written in Arabic numerals.

    b. Hands next to each other turn in opposite directions. For this reason the dials in the ones', hundreds', and ten thousands' places are numbered clockwise, while the dials in the tens' and thousands' places are numbered counter-clockwise.

In Example B, the hand on the tens dial points almost directly to the 4. How does one know whether to write 3 or 4 for that dial? In such a case, look at the next dial to the right, and you will see that the hand is just before or just past 0. If that hand is before 0, write the lower digit. (The last two dials in Example B show 39, not 49.) If the hand is past 0, write the higher digit. (If the last hand were between 0 and 1, you would write 40 for the last two dials.) The meter works this way because while each hand makes a complete revolution, the next hand to the left moves only from one digit to the next.

Every month or two, someone from the electric company (or the customer himself) reads the meter. The previous meter reading is subtracted from the current reading to find the number of kilowatt-hours used. The customer is then billed at a certain rate per kilowatt-hour. In Example C below, the rate is $0.0878 per kilowatt-hour.

An electric company often collects a monthly fee in addition to the rate charged per kilowatt-hour. To find the amount of an electric bill, multiply the number of kilowatt-hours by the rate per kilowatt-hour. Add any monthly fee to the product.

Example C
    Find the number of kilowatt-hours used between the readings in Example A and Example B.

    May reading   16,239 kwh
    April reading − 14,857 kwh
                       1,382 kwh

Example D
    Find the amount of the bill for 1,382 kilowatt-hours at the rate of $0.0878 per kilowatt-hour and a monthly fee of $7.25.

             1,382 kwh
      × $0.0878
        $121.34 (rounded)
    +   7.25
        $128.59

Electric appliances are usually labeled with the number of watts they use. A small heater may use 1,500 watts. To find the number of watt-hours consumed, multiply the number of watts by the number of hours the appliance is used. If the heater runs 4 hours, multiply 4 × 1,500. The result is 6,000 watt-hours or 6 kilowatt-hours.

Example E
    An electric fan uses 80 watts. How many kilowatt-hours of electricity does the fan use if it runs for a full day?

        24
    × 80
  1,920   watt-hours = 1.92 kwh

Example F
    Find the cost of 1.92 kilowatt-hours of electricity at $0.0878 per kilowatt-hour, to the nearest cent.

        1.92
   × 0.0878
  $0.168576 = $0.17

520 Chapter 12 Other Numeration Systems and Final Reviews

CLASS PRACTICE

Write the reading of each meter. Then find the kilowatt-hours of electricity used.

a. 43,018 subtracted from 44,595 = 1,577 kwh

b. 70,215 subtracted from 73,648 = 1,433 kwh

Find the cost of these amounts of electricity.

c. 728 kwh at $0.0745 per kwh $54.24

d. 857 kwh at $0.0817 per kwh $70.02

Calculate the watt-hours used in each case.

e. 55 watts for 112 hours
   6,160 watt-hours

f. 590 watts for 48 hours
   28,320 watt-hours

Change these watt-hours to kilowatt-hours.

g. 15,820 watt-hours   15.82 kwh

h. 36,580 watt-hours   36.58 kwh

WRITTEN EXERCISES

A. *Write the readings of these electric meters. Label your answers with kilowatt-hours (kwh).*

1. 22,853 kwh

2. 7,482 kwh

3. 46,349 kwh

4. 73,580 kwh

B. *Write the reading of each meter. Then find the kilowatt-hours of electricity used.*

5. 18,851 subtracted from 19,332 = 481 kwh

492

4. **To read an electric meter, follow these steps.**

   a. Beginning at the right, write down the digit to which the hand points. If the hand is between two digits, write the lower one. If the hand is between 9 and 0, the lower digit is 9 because the zero then represents 10.

   b. Go left to the next dial, and repeat step 1 until every dial is read.

   Have students read the meters in *Class Practice*. Be sure they understand how to determine the correct reading when a hand seems to point directly to a certain digit (such as 4). Refer to the next dial to the right. If that hand is before the zero, the two dials indicate 39; if the hand is after the zero, the two dials indicate 40.

5. **The number of kilowatt-hours used is determined by subtracting the previous meter reading from the current one.** Calculate the kilowatt-hours in *Class Practice*.

6. **To calculate the amount of an electric bill, multiply the usage in kilowatt-hours by the rate per kilowatt-hour and add any monthly fee.** Use the exercises below and those in *Class Practice* to demonstrate this point.

   a. 789 kwh at $0.0809 per kwh plus a monthly fee of $6.50   ($70.33)

   b. 832 kwh at $0.0919 per kwh plus a monthly fee of $7.25   ($83.71)

7. **Electric appliances are usually labeled with the number of watts they use.** To find the number of watt-hours consumed, multiply the number of watts by the number of hours of use.

   Find the watt-hours used in each of the following situations.

   a. 75 watts for 24 hours

   (1,800 watt-hours)

   b. 150 watts for 15 hours

   (2,250 watt-hours)

8. **To change watt-hours into kilowatt-hours, divide the number of watt-hours by 1,000.** To divide by 1,000, simply move the decimal point three places to the left.

   Change these watt-hours to kilowatt-hours.

   a. 1,800 watt-hours   (1.8 kwh)

   b. 2,250 watt-hours   (2.25 kwh)

Solutions for Part F

23. (60,592 − 57,874) × $0.0925

24. (32,453 − 23,315) × $0.0875 + 5 × $5.75

25. 7 × 2,500 ÷ 1,000

26. 240 × 10 × 6 ÷ 1,000

6. 31,178 subtracted from 32,274 = 1,096 kwh
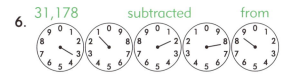 

7. 36,744 subtracted from 38,136 = 1,392 kwh
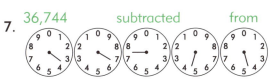 

8. 78,403 subtracted from 81,553 = 3,150 kwh

C. Find the cost of these amounts of electricity.

9. 822 kwh at $0.0798 per kwh $65.60
10. 915 kwh at $0.0756 per kwh $69.17
11. 215 kwh at $0.0696 per kwh $14.96
12. 684 kwh at $0.0746 per kwh $51.03

D. Calculate the watt-hours used in each case.

13. 40 watts for 168 hours   6,720 watt-hours
14. 50 watts for 24 hours   1,200 watt-hours
15. 150 watts for 96 hours
    14,400 watt-hours
16. 225 watts for 48 hours
    10,800 watt-hours

E. Change these watt-hours to kilowatt-hours.

17. 12,400 watt-hours   12.4 kwh
18. 15,780 watt-hours   15.78 kwh
19. 42,630 watt-hours   42.63 kwh
20. 68,800 watt-hours   68.8 kwh
21. 105,750 watt-hours   105.75 kwh
22. 100,350 watt-hours   100.35 kwh

F. Solve these reading problems.

23. The electric meter on the Sensenig farm reads 60,592. Last month at this time it read 57,874. If the rate is $0.0925 per kilowatt-hour, what is the charge for this electricity? $251.42

24. The July reading of the meter at Weaver's Farm Market was 32,453. The previous February reading was 23,315. The rate per kilowatt-hour is $0.0875 plus a monthly charge of $5.75. Find the cost of the electricity for those five months. $828.33

25. Brother Melvin has a 2,500-watt utility heater to keep his milk house warm. How many kilowatt-hours does it take to operate the heater for 7 hours? (Use a decimal in your answer.)   17.5 kwh

26. A large exhaust fan is driven by a $1\frac{1}{2}$-horsepower motor that draws 10 amps at 240 volts. This information can be used to determine watts by applying the formula *volts × amps = watts*. How many kilowatt-hours does the motor use in a 6-hour period?   14.4 kwh

27. Beside the altar in Solomon's temple stood 10 basins that held water for washing (1 Kings 7:38). The diameter of each basin was 4 cubits (6 feet) at the top. Find the circumference at the top of each laver, to the nearest tenth.   **12.6 cubits or 18.8 feet**

28. The millstones used by Hebrew women to grind grain consisted of two stones, each about 18 to 24 inches in diameter. There was a handle to turn the upper stone, which rotated on the stationary lower stone (Matthew 24:41). If the handle was 10 inches from the center, how far did it move with each rotation? (Use a decimal in your answer.)
62.8 inches

Two women shall be grinding at the mill.

Matthew 24:41

REVIEW EXERCISES

G. Simplify these polynomials. *(Lessons 139, 151)*

29. $8m^2n^4 + 3m^2n^4$ **$11m^2n^4$**   **30.** $23a^3b - 6a^3b$ **$17a^3b$**

31. $6p + 3p^2q - 2q^2 - 6p$ **$3p^2q - 2q^2$**   **32.** $5t^2u + 3tu^2 - 7t^2u + tu^2$ **$4tu^2 - 2t^2u$**

H. Multiply these signed numbers. *(Lesson 147)*

33.   +6
     × (−6)
     −36

34.   +5
     × (+8)
     +40

I. Find the interest, compounded quarterly. The decimal from the interest table is given for you. *(Lesson 124)*

35. $6,200 at 8% for $1\frac{1}{2}$ yr.
decimal: 1.1262   **$782.44**

36. $15,100 at 6% for $1\frac{1}{2}$ yr.
decimal: 1.0934   **$1,410.34**

J. Find the circumference of each circle. *(Lesson 91)*

37. $r = 16$ in.   **100.48 in.**   **38.** $r = 6.2$ m   **38.936 m**

K. Find these equivalents, using the tables. *(Lesson 27)*

39. 19 MT = **20.9** tons   **40.** 62 mi. = **99.82** km

27. 3.14 × 6 feet

28. 3.14 × 20-inch diameter of circle traced by handle

T–523 Chapter 12 *Other Numeration Systems and Final Reviews*

LESSON 156

Objective

- To teach *using scientific notation to write very large and very small numbers.

Review

1. *Write equations to solve these reading problems.* (Lesson 152)

   a. In 1842, the first suspension bridge in the United States was built across the Schuylkill River near Philadelphia, Pennsylvania. Its length was 358 feet, which was 8 feet more than 14 times its width. How wide was the bridge?
   $14w + 8 = 358$; $w = 25$ ft.

   b. One of the world's longest suspension bridges is the 853-meter Tacoma Narrows Bridge, which spans Puget Sound. The original bridge collapsed in 1940 after a wind caused the bridge to vibrate at its own natural frequency. Adding 245 to twice its length in meters gives the year when the bridge was rebuilt. What year was that?
   $y = 245 + 2 \cdot 853$; $y = 1951$

2. *Evaluate these expressions.* (Lesson 148)

   a. $2ab - 3b$ if $a = -2$ and $b = -4$ (+28)

   b. $\dfrac{y^2}{3yz}$ if $y = +3$ and $z = -1$ (−1)

3. *Multiply these monomials.* (Lesson 140)

   a. $m^5n^3 \cdot 2m^3n^4$ ($2m^8n^7$)

   b. $6c^3d^5 \cdot 4cd^3$ ($24c^4d^8$)

4. *Use the compound interest formula to find the interest compounded quarterly.* (Lesson 124)

   a. $6,600 at 6% for $\frac{1}{2}$ yr. ($199.49)

   b. $11,800 at 8% for $\frac{1}{2}$ yr. ($476.72)

156. Using Scientific Notation

In the first lesson in this book, you practiced writing large numbers. Some of the largest numbers you wrote were quadrillions, which have six periods (at least 16 digits). Since very large numbers require so many digits, scientists and mathematicians have developed a system called **scientific notation.** This system reduces the size of a numeral by expressing the number as a decimal multiplied by a power of 10.

To write a large number in scientific notation, follow these steps.

1. Move the decimal point left until it comes directly after the digit farthest to the left. Drop any ending zeroes.

2. Write the number found in step 1 as multiplied by 10 to the correct power. To find that power, count the number of places that you moved the decimal point in step 1. In Example A, the exponent is 16 because the decimal point was moved 16 places. In Example B the decimal point was moved 24 places.

> **Example A** $34,000,000,000,000,000 = 3.4 \times 10^{16}$
>
> **Example B** $8,000,000,000,000,000,000,000,000 = 8 \times 10^{24}$

To expand a number written in scientific notation, simply reverse the process. Move the decimal point to the right as many places as the exponent indicates. Use zeroes as place holders, and insert commas as needed.

> **Example C** $2.7 \times 10^{15} = 2,700,000,000,000,000$

Scientific notation is also used for very small decimals, but then the exponent is negative. A negative exponent indicates the reciprocal of a positive exponent. For example, 10^4 equals 10,000 and 10^{-4} equals its reciprocal, 0.0001 ($\frac{1}{10,000}$). A positive exponent indicates how many decimal places to add to the right of the number. A negative exponent indicates how many decimal places to add to the left of the number.

The chart at the right indicates that 10^0 equals 1. How can this be? Consider what you learned in Lesson 141: to divide like terms having exponents, find the difference between the exponents. That is, $n^5 \div n^2 = n^3$.

$10^5 = 100,000$
$10^4 = 10,000$
$10^3 = 1,000$
$10^2 = 100$
$10^1 = 10$
$10^0 = 1$
$10^{-1} = 0.1$ or $\frac{1}{10}$
$10^{-2} = 0.01$ or $\frac{1}{100}$
$10^{-3} = 0.001$ or $\frac{1}{1,000}$
$10^{-4} = 0.0001$ or $\frac{1}{10,000}$
$10^{-5} = 0.00001$ or $\frac{1}{100,000}$

Therefore $10^3 \div 10^3 = 10^0$. This conclusion is supported by using the standard form: $1{,}000 \div 1{,}000 = 1$.

To write a very small decimal in scientific notation, follow these steps.

1. Move the decimal point right until it comes directly after the first nonzero digit. Drop any leading zeroes.

2. Write the number found in step 1 times 10 to the correct negative power. To find that power, count the number of places that you moved the decimal point in step 1. In Example D, the exponent is –10 because the decimal point was moved 10 places. In Example E the decimal point was moved 14 places.

Example D $\quad 0.00000000026 = 2.6 \times 10^{-10}$

Example E $\quad 0.0000000000000375 = 3.75 \times 10^{-14}$

To expand a decimal written in scientific notation, simply reverse the process. Move the decimal point to the left as many places as the negative exponent indicates. Use zeroes as place holders.

Example F $\quad 2.1 \times 10^{-11} = 0.000000000021$

CLASS PRACTICE

Write these large numbers, using scientific notation.
- a. 4,800,000,000,000 4.8×10^{12}
- b. 7,650,000,000,000,000,000 7.65×10^{18}
- c. 3,500,000,000,000,000,000 3.5×10^{18}
- d. 350,000,000,000,000,000,000 3.5×10^{20}

Expand these numbers expressed in scientific notation.
- e. 6.5×10^7 65,000,000
- f. 9.7×10^4 97,000
- g. 5.2×10^8 520,000,000
- h. 7.9×10^{11} 790,000,000,000

Use negative exponents to express these decimals.
- i. 0.00001 10^{-5}
- j. 0.0000001 10^{-7}

Write these decimals, using scientific notation.
- k. 0.000065 6.5×10^{-5}
- l. 0.00000071 7.1×10^{-7}
- m. 0.0000000000065 6.5×10^{-12}
- n. 0.00000000000000000778 7.78×10^{-18}

Expand these decimals expressed in scientific notation.
- o. 3.6×10^{-5} 0.000036
- p. 4.12×10^{-9} 0.00000000412
- q. 7.9×10^{-14} 0.000000000000079
- r. 6.95×10^{-16} 0.000000000000000695

5. *Find the area of each figure.* (Lesson 92)

   a. Square
      $s = 36$ ft.
      $a = (1{,}296$ sq. ft.$)$

   b. Rectangle
      $l = 16$ in.
      $w = 12$ in.
      $a = (192$ sq. in.$)$

6. *Find the equivalent temperatures, to the nearest degree.* (Lesson 28)

   a. $56°C = \underline{}°F$ $(133°)$
   b. $72°F = \underline{}°C$ $(22°)$

Introduction

Write the number one trillion (1,000,000,000,000) on the board. Ask the students, "What exponent can you use with 10 to have it equal one trillion?" One trillion is equal to 10^{12}.

Now write 1,200,000,000,000 on the board, and ask the students to write a multiplication problem using exponents to express this value. The number 1,200,000,000,000 as written in this manner is 1.2×10^{12}.

Using this method to write large numbers is called scientific notation.

Teaching Guide

1. **Scientific notation reduces the size of a numeral by expressing the number as a decimal multiplied by a power of 10.** To write large numbers in scientific notation, follow these steps.

   (1) Move the decimal point left until it comes directly after the digit farthest to the left. Drop any ending zeroes.

   (2) Write the number found in step 1 as multiplied by 10 to the correct power. To find that power, count the number of places that you moved the decimal point in step 1.

   a. $1,000,000,000 = 1 \times 10^9$

   b. $1,200,000,000,000 = 1.2 \times 10^{12}$

2. **To expand a number written in scientific notation, simply reverse the process.** Move the decimal point to the right as many places as the exponent indicates. Use zeroes as place holders, and insert commas as needed.

   a. $2.4 \times 10^6 = 2,400,000$

   b. $3.8 \times 10^8 = 380,000,000$

3. **Scientific notation is also used for very small decimals, but then the exponent is negative.** A negative exponent indicates the reciprocal of a positive exponent. Illustrate with the list in the pupil's lesson.

   a. $0.1 = 10^{-1}$

   b. $0.01 = 10^{-2}$

   c. $0.00000001 = 10^{-8}$

   d. $0.000000000000001 = 10^{-15}$

4. **To write a very small decimal in scientific notation, follow these steps.**

   (1) Move the decimal point right until it comes directly after the first nonzero digit. Drop any leading zeroes.

   (2) Write the number found in step 1 times 10 to the correct negative power. To find that power, count the number of places that you moved the decimal point in step 1.

   a. $0.000000004 = 4 \times 10^{-9}$

   b. $0.0000000000000007 = 7 \times 10^{-16}$

5. **To expand a decimal written in scientific notation, simply reverse the process.** Move the decimal point to the left as many places as the negative exponent indicates. Use zeroes as place holders.

   a. $3 \times 10^{-8} = 0.00000003$

   b. $8 \times 10^{-12} = 0.000000000008$

WRITTEN EXERCISES

A. *Write these large numbers, using scientific notation.*

1. 3,200,000,000 3.2×10^9
2. 4,500,000,000,000 4.5×10^{12}
3. 27,000,000,000,000 2.7×10^{13}
4. 620,000,000,000,000 6.2×10^{14}
5. 70,000,000,000,000,000 7×10^{16}
6. 8,000,000,000,000,000 8×10^{15}
7. 133,000,000,000,000,000 1.33×10^{17}
8. 43,500,000,000,000,000 4.35×10^{16}

B. *Expand these numbers expressed in scientific notation.*

9. 2.1×10^6 2,100,000
10. 4.1×10^7 41,000,000
11. 5.3×10^4 53,000
12. 5.2×10^9 5,200,000,000

C. *Use negative exponents to express these decimals.*

13. 0.00001 10^{-5}
14. 0.000001 10^{-6}
15. 0.000000001 10^{-9}
16. 0.00000000001 10^{-11}

D. *Write these decimals, using scientific notation.*

17. 0.00045 4.5×10^{-4}
18. 0.0000051 5.1×10^{-6}
19. 0.000031 3.1×10^{-5}
20. 0.00000012 1.2×10^{-7}
21. 0.0000053 5.3×10^{-6}
22. 0.000000039 3.9×10^{-8}

E. *Expand these decimals expressed in scientific notation.*

23. 1.1×10^{-4} 0.00011
24. 3.12×10^{-5} 0.0000312
25. 2.7×10^{-6} 0.0000027
26. 8.1×10^{-11} 0.000000000081

F. *Solve these reading problems. Use equations for numbers 31 and 32.*

27. After Saul became king, he set up a standing army with a core consisting of 3,000 chosen men (1 Samuel 13:2). Express this number of men, using scientific notation. 3×10^3 men

28. King David organized a national militia having 12 divisions of 24,000 soldiers apiece (1 Chronicles 27:1). If each division served for one month, how many men served in the militia in one year? (Use scientific notation in your answer.) 2.88×10^5 men

29. One light-year is the distance that light travels in one year, or about 5,880,000,000,000 miles. Express this distance, using scientific notation. 5.88×10^{12} miles

30. Scientists sometimes measure very small objects with the angstrom (ăng′ strəm). One angstrom equals 0.0000000001 meter, which is the approximate diameter of a hydrogen atom. Express this diameter in terms of a meter, using scientific notation. 10^{-10} meters

31. A tsunami (tsoo nä′ mē) is a huge wave caused by an undersea earthquake. In 1883, tsunamis up to 130 feet high destroyed more than 150 villages on an island of Indonesia. The number of lives lost was about 240 times the number of villages destroyed. How many people drowned? $p = 150 \times 240$ $p = 36{,}000$ people

32. The highest tsunami on record struck Alaska on March 28, 1964. It was 40 feet less than twice as high as the waves described in problem 31 above. How high was this tsunami? $t = 2(130) - 40$ $t = 220$ ft.

REVIEW EXERCISES

G. Write the reading of each meter. *(Lesson 155)*

33. 12,953 kwh

34. 14,830 kwh

H. Evaluate these expressions. *(Lesson 148)*

35. $3rs - 4s$ -39
$r = -3; s = 3$

36. $2\pi r^2 + 2\pi rh$ 452.16
$r = 4; h = 14$

I. Multiply these monomials. *(Lesson 140)*

37. $g^3h^2 \cdot 4g^2h^2$ $4g^5h^4$

38. $5t^4u^2 \cdot 6t^2u^3$ $30t^6u^5$

J. Use the compound interest formula to find the total principal and interest if interest is compounded quarterly. *(Lesson 124)*

39. $7,600 at 16% for ½ yr. $8,220.16

40. $15,000 at 12% for ½ yr. $15,913.50

K. Write these formulas. *(Lesson 92)*

41. Area of a square $a = s^2$

42. Area of a rectangle $a = lw$

L. Find the equivalent temperatures, to the nearest degree. *(Lesson 28)*

43. 63°C = 145 °F

44. 76°F = 24 °C

LESSON 157

Objectives

- To review the base ten numeration system.
- To teach that *numeration systems can have bases other than ten.
- To teach *how to read numbers in systems with other bases.

Review

1. *Write these numbers, using scientific notation.* (Lesson 156)

   a. 6,400,000,000,000 (6.4×10^{12})

   b. 7,800,000,000,000,000,000

   (7.8×10^{18})

2. *Find the missing numbers for this table, to the nearest whole number.* (Lesson 149)

   Lateral areas of cylinders 3 feet high

Formula: $a_\ell = 2\pi rh$				
r	2	3	4	5
a_ℓ	(38)	(57)	(75)	(94)

3. *Divide these monomials.* (Lesson 141)

   a. $t^5 \div t^2$ (t^3)   b. $y^7 \div y^5$ (y^2)

4. *Calculate each sales tax or property tax.* (Lesson 125)

	Sale	Tax Rate	Tax
a.	$62.45	5%	($3.12)
	Assessment	Tax Rate	Tax
b.	$45,000	19 mills	($855)

5. *Find the area of each figure.* (Lesson 93)

   a. Parallelogram: $b = 26$ in.; $h = 18$ in.
   (468 sq. in.)

   b. Triangle: $b = 125$ ft.; $h = 95$ ft.
   (5,937.5 sq. ft.)

6. *Find the missing parts.* (Lesson 29)

   a. $d = 700$ mi.; $r = 56$ m.p.h.; $t =$ ___
   ($12\frac{1}{2}$ hr.)

   b. $d =$ ___; $r = 240$ km/h; $t = 4\frac{1}{4}$ hr.
   (1,020 km)

157. The Concept of Base in Numeration

The numeration system with which we are familiar is used throughout the world. With this system we can use ten digits (0, 1, 2, 3, 4, 5, 6, 7, 8, 9) to write any number, large or small. It is called the decimal (tens) system because its base is ten.

Base is an important concept in place-value numeration. The base value is the point at which the quantity in one place is regrouped, and the new group is recorded as 1 in the next place to the left. Study the following illustration.

7	///////	The digit 7 indicates this quantity.
+ 6	//////	The digit 6 indicates this quantity.
?	/////////////	The decimal system has no digit for this quantity.
13	(//////////) ///	Since the base is ten, the quantity is regrouped as shown.

There is 1 of the base group (ten), and 3 more.

The process is repeated with the larger values. When a quantity reaches ten tens, a digit is written in the hundreds' place. When a quantity reaches ten hundreds, a digit is written in the thousands' place; and so forth. This system of tens apparently developed because people used their ten fingers for counting. In fact, the original meaning of *digit* is "finger."

It is possible to use numbers other than ten as the base for a number system. Consider: What if people had eight or twelve fingers? Then the problem illustrated above might be written as shown below.

Base Eight System

```
   7      ///////
 + 6      //////
  15    (////////) /////
```
(1 group of eight, and 5 more)

Base Twelve System

```
   7      ///////
 + 6      //////
  11    (////////////) /
```
(1 group of twelve, and 1 more)

Any number higher than 1 can be the base of a numeration system. The base indicates the point at which regrouping occurs as well as the number of digits in that system. For example, the base two system uses only two digits: 0 and 1. The base six system uses six digits: 0, 1, 2, 3, 4, and 5.

Base	Digits Used	Example
two	0, 1	101_{two}
four	0, 1, 2, 3	21_{four}
six	0, 1, 2, 3, 4, 5	13_{six}
eight	0, 1, 2, 3, 4, 5, 6, 7	24_{eight}
ten	0, 1, 2, 3, 4, 5, 6, 7, 8, 9	17_{ten}

Chapter 12 Other Numeration Systems and Final Reviews

The base of a number is indicated by a subscript word after the numeral, such as 4_{six}; this is read "four, base six." To read a number with more than one digit, say each digit in sequence and then name the base. Read 653_{eight} as "six, five, three, base eight." If a number has no subscript word, its base is understood to be ten.

101_{two} = "one, zero, one, base two"
356_{seven} = "three, five, six, base seven"

Counting in a system other than base ten is interesting. Below are the numbers from 1 to 20 as you would count in various systems.

Base four: 1, 2, 3, 10, 11, 12, 13, 20
Base six: 1, 2, 3, 4, 5, 10, 11, 12, 13, 14, 15, 20
Base eight: 1, 2, 3, 4, 5, 6, 7, 10, 11, 12, 13, 14, 15, 16, 17, 20
Base ten: 1, 2, 3, 4, 5, 6, 7, 8, 9, 10, 11, 12, 13, 14, 15, 16, 17, 18, 19, 20

CLASS PRACTICE

List the digits used in each numeration system.

a. base four   0, 1, 2, 3
b. base seven   0, 1, 2, 3, 4, 5, 6

Read these numbers correctly.

c. 314_{five} three, one, four, base five
d. 152_{six} one, five, two, base six
e. $4,317_{eight}$ four, three, one, seven, base eight
f. $5,304_{seven}$ five, three, zero, four, base seven

Count in the following bases.   (See facing page.)

g. from 1_{four} to 21_{four}
h. from 5_{seven} to 20_{seven}
i. from 15_{six} to 30_{six}
j. from 1_{two} to 111_{two}
k. from 13_{five} to 30_{five}
l. from 1_{three} to 22_{three}

WRITTEN EXERCISES

A. *Write the digits used in each numeration system.*

1. base six   0, 1, 2, 3, 4, 5
2. base five   0, 1, 2, 3, 4
3. base two   0, 1
4. base nine   0, 1, 2, 3, 4, 5, 6, 7, 8

B. *Write these numbers, using words.*

5. 111_{two} one, one, one, base two
6. $2,743_{eight}$ two, seven, four, three, base eight
7. $2,301_{four}$ two, three, zero, one, base four
8. $5,314_{six}$ five, three, one, four, base six

C. *Count in the following bases.* (See facing page.)

9. from 1_{six} to 14_{six}
10. from 1_{four} to 21_{four}
11. from 15_{seven} to 25_{seven}
12. from 6_{nine} to 15_{nine}
13. from 20_{three} to 111_{three}
14. from 20_{five} to 32_{five}

Introduction

We are familiar with the base ten system. In this system, the age of most eighth graders is written as 13, 14, or 15. But how would you write your age in base 7? Thirteen-year-olds would write 16_{seven}; fourteen-year-olds would write 20_{seven}; and fifteen-year-olds would write 21_{seven}.

Teaching Guide

1. **The numeration system we use is a base ten system.** It has the ten digits from 0 to 9. The base of a number system is the point at which the quantity in one place is regrouped, and the new group is recorded in the next place to the left. For example, when the quantity in the ones' place is twelve, it is regrouped as 1 ten and 2 ones (12).

    The base of a number tells how much value each place has. The first place in any system is the ones' place. The second place has the same value as the base; and succeeding places have the value of the base raised to the second power, the third power, and so on..

2. **The base of a number is indicated by a subscript word after the numeral.** (Words are used rather than numerals because a word such as *twelve* always indicates a certain value, whereas the value of a numeral such as 12 depends on the base.) To read numbers with bases other than ten, say each digit in sequence and then give the base. For example, the number $1E_{twelve}$ would be read "one, E, base twelve."

3. **Counting in any numeration system depends on the base of the system.** Practice counting in bases other than base ten. Count around the class.

Answers for CLASS PRACTICE g–l

g. 1, 2, 3, 10, 11, 12, 13, 20, 21
h. 5, 6, 10, 11, 12, 13, 14, 15, 16, 20
i. 15, 20, 21, 22, 23, 24, 25, 30
j. 1, 10, 11, 100, 101, 110, 111
k. 13, 14, 20, 21, 22, 23, 24, 30
l. 1, 2, 10, 11, 12, 20, 21, 22

Answers for Part C

9. 1, 2, 3, 4, 5, 10, 11, 12, 13, 14
10. 1, 2, 3, 10, 11, 12, 13, 20, 21
11. 15, 16, 20, 21, 22, 23, 24, 25
12. 6, 7, 8, 10, 11, 12, 13, 14, 15
13. 20, 21, 22, 100, 101, 102, 110, 111
14. 20, 21, 22, 23, 24, 30, 31, 32

Further Study

Although the base ten numeration system is the most common one, it is not the only system used in history. The Sumerians and Babylonians used a sexagesimal system based on the number sixty. It featured a type of place-value notation. This system is reflected in the fact that a circle is divided into 360 degrees, each degree into 60 minutes, and each minute into 60 seconds.

The Romans made limited use of the duodecimal (base twelve) number system. This is the reason for twelve inches in a foot, twelve Roman ounces in a pound, twelve items in a dozen, and twelve dozen in a gross. Interestingly, the word *dozen* comes from the Latin word for "twelve"; and the words *ounce* and *inch* come from the Latin words for "twelfth."

Solutions for Part D

15. $1{,}777 \times 0.3$
16. 67×0.84
17. $(\$15.95 + \$17.95) \times 1.06$
18. $\$139.50 \times 1.06 \div 10$

D. Solve these reading problems.

15. The Siloam tunnel brings water from the Gihon spring into the city of Jerusalem. It was built by King Hezekiah's workers (2 Kings 20:20), who cut through 1,777 feet of solid rock. Express this distance in meters, to the nearest whole number. (1 foot = 0.3 meter)   533 meters

16. The Pool of Siloam, which is fed by the Gihon spring, has an area of about 67 square yards. Express this surface area in square meters, to the nearest whole number. (1 square yard = 0.84 square meter)   56 square meters

17. At Clearton Bookstore, Fred bought a *Strong's Concordance* for $15.95 and a *Nave's Topical Bible* for $17.95. Calculate the total cost of these books, including 6% sales tax. $35.93

18. Clearton Bookstore offers a discount for an order of 10 or more *Strong's Concordances*. The price of 10 copies is $139.50 plus 6% sales tax. To the nearest cent, find the cost of one concordance at this rate, including sales tax. $14.79

REVIEW EXERCISES

E. Write these numbers, using scientific notation. *(Lesson 156)*

19. 7,900,000,000,000 7.9×10^{12} 20. 2,600,000,000,000,000,000 2.6×10^{18}

F. Write the reading of each meter. *(Lesson 155)*

21. 38,940 kwh   22. 16,552 kwh

G. Find the missing numbers for this table. *(Lesson 149)*

Formula: $a = 3b^2$				
b	2	4	6	8
a	23. 12	24. 48	25. 108	26. 192

H. Divide these monomials. *(Lesson 141)*

27. $h^{11} \div h^8$ h^3 28. $q^{10} \div q^6$ q^4

I. Calculate each sales tax or property tax. *(Lesson 125)*

29. Sale: $68.15 $4.77   30. Assessment: $64,200 $1,348.20
    Tax rate: 7%   Tax rate: 21 mills

J. Write these formulas. *(Lesson 93)*

31. Area of a parallelogram $a = bh$ 32. Area of a triangle $a = \frac{1}{2}bh$

K. Find the missing parts. *(Lesson 29)*

33. d = 1,560 mi.; r = 60 m.p.h.; t = __26__ hr.

34. d = 460 km; r = __92__ km/h; t = 5 hr.

158. The Duodecimal (Base Twelve) Numeration System

Most of the numeration systems introduced in Lesson 157 have bases of ten or less. One example given was in the base twelve system, which is also called the **duodecimal system**. This system uses twelve digits. But how can you express the values of 10 and 11 with single digits? One way is to use T for "ten" and E for "eleven."

Counting in base twelve: 1, 2, 3, 4, 5, 6, 7, 8, 9, T, E, 10

There are twelve inches in 1 foot and twelve items in 1 dozen. So you have actually been working with base twelve numeration for quite some time—though you probably have not thought of it in that way. Compound measures of feet and inches, as well as dozens and units, can easily be written as base 12 numbers. Consider the following examples.

$$2 \text{ feet } 4 \text{ inches} = 2 \text{ twelves and } 4 \text{ more} = 24_{twelve}$$
$$6 \text{ feet } 10 \text{ inches} = 6 \text{ twelves and } T \text{ more} = 6T_{twelve}$$
$$3 \text{ dozen and } 7 = 3 \text{ twelves and } 7 \text{ more} = 37_{twelve}$$
$$4 \text{ dozen and } 11 = 4 \text{ twelves and } E \text{ more} = 4E_{twelve}$$

The duodecimal system uses place value in the same way that the decimal system does. However, the value of each place is a power of twelve rather than of ten as in the base ten system. The following example shows that 135_{twelve} equals $(1 \times 12^2) + (3 \times 12^1) + (5 \times 12^0)$.

	twelve twelves' place (12^2)	twelves' place (12^1)	ones' place (12^0)
$135_{twelve} =$	1	3	5

Base twelve numbers have little meaning for us unless we change them to an equivalent value in base ten. Actually, you make this change when you convert a measure such as 7 feet to 84 inches. Seven feet (70_{twelve}) is the same length as 84 inches (84_{ten}).

To express a base twelve numeral in the base ten system, multiply each place value in base twelve by its equivalent value in base ten. The place farthest to the right is multiplied by 1 (12^0), the second place by 12 (12^1), the third place by 144 (12^2), and the fourth place by 1,728 (12^3).

Example A
Change 345_{twelve} to base ten.
$3 \times 12^2 = 432$
$4 \times 12^1 = 48$
$5 \times 12^0 = \underline{5}$
Total value: 485_{ten}

Example B
Change $2,6TE_{twelve}$ to base ten.
Remember that $T = 10_{ten}$ and $E = 11_{ten}$.
$2 \times 12^3 = 3,456$
$6 \times 12^2 = 864$
$10 \times 12^1 = 120$
$11 \times 12^0 = \underline{11}$
Total value: $4,451_{ten}$

LESSON 158

Objectives

- To introduce *the base twelve numeration system, also known as the duodecimal system.

- To teach *how to express base twelve numbers as base ten numbers, and base ten numbers as base twelve numbers. (The same methods, of course, will work for converting to and from numeration systems with any base.)

Review

1. *Count in the following bases.* (Lesson 157)

    a. from 21_{five} to 41_{five} (21, 22, 23, 24, 30, 31, 32, 33, 34, 40, 41)

    b. from 1_{six} to 22_{six} (1, 2, 3, 4, 5, 10, 11, 12, 13, 14, 15, 20, 21, 22)

2. *Write these numbers, using scientific notation.* (Lesson 156)

    a. 220,000,000,000 (2.2×10^{11})

    b. 70,000,000,000,000,000 (7×10^{16})

3. *Calculate the net profit in each exercise.* (Lesson 126)

	Income	Cost of Goods	Overhead	Net Profit
a.	$425	$195	$75	($155)
b.	$526	$319	$125	($82)

4. *Find the area of each trapezoid.* (Lesson 94)

    a. $h = 12$ in.
       $b_1 = 14$ in.
       $b_2 = 18$ in.
       $a = $ (192 sq. in.)

    b. $h = 78$ ft.
       $b_1 = 85$ ft.
       $b_2 = 66$ ft.
       $a = $ (5,889 sq. ft.)

5. *Find these equivalents.* (Lesson 30)

    a. 250 gerahs = ___ oz. (5)

    b. 19 spans = ___ in. (171)

Introduction

Ask the students if they know how many items are in a gross. A gross is 144. For example, there are 144 pencils in a gross. You might even bring a gross of pencils to show the class.

Dozens and grosses are helpful for understanding the base twelve numeration system. Dozens take the place of tens, and grosses take the place of hundreds.

Teaching Guide

1. **The base twelve numeration system has a base of twelve.** It is also known as the duodecimal system, from *duo* (two) and *deci* (ten).

    The base twelve system has twelve digits, including T for "ten" and E for "eleven." (Capital letters are used for the sake of appearance as well as clarity. A numeral such as $3t4e$ could be mistaken for an algebraic expression.)

2. **Base twelve place value is a system of twelves.** The first place on the right is the ones' place. The second place has a value of twelve; the third place has a value of twelve to the second power (144); and so on. Discuss the meaning of 135_{twelve} as shown in the lesson.

3. **The base twelve system is already somewhat familiar because we count various things by twelves.**

Two examples are feet and dozens. Illustrate with the following examples.

a. 3 dozen and 2 = 32_{twelve}
b. 4 dozen and 9 = 49_{twelve}
c. 9 dozen and 11 = $9E_{twelve}$
d. 6 dozen and 10 = $6T_{twelve}$
e. 8 feet 11 inches = $8E_{twelve}$
f. 10 feet 11 inches = TE_{twelve}

4. **To express a base twelve numeral in the base ten system, multiply each place value in base twelve by its equivalent value in base ten.** The first place from the right is multiplied by 1 (12^0), the second place by twelve (12^1), the third by 144 (12^2), the fourth by 1,728 (12^3).

Change the following base twelve numbers to base ten.

a. 279_{twelve}
$2 \times 12^2 = 288$
$7 \times 12^1 = 84$
$9 \times 12^0 = \underline{9}$
Total value: 381_{ten}

b. $4,196_{twelve}$
$4 \times 12^3 = 6,912$
$1 \times 12^2 = 144$
$9 \times 12^1 = 108$
$6 \times 12^0 = \underline{6}$
Total value: $7,170_{ten}$

c. $ET9_{twelve}$
$E (11) \times 12^2 = 1,584$
$T (10) \times 12^1 = 120$
$9 \times 12^0 = \underline{9}$
Total value: $1,713_{ten}$

d. T,TTT_{twelve}
$10 \times 12^3 = 17,280$
$10 \times 12^2 = 1,440$
$10 \times 12^1 = 120$
$10 \times 12^0 = \underline{10}$
Total value: $18,850_{ten}$

5. **To express a base ten number in the base twelve system, follow the steps below.**

(1) Divide the base ten number by twelve. Write the quotient above the number and the remainder to the right. If the remainder is ten, write T. If it is eleven, write E.

(2) Repeat step 1 with the new quotient until you have only a remainder.

(3) Write the remainders in order, starting with the last one. This is the base twelve number.

Change the following base ten numbers to base twelve.

a. 665

$0 R\ 4$
$12\overline{)4} R\ 7$
$12\overline{)55} R\ 5$
$12\overline{)665}$
$665_{ten} = 475_{twelve}$

b. 20,615

$0 R\ E$
$12\overline{)11} R\ E$
$12\overline{)143} R\ 1$
$12\overline{)1,717} R\ E$
$12\overline{)20,615}$
$20,615_{ten} = E,E1E_{twelve}$

To express a base ten number as a base twelve number, use the following steps.

1. Divide the base ten number by twelve. Write the quotient above the number and the remainder to the right. If the remainder is ten, write T. If it is eleven, write E.
2. Repeat step 1 with the new quotient until you have only a remainder.
3. Write the remainders in order, starting with the last one. This is the base twelve number.

Example C
Change 777_{ten} to base twelve.
$$\begin{array}{r} 0 \quad R\;5 \\ 12\overline{)5} \quad R\;4 \\ 12\overline{)64} \quad R\;9 \\ 12\overline{)777} \end{array}$$

$777_{ten} = 549_{twelve}$

Example D
Change $2{,}435_{ten}$ to base twelve.
$$\begin{array}{r} 0 \quad R\;1 \\ 12\overline{)1} \quad R\;4 \\ 12\overline{)16} \quad R\;T \\ 12\overline{)202} \quad R\;E \\ 12\overline{)2{,}435} \end{array}$$

$2{,}435_{ten} = 1{,}4TE_{twelve}$

CLASS PRACTICE

Use the base twelve number system to write these numbers.

a. four dozen and four 44_{twelve} b. eight dozen and eleven $8E_{twelve}$

c. three feet ten inches $3T_{twelve}$ d. six feet 60_{twelve}

Change these base twelve numbers to base ten.

e. 87_{twelve} 103   f. $T9_{twelve}$ 129   g. $68E_{twelve}$ 971

Change these base ten numbers to base twelve.

h. 520 374_{twelve} i. 154 $10T_{twelve}$ j. 1,056 740_{twelve}

WRITTEN EXERCISES

A. *Use the base twelve number system to write these numbers.*

1. one dozen and seven 17_{twelve} 2. three dozen and five 35_{twelve}
3. six feet 60_{twelve} 4. nine dozen 90_{twelve}
5. eight feet ten inches $8T_{twelve}$ 6. five feet eleven inches $5E_{twelve}$

B. *Change these base twelve numbers to base ten.*

7. 49_{twelve} 57   8. 60_{twelve} 72
9. 84_{twelve} 100   10. 494_{twelve} 688

C. *Change these base ten numbers to base twelve.*

11. 67 57_{twelve} 12. 268 $1T4_{twelve}$ 13. 625 441_{twelve}
14. 1,146 $7E6_{twelve}$ 15. 1,727 EEE_{twelve} 16. 430 $2ET_{twelve}$

532   Chapter 12   Other Numeration Systems and Final Reviews

D. Solve these reading problems. Use equations to solve exercises 21 and 22.

17. The *Hindenburg* was a huge German airship that exploded at Lakehurst, New Jersey, in 1937. It was 578$_{twelve}$ feet long and held 7,000,000 cubic feet of hydrogen. Express its length with a base ten numeral.   **812 feet**

18. The *Hindenburg* was propelled by 4 diesel engines that enabled the airship to travel as fast as 71$_{twelve}$ miles per hour. Express that speed with a base ten numeral.   **85 miles per hour**

19. During the Indian removal of 1838, General Winfield Scott and his troops forced many Cherokee Indians from the East to move to new lands west of the Mississippi. The group started with more than 14,000 Indians, but an estimated 4,000 of them died along the way. Use a base twelve numeral to tell how many survived the journey.   **5,954$_{twelve}$ Indians**

20. The Trail of Tears traveled by these Indians was nearly 800 miles long. What is this distance in base 12?   **568$_{twelve}$ miles**

21. The Steiners spent 3 days traveling the 1,380 miles to Grandfather Steiner's. On the first day they traveled 15 miles more than twice as far as they traveled on the second day, and on the third day they traveled 135 miles less than twice as far as they did on the second day. How many miles did they travel on the second day?
    $(2s + 15) + s + (2s - 135) = 1,380$ **$s = 300$ miles**

22. The sum of three consecutive odd numbers is 51. What is the smallest number? (Hint: Each consecutive odd number is 2 more than the previous odd number.)
    $n + (n + 2) + (n + 4) = 51$ **$n = 15$**

REVIEW EXERCISES

E. Write these numbers, using words. *(Lesson 157)*

23. 121$_{three}$ **one, two, one, base three**   24. 1,235$_{six}$ **one, two, three, five, base six**

F. Write these numbers, using scientific notation. *(Lesson 156)*

25. 35,000,000,000   **3.5×10^{10}**   26. 86,000,000,000,000,000   **8.6×10^{16}**

G. Calculate the net profit in each exercise. *(Lesson 126)*

27. Sales: $624; cost of goods: $342; overhead: $98   **$184**
28. Sales: $2,482; cost of goods: $997; overhead: $483   **$1,002**

H. Find the area of each trapezoid. *(Lesson 94)*

29. $h = 42$ ft.   **2,520 sq. ft.**   30. $h = 100$ ft.   **8,800 sq. ft.**
    $b_1 = 61$ ft. $b_1 = 84$ ft.
    $b_2 = 59$ ft. $b_2 = 92$ ft.

I. Find these equivalents. *(Lesson 30)*

31. 560 Roman pounds = **420** lb.   32. 19 cubits = **874** cm

Further Study

Some mathematicians have recommended use of the duodecimal system of numeration rather than the decimal system. Their reasoning is that twelve has more factors than ten (2, 3, 4, and 6 as opposed to only 2 and 5); and for this reason, calculations would be greatly simplified. That may be logical; but the base ten system is so deeply entrenched, and a worldwide changeover would involve so many complications, that the idea is completely impractical at this point in history.

The duodecimal system would produce some interesting results in relation to measures. A square foot would equal 100_{twelve} square inches; a cubic foot would equal $1,000_{twelve}$ cubic inches; and a troy pound would equal 10 troy ounces. But rather than changing the number system to better match a system of measures based on twelve, most of the world has adopted a system of measures based on ten (the metric system) to better match the base ten number system.

The base twelve numeration system can be extended to the right of the decimal point (which should probably be called the "duodecimal point" in base twelve). See the following examples.

$0.1 = 12^{-1} = \frac{1}{12}$
$0.01 = 12^{-2} = \frac{1}{144}$
$0.001 = 12^{-3} = \frac{1}{1,728}$

It is interesting to see what some of the common decimal-fraction equivalents are in base twelve.

$\frac{1}{2}_{twelve} = 0.6_{twelve}$

$\frac{1}{4}_{twelve} = 0.3_{twelve}$

$\frac{1}{3}_{twelve} = 0.4_{twelve}$

$\frac{1}{5}_{twelve} = 0.24_{twelve}$

$\frac{1}{6}_{twelve} = 0.2_{twelve}$

$\frac{1}{8}_{twelve} = 0.16_{twelve}$

LESSON 159

Objectives

- To introduce *the base two numeration system, also known as the binary system.
- To teach *how to express base two numbers as base ten numbers, and base ten numbers as base two numbers.

Review

1. *Count to 20_{twelve} in the base twelve system.* (Lesson 157)

   (1, 2, 3, 4, 5, 6, 7, 8, 9, T, E, 10, 11, 12, 13, 14, 15, 16, 17, 18, 19, 1T, 1E, 20)

2. *Simplify these polynomials.* (Lesson 151)

   a. $2bc + b^2c - 2bc - b^4c + b^2c$
   $(-b^4c + 2b^2c)$

   b. $9m - 5n + n - 6$ $(9m - 4n - 6)$

3. *Calculate the missing percents on this income statement.* (Lesson 127)

   Tucson Lawn & Garden
   Income Statement
   For the Year Ending
   December 31, 20—

	Amount	Percent of Sales
Total sales	$88,000	100%
Cost of goods	46,000	(52%)
Gross profit	42,000	(48%)
Overhead	19,000	(22%)
Net profit	23,000	(26%)

4. *Find the area of each circle.* (Lesson 95)

   a. $r = 25$ in. $a = $ (1,962.5 sq. in.)

   b. $r = 6.5$ m $a = $ (132.665 m²)

   c. $r = 64$ ft. $a = $ (12,861.44 sq. ft.)

5. *Solve these reading problems, which contain extra information.* (Lesson 31)

   a. In 1998, beef farmers received about 24 cents of every dollar that consumers spent for beef. That figure was closer to 50 cents in the 1950s. If a consumer spent $500 for beef in 1998, how much of that money did farmers receive? ($120)

   b. Beef cattle during the summer of 1998 sold for as little as 58 cents per pound (live weight). The price had formerly been as high as 80 cents per pound. What would be the selling price of a 1,350-pound steer at 62 cents per pound? ($837)

159. The Binary (Base Two) Numeration System

The base two numeration system, also known as the **binary system** (bī′ nə rē), uses just two digits: 0 and 1. In this system, every number must be expressed with only those two digits. This means that even fairly small numbers will have many places. The following table shows the numbers from 1 to 10,000 in the binary system, along with their corresponding base ten values.

Decimal	Binary	Decimal	Binary
1	1	9	1,001
2	10	10	1,010
3	11	11	1,011
4	100	12	1,100
5	101	13	1,101
6	110	14	1,110
7	111	15	1,111
8	1,000	16	10,000

Changing binary numerals to decimal numerals is done by the same method given in Lesson 158 for base twelve conversions. To express a base two numeral in base ten, multiply each base two place value by its base ten value. Thus, the first place on the right is multiplied by 1 (2^0), the second place by two (2^1), the third place by 4 (2^2), the fourth place by 8 (2^3), and so on.

Example A
Change $1,111_{two}$ to base ten.
(This numeral is read
   "one, one, one, one, base two.")

$1 \times 2^3 = 8$
$1 \times 2^2 = 4$
$1 \times 2^1 = 2$
$1 \times 2^0 = \underline{1}$
Total value: 15_{ten}

Example B
Change $1,101,011_{two}$ to base ten.

$1 \times 2^6 = 64$
$1 \times 2^5 = 32$
$0 \times 2^4 = 0$
$1 \times 2^3 = 8$
$0 \times 2^2 = 0$
$1 \times 2^1 = 2$
$1 \times 2^0 = \underline{1}$
Total value: 107_{ten}

534 *Chapter 12 Other Numeration Systems and Final Reviews*

To express a base ten number as a base two number, use the following steps.

1. Divide the base ten number by two. Write the quotient above the number and the remainder to the right.
2. Repeat step 1 with the new quotient until you have only a remainder.
3. Write the remainders in order, starting with the last one. This is the base two number.

Example C
Change 38_{ten} to base two.

$$\begin{array}{ll} 0 & R\ 1 \\ 2\overline{)1} & R\ 0 \\ 2\overline{)2} & R\ 0 \\ 2\overline{)4} & R\ 1 \\ 2\overline{)9} & R\ 1 \\ 2\overline{)19} & R\ 0 \\ 2\overline{)38} & \end{array}$$

$38_{ten} = 100{,}110_{two}$

Example D
Change 125_{ten} to base two.

$$\begin{array}{ll} 0 & R\ 1 \\ 2\overline{)1} & R\ 1 \\ 2\overline{)3} & R\ 1 \\ 2\overline{)7} & R\ 1 \\ 2\overline{)15} & R\ 1 \\ 2\overline{)31} & R\ 0 \\ 2\overline{)62} & R\ 1 \\ 2\overline{)125} & \end{array}$$

$125_{ten} = 1{,}111{,}101_{two}$

Computers use binary numeration as their basic language. They have thousands of switches that are either on (representing 1) or off (representing 0); and they use these switches to perform calculations. Numbers are recorded on a magnetic disk or tape by using a magnetized dot for 1 and its absence for 0. When a calculation is complete, the binary numerals go through special circuits that convert them to decimal numerals to be printed out or displayed on a screen.

CLASS PRACTICE

Change these binary (base two) numbers to base ten.

a. 111_{two} 7   b. $10{,}011_{two}$ 19   c. $1{,}101{,}101_{two}$ 109

Write these base ten numbers as binary numbers.

d. 15 $1{,}111_{two}$ e. 86 $1{,}010{,}110_{two}$ f. 105 $1{,}101{,}001_{two}$

WRITTEN EXERCISES

A. *Change these binary (base two) numbers to base ten.*

1. 110_{two} 6   2. $11{,}101_{two}$ 29
3. $11{,}001_{two}$ 25   4. $10{,}101_{two}$ 21
5. $101{,}111_{two}$ 47   6. $110{,}010_{two}$ 50
7. $1{,}110{,}111_{two}$ 119   8. $1{,}101{,}010_{two}$ 106
9. $11{,}101{,}100_{two}$ 236   10. $10{,}111{,}100_{two}$ 188

Lesson 159 T–534

Introduction

Review the base twelve number system.

a. Change $2{,}3E7_{twelve}$ to base ten.
$2 \times 12^3 = 3{,}456$
$3 \times 12^2 = 432$
$E \times 12^1 = 132$
$7 \times 12^0 = \underline{7}$
Total value: $4{,}027_{ten}$

b. Change 971_{ten} to base twelve.

$$\begin{array}{r} 0 \\ 12\overline{)6} \\ 12\overline{)80} \\ 12\overline{)971} \end{array} \begin{array}{l} R\ 6 \\ R\ 8 \\ R\ E \end{array}$$

$971_{ten} = 68E_{twelve}$

Teaching Guide

1. **The base two numeration system, or binary numeration system, uses just two digits: 0 and 1.** Every number in this system must be expressed with only those two digits. As a result, even fairly small numbers will have many places.

2. **Changing a binary number to base ten is done by the same method as that for base twelve conversions.** To express a base two numeral in base ten, multiply each base two place value by its base ten value. Thus, the first place on the right is multiplied by 1 (2^0), the second place by two (2^1), the third place by 4 (2^2), the fourth place by 8 (2^3), and so on.

   a. Change $10{,}111_{two}$ to base ten.
   $1 \times 2^4 = 16$
   $0 \times 2^3 = 0$
   $1 \times 2^2 = 4$
   $1 \times 2^1 = 2$
   $1 \times 2^0 = \underline{1}$
   Total value: 23_{ten}

   b. Change $1{,}011{,}011_{two}$ to base ten.
   $1 \times 2^6 = 64$
   $0 \times 2^5 = 0$
   $1 \times 2^4 = 16$
   $1 \times 2^3 = 8$
   $0 \times 2^2 = 0$
   $1 \times 2^1 = 2$
   $1 \times 2^0 = \underline{1}$
   Total value: 91_{ten}

   c. Change $11{,}101{,}111_{two}$ to base ten.
   (239_{ten})

   d. Change $111{,}011{,}111_{two}$ to base ten.
   (479_{ten})

3. **To express a base ten number as a binary number, follow the steps below.**

   (1) Divide the base ten number by two. Write the quotient above the number and the remainder to the right.

   (2) Repeat step 1 with the new quotient until you have only a remainder.

   (3) Write the remainders in order, starting with the last one. This is the base two number.

   a. Change 14 to base two.
   $$\begin{array}{ll} 0 & R\ 1 \\ 2\overline{)1} & R\ 1 \\ 2\overline{)3} & R\ 1 \\ 2\overline{)7} & R\ 0 \\ 2\overline{)14} & \end{array}$$
   $14_{ten} = 1{,}110_{two}$

   b. Change 25 to base two.
   $$\begin{array}{ll} 0 & R\ 1 \\ 2\overline{)1} & R\ 1 \\ 2\overline{)3} & R\ 0 \\ 2\overline{)6} & R\ 0 \\ 2\overline{)12} & R\ 1 \\ 2\overline{)25} & \end{array}$$
   $25_{ten} = 11{,}001_{two}$

   c. Change 40 to base two.
   $(101{,}000_{two})$

   d. Change 75 to base two.
   $(1{,}001{,}011_{two})$

Further Study

1. Every even number in base ten ends with 0 in binary. Every odd number in base ten ends with 1 in binary.

2. Binary decimals follow the same principle as other decimals. Some of the results are rather interesting when expressed in base ten.

   $0.1 = 2^{-1}$
   $0.01 = 2^{-2}$
   $0.001 = 2^{-3}$
   $0.0001 = 2^{-4}$
   $0.00001 = 2^{-5}$
   $0.000001 = 2^{-6}$

   $0.1_{two} = \frac{1}{2}_{ten} \quad = 1 \times 2^{-1} = 0.5_{ten}$

   $0.11_{two} = \frac{3}{4}_{ten} \quad = 1 \times 2^{-1} = 0.5$
   $\phantom{0.11_{two} = \frac{3}{4}_{ten} \quad } + 1 \times 2^{-2} = \underline{0.25}$
   Total value: 0.75_{ten}

   $0.111_{two} = \frac{7}{8}_{ten} \quad = 1 \times 2^{-1} = 0.5$
   $\phantom{0.111_{two} = \frac{7}{8}_{ten} \quad } + 1 \times 2^{-2} = 0.25$
   $\phantom{0.111_{two} = \frac{7}{8}_{ten} \quad } + 1 \times 2^{-3} = \underline{0.125}$
   Total value: 0.875_{ten}

   $0.1111_{two} = \frac{15}{16}_{ten} = 1 \times 2^{-1} = 0.5$
   $\phantom{0.1111_{two} = \frac{15}{16}_{ten} } + 1 \times 2^{-2} = 0.25$
   $\phantom{0.1111_{two} = \frac{15}{16}_{ten} } + 1 \times 2^{-3} = 0.125$
   $\phantom{0.1111_{two} = \frac{15}{16}_{ten} } + 1 \times 2^{-4} = \underline{0.0625}$
   Total value: 0.9375_{ten}

   $0.11111_{two} = \frac{31}{32}_{ten} = 1 \times 2^{-1} = 0.5$
   $\phantom{0.11111_{two} = \frac{31}{32}_{ten} } + 1 \times 2^{-2} = 0.25$
   $\phantom{0.11111_{two} = \frac{31}{32}_{ten} } + 1 \times 2^{-3} = 0.125$
   $\phantom{0.11111_{two} = \frac{31}{32}_{ten} } + 1 \times 2^{-4} = 0.0625$
   $\phantom{0.11111_{two} = \frac{31}{32}_{ten} } + 1 \times 2^{-5} = \underline{0.03125}$
   Total value: 0.96875_{ten}

B. Write these base ten numbers as binary numbers.

11. 9	1,001₍two₎	12. 11	1,011₍two₎
13. 14	1,110₍two₎	14. 23	10,111₍two₎
15. 33	100,001₍two₎	16. 48	110,000₍two₎
17. 55	110,111₍two₎	18. 65	1,000,001₍two₎
19. 122	1,111,010₍two₎	20. 115	1,110,011₍two₎

C. Solve these reading problems. Be careful, for some of them contain extra information.

21. How would a 13-year-old boy express his age in base 2?
    1,101$_{two}$

22. Express the number of inches in a foot as a base 2 number.
    1,100$_{two}$

23. Goliath's coat of mail weighed 5,000 shekels of silver, or about 10,011,100$_{two}$ pounds (1 Samuel 17:5). Express this weight in base ten.
    156 pounds

24. The head of Goliath's spear weighed 600 shekels of iron (1 Samuel 17:7), or about 10,011$_{two}$ pounds. What was the weight in base ten?
    19 pounds

25. Thomas Edison registered a total of 1,300 patents for his inventions. His "invention factory" at Newark, New Jersey, operated for 6 years before he moved to Menlo Park, New Jersey. During that time, Edison received nearly 200 patents. To the nearest whole number, what was the average number of patents Edison received per year at Newark?
    33 patents   (200 ÷ 6)

26. In 1882, Edison received 141 patents. That amounted to one patent nearly every three days. From 1869 through 1910, he averaged 1 patent every 2 weeks. How many patents did he receive in that 42-year period?
    1,092 patents   (42 × 52 ÷ 2)

incandescent light

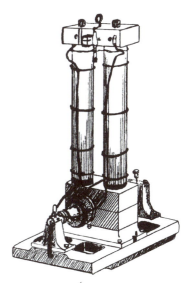
dynamo

phonograph

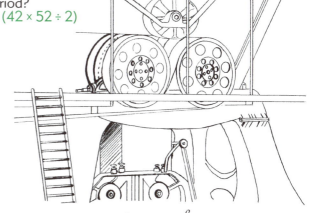
stone crusher

536 Chapter 12 Other Numeration Systems and Final Reviews

REVIEW EXERCISES

D. Express each number in the numeration system indicated. *(Lesson 158)*

27. six dozen and four in base ten   76

28. eight dozen and ten in base twelve   8T$_{twelve}$

E. Simplify these polynomials. *(Lesson 151)*

29. $3rs + r^3s - 2rs - r^3s$ rs

30. $4x - 2v + v - 2$ $-v + 4x - 2$

F. Calculate the missing percents on this income statement. *(Lesson 127)*

Zook Clothing Store
Income Statement
For the Year Ending
December 31, 20—

	Amount	Percent of Sales
Total sales	$83,000	100%
Cost of goods	50,000	31. 60%
Gross profit	33,000	32. 40%
Overhead	19,000	33. 23%
Net profit	14,000	34. 17%

G. Find the area of each circle. *(Lesson 95)*

35. $r = 32$ in.   3,215.36 sq. in.   36. $r = 210$ m   138,600 m²
    $\pi = 3.14$ $\pi = 3\frac{1}{7}$

T-537 Chapter 12 *Other Numeration Systems and Final Reviews*

LESSON 160

Objective

- To give more practice with base twelve and base two numeration.

Review

1. Give Lesson 160 Quiz (Scientific Notation).

2. *Express each number in the decimal system.* (Lessons 158, 159)
    a. $1,001_{two}$ (9)
    b. $1,100,111_{two}$ (103)
    c. $65T_{twelve}$ (934)
    d. $E2T_{twelve}$ (1,618)

3. *Write these numbers, using scientific notation.* (Lesson 156)
    a. 580,000,000,000 (5.8×10^{11})
    b. 25,000,000,000,000,000,000
        (2.5×10^{19})

4. *Compare each set, and write < or > between them.* (Lesson 144)
    a. $+8$ ($\geq$) -5
    b. $+3$ ($\geq$) 0
    c. $|-6|$ ($\geq$) $|+5|$
    d. $|+6|$ ($\leq$) $|-7|$

5. *Write equations to solve these reading problems.* (Lesson 138)
    a. Laura is 1 year older than Marie, and Rose is 2 years older than Marie. The sum of their ages is 45. How old is Marie?
        ($m + [m + 1] + [m + 2] = 45$
        $m = 14$ yr.)
    b. The sum of two consecutive odd numbers is 44. Find the smaller number.
        ($n + [n + 2] = 44; n = 21$)

160. Review of Base Twelve and Base Two Numeration

The last two lessons worked with the duodecimal (base twelve) and the binary (base two) number systems. You practiced converting numbers in those two bases to and from the decimal system.

Here is a review of changing numbers from another base to the decimal system.

Base twelve: Multiply the first place on the right by 1 (12^0), the second place by 12 (12^1), the third place by 144 (12^2), and the fourth place by 1,728 (12^3). See Example A.

Base two: Multiply the first place on the right by 1 (2^0), the second place by 2 (2^1), the third place by 4 (2^2), the fourth place by 8 (2^3), and so on. See Example B.

Converting a decimal number to another base is done by following the steps below. See Examples C and D.

1. Divide the base ten number by the new base. Write the quotient above the number and the remainder to the right.

2. Repeat step 1 with the new quotient until you have only a remainder.

3. Write the remainders in order, starting with the last one. This is the number in the new base. Identify the base with a subscript word.

Example A

Change $3,E7T_{twelve}$ to base ten.

$3 \times 12^3 = 5,184$
$11 \times 12^2 = 1,584$
$7 \times 12^1 = 84$
$10 \times 12^0 = 10$
Total value: $6,862_{ten}$

Example B

Change $1,111,111_{two}$ to base ten.
$1 \times 2^6 = 64$
$1 \times 2^5 = 32$
$1 \times 2^4 = 16$
$1 \times 2^3 = 8$
$1 \times 2^2 = 4$
$1 \times 2^1 = 2$
$1 \times 2^0 = 1$
Total value: 127_{ten}

Example C

Change $7,288_{ten}$ to base twelve.

```
         0    R 4
     12)4     R 2
    12)50     R 7
   12)607     R 4
  12)7,288
```

$7,288_{ten} = 4,274_{twelve}$

Example D

Change 133$_{ten}$ to base two.

```
           0    R 1
         2)1    R 0
         2)2    R 0
         2)4    R 0
         2)8    R 0
        2)16    R 1
        2)33    R 0
        2)66    R 1
       2)133
```

133$_{ten}$ = 10,000,101$_{two}$

CLASS PRACTICE

Change these binary numbers to decimal numbers.

a. 101$_{two}$ 5   b. 11,100$_{two}$ 28

Change these decimal numbers to binary numbers.

c. 22   10,110$_{two}$ d. 55   110,111$_{two}$

Change these duodecimal numbers to decimal numbers.

e. 87$_{twelve}$ 103   f. E3T$_{twelve}$ 1,630

Change these decimal numbers to duodecimal numbers.

g. 42   36$_{twelve}$ h. 1,511   T5E$_{twelve}$

WRITTEN EXERCISES

A. *Change these binary numbers to decimal numbers.*

1. 111$_{two}$ 7   2. 1,101$_{two}$ 13
3. 10,101$_{two}$ 21   4. 110,101$_{two}$ 53
5. 1,000,011$_{two}$ 67   6. 10,111,111$_{two}$ 191

B. *Change these decimal numbers to binary numbers.*

7. 6   110$_{two}$ 8. 11   1,011$_{two}$
9. 24   11,000$_{two}$ 10. 31   11,111$_{two}$

C. *Change these duodecimal numbers to decimal numbers.*

11. 47$_{twelve}$ 55   12. 17T$_{twelve}$ 238
13. 69$_{twelve}$ 81   14. 91$_{twelve}$ 109
15. 2E6$_{twelve}$ 426   16. ETT$_{twelve}$ 1,714

Introduction

The main purpose of today's lesson is to summarize and review the conversion methods taught in the last three lessons. The examples and exercises in the lesson should help to rivet these concepts in students' minds.

Teaching Guide

1. **To change a number from another base to a decimal number, multiply each place value by its base ten value.**

    Change the following duodecimal numbers to base ten.

    a. 275_{twelve} (377)

    b. $48E_{twelve}$ (683)

    Change the following binary numbers to base ten.

    c. $1,111,110_{two}$ (126)

    d. $1,101,111_{two}$ (111)

2. **To change a decimal number to another base, divide by the base to which the number is being changed.**

    Change the following decimal numbers to duodecimal numbers.

    a. 1,727 (EEE_{twelve})

    b. 18,993 ($T,ET9_{twelve}$)

    Change the following decimal numbers to binary numbers.

    c. 85 ($1,010,101_{two}$)

    d. 157 ($10,011,101_{two}$)

Further Study

The methods exercised in this lesson work for conversions between any numeration systems. Below are examples showing these same methods applied to base six and to hexadecimal (base sixteen) numeration. The digits 0, 1, 2, 3, 4, 5, 6, 7, 8, 9, A, B, C, D, E, F are used in the hexadecimal system.

a. Change 453_{six} to base ten.
$4 \times 6^2 = 144$
$5 \times 6^1 = 30$
$3 \times 6^0 = \underline{3}$
Total value: 177

b. Change $48E_{sixteen}$ to base ten.
$4 \times 16^2 = 1{,}024$
$8 \times 16^1 = 128$
$E \times 16^0 = \underline{14}$
Total value: 1,166

c. Change 6,948 to base six.

$0 \quad\quad R\,5$
$6\overline{)5} \quad\quad R\,2$
$6\overline{)32} \quad\quad R\,1$
$6\overline{)193} \quad\quad R\,0$
$6\overline{)1{,}158} \quad R\,0$
$6\overline{)6{,}948}$

$6{,}948_{ten} = 52{,}100_{six}$

d. Change 48,851 to base sixteen.

$0 \quad\quad R\,11 = B$
$16\overline{)11} \quad\quad R\,14 = E$
$16\overline{)190} \quad\quad R\,13 = D$
$16\overline{)3{,}053} \quad R\,3 = 3$
$16\overline{)48{,}851}$

$48{,}851_{ten} = B{,}ED3_{sixteen}$

D. Change these decimal numbers to duodecimal numbers.

17. 79   67$_{twelve}$
18. 128   T8$_{twelve}$
19. 135   E3$_{twelve}$
20. 1,706   ET2$_{twelve}$

E. Solve these reading problems. Write equations for numbers 25 and 26.

21. After Solomon's kingdom was divided, the kingdom of Israel was governed by 19 kings until the Assyrian conquest. Although its kingdom lasted much longer than did Israel's, Judah was governed by only one more ruler than was Israel. Using base two, give the number of rulers in the kingdom of Judah.   10,100$_{two}$ rulers

22. It is commonly accepted that Israel fell to Assyria in 722 B.C., and that Judah fell to Babylon in 587 B.C. Using the binary system, express the interval between these years.   10,000,111$_{two}$ years   (722 – 587 = 135; Change to base two.)

23. The Kurtz family raises 10 acres of Jumbo Nantes carrots on their Alberta farm. These carrots may grow as long as 10,010$_{two}$ inches. Express this length in base ten.   18 inches

24. The Kurtzes clean and bag the carrots and other produce in a shed with a floor area of 2,T88$_{twelve}$ square feet. What is this area in base ten?   5,000 square feet

25. In a certain paragraph, the letter e is used 4 more times than a. The two letters are used a total of 50 times. How many times does a occur in the paragraph? $a + (a + 4) = 50$ $a = 23$ times

26. A history test has four parts. Part B has twice as many points as Part A. Parts C and D together have 14 points. The test has a total of 38 points. How many points are in Part A? $a + 2a + 14 = 38$ $a = 8$ points

REVIEW EXERCISES

F. Change these numbers as indicated. *(Lessons 158, 159)*

27. 1,101$_{two}$ to base ten   13
28. 101$_{ten}$ to base two   1,100,101$_{two}$
29. E2E$_{twelve}$ to base ten   1,619
30. 262$_{ten}$ to base twelve   19T$_{twelve}$

G. Write these numbers, using scientific notation. *(Lesson 156)*

31. 810,000,000,000 8.1×10^{11}
32. 44,000,000,000,000,000 4.4×10^{16}

H. Compare each set of numbers, and write < or > between them. *(Lesson 144)*

33. +6 $>$ −4
34. −8 $<$ −6
35. |−4| $>$ |+2|
36. |+7| $>$ |−6|

161. Final Review of Basic Mathematical Operations and Measures

A. Copy and solve these problems, labeling the parts as indicated. *(Lessons 4, 6, 8, 11)*

1. 671   addend
    + 624   addend
    1,295   sum

2. 7,945   addend
    + 6,410   addend
    14,355   sum

3. 8,634   minuend
    − 6,568   subtrahend
    2,066   difference

4. 34,272   minuend
    − 18,567   subtrahend
    15,705   difference

5. 765   multiplicand
    × 65   multiplier
    49,725   product

6. 748   multiplicand
    × 94   multiplier
    70,312   product

7. divisor 56)6,845 dividend   quotient 122 R 13 remainder

8. divisor 97)8,723 dividend   quotient 89 R 90 remainder

B. Write these English equivalents. *(Lessons 17–20)*

9. 1 fl. oz. = __2__ tbsp.
10. 1 cup = __8__ fl. oz.
11. 1 qt. = __$\frac{1}{8}$__ pk.
12. 1 sq. yd. = __9__ sq. ft.
13. 9 ft. 4 in. = __$9\frac{1}{3}$__ ft.
14. 8 lb. 6 oz. = __$8\frac{3}{8}$__ lb.
15. 3 squares = __300__ sq. ft.
16. 500 sq. ft. = __5__ squares

C. Write the equivalents of these metric and Bible measures. *(Lessons 21–24, 30)*

17. 3,750 l = __3.75__ kl
18. 950 l = __0.95__ kl
19. 2,200 cm² = __0.22__ m²
20. 91 ha = __0.91__ km²
21. 95 shekels = __38__ oz.
22. 32 talents = __1,088__ kg

D. Write these English/metric equivalents. *(Lesson 27)*

23. 1 ft. = __0.3__ m
24. 1 m = __3.28__ ft.
25. 1 km = __0.62__ mi.
26. 1 a. = __0.4__ ha
27. 32 m = ____ ft.   104.96
28. 235 km = __145.7__ mi.
29. 14 ha = __35__ a.
30. 78 a. = __31.2__ ha

LESSON 161

Objective

- To give a year-end review of Chapters 1 and 2. (The mental math in these chapters is reviewed in Lesson 162.)

Teaching Guide

1. Lesson 161 reviews the material taught in Chapters 1, 2 (Lessons 1–31) except for the mental math. For pointers on using review lessons, see *Teaching Guide* for Lesson 98.

2. The table below shows new concepts taught in Chapters 1, 2 and reviewed in this lesson.

3. Reading problem review. (This section in these review lessons names the reading problem skills that are taught in the chapters being reviewed and in the current lesson.)

    Choosing the correct operation (Lesson 14).

    Choosing the necessary information (Lesson 31).

Lesson number and new concept	Exercises in Lesson 161
1—Numbers to the quadrillions' place.	45, 46
4—Commutative and associative laws.	55, 56
10—Distributive law.	57, 58
19—Memorizing that 1 square = 100 square feet.	15, 16
28—Memorizing the Fahrenheit/Celsius conversion formulas.	39, 40

Answers for Exercises 45 and 46

45. Twenty-three quadrillion, nine hundred seventy-five trillion, nine billion

46. Eight hundred seventy-eight quadrillion, six hundred trillion

E. Solve these problems involving English measures. *(Lessons 25, 26)*

31. 7 ft. 6 in.
   + 8 ft. 8 in.
   ‾‾‾‾‾‾‾‾‾‾
     16 ft. 2 in.

32. 7 pk. 1 qt.
   − 6 pk. 3 qt.
   ‾‾‾‾‾‾‾‾‾‾
        6 qt.

33. 3 yd. 1 ft.
   ×   8
   ‾‾‾‾‾‾‾‾‾‾
     26 yd. 2 ft.

34. 8 lb. 14 oz.
   ×   6
   ‾‾‾‾‾‾‾‾‾‾
     53 lb. 4 oz.

35. $7 \overline{)16 \text{ yd. } 1 \text{ ft.}}$ = 2 yd. 1 ft.

36. $15 \overline{)64 \text{ lb. } 11 \text{ oz.}}$ = 4 lb. 5 oz.

F. Solve these problems involving metric measures. *(Lesson 25)*

37. 40 cm + 80 mm = __480__ mm

38. 2.9 kg − 250 g = __2,650__ g

G. Find these temperature equivalents, to the nearest degree. *(Lesson 28)*

39. 23°C = __73__ °F

40. 226°F = __108__ °C

H. Write the formulas. Find the missing parts, using a fraction for any remainder. *(Lesson 29)*

41. Write the formula for finding time when rate and distance are known. $t = \dfrac{d}{r}$

42. Write the formula for finding rate when time and distance are known. $r = \dfrac{d}{t}$

43. d = 975 ft.; r = __$97\frac{1}{2}$__ ft. per sec.; t = 10 sec.

44. d = __660__ m; r = 22 m/sec.; t = 30 sec.

I. Do these exercises with numerals. *(Lessons 1–3)*

45. Write 23,975,009,000,000,000 using words.   (See facing page.)
46. Write 878,600,000,000,000,000 using words.
47. Write the value of the 7 in 207,316,000,000,000.   seven trillion
48. Write the value of the 9 in 200,900,000,111,000,000.   nine hundred trillion
49. Round 3,696,434 to the nearest million.   4,000,000
50. Round 297,777,111 to the nearest ten million.   300,000,000
51. Write $\overline{\text{MC}}$ICLXXII as an Arabic numeral.   1,101,172
52. Write $\overline{\text{MD}}$CMXIV as an Arabic numeral.   1,500,914
53. Write 4,717 as a Roman numeral. $\overline{\text{IV}}$DCCXVII
54. Write 131,000 as a Roman numeral. $\overline{\text{CXXXI}}$

J. Write the correct word or number for each blank. *(Lessons 4, 10)*

55. The equation 6 × (4 × 3) = (6 × 4) × 3 is an example of the ____ law.   associative
56. The equation 7 + 5 = 5 + 7 is an example of the ____ law.   commutative
57. (6 × 9) + (8 × 9) = __14__ × 9
58. (15 × 8) + (__8__ × 8) = 23 × 8

K. Do these exercises related to multiplication and division. *(Lessons 8, 12)*

59. Estimate the product of 345 × 616.   180,000
60. Estimate the product of 278 × 319.   90,000
61. Is 26,217,344 divisible by 4?   yes
62. Is 3,619,539 divisible by 3?   yes

L. Follow the directions for these exercises with measures. *(Lessons 19–21, 24)*

63. Write the abbreviation for *square inch* by memory.   sq. in.
64. Write the abbreviation for *square millimeter* by memory.   mm^2
65. Write the metric prefix that means 0.001.   milli-
66. Write the metric prefix that means 0.01.   centi-
67. 2:00 A.M. in Newfoundland = ____ in Los Angeles, California.   9:30 P.M.
68. 9:00 A.M. in Phoenix, Arizona = ____ in St. Johns, Newfoundland.   12:30 P.M.

M. Write a number sentence for each problem, and find the answer. *(Lesson 14)*

69. The original cost to build the Erie Canal was $7,143,789. In 1895, New York authorized a sum $1,856,211 greater than that for improving the canal. How much money was authorized to improve the canal in 1895? $m = \$7{,}143{,}789 + \$1{,}856{,}211 = \$9{,}000{,}000$

70. People using the Erie Canal traveled about $1\frac{1}{2}$ miles per hour, or 36 miles per day. To the nearest whole day, how long did it take them to travel the entire 363-mile length of the canal? $d = 363 \div 36 = 10$ days

71. When toll charges were abolished in 1882, a total of $121,461,891 in tolls had been collected. That was $20,461,891 more than the amount authorized in 1903 to make the canal part of a larger modern waterway. How much money was authorized in 1903 for the purpose stated? $m = \$121{,}461{,}891 - \$20{,}461{,}891 = \$101{,}000{,}000$

72. The Erie Canal in the state of New York was originally 363 miles long. In 1918, the Erie Canal was joined with three shorter canals to form the New York State Barge Canal System, which is 161 miles longer than the original Erie Canal. How long is the New York Barge Canal System? $c = 363 + 161 = 524$ miles

N. Write the facts you will use to solve each problem. Then find the answer. *(Lesson 31)*

73. One cup of skim milk contains 130 milligrams of sodium, 12 milligrams of sugar, and 8 milligrams of protein. How much sugar is in a quart of skim milk?
    Fact: 12 mg, (Also accept 1 qt. = 4c.); Answer: 48 mg

74. A box of corn cereal will serve 14 people with 1 cup each, which is equal to about 1.1 ounces or 31 grams. Find how many ounces of cereal are in the box, to nearest ounce.
    Facts: 14 people, 1.1 oz.; Answer: 15 oz.

LESSON 162

Objective

- To give a year-end review of mental math and of Chapter 3.

Teaching Guide

1. Lesson 162 reviews the mental math taught in Chapters 1 and 2, along with the material in Chapter 3 (Lessons 34–44). For pointers on using review lessons, see *Teaching Guide* for Lesson 98.
2. The table below shows new concepts taught in Chapter 3 and reviewed in this lesson.
3. Reading problem review.

    Deciding what information is missing (Lesson 44).

Lesson number and new concept	Exercises in Lesson 162
34—Memorizing the prime factors from 2 to 19.	57, 58
34—Using exponents to express prime factors.	59, 60
40—Simplifying complex fractions.	55, 56
42—Mentally dividing whole numbers by fractions.	33–36

162. Final Review of Mental Calculation, Factoring, and Fractions

A. Find the answers without copying the problems. *(Lessons 5, 9, 11)*

1. 6,696 + 496   7,192
2. 7,467 + 979   8,446
3. 4 × 43,374   173,496
4. 7 × 45,483   318,381
5. 8)$323.36 $40.42
6. 4)$643.52 $160.88

B. Solve mentally, following the directions for each set.

Use sets of addends or factors that are multiples of 10. *(Lessons 4, 9)*

7. 5 + 4 + 6 + 7 + 3 + 9 + 2 + 8   44
8. 6 + 5 + 5 + 9 + 5 + 1 + 6 + 5   42
9. 4 × 5 × 8 × 5   800
10. 8 × 2 × 9 × 5   720
11. 25 × 20   500
12. 11 × 60   660

Calculate from left to right. *(Lessons 5, 7, 9, 10)*

13. 567 + 982   1,549
14. 976 + 975   1,951
15. 526 − 349   177
16. 934 − 378   556
17. 3,872 × 1,000   3,872,000
18. 95,457 × 10,000   954,570,000
19. 6 × 267   1,602
20. 4 × 1,343   5,372

Apply the distributive law. *(Lesson 10)*

21. 15 × 14   210
22. 24 × 31   744

Use the double-and-divide method. *(Lesson 10)*

23. 28 × 15   420
24. 35 × 24   840

Use the divide-and-divide method. *(Lesson 12)*

25. 216 ÷ 18   12
26. 288 ÷ 24   12

Use the double-and-double method. *(Lesson 13)*

27. 24,500 ÷ 5,000   4.9
28. 32,000 ÷ 500   64

Use the quadruple-and-quadruple method. *(Lesson 13)*

29. 6,000 ÷ 250   24
30. 10,000 ÷ 250   40

C. Solve these fraction problems mentally. *(Lessons 38, 42)*

31. $\frac{5}{8}$ of 24   15
32. $\frac{5}{6}$ of 36   30
33. 12 ÷ $\frac{1}{3}$ 36
34. 8 ÷ $\frac{1}{7}$ 56
35. 6 ÷ $\frac{6}{7}$ 7
36. 8 ÷ $\frac{4}{5}$ 10

D. Do these calculations with fractions. (Lessons 37–41)

37. $\frac{5}{8}$
   $+\frac{3}{5}$
   $1\frac{9}{40}$

38. $\frac{5}{6}$
   $+\frac{5}{8}$
   $1\frac{11}{24}$

39. $\frac{11}{12}$
   $-\frac{2}{3}$
   $\frac{1}{4}$

40. $\frac{7}{16}$
   $-\frac{5}{12}$
   $\frac{1}{48}$

41. $\frac{5}{8}$ of 26 $16\frac{1}{4}$
42. $\frac{3}{8}$ of 23 $8\frac{5}{8}$
43. $1\frac{1}{4} \times 1\frac{4}{5}$ $2\frac{1}{4}$
44. $1\frac{1}{4} \times 2\frac{1}{2}$ $3\frac{1}{8}$
45. $23 \div \frac{1}{6}$ 138
46. $22 \div \frac{1}{3}$ 66
47. $\frac{5}{9} \div \frac{5}{8}$ $\frac{8}{9}$
48. $\frac{3}{4} \div \frac{2}{3}$ $1\frac{1}{8}$
49. $3\frac{3}{4} \div 1\frac{4}{5}$ $2\frac{1}{12}$
50. $3\frac{1}{8} \div 2\frac{1}{2}$ $1\frac{1}{4}$
51. 8 is $\frac{2}{5}$ of 20
52. 15 is $\frac{3}{7}$ of 35

E. Solve these problems by vertical multiplication. (Lesson 38)

53. 24
   $\times 2\frac{3}{4}$
   66

54. 30
   $\times 3\frac{4}{5}$
   114

F. Use division to simplify these complex fractions. (Lesson 40)

55. $\frac{\frac{5}{6}}{\frac{5}{9}}$ $1\frac{1}{2}$

56. $\frac{\frac{3}{4}}{\frac{9}{10}}$ $\frac{5}{6}$

G. Do these exercises with factoring and fractions. (Lessons 34–36, 39)

57. Is the number 13 prime or composite?   prime
58. Is the number 15 prime or composite?   composite
59. Divide by primes to find the prime factors of 32. Use exponents when possible. $32 = 2^5$
60. Divide by primes to find the prime factors of 54. Use exponents when possible. $54 = 2 \times 3^3$
61. Find the greatest common factor of 65 and 91.   13
62. Find the greatest common factor of 42 and 63.   21
63. Find the lowest common multiple of 15 and 20.   60
64. Find the lowest common multiple of 18 and 20.   180
65. Compare $\frac{3}{8}$ and $\frac{2}{5}$, and write < or > between them. $\frac{3}{8} < \frac{2}{5}$
66. Compare $\frac{2}{7}$ and $\frac{5}{16}$, and write < or > between them. $\frac{2}{7} < \frac{5}{16}$
67. What kind of fraction is $\frac{5}{8}$?   proper
68. What kind of fraction is $\frac{8}{5}$?   improper
69. Expand $\frac{2}{3}$ by multiplying both terms by 8. $\frac{16}{24}$
70. Expand $\frac{5}{6}$ by multiplying both terms by 50. $\frac{250}{300}$

Lesson 162 T-544

71. Reduce $\frac{45}{54}$ to lowest terms. $\frac{5}{6}$

72. Reduce $\frac{35}{56}$ to lowest terms. $\frac{5}{8}$

73. What is the reciprocal of $2\frac{1}{5}$? $\frac{5}{11}$

74. What is the reciprocal of $\frac{7}{16}$? $\frac{16}{7}$

H. Tell what missing information is needed to solve each problem. *(Lesson 44)*

75. In 2 Kings 20:6, God granted an additional 15 years to Hezekiah's life because of Hezekiah's prayer. How old was Hezekiah when he died? age at time of sickness

76. Mark 2:1–4 tells of four men who brought a man sick of the palsy on his bed to Jesus. If the sick man weighed 150 pounds, what was the total weight carried by each bearer? weight of the bed

77. The Panama Canal was built from 1907 to 1914, with more than 43,000 people working on the project at its peak. The French had tried to build a canal in Panama before the United States took up the challenge, but they gave up after 40,000 of their workmen died of malaria and yellow fever. How many more workers did the French lose than did the Americans? number of workers lost by Americans

78. The United States built the Panama Canal at an initial cost of $366,650,000. That is nearly 3 times the cost of building the Suez Canal, which is much longer. The Panama Canal is nearly 40 miles long. How much shorter is the Panama Canal than the Suez Canal? length of the Suez Canal

79. Some fruit growers use huge fans to keep the air moving and thus protect fruit blossoms from frost. One such fan is mounted on top of a 35-foot mast, its 2 propellers spanning 18 feet. How far does the outer tip of a propeller travel in one minute? number of revolutions per minute

80. A large V-8 engine powers this 18-foot fan (number 79). The engine rests on a concrete pad 3 feet thick. One week in February the engine ran an average of 5.5 hours each night. How much fuel did it use in that time? amount of fuel used each hour

163. Final Review of Decimals, Ratios, and Proportions

A. Do these calculations with decimals. *(Lessons 47–49, 51)*

1. 3.8 + 2.7 = 6.5
2. 3.3 + 1.4016 = 4.7016
3. 6.7 − 4.792 = 1.908
4. 6.01 − 4.4979 = 1.5121
5. 3.1725 × 0.06 = 0.19035
6. 1.4972 × 0.035 = 0.052402
7. 0.6)2.76 = 4.6
8. 0.8)1.26 = 1.575
9. 12 + 15.74 + 14.8   42.54
10. 2.104 + 6.8 + 2.53 + 1.6889   13.1229
11. 0.79 − 0.6984   0.0916
12. 0.7 − 0.03991   0.66009
13. 15.6 × 0.53   8.268
14. 15.6 × 2.009   31.3404
15. 32.8 × 3.603   118.1784
16. 1.66 × 2.015   3.3449

B. Solve these problems mentally by moving decimal points or by using fractions. *(Lessons 48, 49, 51)*

17. 12.943 × 100   1,294.3
18. 3.01 × 1,000   3,010
19. 57.24 ÷ 1,000   0.05724
20. 221.4 ÷ 1,000   0.2214
21. 0.6 of 55   33
22. 0.8 of 45   36
23. 0.375 of 32   12
24. 0.625 of 24   15

C. Do these exercises with decimals. *(Lessons 47, 50)*

25. Write 0.65 as a common fraction in lowest terms. $\frac{13}{20}$
26. Write 0.48 as a common fraction in lowest terms. $\frac{12}{25}$
27. Give the place value of the digit farthest to the right in 0.312121.   millionths
28. Give the place value of the digit farthest to the right in 0.21223.   hundred-thousandths
29. Write 4.0076, using words.   Four and seventy-six ten-thousandths
30. Write 19.000101, using words.   Nineteen and one hundred one millionths
31. Compare the following numbers, and place < or > between them.
    0.00302 _<_ 0.02188
32. Compare the following numbers, and place < or > between them.
    2.40888 _<_ 2.408881
33. Round 3.67899 to the nearest thousandth.   3.679
34. Round 4.182999 to the nearest ten-thousandth.   4.1830

LESSON 163

Objectives

- To give a year-end review of Chapter 4.

Teaching Guide

1. Give Lesson 163 Quiz (The Duodecimal System).

2. Lesson 163 reviews the material taught in Chapter 4 (Lessons 47–57). For pointers on using review lessons, see *Teaching Guide* for Lesson 98.

3. The table below shows new concepts taught in Chapter 4 and reviewed in this lesson.

4. Reading problem review.

    Using direct and inverse proportions (Lesson 55).

    Be sure to discuss reading problems that involve direct proportions, direct proportions in which one part must be calculated, and inverse proportions.

Lesson number and new concept	Exercises in Lesson 163
52—Expressing ratios with either the antecedent or the consequent as 1 and the other part as a fraction or a decimal.	43, 44
54—Inverse proportions, and reading problems to be solved by them.	40, 53, 57
57—Using proportions to determine lengths on a blueprint.	47, 48

T-547 Chapter 12 *Other Numeration Systems and Final Reviews*

An Ounce of Prevention

This lesson contains enough reading problems involving proportions that you could make a separate lesson out of the portion on reading problems if your students need the extra practice. If you do that, you might assign exercises 1–48 for the first day's lesson. On the second day, you could review the examples given in Lessons 53–55 and assign all the reading problems.

Proportions for Part F

45. $\dfrac{\text{scale in.}}{\text{actual mi.}} \ \dfrac{1}{16} = \dfrac{1\frac{3}{4}}{n} \ \dfrac{\text{scale in.}}{\text{actual mi.}}$ $\qquad n = 28$ mi.

46. $\dfrac{\text{scale in.}}{\text{actual mi.}} \ \dfrac{1}{16} = \dfrac{5\frac{3}{8}}{n} \ \dfrac{\text{scale in.}}{\text{actual mi.}}$ $\qquad n = 86$ mi.

47. $\dfrac{\text{scale in.}}{\text{actual ft.}} \ \dfrac{1}{4} = \dfrac{n}{24} \ \dfrac{\text{scale in.}}{\text{actual ft.}}$ $\qquad n = 6$ in.

48. $\dfrac{\text{scale in.}}{\text{actual ft.}} \ \dfrac{1}{4} = \dfrac{n}{15} \ \dfrac{\text{scale in.}}{\text{actual ft.}}$ $\qquad n = 3\frac{3}{4}$ in.

Proportions for Part G

49. $\dfrac{\text{cinnamon}}{\text{myrrh}} \ \dfrac{250}{500} = \dfrac{n}{600} \ \dfrac{\text{cinnamon}}{\text{myrrh}}$ $\qquad n = 300$ shekels

50. $\dfrac{\text{calamus}}{\text{total}} \ \dfrac{1}{6} = \dfrac{n}{24} \ \dfrac{\text{calamus}}{\text{total}}$ $\qquad n = 4$ pounds

51. $\dfrac{\text{lunch and recesses}}{\text{total}} \ \dfrac{5}{39} = \dfrac{n}{390} \ \dfrac{\text{lunch and recesses}}{\text{total}}$ $\qquad n = 50$ minutes

52. $\dfrac{\text{cups cream}}{\text{gallons milk}} \ \dfrac{2}{1} = \dfrac{5\frac{1}{2}}{n} \ \dfrac{\text{cups cream}}{\text{gallons milk}}$ $\qquad n = 2\frac{3}{4}$ gallons

35. Express $\frac{5}{16}$ as a decimal.   0.3125

36. Express $\frac{23}{40}$ as a decimal.   0.575

37. Solve $42.6 \div 0.17$, to the nearest ten-thousandth.   250.5882

38. Solve $31.71 \div 7.3$, to the nearest ten-thousandth.   4.3438

D. Write *direct* or *inverse* to tell what type each proportion is. *(Lesson 52)*

39. $\dfrac{\text{bushels processed Mon.}}{\text{quarts canned Mon.}} \ \dfrac{3}{75} = \dfrac{5}{n} \ \dfrac{\text{bushels processed Tue.}}{\text{quarts canned Tue.}}$ direct

40. $\dfrac{\text{size of drive gear}}{\text{size of driven gear}} \ \dfrac{4}{7} = \dfrac{15}{n} \ \dfrac{\text{r.p.m. of driven gear}}{\text{r.p.m. of drive gear}}$ inverse

E. Write a ratio in lowest terms for each statement. *(Lesson 52)*

41. David has memorized 8 of the 12 verses in Isaiah 53.   2:3

42. The van traveled 340 miles on 20 gallons of gasoline.   17:1

43. Larry plowed 12 acres of a 15-acre field. (Write this ratio so that the antecedent is 1 and the consequent is a decimal.)   1 to 1.25

44. The Martins are planting 52 acres of corn on their 80-acre farm. (Write this ratio so that the antecedent is a decimal and the consequent is 1.)   0.65 to 1

F. Use proportions to do these exercises with maps and blueprints. *(Lessons 56, 57)*

Find the distance represented by each length if 1 in. = 16 mi.

45. $1\frac{3}{4}$ in. $n = 28$ mi.   46. $5\frac{3}{8}$ in. $n = 86$ mi.

Find the length needed to represent each distance if 1 in. = 4 ft.

47. 24 ft. $n = 6$ in.   48. 15 ft. $n = 3\frac{3}{4}$ in.

G. Use direct proportions to solve these reading problems. Be careful, for one part of a proportion may need to be calculated. *(Lesson 53)*

49. God gave a special recipe for the holy anointing oil (Exodus 30:23–25). It was to contain 250 shekels of sweet cinnamon for every 500 shekels of pure myrrh. How much cinnamon would be in a mixture containing 600 shekels of myrrh?   300 shekels

50. The holy anointing oil (number 49 above) also contained sweet calamus and cassia. The oil consisted of 2 parts myrrh, 1 part cinnamon, 1 part calamus, and 2 parts cassia. How many pounds of calamus would have been in 24 pounds of the anointing oil?
                                                                                             4 pounds

51. For every 5 minutes that the upper grade students spend at lunch or recess, they spend 34 minutes in study or other school activities. If the school day is $6\frac{1}{2}$ hours long (390 minutes), how much time do they have each day for lunch and recesses?   50 minutes

52. When Jennifer skims cream off the milk from the two family cows, she usually gets 2 cups from every gallon of milk. How many gallons does it take to produce $5\frac{1}{2}$ cups of cream? $2\frac{3}{4}$ gallons

H. Solve these reading problems by using inverse proportions. *(Lesson 54)*

53. Last week Karla cleaned several rooms in 2 hours. This week, with her sisters helping, the same cleaning took only 40 minutes. If they all worked at the same rate, how many girls did the cleaning this week?   3 girls

54. Michael weighs 60 pounds and sits 6 feet from the fulcrum of a seesaw. Matthew balances him if he sits $7\frac{1}{2}$ feet from the fulcrum. How much does Matthew weigh?   48 pounds

I. Solve these reading problems by using proportions. This section includes a mixture of proportion types. *(Lessons 53–55)*

55. Two families traveled the same distance to the same destination, but they came from opposite directions. The Hurst family traveled 4 hours at an average speed of 50 miles per hour. The Bender family traveled 3.2 hours. What was the Bender family's average speed?   62.5 miles per hour

56. If a train covers the distance between two cities in 8 hours at 60 miles per hour, how long will it take to cover the same distance at 80 miles per hour?   6 hours

57. If a 7-bottom plow can turn over 30 acres of soil in 5 hours, how long would it take to do the same acreage with a 4-bottom plow driving the same speed? $8\frac{3}{4}$ hours

58. The ratio of hogs to steers on the Groff farm is 7 to 2. If the total number of hogs and steers is 378, how many steers are there?   84 steers

59. In Exodus 34:21, God commanded man to rest 1 day out of every 7. If all those days of rest have been kept and if 6,000 years have passed since Creation, how many years of rest is that? (Answer to the nearest whole number.)   857 years

60. In response to Jethro's advice, Moses established a court system that divided his responsibility among many able men (Exodus 18). According to verse 25, Moses chose "rulers of thousands, rulers of hundreds, rulers of fifties, and rulers of tens." If the Israelites numbered 2,000,000 at this time, how many men served as rulers of fifties?   40,000 men

Proportions for Part H

53. $\dfrac{\text{workers (1)}}{\text{workers (2)}} \quad \dfrac{1}{n} = \dfrac{40}{120} \quad \dfrac{\text{minutes (2)}}{\text{minutes (1)}}$ $n = 3$ girls

54. $\dfrac{\text{pounds on side 1}}{\text{pounds on side 2}} \quad \dfrac{60}{n} = \dfrac{7\frac{1}{2}}{6} \quad \dfrac{\text{feet from fulcrum (2)}}{\text{feet from fulcrum (1)}}$ $n = 48$ pounds

Proportions for Part I

55. $\dfrac{\text{m.p.h. (1)}}{\text{m.p.h. (2)}} \quad \dfrac{50}{n} = \dfrac{3.2}{4} \quad \dfrac{\text{hours (2)}}{\text{hours (1)}}$ $n = 62.5$ miles per hour

56. $\dfrac{\text{m.p.h. (1)}}{\text{m.p.h. (2)}} \quad \dfrac{60}{80} = \dfrac{n}{8} \quad \dfrac{\text{hours (2)}}{\text{hours (1)}}$ $n = 6$ hours

57. $\dfrac{\text{bottoms (1)}}{\text{bottoms (2)}} \quad \dfrac{7}{4} = \dfrac{n}{5} \quad \dfrac{\text{hours (2)}}{\text{hours (1)}}$ $n = 8\frac{3}{4}$ hours

58. $\dfrac{\text{steers}}{\text{total}} \quad \dfrac{2}{9} = \dfrac{n}{378} \quad \dfrac{\text{steers}}{\text{total}}$ $n = 84$ steers

59. $\dfrac{\text{rest}}{\text{total}} \quad \dfrac{1}{7} = \dfrac{n}{6{,}000} \quad \dfrac{\text{rest}}{\text{total}}$ $n = 857$ years

60. $\dfrac{\text{rulers}}{\text{Israelites}} \quad \dfrac{1}{50} = \dfrac{n}{2{,}000{,}000} \quad \dfrac{\text{rulers}}{\text{Israelites}}$ $n = 40{,}000$ men

Chapter 12 Other Numeration Systems and Final Reviews

LESSON 164

Objective

- To give a year-end review of Chapter 5 and the first part of Chapter 6.

Teaching Guide

1. Lesson 164 reviews the material taught in Lessons 60–74. For pointers on using review lessons, see *Teaching Guide* for Lesson 98.
2. The table below shows new concepts taught in Lessons 60–74 and reviewed in this lesson.
3. Reading problem review.

    Using sketches (Lesson 70).

Lesson number and new concept	Exercises in Lesson 164
60—Changing percents to ratios.	9–12
66—Finding the rate of increase when the rate is greater than 100%.	40
67—Finding the base when the rate is a fractional part of a percent.	43, 44
67—Finding the base when the rate is more than 100%.	45, 46
73—Using *arithmetic mean* instead of *average*.	53–60
74—Finding the median.	61, 62
74—Finding the mode.	63, 64

164. Final Review of Percents and Statistics

A. *Express these numbers as indicated.* *(Lesson 60)*

As decimals

1. 62%   0.62
2. 34%   0.34
3. 535%   5.35
4. 2,174% 21.74
5. $\frac{1}{2}$%   0.005
6. $\frac{1}{5}$%   0.002

As fractions in lowest terms

7. 55% $\frac{11}{20}$
8. 61% $\frac{61}{100}$

As ratios in which the consequent is 100

9. 23%   23:100
10. 36%   36:100

As ratios reduced to lowest terms

11. 42%   21:50
12. 48%   12:25

As percents

13. $\frac{14}{20}$ 70%
14. 0.6   60%
15. $\frac{4}{5}$ (Be sure you still know the fraction–percent equivalents by memory.)   80%
16. $\frac{3}{8}$ (Be sure you still know the fraction–percent equivalents by memory.) $37\frac{1}{2}$%

B. *Identify the given numbers as* base, rate, *or* percentage. *(Lesson 61)*

17. 40% of 19 = 7.6   7.6 is the ___.   percentage
18. 125% of 36 = 45   36 is the ___.   base
19. 22% of 56 = 12.32   22% is the ___.   rate
20. 105% of 62 = 65.1   65.1 is the ___.   percentage

C. *Find these percentages.* *(Lessons 61, 62)*

21. 18 × 42%   7.56
22. 46 × 8%   3.68
23. 125% of 90   112.5
24. 135% of 68   91.8
25. $\frac{1}{2}$% of 250   1.25
26. $\frac{1}{4}$% of 450   1.125
27. $2\frac{3}{8}$% of 410   9.7375
28. $4\frac{1}{4}$% of 320   13.6

D. *Find the amount of increase or decrease, to the nearest cent. Also find the new price.* *(Lesson 63)*

29. $4.49 increased by 28% $1.26; $5.75
30. $28.37 increased by 34% $9.65; $38.02
31. $24.95 decreased by 15% $3.74; $21.21
32. $14.95 decreased by 41% $6.13; $8.82

550 Chapter 12 *Other Numeration Systems and Final Reviews*

E. Find the new amount after each increase or decrease, to the nearest cent. *(Lesson 64)*

33. $32.95 increased by 35% $44.48
34. $17.25 decreased by 26% $12.77
35. $155.95 decreased by 6% $146.59
36. $195.15 increased by 8% $210.76

F. Find the missing rates, to the nearest whole percent. *(Lesson 65)*

37. 8 is 62% of 13
38. 6 is 35% of 17

G. Find the rate of each change, to the nearest whole percent. Include the label *increase* **or** *decrease*. *(Lesson 66)*

39. year 1 enrollment, 26; year 2 enrollment, 15   42% decrease
40. birth weight, 8 pounds; weight after 1 year, 21 pounds   163% increase

H. Find the base in each problem. *(Lesson 67)*

41. 18 is 30% of 60
42. 44 is 40% of 110
43. 6 is $2\frac{1}{2}$% of 240
44. 14 is $8\frac{3}{4}$% of 160
45. 75 is 150% of 50
46. 96 is 320% of 30

I. Find the missing parts. *(Lesson 68)*

47. sales: $1,316; rate: 9%
    commission: $118.44
48. sales: $1,475; rate: 7%
    commission: $103.25

J. Solve these percent problems mentally. *(Lesson 69)*

49. 8 is $16\frac{2}{3}$% of 48
50. 15 is 60% of 25
51. $\frac{1}{2}$% of 1,800 = 9
52. $\frac{3}{4}$% of 1,600 = 12

K. Find the arithmetic mean of each set as directed. *(Lesson 73)*

Express any remainder as a fraction.

53. 36, 38, 32, 35, 34   35
54. 80, 84, 81, 73, 82   80
55. 138, 151, 139, 123 $137\frac{3}{4}$
56. 464, 462, 426, 489 $460\frac{1}{4}$

Round to the nearest tenth.

57. 47, 48, 52, 51, 57, 56   51.8
58. 67, 69, 58, 71, 65, 63   65.5

Round to the nearest whole percent.

59. 97%, 98%, 94%, 98%, 93%   96%
60. 93%, 84%, 98%, 87%, 90%   90%

L. Find the median of each set. *(Lesson 74)*

61. 215, 199, 217, 216, 399, 115, 269, 252, 214, 235 $216\frac{1}{2}$
62. 315, 326, 310, 416, 305, 499, 375, 368, 389, 325, 378, 391 $371\frac{1}{2}$

Lesson 164 T-550

Sketches for Part N

65.

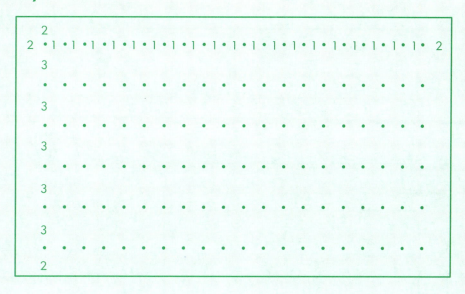

66.

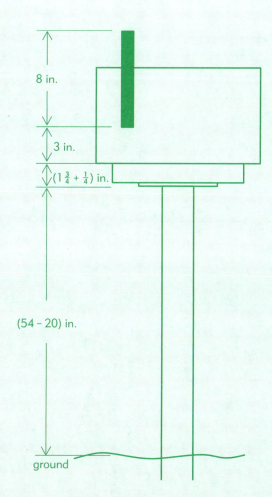

67.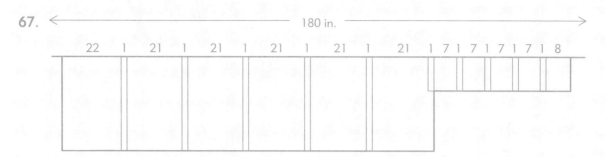

68.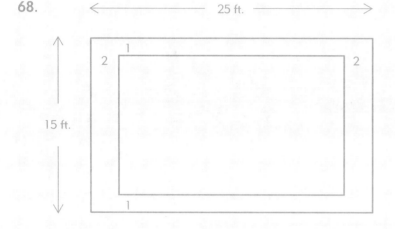

69.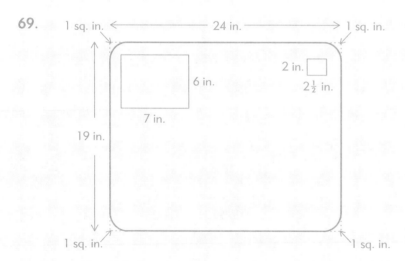

70.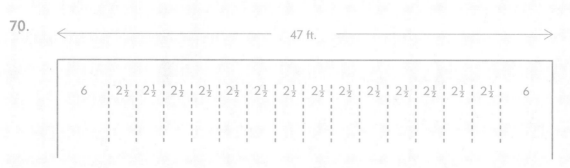

M. Find the mode(s) of each set. *(Lesson 74)*

63. 277, 276, 275, 279, 277, 272, 273, 275, 274, 277, 269, 288   277
64. 122, 126, 127, 122, 123, 124, 122, 125, 126, 127, 126, 123   122, 126

N. Draw a sketch for each problem, and use it to find the solution. *(Lesson 70)*
(See page T–550 for sketches.)

65. The Warrens planted a strawberry patch with 3 feet between the rows and a 2-foot border all around. There were 6 rows of 20 plants each, and the plants were 1 foot apart. What were the dimensions of the strawberry patch?   19 feet by 23 feet

66. Brother Gerald put up a new mailbox at the end of his driveway. He used a metal post $4\frac{1}{2}$ feet long with a $\frac{1}{4}$-inch metal plate welded to the top. He bolted a board $1\frac{3}{4}$ inches thick to the plate and then screwed the mailbox (8 inches high) to the board. An 8-inch-long flag was attached to the side of the mailbox 3 inches up from its base. When the flag is up, how far above the ground does it reach if the post extends 20 inches into the ground?   47 inches

67. Heidi has set up a 15-foot temporary clothesline for drying extra towels and washcloths. She hangs up 6 towels each 23 inches wide and 5 washcloths each 9 inches square. If adjoining pieces overlap by 1 inch, how much of the line remains empty?   7 inches

68. On a concrete pad 25 feet by 15 feet, Brian dumped a load of mulch that covered the pad except for 1 foot along each 25-foot side and 2 feet along each 15-foot end. How many square feet did the mulch cover?   273 square feet

69. On Sharon's desktop is a schedule measuring 6 inches by 7 inches and a name card 2 inches by $2\frac{1}{2}$ inches. The lid itself is 24 inches by 19 inches, but since the corners are rounded, it has 1 square inch less per corner than if it were a perfect rectangle. How much surface area is not covered by the schedule and the name card?   405 square inches

70. The Shirks' garden is 47 feet wide. If there is a 6-foot border on each side and the rows are $2\frac{1}{2}$ feet apart, how many rows are in the garden?   15 rows

552 Chapter 12 Other Numeration Systems and Final Reviews

165. Final Review of Plane Geometry and of Graphs

A. Write the formulas for these facts. Be sure you know them all by memory. *(Lessons 90–95)*

1. Perimeter of a square — $p = 4s$
2. Perimeter of a rectangle — $p = 2(l + w)$
3. Circumference of a circle when the radius is known — $c = 2\pi r$
4. Circumference of a circle when the diameter is known — $c = \pi d$
5. Area of a square — $a = s^2$
6. Area of a rectangle — $a = lw$
7. Area of a parallelogram — $a = bh$
8. Area of a triangle — $a = \frac{1}{2}bh$
9. Area of a trapezoid — $a = \frac{1}{2}h(b_1 + b_2)$
10. Area of a circle — $a = \pi r^2$

B. Do these exercises with geometric symbols. *(Lesson 85)*

11. Use symbols to write "line segment AB." $\overline{AB}$
12. Use symbols to write "line CD is perpendicular to line EF." $\overleftrightarrow{CD} \perp \overleftrightarrow{EF}$
13. Draw the figure indicated by this expression: $\overleftrightarrow{GH} \parallel \overleftrightarrow{JK}$.
14. Draw the figure indicated by this expression: $\angle KLM$.

(See facing page.)

C. Give the correct term for each blank or description. *(Lessons 85–87, 89)*

15. Name a polygon with five straight sides.   pentagon
16. Name a polygon with nine straight sides.   nonagon
17. This is a(n) ____ angle.   reflex

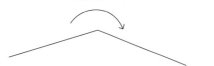

18. This is a(n) ____ angle.   acute

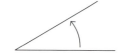

Lesson 165   T–552

LESSON 165

Objective

- To give a year-end review of the latter part of Chapter 6 and of Chapter 7.

Teaching Guide

1. Lesson 165 reviews the material taught in Lessons 75–97. For pointers on using review lessons, see *Teaching Guide* for Lesson 98.

2. The table below shows new concepts taught in Lessons 75–97 and reviewed in this lesson.

3. Reading problem review.

   Solving multistep problems (Lesson 97).

Lesson number and new concept	Exercises in Lesson 165
78—Making double-line graphs.	65–76
85—Identifying various polygons.	15, 16
86—Identifying reflex angles.	17
88—Constructing triangles with different sets of information.	33, 34

Sample Figures for Exercises 13 and 14

13.

14.

```
            K
           /|
          / |
         /  |
        /   |
       L----M
```

(Accept any size angle.)

T-553   Chapter 12   *Other Numeration Systems and Final Reviews*

Model Constructions for Part E

31.

32.

33.
30°   80°
2 in.

34.
2 in.
125°
2 in.

35.

36.

19. A triangle with angles of 46°, 51°, and 83° is called (obtuse, acute, right).　　acute

20. A triangle with angles of 53°, 17°, and 110° is called (obtuse, acute, right).　　obtuse

21. A triangle with two sides of 6 inches and one side of 4 inches is called　　isosceles
(scalene, equilateral, isosceles).

22. A triangle with sides of 5 inches, 6 inches, and 7 inches is called　　scalene
(scalene, equilateral, isosceles).

23. A line from the outer edge to the center of the circle is the ___.　　radius

24. One-half the distance around a circle is a ___.　　semicircle

25. Triangles with the same shape and size are (similar, congruent).　　congruent

26. Triangles with the same shape but different sizes are (similar, congruent).　　similar

D. Supply the correct number of degrees. *(Lessons 86, 87)*

27. The complementary angle to a 63° angle has ___°.　　27°

28. The supplementary angle to a 75° angle has ___°.　　105°

29. If two angles of a triangle have 18° and 38°, the third angle has ___°.　　124°

30. If two angles of a triangle have 86° and 91°, the third angle has ___°.　　3°

E. Draw these geometric constructions. *(Lessons 86, 88)*
(See facing page for model constructions.)

31. Draw a 45° angle.

32. Draw a 120° angle.

33. Construct a triangle with these dimensions. Label the given dimensions. 30°, 2 in., 80°

34. Construct a triangle with these dimensions. Label the given dimensions. 2 in., 125°, 2 in.

35. Use a compass to draw a circle with a 2-inch radius.

36. Use a compass to draw a circle with a 3-inch diameter.

F. Find the perimeter or circumference of each figure. *(Lessons 90, 91)*

37. A polygon with sides 13 in., 7 in., 8 in., 6 in., 6 in., 9 in., and 8 in.　　57 in.

38. A square with sides of 48 cm　　192 cm

39. A rectangle 21 inches long and 15 inches wide　　72 in.

40. An equilateral triangle with 9-inch sides　　27 in.

41. A circle with a radius of 6 centimeters　　37.68 cm

42. A circle with a diameter of 75 centimeters　　235.5 cm

554 Chapter 12 *Other Numeration Systems and Final Reviews*

G. Measure the angles in these triangles. Check your work by making sure the sum of the three angles is 180°. *(Lesson 87)*

43.

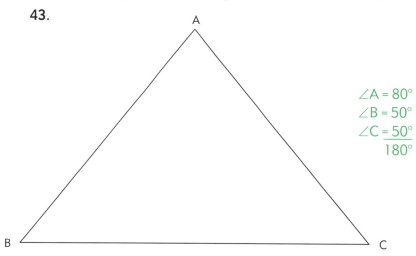

∠A = 80°
∠B = 50°
∠C = 50°
―――
180°

44.

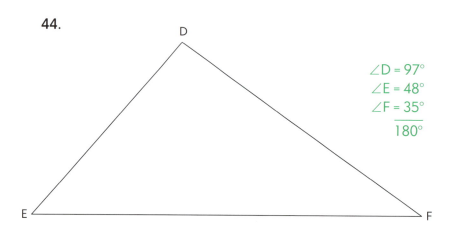

∠D = 97°
∠E = 48°
∠F = 35°
―――
180°

H. Find the areas of the following figures. *(Lessons 92–95)*

Rectangles

45. l = 7.5 m   43.125 m²   **46.** $l = 5\frac{1}{2}$ ft. $26\frac{1}{8}$ sq. ft.
w = 5.75 m $w = 4\frac{3}{4}$ ft.

Squares

47. s = 12.5 m   156.25 m²   **48.** $s = 4\frac{3}{4}$ ft. $22\frac{9}{16}$ sq. ft.

Parallelograms

49. b = 6.4 m   30.4 m²   **50.** $b = 5\frac{1}{4}$ in. $14\frac{7}{16}$ sq. in.
h = 4.75 m $h = 2\frac{3}{4}$ in.

Triangles

51. b = 12 ft.   60 sq. ft.   **52.** b = 4.7 cm   8.93 cm²
h = 10 ft. h = 3.8 cm

T–555 Chapter 12 *Other Numeration Systems and Final Reviews*

Answers for Exercise 59

59. Florida, Georgia, Louisiana, Alabama, South Carolina (in that order)

Graphs for Part J

61–66.   67–72.

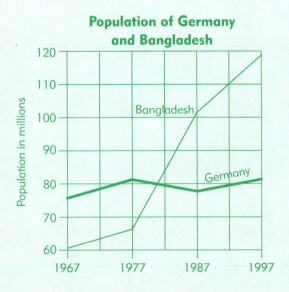

Trapezoids

53. $h = 6$ in.   27 sq. in.   54. $h = 8$ cm   44 cm²
    $b_1 = 5$ in. $b_1 = 5$ cm
    $b_2 = 4$ in. $b_2 = 6$ cm

Circles

55. $r = 32$ in. ($\pi = 3\frac{1}{7}$) $3{,}218\frac{2}{7}$ sq. in.   56. $r = 44$ in. ($\pi = 3.14$) $6{,}079.04$ sq. in.

I. Use the given data to do the exercises on preparing bar graphs. You do not need to draw the graphs. (Lesson 77)

Annual Sales of Strahm's Produce

57. Choose the most reasonable interval for the vertical scale of the graph. **$10,000**
    $500; $1,000; $10,000; $100,000

58. List the numbers that you would place along the horizontal scale. **1, 2, 3, 4, 5**

Year	Sales
1	$125,012
2	115,977
3	131,870
4	145,212
5	150,111

Population of Five Southern States in 1995

59. List the states in the correct order to rank the data from largest to smallest. **(See facing page.)**

60. Choose the most reasonable interval for the vertical scale of the graph.
    50,000; 100,000; 1,000,000   **1,000,000**

State	Estimated Population
Alabama	4,040,389
Florida	12,938,071
Georgia	6,478,149
Louisiana	4,220,164
South Carolina	3,486,310

Source: U.S. Census Bureau, 1990

J. Construct double-line graphs from the following sets of data. Choose the most logical scale from the suggested intervals. (Lesson 78) (See facing page for graphs.)

61–66. Suggested intervals for vertical scale: **(100,000)**; 200,000; 500,000

Population of Pittsburgh, Pennsylvania, and Phoenix, Arizona

Cities	1950	1960	1970	1980	1990
Pittsburgh	676,806	604,332	520,089	423,959	369,879
Phoenix	106,818	439,170	584,303	789,704	983,403

Source: U.S. Census Bureau, 1990

67–72. Suggested intervals for vertical scale: **(10,000,000)**; 50,000,000; 100,000,000

Population of Germany and Bangladesh

Country	1967	1977	1987	1997
Germany	75,400,000	80,566,000	77,536,000	81,640,000
Bangladesh	60,500,000	66,310,000	101,992,000	118,157,000

Source: *World Book Encyclopedia*

K. Tell how many symbols should be used for these states in a picture graph, to the nearest half symbol. You do not need to draw the graph. *(Lesson 76)*

States With Greatest Harvested Acreage in 1996

73. Iowa   12
74. Illinois $11\frac{1}{2}$
75. Kansas $10\frac{1}{2}$
76. Minnesota   10

Iowa .... 24,057,000 acres
Illinois .... 23,183,000 acres
North Dakota .... 22,237,000 acres
Kansas .... 20,899,000 acres
Minnesota .... 19,587,000 acres

Source: *1998 World Almanac*
Scale: 1 symbol = 2,000,000 acres

L. Solve these multistep reading problems. *(Lesson 97)*

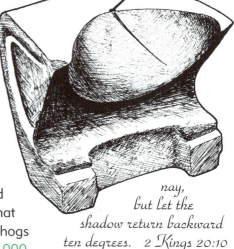

It is a light thing for the shadow to go down ten degrees: nay, but let the shadow return backward ten degrees. 2 Kings 20:10

77. In Isaiah 38:8, God confirmed His promise of longer life to Hezekiah by moving the shadow on the sundial back 10 degrees. Scholars think the degrees represented hours, half hours, or quarter hours. If they were quarter hours, how many hours and minutes did the sun retreat? $2\frac{1}{2}$ hours

78. In Mark 5 we read that Jesus cast many evil spirits out of the Gadarene demoniac. The spirits entered a herd of about 2,000 swine, which rushed down a steep place and drowned in the sea. Suppose that half of these hogs averaged 100 pounds each and the rest averaged twice that much. What was the loss to the owners if the hogs were worth about $0.25 per pound? $75,000

79. One year the price for a gallon of gasoline rose from $1.14 to $1.46 in about six months. How much more did it cost to buy 15 gallons of fuel at the higher price than at the lower price? $4.80

80. When a mason laid blocks for an addition to the barn, Marlin calculated that he laid an average of $1\frac{1}{2}$ blocks per minute. There will be 12 rows of blocks with $21\frac{1}{2}$ blocks in each row except for a window opening that is 3 blocks wide and 4 blocks high. To the nearest quarter hour, how long should it take for the mason to lay all the blocks? $2\frac{3}{4}$ hours

81. Kathleen ordered 3 pounds of Incredible sweet corn seed at $4.95 per pound, 2 packets of cantaloupe seeds at $6.25 per packet, and a 50-pound bag of seed potatoes for $9.00. Calculate her bill, including a 6% sales tax. $38.53

82. Most lightning strokes consist of a downward leader stroke and a return stroke. The downward stroke varies in speed from 100 to 1,000 miles per second. The return stroke travels at 87,000 miles per second. If the moon is 239,000 miles from the earth, how long would it take for an object traveling 87,000 miles per second to travel to the moon and back? (Answer to the nearest tenth of a second.)   5.5 seconds

Lesson 165   T–556

Solutions for Part L
77. $10 \times 15 \div 60$
78. $[(\frac{1}{2} \times 2{,}000 \times 100) + (\frac{1}{2} \times 2{,}000 \times 200)] \times \$0.25$
79. $15 \times (\$1.46 - \$1.14)$
80. $(12 \times 21\frac{1}{2} - 3 \times 4) \div 1\frac{1}{2} \div 60$
81. $(3 \times \$4.95 + 2 \times \$6.25 + \$9.00) \times 1.06$
82. $2 \times 239{,}000 \div 87{,}000$

T–557 Chapter 12 *Other Numeration Systems and Final Reviews*

LESSON 166

Objective

- To give a year-end review of Chapter 8 and of the graphs in Lessons 75–80.

Teaching Guide

1. Lesson 166 reviews the material taught in Lessons 100–114 and the graphs in Lessons 75, 79, and 80. For pointers on using review lessons, see *Teaching Guide* for Lesson 98.
2. The table below shows new concepts taught in Chapter 8 and reviewed in this lesson.
3. Reading problem review.

   Using parallel problems (Lesson 114).

Lesson number and new concept	Exercises in Lesson 166
103—Finding the surface area of a pyramid.	4, 17, 18
104—Finding the surface area of a sphere.	5, 19, 20
107—Finding the volume of a square pyramid.	9, 29, 30
108—Finding the volume of a sphere.	10, 31, 32
112—Using the Pythagorean rule to find the hypotenuse.	39
113—Using the Pythagorean rule to find the length of one leg of a right triangle.	40

166. Final Review of Solid Geometry and the Pythagorean Rule

A. Write these formulas. Be sure you know them by memory. *(Lessons 100–108)*

1. Surface area of a cube — $a_s = 6e^2$
2. Surface area of a rectangular solid — $a_s = 2lw + 2wh + 2lh$
3. Surface area of a cylinder — $a_s = 2\pi r^2 + 2\pi rh$
4. Surface area of a square pyramid — $a_s = 4(\frac{1}{2}b\ell) + b^2$
5. Surface area of a sphere — $a_s = 4\pi r^2$
6. Volume of a rectangular solid — $v = lwh$
7. Volume of a cylinder — $v = \pi r^2 h$
8. Volume of a cone — $v = \frac{1}{3}\pi r^2 h$
9. Volume of a square pyramid — $v = \frac{1}{3}lwh$
10. Volume of a sphere — $v = \frac{4}{3}\pi r^3$

B. Find the surface areas of the figures described. *(Lessons 100–104)*

Cubes
11. edge = 5 in.   150 sq. in.
12. edge = 12 in.   864 sq. in.

Rectangular Solids
13. 6 in. by 5 in. by 4 in.   148 sq. in.
14. 8 in. by 7 in. by 5 in.   262 sq. in.

Cylinders
15. radius = 3 in.; height = 4 in.   131.88 sq. in.
16. radius = 4 in.; height = 3 in.   175.84 sq. in.

Square pyramids
17. base = 4 in.; slant height = 4 in.   48 sq. in.
18. base = 6 in.; slant height = 5 in.   96 sq. in.

Spheres
19. radius = 11 in.   1,519.76 sq. in.
20. radius = 17 cm   3,629.84 cm²

558   Chapter 12   *Other Numeration Systems and Final Reviews*

C. Find the volumes of the figures described. Use 3.14 for pi unless otherwise indicated. (*Lessons 105–108*)

Cubes

21. edge = 9 cm   729 cm³
22. edge = 12 in.   1,728 cu. in.

Rectangular Solids

23. 9 cm by 8 cm by 7 cm   504 cm³
24. 15 ft. by 6 ft. by 9 ft.   810 cu. ft.

Cylinders

25. radius = 6 in.; height = 9 in.   1,017.36 cu. in.
26. radius = 12 in.; height = 7 in.; pi = $3\frac{1}{7}$ 3,168 cu. in.

Cones

27. radius = 11 cm; height = 9 cm   1,139.82 cm³
28. radius = 18 m; height = 20 m   6,782.4 m³

Square pyramids

29. side = 8 in.; height = 9 in.   192 cu. in.
30. side = 9 cm; height = 15 cm   405 cm³

Spheres

31. radius = 85 cm (Use 4.18.)   2,567,042.5 cm³
32. radius = 67 cm (Use 4.18.)   1,257,189.34 cm³

D. Do the following exercises with square roots. (*Lessons 109–113*)

33. Use the chart on page 584 to find the square root of 197.   14.036

34. Use the chart on page 584 to find the square root of 179.   13.379

35. The square root of 4,489 is between   60 and 70
    (40 and 50, 50 and 60, 60 and 70, 70 and 80, 80 and 90, 90 and 100).

36. The square root of 7,921 is between   80 and 90
    (40 and 50, 50 and 60, 60 and 70, 70 and 80, 80 and 90, 90 and 100).

37. Extract the square root of 4,489, a perfect square.   67

38. Extract the square root of 7,921, a perfect square.   89

39. Leg *a* of a right triangle is 4 inches, and leg *b* is 5 inches. Use the chart on page 584 to find the hypotenuse, to the nearest tenth.   6.4 inches

40. Leg *a* of a right triangle is 4 centimeters, and the hypotenuse is 10 centimeters. Use the chart on page 584 to find the length of leg *b*, to the nearest tenth.   9.2 centimeters

Lesson 166 T–558

Extractions for Exercises 37 and 38

37.
$$\begin{array}{r}6\ \ 7\\\sqrt{44\!\wedge\!89}\\36\end{array}$$
(20 × 6) 120   8 89
    7 × 127 = 8 89
                     0

38.
$$\begin{array}{r}8\ \ 9\\\sqrt{79\!\wedge\!21}\\64\end{array}$$
(20 × 8) 160   15 21
    9 × 169 = 15 21
                       0

T–559 *Chapter 12 Other Numeration Systems and Final Reviews*

E. **Complete the table for preparation of a circle graph. You do not need to draw the graph.** *(Lesson 79)*

Major Divisions of Time in the Old Testament

Division	Years	Fraction	Decimal	Degrees
Creation to the Flood	1,656	$\frac{1,656}{4,046}$	0.409	41. 147°
The Flood to Entering of Canaan	897	$\frac{897}{4,046}$	0.222	42. 80°
Entering of Canaan to Captivity of Judah	986	$\frac{986}{4,046}$	0.244	43. 88°
Captivity of Judah to Birth of Christ	507	$\frac{507}{4,046}$	0.125	44. 45°
Totals	4,046		1.000	

Source: *The Wonders of Bible Chronology*, by Philip Mauro

F. **Complete the frequency distribution table for a histogram. You do not need to draw the histogram.** *(Lesson 75)*

Sweet Corn Picked by the Lewis Family

Week 1	Dozens	Week 2	Dozens	Week 3	Dozens
Monday	76	Monday	46	Monday	101
Tuesday	88	Tuesday	38	Tuesday	68
Wednesday	65	Wednesday	47	Wednesday	88
Thursday	51	Thursday	95	Thursday	78
Friday	57	Friday	104	Friday	85
Saturday	111	Saturday	93	Saturday	105

Frequency Distribution Table

Intervals	Frequency
More than 25 but not more than 50 dozen	45. 3
More than 50 but not more than 75 dozen	46. 4
More than 75 but not more than 100 dozen	47. 7
More than 100 but not more than 125 dozen	48. 4

G. **Tell how long each segment should be on a rectangle graph 10 centimeters long. You do not need to draw the graph.** *(Lesson 80)*

Students at Laurel Mennonite School

49. Grades 1, 2 — 26 mm — Grades 1, 2 15
50. Grades 3, 4 — 29 mm — Grades 3, 4 17
51. Grades 5–7 — 24 mm — Grades 5–7 14
52. Grades 8–10 — 21 mm — Grades 8–10 12

H. Solve these sets of parallel reading problems. Use your solution to the first problem in each set as a help in solving the second problem. *(Lesson 114)*

53. a. Of the 10 virgins who waited for the bridegroom (Matthew 25), 5 were foolish because they failed to bring extra oil for their lamps. What percent of the virgins were foolish?
    50%

    b. Moses was 3 years younger than his brother Aaron (Exodus 7:7). When Moses was 80 years old, Aaron's age was what percent of Moses' age, to the nearest whole percent?
    104%

54. a. In still air, a typical raindrop falls about 600 feet per minute. What part of a second would it take for a drop to fall 1 foot?
    0.1 second

    b. One of the costliest disasters in the United States resulted when Hurricane Andrew struck Florida and Louisiana in 1992. Winds gusted up to 165 miles per hour. At that speed, what part of a minute would it take for an object to travel 1 mile? (Round to nearest hundredth.)
    0.36 minute

55. a. Two 50-acre farms subdivided into ½-acre plots would make how many plots?
    200 plots

    b. Four $8\frac{3}{4}$-yard pieces of fabric could be divided into how many $1\frac{3}{4}$-yard pieces?
    20 pieces

Solutions for Part H

53. a. $5 \div 10$
b. $(80 + 3) \div 80$

54. a. $60 \div 600$
b. $60 \div 165$

55. a. $2 \times 50 \div \frac{1}{2}$
b. $4 \times 8\frac{3}{4} \div 1\frac{3}{4}$

T-561 *Chapter 12 Other Numeration Systems and Final Reviews*

LESSON 167

Objective

- To give a year-end review of Chapter 9.

Teaching Guide

1. Give Lesson 167 Quiz (The Binary System).
2. Lesson 167 reviews the material taught in Chapter 9. For pointers on using review lessons, see *Teaching Guide* for Lesson 98.
3. Pass out the banking forms that the students need for exercises 1–14. These forms are included with the quiz sheets for this chapter.
4. The following table shows new concepts taught in Chapter 9 and reviewed in this lesson.

Lesson number and new concept	Exercises in Lesson 167
117—Using checking account deposit tickets.	1, 2
118—Writing checks.	3, 4
119—Keeping a check register.	5–10
120—Reconciling a checking account.	11, 12
121—Using savings account deposit tickets.	13, 14
124—Using the compound interest formula, $a = p(1 + r)^n$.	23, 24
124—Using a compound interest table.	25, 26
125—Calculating property tax.	29, 30
126—Working with gross profit, overhead, and net profit.	31–34
127—Profits and expenses as percents of total sales.	37–40

167. Final Review of Finances

(Forms provided in quiz booklet.)

A. Complete deposit tickets for the following deposits. Be sure your calculations are correct. Use the current date. *(Lesson 117)* (See page T–562.)

1. Coins 19.25
   Checks
       69-828 .... 80.35
       73-222 .... 68.17
       82-225 .... 42.49
   Cash received 80.00

2. Currency $95.00
   Coins 49.60
   Checks
       24-362 .... 45.09
       38-701 .... 98.72
       52-660 .... 85.57

B. Write the checks indicated. Use the current year in the date, and sign your own name as the payer. *(Lesson 118)*

	Check number	Date	Payee	Amount	Memo
3.	284	March 25	Leonard White	$90.45	Repair parts
4.	285	March 25	Lewis Mills	$78.36	Calf feed

C. Use this information to fill in check registers. *(Lesson 119)*

5–7.   (See page T–563.)

Balance carried forward: $367.24
Check number 538 on 7/10 to Phipps Garden Center for $21.95
Deposit from sales on 7/10 for $206.57
Check number 539 on 7/11 to Sensenig's Welding for $75.26

8–10.

Balance carried forward: $520.04
Check number 601 on 8/15 to Philip's Appliance for $420.70
Deposit of milk check on 8/15 for $3,217.63
Check number 602 on 8/16 to Gehman Dairy Supply for $95.15

D. Use reconciliation forms to reconcile the following accounts. Write *yes* or *no* to tell whether the balances agree. *(Lesson 120)*

	Ending bank balance		Deposits outstanding	Checks outstanding		Ending register balance	
11.	$468.52	+	$175.62	644.14 – $162.82	481.32	$491.32	no
12.	$585.22	+	$235.40	820.62 – $668.26		$152.36	yes

E. Complete savings deposit tickets for the following deposits. Use your own name and today's date. *(Lesson 121)*   *(See page T–564.)*

13. Account number: 12-98127634
    Items deposited:
        Coins .... $ 79.95
        Currency 314.00
        Checks
            13-491 .... 54.48
            24-503 .... 61.77

14. Account number: 34-95175310
    Items deposited:
        Coins .... $ 79.95
        Currency 83.00
        Checks
            32-945 .... 313.00
            43-841 .... 277.52

F. Find the simple interest. *(Lessons 121, 122)*

15. p = $3,000 $720.00
    r = 6%
    t = 4 yr.

16. p = $485 $116.40
    r = 8%
    t = 3 yr.

17. p = $960 $44.80
    r = 7%
    t = 8 mo.

18. p = $2,500 $516.67
    r = 8%
    t = 2 yr. 7 mo.

G. Use the 360-day year to calculate this interest. *(Lesson 122)*

19. $600 at 6% for 240 days $24
    ($600 × 0.06 × $\frac{2}{3}$)

20. $900 at 8% for 105 days $21
    ($900 × 0.08 × $\frac{7}{24}$)

H. Find the total interest if it is compounded annually. *(Lesson 123)*

21. $2,000 at 7% for 2 years $289.80
    ($140.00 + $149.80)

22. $3,000 at 8% for 2 years $499.20
    ($240.00 + $259.20)

I. Use the compound interest formula to find the total amount of principal plus interest if interest is compounded semiannually. *(Lesson 124)*

23. $3,000 at 12% for 1 year $3,370.80
    ($3,000 × 1.06^2; decimal: 1.1236)

24. $6,000 at 16% for 1 year $6,998.40
    ($6,000 × 1.08^2; decimal: 1.1664)

J. Use the compound interest table on page 418 to find the total interest if it is compounded quarterly. *(Lesson 124)*

25. $2,000 at 5% for 2 years $209.00
    (0.1045 × $2,000)

26. $4,000 at 7% for $2\frac{1}{2}$ years $757.60
    (0.1894 × $4,000)

K. Calculate these property taxes. *(Lesson 125)*

27. Assessment: $70,000; tax rate: 20 mills $1,400

28. Assessment: $80,000; tax rate: 22 mills $1,760

Lesson 167 T–562

Answers for Part A

1.

DEPOSIT TICKET			
[Your name] 56-334/809	CURRENCY		
[Your street address]	COIN	19	25
[Your city and state]	CHECKS		
	69-828	80	35
Account # 36-5551123	73-222	68	17
	82-225	42	49
DATE __[current date]__ 20 ___	SUBTOTAL	210	26
	LESS CASH RECEIVED	80	00
Bryan National Bank	TOTAL DEPOSIT	130	26

2.

DEPOSIT TICKET			
[Your name] 56-334/809	CURRENCY	95	00
[Your street address]	COIN	49	60
[Your city and state]	CHECKS		
	24-362	45	09
Account # 36-5551123	38-701	98	72
	52-660	85	57
DATE __[current date]__ 20 ___	SUBTOTAL	373	98
	LESS CASH RECEIVED		
Bryan National Bank	TOTAL DEPOSIT	373	98

Answers for Part B

3.

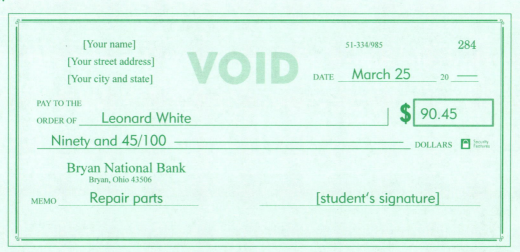

577

T–563 Chapter 12 Other Numeration Systems and Final Reviews

4.

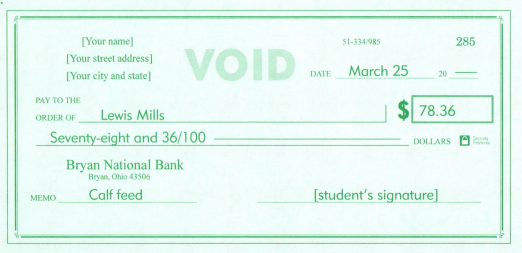

Answers for Part C

5–7.

NUMBER	DATE	DESCRIPTION	✓	FEE	CHECK AMOUNT		DEPOSIT AMOUNT		BALANCE	
		Balance carried forward							$367	24
538	7/10	Phipps Garden Center			21	95			345	29
D	7/10	Deposit—Sales					206	57	551	86
539	7/11	Sensenig's Welding			75	26			476	60

8–10.

NUMBER	DATE	DESCRIPTION	✓	FEE	CHECK AMOUNT		DEPOSIT AMOUNT		BALANCE	
		Balance carried forward							$520	04
601	8/15	Philip's Appliance			420	70			99	34
D	8/15	Deposit—Milk check					3,217	63	3,316	97
602	8/16	Gehman Dairy Supply			95	15			3,221	82

L. Following are two sales receipts with both taxable and nontaxable items. Find the sales tax and the total amount due. *(Lesson 125)*

29.
```
        26.95 T
        19.95 T
        19.01
        24.95 T
Subtotal  90.86
6% Tax    $4.31
Total     $95.17
```
(taxable: $71.85)

30.
```
        57.63 T
        39.95
        58.74 T
        22.35
Subtotal  178.67
7% Tax    $8.15
Total     $186.82
```
(taxable: $116.37)

M. Calculate the gross profit or net profit in each exercise. *(Lesson 126)*

31. Income: $394.42; expenses: $59.95 — $334.47
32. Income: $714.74; expenses: $91.96 — $622.78
33. Sales: $275.96; cost of goods: $141.94; overhead: $71.41 — $62.61
34. Sales: $621.96; cost of goods: $385.74; overhead: $181.74 — $54.48

N. Calculate the loss in each exercise. *(Lesson 126)*

35. Sales: $949.47; cost of goods: $839.78; overhead: $205.87 — $96.18
36. Sales: $809.75; cost of goods: $616.89; overhead: $271.97 — $79.11

O. Calculate the missing percents on this income statement. *(Lesson 127)*

Rayburn Mower Sales and Service
Income Statement
For the Year Ending
December 31, 20—

	Amount	Percent of Sales
Total sales	$41,000	100%
Cost of goods	19,000	37.46%
Gross profit	22,000	38.54%
Overhead	7,000	39.17%
Net profit	15,000	40.37%

P. Solve these problems mentally. *(Lesson 129)*

41. 148 + 229   377   42. 315 + 289   604
43. 284 − 148   136   44. 267 − 139   128
45. 9 × 58   522   46. 8 × 63   504
47. 26 × 50   1,300   48. 38 × 50   1,900
49. 288 ÷ 24   12   50. 216 ÷ 18   12
51. 4,200 ÷ 50   84   52. 3,600 ÷ 50   72

Q. Solve these reading problems.

53. Using the 360-day year, find the interest on $3,000 at 6% for 120 days. (Solve mentally if you can.) $60

54. Using the 360-day year, find the interest on $6,000 at 6% for 180 days. (Solve mentally if you can.) $180

55. A residence is assessed at $122,000 in a town with a tax rate of 20 mills per year. Half of the annual tax is due by May. What is the amount due on May 1? $1,220

56. A commercial building is assessed at $245,000 where the annual rate is 19 mills. Half of the annual tax is due by November. What is the amount due on November 1? $2,327.50

57. A business with sales of $95,000 had a gross profit of $38,000. What percent was the gross profit of the sales, to the nearest whole percent? 40%

58. The same business had a net profit of $14,000 on its sales of $95,000. What percent was the net profit of the sales, to the nearest whole percent? 15%

Lesson 167 T–564

Answers for Part E

13.

SAVINGS DEPOSIT			
DATE [current date] 20 ___	CASH	393	95
	CHECKS		
[student's name]	13-491	54	48
Account name	24-503	61	77
12-98127634	SUBTOTAL	510	20
Account number	LESS CASH RECEIVED		
The Longville Bank	**TOTAL DEPOSIT**	510	20

14.

SAVINGS DEPOSIT			
DATE [current date] 20 ___	CASH	162	95
	CHECKS		
[student's name]	32-945	313	00
Account name	43-841	277	52
34-95175310	SUBTOTAL	753	47
Account number	LESS CASH RECEIVED		
The Longville Bank	**TOTAL DEPOSIT**	753	47

Solutions for Part Q

53. $\$3{,}000 \times 0.06 \times \frac{1}{3}$

54. $\$6{,}000 \times 0.06 \times \frac{1}{2}$

55. $\$122{,}000 \times 0.020 \times \frac{1}{2}$

56. $\$245{,}000 \times 0.019 \times \frac{1}{2}$

57. $38{,}000 \div 95{,}000$

58. $14{,}000 \div 95{,}000$

T-565 *Chapter 12 Other Numeration Systems and Final Reviews*

LESSON 168

Objective

- To give a year-end review of Chapters 10 and 11.

Teaching Guide

1. Lesson 168 reviews the material taught in Chapters 10 and 11 (Lessons 132–152.) For pointers on using review lessons, see *Teaching Guide* for Lesson 98.
2. The table below shows new concepts taught in Chapters 10 and 11 that are reviewed in this lesson.
3. Reading problem review.
   Writing equations (Lesson 138).

Lesson number and new concept	Exercises in Lesson 168
132—Algebraic terms: monomial, polynomial, coefficient.	1–6
133—Use of brackets.	9, 10
135—Adding and subtracting monomials.	12, 13
139—Using exponents with literal numbers.	14, 15, 18–21
140—Multiplying numbers with exponents.	16, 17
141—Finding square roots of numbers with exponents.	22, 23
144—Comparing the values and absolute values of signed numbers.	28, 29
147—Multiplying more than two signed numbers.	40, 41
148—Substituting literal numbers with negative values.	44, 45
150—Making graphs based on formulas that include exponents to the third power.	56–61

168. Final Review of Algebra

A. Do these exercises on algebraic terms. *(Lesson 132)*

1. Is $5b^3c^4de$ a monomial or a polynomial? — monomial
2. Is $c + d$ a monomial or a polynomial? — polynomial
3. Is $7e + 3f$ a monomial, a binomial, or a trinomial? — binomial
4. Is $6g^2h + 7g^3h - 3g$ a monomial, a binomial, or a trinomial? — trinomial
5. Write the numerical coefficient in the monomial $4pq$. — 4
6. Write the literal coefficients in the monomial $7mn$. — mn

B. Name the operation that will solve each equation. *(Lessons 136, 137)*

7. $8s = 3$ (addition, subtraction, multiplication, division) — division
8. $x + 8 = 3$ (addition, subtraction, multiplication, division) — subtraction

C. Simplify these expressions. *(Lessons 133, 139)*

9. $2 \cdot 4[7 - (2 + 3)]$ — 16
10. $3[\frac{12}{3}(6 - 3)]$ — 36
11. $v^3w + v^2w + 3v^3w$ — $4v^3w + v^2w$
12. $7y^2 + 6y^2$ — $13y^2$
13. $8a^2b^4 - 3a^2b^4$ — $5a^2b^4$

D. Do these exercises with exponents. *(Lessons 139–141)*

14. Write x^4 in expanded form. — $x \cdot x \cdot x \cdot x$
15. Write y^3 in expanded form. — $y \cdot y \cdot y$
16. Multiply $2^3 \cdot 2^3$, and write the product in standard form. — 64
17. Multiply $3^2 \cdot 3^3$, and write the product in standard form. — 243
18. Simplify $c^3de^2 \cdot c^2d^3e$, using exponents. — $c^5d^4e^3$
19. Simplify $x^4y^2z \cdot y^3z^4$, using exponents. — $x^4y^5z^5$
20. Solve $m^5 \div m^3$. — m^2
21. Solve $n^4 \div n^2$. — n^2
22. Find $\sqrt{h^8}$. — h^4
23. Find $\sqrt{k^{12}}$. — k^6

E. Follow the directions, using signed numbers. *(Lesson 144)*

24. Use a signed number to show a 3,000-foot decrease in altitude.   −3,000 feet
25. Use a signed number to show a $3,400 profit.   +$3,400
26. Write the absolute value of −7.   7
27. Write the absolute value of +8.   8
28. Compare these numbers, and write < or > between them.   −5 _>_ −6
29. Compare these numbers, and write < or > between them. |+8| _>_ |−2|

F. Simplify these polynomials by combining like terms. *(Lesson 151)*

30. $-2vw - y^2z - 3vw - y^2z - 6$ $-5vw - 2y^2z - 6$
31. $3cd + 2g^2h - 5cd - 3cd - 4g^2h$ $-5cd - 2g^2h$

G. Solve these equations, showing each step in your work. *(Lessons 136, 137)*

32. $3x + 2 = 17$ $x = 5$
33. $n - 5 = 8$ $n = 13$ (See facing page for steps.)
34. $n + 6 = 18$ $n = 12$
35. $5 - 3 = \frac{n}{7}$ $14 = n$

H. Do these calculations with signed numbers. *(Lessons 145–147)*

36. $-5 + (-7) = -12$
37. $-6 + (-8) = -14$
38. $-6 - (+3) = -9$
39. $+7 - (-12) = +19$
40. $-1(+2)(-2)(+2)(+3)(-2)$ -48
41. $+1(-2)(-2)(+1)(-2)(-1)$ $+8$
42. $+18 \div (-3)$ -6
43. $-42 \div (-6)$ $+7$

I. Evaluate these expressions. *(Lesson 148)*

44. $2m - 4n$ if $m = 4$ and $n = -4$ $+24$
45. $2p - 4r$ if $p = -3$ and $r = 2$ -14

J. Find the missing numbers for this table, to the nearest whole number. *(Lesson 149)*

Formula: $a = \pi r^2$	Areas of circles			
r	20	21	22	23
a	46. 1,256	47. 1,385	48. 1,520	49. 1,661

K. Prepare graphs from these tables. *(Lesson 150)*

50–55.   (See facing page.)

Formula: $p = \frac{i}{0.05}$						
i	$0	$10	$20	$30	$40	$50	$60
p	$0	$200	$400	$600	$800	$1,000	$1,200

Lesson 168 T–566

Solutions for Part G

32. $3x + 2 - 2 = 17 - 2$
$3x = 15$
$\dfrac{3x}{3} = \dfrac{15}{3}$
$x = 5$

33. $n - 5 + 5 = 8 + 5$
$n = 13$

34. $n + 6 - 6 = 18 - 6$
$n = 12$

35. $2 = \dfrac{n}{7}$
$2 \cdot 7 = \dfrac{n}{7} \cdot 7$
$14 = n$

Graphs for Part K

50–55.

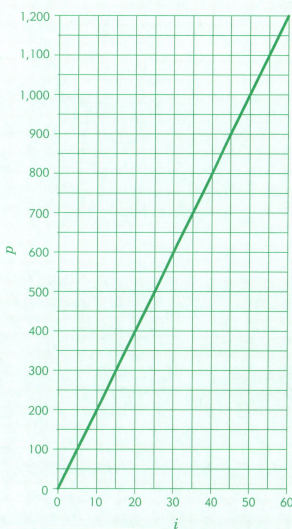

56–61.

$a = 5b^3$

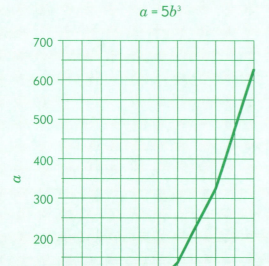

56–61.

Formula: $a = 5b^3$						
b	0	1	2	3	4	5
a	0	5	40	135	320	625

L. Use equations to solve these reading problems. *(Lessons 137, 152)*

62. James and Wilbur stacked 1,260 bales of straw in the barn. James stacked twice as many as Wilbur did. How many did Wilbur stack?
    $w + 2w = 1,260$ $\qquad$ $w = 420$ bales

63. It took Curtis twice as much time to cut the alfalfa hay as to rake it. The baling took the same length of time as the cutting. The total time he spent working with the hay was $7\frac{1}{2}$ hours. How long did it take to do the raking?
    $r + 2r + 2r = 7\frac{1}{2}$ $\qquad$ $r = 1\frac{1}{2}$ hours

64. The Millers' turkey barn is 5 times as long as it is wide. The combined length and width is 240 feet. How wide is the barn?
    $w + 5w = 240$ $\qquad$ $w = 40$ feet

65. Samuel found that one of the living room windows was 2 times as high as it was wide. The total height and width was 72 inches. How wide was the window?
    $w + 2w = 72$ $\qquad$ $w = 24$ inches

66. Man knows 3,000 more species of birds than of reptiles. The total number of species in the two groups is 15,000. How many species of reptiles are known?
    $r + (r + 3,000) = 15,000$ $\qquad$ $r = 6,000$ species

67. Man knows 47,500 species of vertebrates. There are twice as many species of mammals as of amphibians. The sum of the species of mammals and amphibians is 40,000 less than the total number of species of vertebrates. How many species of amphibians are known?
    $a + 2a = 47,500 - 40,000$ $\qquad$ $a = 2,500$ species

568   Chapter 12   Other Numeration Systems and Final Reviews

169. Review of Chapter 12 and Final Review

A. **Write the reading of each meter. Then find the kilowatt-hours of electricity used.** *(Lesson 155)*

1. 23,353   subtracted   from   25,518 = 2,165 kwh

2. 49,879   subtracted   from   51,089 = 1,210 kwh

 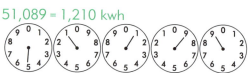

B. **Write these numbers, using scientific notation.** *(Lesson 156)*

3. 3,000,000,000 3×10^9
4. 43,500,000,000,000 4.35×10^{13}
5. 0.000022 2.2×10^{-5}
6. 0.00000013 1.3×10^{-7}

C. **Write these base twelve numbers as base ten numbers.** *(Lesson 158)*

7. 25_{twelve} 29
8. 105_{twelve} 149
9. TEE_{twelve} 1,583
10. ETE_{twelve} 1,715

D. **Write these base ten numbers as base twelve numbers.** *(Lesson 158)*

11. 73 61_{twelve}
12. 138 $E6_{twelve}$
13. 1,726 EET_{twelve}
14. 1,698 $E96_{twelve}$

E. **Write these binary numbers as base ten numbers.** *(Lesson 159)*

15. 11_{two} 3
16. 111_{two} 7
17. 100_{two} 4
18. $1,010_{two}$ 10

F. **Write these base ten numbers as binary numbers.** *(Lesson 159)*

19. 5 101_{two}
20. 307 $100,110,011_{two}$
21. 16 $10,000_{two}$
22. 190 $10,111,110_{two}$

G. **Do these Final Review exercises.**

CHAPTER 1

Write each number, using words. *(Lesson 1)*

23. 3,450,000,000,000,000   Three quadrillion, four hundred fifty trillion
24. 7,000,900,000,000,000   Seven quadrillion, nine hundred billion

LESSON 169

Objectives

- To review the first part of Chapter 12 (Lessons 155–160), along with the new concepts taught in this Grade 8 course, in preparation for the final test.

Teaching Guide

1. The first part of Lesson 169 reviews material from the first part of Chapter 12 (Lessons 155–160). For pointers on using review lessons, see *Teaching Guide* for Lesson 98.

2. Be sure to review the new concepts taught in this chapter, shown on the chart below.

3. The remainder of Lesson 169 is a general review of new material taught in this text. Be sure to discuss any concepts that caused significant problems during the year.

Lesson number and new concept	Exercises in Lesson 169
155—Reading numbers shown by dials, as on some electric meters.	1, 2
156—Using scientific notation to write very large and very small numbers.	3–6
157—Concept that numeration systems can have bases other than ten.	7–22
158—Working with the duodecimal system.	7–14
159—Working with the binary system.	15–22

Name the mathematical law illustrated by each equation. (Lessons 4, 10)

25. 7 × 6 = 6 × 7   commutative law

26. 23 × 15 = 20 × 15 + 3 × 15   distributive law

CHAPTER 2

Express the Fahrenheit temperature as Celsius, and the Celsius as Fahrenheit, to the nearest degree. (Lesson 28)

27. 45°F   7°C

28. 41°C   106°F

CHAPTER 3

List the prime factors for each number. Use exponents for repeated factors. (Lesson 34)

29. 40 $40 = 2^3 \times 5$

30. 180 $180 = 2^2 \times 3^2 \times 5$

Simplify each complex fraction. (Lesson 40)

31. $\dfrac{\frac{5}{6}}{\frac{7}{8}}$ $\dfrac{20}{21}$ **32.** $\dfrac{\frac{3}{4}}{\frac{5}{8}}$ $1\frac{1}{5}$

CHAPTER 4

Express these ratios as indicated. (Lesson 52)

33. 14:5 antecedent: 1, consequent: a decimal rounded to the nearest tenth   1 to 0.4

34. 18:7 consequent: 1, antecedent: in fraction form $2\frac{4}{7}$ to 1

Choose the correct word. (Lesson 55)

35. The number of workers on a job varies (directly, inversely) in relation to the days needed to complete the job.   inversely

36. The number of hours needed to complete a journey varies (directly, inversely) in relation to the distance.   directly

Find the actual distance represented by these measurements on a scale drawing. (Lesson 57)

37. scale = 1:40; length = $4\frac{1}{4}$ inches   170 inches

38. scale = 5:2; length = $3\frac{3}{4}$ inches $1\frac{1}{2}$ inches

CHAPTER 5

Express each fraction as a percent, to the nearest whole percent. (Lesson 65)

39. $\frac{13}{15}$ 87%   **40.** $\frac{5}{7}$ 71%

Find the rate of increase or decrease. Include the label increase *or* decrease. *(Lesson 66)*

41. base: 30; new amount: 75   150% increase

42. base: 120; new amount: 42   65% decrease

Find the base amounts. (Lesson 67)

43. 15 is ¾% of 2,000

44. 16 is ½% of 3,200

45. 30 is 125% of 24.

46. 18 is 120% of 15.

CHAPTER 6

Find the arithmetic mean of each set. (Lesson 73)

47. 34, 36, 38, 21, 85, 42   42⅔

48. 78, 96, 65, 79, 76, 89   80½

Find the median of each set. (Lesson 74)

49. 34, 36, 38, 21, 85, 42   37

50. 78, 96, 65, 79, 76, 89   78½

Find the mode(s) of each set. (Lesson 74)

51. 22, 23, 24, 23, 25, 24, 26, 25, 26, 24, 21, 23, 22, 21, 23, 24   23, 24

52. 78, 76, 78, 76, 78, 75, 78, 79, 79, 73, 76, 75, 78, 76, 75, 77   78

CHAPTER 7

Name the plane figures described. (Lesson 85)

53. Any plane figure with all straight sides   polygon

54. A plane figure with nine straight sides   nonagon

Name the type of angle for each size. (Lesson 86)

55. A 165° angle   obtuse

56. A 255° angle   reflex

CHAPTER 8

Find the surface area of each square pyramid. (Lesson 103)

57. base: 5 in.; slant height: 6 in.   85 sq. in.

58. base: 6 in.; slant height: 7 in.   120 sq. in.

Find the surface area of each sphere. (Lesson 104)

59. 2-inch radius   50.24 sq. in.

60. 3-inch radius   113.04 sq. in.

Find the volume of each square pyramid. (Lesson 107)

61. base: 5 in.; height: 6 in.   50 cu. in.

62. base: 6 in.; height: 7 in.   84 cu. in.

Find the volume of each sphere. Use 4.18. (Lesson 108)

63. 6-inch radius   902.88 cu. in.

64. 8-inch radius   2,140.16 cu. in.

Use the table of square roots on page 584 to find the missing measure for each triangle. (Lessons 112, 113)

65. leg a = 6 in.; leg b = 7 in.; hypotenuse = ___ to the nearest tenth   9.2 in.

66. leg a = 7 in.; hypotenuse = 14 in.; leg b = ___ to the nearest tenth   12.1 in.

T–571 *Chapter 12 Other Numeration Systems and Final Reviews*

CHAPTER 9

Find the total amount of principal plus interest. (Lesson 124)

67. $6,000 for 1 year at 12% compounded semiannually. Use the compound interest formula. ($6,000 × 1.06^2; decimal: 1.1236) $6,741.60

68. $8,000 for $2\frac{1}{2}$ years at 7% compounded quarterly. Use the compound interest table on page 418. (1.1894 × $8,000) $9,515.20

Calculate the property tax. (Lesson 125)

69. Assessment: $115,000; tax rate: 24 mills $2,760
70. Assessment: $145,000; tax rate: 21 mills $3,045

Find the profit as indicated. (Lesson 126)

71. Sales: $45,350; cost of goods: $32,670; gross profit: $12,680
72. Sales: $23,500; cost of goods: $13,500; overhead: $7,800; net profit: $2,200

Find what percent the gross profit is of the sales. (Lesson 127)

73. Sales: $675; cost of goods: $513 24%
74. Sales: $2,450; cost of goods: $1,666 32%

CHAPTER 10

Choose the correct name for each expression. (Lesson 132)

75. 3c + 5d (monomial, binomial, trinomial) binomial
76. 4x + 2y − 3z (monomial, polynomial). polynomial

Simplify these expressions. (Lessons 133, 140, 141)

77. 100 − 3[3 + 2(5 + 6)]   25
78. 80 − 5[39 − 4(28 − 22)]   5
79. $4cd^2 \cdot 2^2c^2d^2$ $16c^3d^4$
80. $4s^3t^2 \cdot 2^3s^2t^3$ $32s^5t^5$
81. $n^6 \div n^2$ n^4
82. $x^9 \div x^5$ x^4

Find these square roots. (Lesson 141)

83. $\sqrt{m^8}$ m^4
84. $\sqrt{x^{12}}$ x^6

CHAPTER 11

Compare the integers and write < or > between them. (Lesson 144)

85. $|-8|$ _>_ $|+7|$
86. −3 _<_ −2

Solve these multiplications. (Lesson 147)

87. −2(−7)(+1)(−5)(+2)(−1)   +140
88. +1(−3)(−2)(+3)(−2)(+1)   −36

Evaluate each expression with the values given. (Lesson 148)

89. $\frac{4x}{3y}$ ($x = -3$ and $y = -2$)   +2

90. $\frac{m-n}{n-m}$ ($m = -3$ and $n = -5$)   −1

Simplify these expressions. (Lesson 151)

91. $3a^2b + 2a^2c - 6a^2b$ $-3a^2b + 2a^2c$

92. $2d^3e - 6d^3e - 6d^2e + 7d$ $-4d^3e - 6d^2e + 7d$

H. Use equations to solve these reading problems. (Lessons 138, 152)

93. The sum of three consecutive numbers is 63. What is the smallest number?
  $n + (n + 1) + (n + 2) = 63$ $n = 20$

94. Lewis is twice as old as Marlin. Father, who is 41, is 20 years older than the sum of the boys' ages. How old is Marlin? $m + 2m + 20 = 41$ $m = 7$ years

95. According to one study, the average American generated nearly $5\frac{1}{2}$ pounds of consumer waste per day in 1988. That year 6 Americans produced as much waste as did 11 Americans in 1960. What was the average waste per day for an American of 1960? $11w = 6 \cdot 5\frac{1}{2}$ $w = 3$ pounds

96. One year the amount of municipal wastes in America was estimated to be 50 million tons less than twice the amount of industrial wastes. Municipal and industrial wastes together amounted to 370 million tons. What was the amount of industrial wastes? $i + (2i - 50) = 370$ $i = 140$ million tons

97. Black ironwood is the heaviest known wood, and balsa is one of the lightest. One cubic foot of black ironwood weighs 3 pounds more than 15 times as much as a cubic foot of balsa. A cubic foot of ironwood plus a cubic foot of balsa stacked together weigh 99 pounds. What does a cubic foot of balsa weigh? $b + (15b + 3) = 99$ $b = 6$ pounds

98. The thickness of the Hoover Dam at its crest, added to its height, is 771 feet. The height is 6 feet more than 16 times the crest thickness. What is the crest thickness of the Hoover Dam? $c + (16c + 6) = 771$ $c = 45$ feet

170. Final Test

LESSON 170

Objective

- To test the students' mastery of the concepts taught in this Grade 8 course.

Teaching Guide

1. Correct Lesson 169.
2. Review any areas of special difficulty.
3. Administer the test. For pointers on giving tests, see *Teaching Guide* for Lesson 99.

QUIZZES and SPEED TESTS

Answer Key

He maketh my feet like hinds' feet:
and setteth me upon my high places.
2 Samuel 22:34

Lesson 86 ~ Speed Test (Time: 1 minute) *Basic Math Facts*

1. 9+5=14, 7+0=7, 3+7=10, 2+5=7, 6+4=10, 0+6=6, 7+9=16, 2+4=6, 2+5=7, 7+7=14

2. 3+3=6, 1+9=10, 7+6=13, 6+9=15, 2+0=2, 9+9=18, 5+8=13, 4+0=4, 8+7=15, 0+9=9

3. 7+3=10, 8+5=13, 8+7=15, 7+3=10, 2+5=7, 8+1=9, 2+3=5, 6+8=14, 4+0=4, 3+5=8

4. 3−3=0, 9−5=4, 4−2=2, 5−1=4, 7−5=2, 6−1=5, 4−3=1, 8−2=6, 2−0=2, 7−6=1

5. 14−8=6, 5−5=0, 13−5=8, 17−8=9, 5−3=2, 15−6=9, 8−3=5, 7−2=5, 4−0=4, 16−9=7

6. 3×7=21, 12×8=96, 7×8=56, 6×7=42, 2×2=4, 9×5=45, 5×2=10, 4×6=24, 8×9=72, 12×2=24

7. 7×0=0, 11×8=88, 9×7=63, 6×8=48, 12×3=36, 8×1=8, 10×0=0, 6×5=30, 12×4=48, 11×6=66

8. 3×8=24, 11×10=110, 9×6=54, 12×5=60, 12×7=84, 3×6=18, 1×0=0, 11×12=132, 4×5=20, 12×6=72

9. 66÷11=6, 9÷1=9, 72÷8=9, 15÷3=5, 120÷10=12, 36÷9=4, 36÷6=6

10. 16÷4=4, 32÷8=4, 108÷12=9, 20÷10=2, 28÷7=4, 88÷11=8, 16÷8=2

11. 40÷4=10, 24÷6=4, 48÷4=12, 0÷3=0, 27÷9=3, 84÷7=12

Lesson 88 ~ Quiz

Name _____

Date _____

Geometric Figures

Score _____

A. Write each expression, using symbols.

1. angle LMN = ____∠LMN____
2. ray QR = ____$\overrightarrow{QR}$____
3. line segment TU = ____$\overline{TU}$____
4. line IJ = ____$\overleftrightarrow{IJ}$____
5. line FG is parallel to line HI = ____$\overleftrightarrow{FG} \parallel \overleftrightarrow{HI}$____
6. line JK is perpendicular to line LM = ____$\overleftrightarrow{JK} \perp \overleftrightarrow{LM}$____

B. Draw the following polygons.

7. octagon (8 sides, equal or unequal)

8. decagon (10 sides, equal or unequal)

C. Measure the following angles. Answers are multiples of 5 degrees.

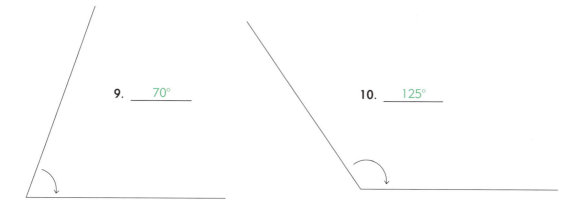

9. ____70°____

10. ____125°____

Lesson 91 ~ Quiz Name _____

  Date _____

Geometric Symbols and Figures Score _____

A. Write these expressions, using symbols.

1. line segment ST = _____$\overline{ST}$_____
2. angle MNO = _____$\angle MNO$_____
3. line DE is perpendicular to line FG = _____$\overleftrightarrow{DE} \perp \overleftrightarrow{FG}$_____
4. line HI is parallel to line JK = _____$\overleftrightarrow{HI} \parallel \overleftrightarrow{JK}$_____

B. Write what kind of angle each one is.

5. ____straight____
6. ____reflex____

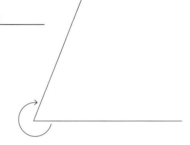

C. Name each triangle according to the type of angle.

7. ____right____
8. ____obtuse____

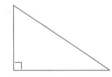

D. Identify each pair of triangles as similar or congruent, using symbols.

9. ____$\triangle ABC \cong \triangle DEF$____
10. ____$\triangle GHI \sim \triangle JKL$____

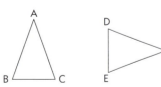
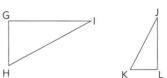

Lesson 93 ~ Quiz   Name _____

                          Date _____

Perimeter and Circumference   Score _____

A. Find the perimeter of each figure.

Rectangles

1. $l = 16$ m
   $w = 7.5$ m

   $p =$ __47 m__

2. $l = 18$ ft.
   $w = 11$ ft.

   $p =$ __58 ft.__

3. $l = 9\frac{1}{2}$ in.
   $w = 5\frac{2}{3}$ in.

   $p =$ __$30\frac{1}{3}$ in.__

Squares

4. $s = 14.2$ cm

   $p =$ __56.8 cm__

5. $s = 2.3$ m

   $p =$ __9.2 m__

6. $s = 10\frac{1}{4}$ ft.

   $p =$ __41 ft.__

B. Find the circumference of each circle.

7. $d = 25$ m
   $\pi = 3.14$

   $c =$ __78.5 m__

8. $r = 5.5$ cm
   $\pi = 3.14$

   $c =$ __34.54 cm__

9. $d = 5\frac{1}{4}$ in.
   $\pi = 3\frac{1}{7}$

   $c =$ __$16\frac{1}{2}$ in.__

10. $r = 7$ in.
    $\pi = 3\frac{1}{7}$

    $c =$ __44 in.__

Lesson 97 ~ Quiz

Name _____

Date _____

Area

Score _____

Find the area of each figure.

Triangles

1. $b = 16$ m
   $h = 11.2$ m

   $a =$ __89.6 m²__

2. $b = 5\frac{1}{2}$ in.
   $h = 13$ in.

   $a =$ __$35\frac{3}{4}$ sq. in.__

Rectangles

3. $l = 23.4$ cm
   $w = 18.1$ cm

   $a =$ __423.54 cm²__

4. $l = 12$ ft.
   $w = 4\frac{1}{2}$ ft.

   $a =$ __54 sq. ft.__

Squares

5. $s = 14$ in.

   $a =$ __196 sq. in.__

6. $s = 5.8$ km

   $a =$ __33.64 km²__

Parallelograms

7. $b = 31$ ft.
   $h = 19$ ft.

   $a =$ __589 sq. ft.__

8. $b = 26.4$ cm
   $h = 11.9$ cm

   $a =$ __314.16 cm²__

Trapezoids

9. $h = 8$ in.
   $b_1 = 11\frac{1}{2}$ in.
   $b_2 = 15$ in.

   $a =$ __106 sq. in.__

10. $h = 34.1$ m
    $b_1 = 40$ m
    $b_2 = 52.8$ m

    $a =$ __1,582.24 m²__

Circles

11. $r = 18$ cm
    $\pi = 3.14$

    $a =$ __1,017.36 cm²__

12. $d = 28$ in.
    $\pi = 3\frac{1}{7}$

    $a =$ __616 sq. in.__

Lesson 100 ~ Quiz

Name _____

Date _____

Finding Area

Score _____

Write the formula for each area. Then find the area of the figure described.

Triangle: b = 10 cm   h = 15 cm

1. Formula: ___$a = \frac{1}{2}bh$___

2. a = ___75 cm²___

Rectangle: l = 22 in.   w = 18 in.

3. Formula: ___$a = lw$___

4. a = ___396 sq. in.___

Square: s = 17 mi.

5. Formula: ___$a = s^2$___

6. a = ___289 sq. mi.___

Parallelogram: b = 26 ft.   h = 18 ft.

7. Formula: ___$a = bh$___

8. a = ___468 sq. ft.___

Trapezoid: h = 15 m b_1 = 12 m b_2 = 16 m

9. Formula: ___$a = \frac{1}{2}h(b_1 + b_2)$___

10. a = ___210 m²___

Circle: r = 15 cm   π = 3.14

11. Formula: ___$a = \pi r^2$___

12. a = ___706.5 cm²___

Lesson 102 ~ Speed Test Preparation

Addition

1. 9   7   3   2   6   0   6   2   2   0
   +0 +1 +7 +8 +3 +1 +0 +7 +8 +6
   ─   ─   ── ── ─   ─   ─   ─   ── ─
   9   8   10 10 9   1   6   9   10 6

2. 8   6   2   1   5   1   5   1   1   1
   +3 +5 +7 +3 +5 +1 +3 +8 +0 +5
   ── ── ─   ─   ── ─   ─   ─   ─   ─
   11 11 9   4   10 2   8   9   1   6

3. 1   2   9   5   8   5   7   7   2   2
   +3 +5 +7 +3 +5 +1 +3 +8 +0 +5
   ─   ─   ── ─   ── ─   ── ── ─   ─
   4   7   16 8   13 6   10 15 2   7

4. 3   1   7   6   2   9   5   4   8   0
   +3 +5 +7 +3 +5 +1 +3 +8 +0 +5
   ─   ─   ── ─   ─   ── ─   ── ─   ─
   6   6   14 9   7   10 8   12 8   5

5. 3   9   6   9   0   9   8   0   7   9
   +7 +8 +8 +7 +2 +8 +2 +6 +4 +2
   ── ── ── ── ─   ── ── ─   ── ──
   10 17 14 16 2   17 10 6   11 11

6. 7   8   8   7   2   8   2   6   4   2
   +3 +9 +6 +9 +0 +9 +8 +0 +7 +9
   ── ── ── ── ─   ── ── ─   ── ──
   10 17 14 16 2   17 10 6   11 11

7. 3   9   0   5   1   6   4   8   2   7
   +6 +5 +2 +0 +8 +4 +7 +1 +9 +3
   ─   ── ─   ─   ─   ── ── ─   ── ──
   9   14 2   5   9   10 11 9   11 10

8. 4   5   8   4   3   3   1   0   4   5
   +0 +1 +7 +8 +3 +1 +0 +7 +8 +6
   ─   ─   ── ── ─   ─   ─   ─   ── ──
   4   6   15 12 6   4   1   7   12 11

9. 8   6   9   4   2   4   5   3   8   6
   +5 +0 +7 +5 +4 +6 +0 +4 +5 +6
   ── ─   ── ─   ─   ── ─   ─   ── ──
   13 6   16 9   6   10 5   7   13 12

10. 1   9   2   8   3   7   4   6   5   9
   +0 +9 +2 +8 +3 +7 +4 +6 +7 +8
   ─   ── ─   ── ─   ── ─   ── ── ──
   1   18 4   16 6   14 8   12 12 17

Lesson 103 ~ Speed Test (Time: 1 minute) *Addition*

1. 9+8=17, 7+1=8, 3+7=10, 2+2=4, 6+6=12, 0+4=4, 6+5=11, 2+9=11, 2+8=10, 0+9=9

2. 3+5=8, 1+0=1, 7+7=14, 6+5=11, 2+4=6, 9+6=15, 5+0=5, 4+4=8, 8+5=13, 0+6=6

3. 7+3=10, 8+9=17, 8+6=14, 7+9=16, 2+0=2, 8+9=17, 2+8=10, 6+0=6, 4+7=11, 2+9=11

4. 3+3=6, 9+5=14, 0+7=7, 5+3=8, 1+5=6, 6+1=7, 4+3=7, 8+8=16, 2+0=2, 7+5=12

5. 1+3=4, 9+5=14, 2+7=9, 8+3=11, 3+5=8, 7+1=8, 4+3=7, 6+8=14, 5+0=5, 9+5=14

6. 8+3=11, 6+7=13, 9+7=16, 4+3=7, 2+5=7, 4+1=5, 5+3=8, 3+8=11, 8+0=8, 6+5=11

7. 4+7=11, 5+8=13, 8+8=16, 4+9=13, 3+2=5, 3+8=11, 1+2=3, 0+6=6, 4+4=8, 5+2=7

8. 3+6=9, 9+5=14, 6+2=8, 9+0=9, 0+8=8, 9+4=13, 8+7=15, 0+1=1, 7+9=16, 9+3=12

9. 1+0=1, 2+9=11, 9+3=12, 5+8=13, 8+3=11, 5+7=12, 7+4=11, 7+6=13, 2+7=9, 2+8=10

10. 9+8=17, 7+1=8, 3+7=10, 2+2=4, 6+6=12, 0+4=4, 6+5=11, 2+9=11, 2+8=10, 0+9=9

Lesson 106 ~ Quiz　　　　　　　　　　　　Name _____

　　　　　　　　　　　　　　　　　　　　　Date _____

Finding Surface Area　　　　　　　　　　Score _____

Write the formula for each surface area. Then find the surface area of the figure described.

　　Cube:　　$e = 12$ in.

1. Formula: ___$a_s = 6e^2$___

2. $a_s =$ ___864 sq. in.___

　　Rectangular solid:　　$l = 5$ in.　　$w = 4$ in.　　$h = 8$ in.

3. Formula: ___$a_s = 2lw + 2wh + 2lh$___

4. $a_s =$ ___184 sq. in.___

　　Cylinder:　　$r = 4$ in.　　$h = 12$ in.　　$\pi = 3.14$

5. Formula: ___$a_s = 2\pi r^2 + 2\pi rh$___

6. $a_s =$ ___401.92 sq. in.___

　　Square pyramid:　　$b = 7$ cm　　$\ell = 5$ cm

7. Formula: ___$a_s = 4(\frac{1}{2}b\ell) + b^2$___

8. $a_s =$ ___119 cm²___

　　Sphere:　　$r = 10$ in.　　$\pi = 3.14$

9. Formula: ___$a_s = 4\pi r^2$___

10. $a_s =$ ___1,256 sq. in.___

Lesson 109 ~ Quiz Name _____

Date _____

Finding Volume Score _____

Write the formula for each volume. Then find the volume of the figure described.

Cube: e = 12 in.

1. Formula: ____$v = e^3$____

2. v = ____1,728 cu. in.____

Rectangular solid: l = 5 in.   w = 4 in.   h = 8 in.

3. Formula: ____$v = lwh$____

4. v = ____160 cu. in.____

Cylinder: r = 4 in.   h = 12 in.   π = 3.14

5. Formula: ____$v = \pi r^2 h$____

6. v = ____602.88 cu. in.____

Cone: r = 5 cm   h = 6 cm   π = 3.14

7. Formula: ____$v = \frac{1}{3}\pi r^2 h$____

8. v = ____157 cm³____

Square pyramid: s = 7 ft.   h = 5 ft.

9. Formula: ____$v = \frac{1}{3}lwh$____

10. v = ____$81\frac{2}{3}$ cu. ft.____

Sphere: r = 10 m   (Use 4.18.)

11. Formula: ____$v = \frac{4}{3}\pi r^3$____

12. v = ____4,180 m³____

Lesson 115 ~ Quiz Name _____

 Date _____

The Pythagorean Rule Score _____

A. Write the formula for finding each fact.

1. The hypotenuse of a right triangle. $c^2 = a^2 + b^2$

2. The length of leg a when the hypotenuse and leg b are known. $a^2 = c^2 - b^2$

B. Find the hypotenuse of each right triangle, to the nearest tenth. Use the table on the back of this sheet.

3. leg a = 5 in.; leg b = 7 in.; hypotenuse = ___8.6 in.___

4. leg a = 8 in.; leg b = 5 in.; hypotenuse = ___9.4 in.___

5. leg a = 9 in.; leg b = 10 in.; hypotenuse = ___13.5 in.___

6. leg a = 8 in.; leg b = 8 in.; hypotenuse = ___11.3 in.___

C. Find the length of leg a in each triangle, to the nearest tenth. Use the table on the back of this sheet.

7. leg b = 3 in.; hypotenuse = 5 in.; leg a = ___4 in.___

8. leg b = 5 in.; hypotenuse = 8 in.; leg a = ___6.2 in.___

9. leg b = 11 in.; hypotenuse = 15 in.; leg a = ___10.2 in.___

10. leg b = 8 in.; hypotenuse = 12 in.; leg a = ___8.9 in.___

Lesson 118 ~ Quiz　　　　　　　　　　Name _____

　　　　　　　　　　　　　　　　　　　　Date _____

Surface Area and Volume　　　　　　　　Score _____

A. Find the surface area of each figure. Use 3.14 for pi.

1. *Rectangular solid*
   l = 2 in.
   w = 3 in.
   h = 4 in.

   __52 sq. in.__

2. *Cube*
   e = 2 in.

   __24 sq. in.__

3. *Square pyramid*
   b = 2 in.
   ℓ = 2 in.

   __12 sq. in.__

4. *Cylinder*
   r = 1 in.
   h = 2 in.

   __18.84 sq. in.__

5. *Sphere*
   r = 2 in.

   __50.24 sq. in.__

B. Find the volume of each figure. Use 3.14 for pi.

6. *Rectangular solid*
   l = 2 in.
   w = 3 in.
   h = 4 in.

   __24 cu. in.__

7. *Cube*
   e = 2 in.

   __8 cu. in.__

8. *Square pyramid*
   b = 2 in.
   h = 6 in.

   __8 cu. in.__

9. *Cylinder*
   r = 1 in.
   h = 2 in.

   __6.28 cu. in.__

10. *Sphere* (Use 4.18.)
    r = 2 in.

    __33.44 cu. in.__

Lesson 122 ~ Speed Test   (Time: 3 minutes)   *Horizontal Addition*

Solve these horizontal addition problems by writing only the answers.

1. 345 + 732 = __1,077__   2. 512 + 647 = __1,159__

3. 278 + 875 = __1,153__   4. 486 + 399 = __885__

5. 1,154 + 3,651 = __4,805__   6. 2,614 + 5,481 = __8,095__

7. 7,684 + 4,382 = __12,066__   8. 8,381 + 6,375 = __14,756__

9. 17,482 + 19,612 = __37,094__ 10. 27,142 + 75,763 = __102,905__

Lesson 126 ~ Quiz Name _____

 Date _____

Calculating Interest Score _____

A. Calculate the simple interest.

1. $p = \$2{,}000$
   $r = 5\%$
   $t = 2$ yr.

   $i =$ __$200.00__

2. $p = \$3{,}000$
   $r = 6\%$
   $t = 4$ yr.

   $i =$ __$720.00__

3. $p = \$4{,}000$
   $r = 6\%$
   $t = 6$ mo.

   $i =$ __$120.00__

4. $p = \$5{,}000$
   $r = 5\%$
   $t = 210$ days

   $i =$ __$145.83__

5. $p = \$2{,}000$
   $r = 5\frac{1}{2}\%$
   $t = 2$ yr.

   $i =$ __$220.00__

6. $p = \$5{,}000$
   $r = 6\frac{1}{2}\%$
   $t = 3$ yr.

   $i =$ __$975.00__

B. Use the compound interest formula to find the total principal plus interest.

7. $p = \$1{,}000$
   $r = 6\%$ compounded annually
   $t = 2$ yr.

   $a =$ __$1,123.60__

8. $p = \$2{,}000$
   $r = 8\%$ compounded semiannually
   $t = 1$ yr.

   $a =$ __$2,163.20__

C. Find the compound interest by using the compound interest table. (Decimals from the table are included in the exercises.)

9. $p = \$2{,}000$
   $r = 8\%$ compounded semiannually
   $t = 2$ yr.
   Decimal: 1.1699

   $i =$ __$339.80__

10. $p = \$3{,}000$
    $r = 6\%$ compounded quarterly
    $t = 2\frac{1}{2}$ yr.
    Decimal: 1.1605

    $i =$ __$481.50__

Lesson 129 ~ Quiz Name _____

Sales Tax, Property Tax, and Profit Date _____

Score _____

A. Calculate the sales tax.

1. Sale: $56.00
   Tax rate: 6%

   Tax: ____$3.36____

2. Sale: $75.50
   Tax rate: 5%

   Tax: ____$3.78____

B. Calculate the property tax.

3. Assessment: $95,000
   Tax rate: 23 mills

   Tax: ____$2,185____

4. Assessment: $125,000
   Tax rate: 22 mills

   Tax: ____$2,750____

C. Calculate the profit.

5. Income: $2,500
   Expenses: $776

   Profit: ____$1,724____

6. Income: $3,257
   Expenses: $975

   Profit: ____$2,282____

D. Calculate the gross profit.

7. Sales: $876.98
   Cost of goods: $583.72

   Gross profit: ____$293.26____

8. Sales: $1,817.42
   Cost of goods: $1,274.98

   Gross profit: ____$542.44____

E. Calculate the net profit or loss. Label your answers *profit* or *loss*.

9. Sales: $894.14
   Cost of goods: $626.71
   Overhead: $311.67

   ____$44.24 loss____

10. Sales: $2,011.41
    Cost of goods: $1,482.14
    Overhead: $357.14

    ____$172.13 profit____

Lesson 132 ~ Quiz

Name _____

Date _____

Calculating Interest

Score _____

A. Find the simple interest on each loan. Calculate mentally if you can.

1. p = $2,000
   r = 10%
   t = 3 yr.

   i = ___$600___

2. p = $5,000
   r = 5%
   t = 5 yr.

   i = ___$1,250___

3. p = $10,000
   r = 4%
   t = 4 yr.

   i = ___$1,600___

4. p = $6,000
   r = 6%
   t = 3 yr.

   i = ___$1,080___

5. p = $6,000
   r = 8%
   t = 3 yr.

   i = ___$1,440___

6. p = $8,000
   r = 5%
   t = 4 yr.

   i = ___$1,600___

B. Use the compound interest formula to find the amount of principal plus interest.

7. p = $1,000
   r = 10% compounded semiannually
   t = 1 yr.

   a = ___$1,102.50___

8. p = $2,000
   r = 6% compounded semiannually
   t = 1 yr.

   a = ___$2,121.80___

C. Find the compound interest by using the compound interest table. (Decimals from the table are included in the exercises.)

9. p = $10,000
   r = 8% compounded semiannually
   t = $2\frac{1}{2}$ yr.
   Decimal: 1.2167

   i = ___$2,167.00___

10. p = $6,000
    r = 7% compounded quarterly
    t = 2 yr.
    Decimal: 1.1489

    i = ___$893.40___

Lesson 136 ~ Quiz

Name _____

Date _____

Order of Operations

Score _____

Solve each problem, using the correct order of operations.

1. $4 + 3 \cdot 6 =$ ___22___

2. $12 - 2 \cdot 3 =$ ___6___

3. $7 \cdot 6 - 8 - 2 =$ ___32___

4. $15 \div 3 + 2 \cdot 6 =$ ___17___

5. $3(4 + 2)(4 - 1) =$ ___54___

6. $7 + 2(5 + 1)(5 + 1) =$ ___79___

7. $70 - 3^2(3 + 4) =$ ___7___

8. $4^2(8 - 5) - \frac{40}{20} =$ ___46___

9. $20 - [2^3 - (5 - 3)] =$ ___14___

10. $5[3^3 - (8 - 6)] =$ ___125___

Lesson 141 ~ Quiz   Name _____

                                   Date _____

Solving Equations   Score _____

Solve these equations. Write answers in the form *n* = 4.

1. $8 + n = 23$ *n* = 15   2. $25 - n = 16$ *n* = 9

3. $n - 5 = 7^2$ *n* = 54   4. $t - 5 = 4 \cdot 2$ *t* = 13

5. $w + 3w - 4 = 8$ *w* = 3   6. $15 - 2z = 7$ *z* = 4

7. $\frac{s}{7} + 3 = 9$ *s* = 42   8. $\frac{a}{4} - 3 = 8 - 2(3)$ *a* = 20

9. $2x + 3x - 12 = 8$ *x* = 4   10. $n + (n + 1) + (n + 2) = 21$ *n* = 6

Lesson 145 ~ Quiz

Name _____

Date _____

Multiplying and Dividing With Exponents

Score _____

A. Write these products with exponents.

1. $a^2 \cdot a^2 = $ a^4

2. $b^4 \cdot b^3 = $ b^7

3. $c^3 \cdot c^2 = $ c^5

4. $d^5 \cdot d^4 = $ d^9

5. $e^3 \cdot e^3 \cdot e^2 = $ e^8

6. $f^6 \cdot f^2 \cdot f^3 = $ f^{11}

7. $g^4 \cdot g^9 \cdot g = $ g^{14}

8. $h^6 \cdot h \cdot h^3 = $ h^{10}

9. $h^3 k^2 \cdot h^2 k^3 = $ $h^5 k^5$

10. $k^4 m^2 \cdot k^3 m^2 = $ $k^7 m^4$

B. Write the answers with exponents.

11. $n^7 \div n^4 = $ n^3

12. $p^7 \div p^3 = $ p^4

13. $q^6 \div q^4 = $ q^2

14. $r^7 \div r^5 = $ r^2

15. $s^7 \div s^3 = $ s^4

16. $t^9 \div t = $ t^8

17. $u^8 \div u^2 = $ u^6

18. $v^6 \div v^2 = $ v^4

19. $w^5 \div w^2 = $ w^3

20. $x^4 \div x^2 = $ x^2

Lesson 147 ~ Quiz Name _____

Date _____

Adding and Subtracting Signed Numbers   Score _____

Add or subtract these signed numbers.

1. +3
    + (−5)
    ─────
    −2

2. −1
    + (−7)
    ─────
    −8

3. +4
    + (+8)
    ─────
    +12

4. −7
    + (+5)
    ─────
    −2

5. +7
    + (−3)
    ─────
    +4

6. −8
    + (+6)
    ─────
    −2

7. +10
    + (−12)
    ─────
    −2

8. +2
    + (+15)
    ─────
    +17

9. +11
    + (−11)
    ─────
    0

10. −17
    + (−10)
    ─────
    −27

11. +12
    − (+11)
    ─────
    +1

12. −5
    − (−5)
    ─────
    0

13. +4
    − (−6)
    ─────
    +10

14. −2
    − (−6)
    ─────
    +4

15. +3
    − (+7)
    ─────
    −4

16. −6
    − (+4)
    ─────
    −10

17. +6
    − (−2)
    ─────
    +8

18. −7
    − (+5)
    ─────
    −12

19. +11
    − (+13)
    ─────
    −2

20. +3
    − (+13)
    ─────
    −10

Lesson 151 ~ Quiz

Name _____

Date _____

Calculation With Signed Numbers

Score _____

Solve these problems.

1. -5
    $+(-7)$
    $\overline{-12}$

2. -4
    $+(-9)$
    $\overline{-13}$

3. -6
    $+(+3)$
    $\overline{-3}$

4. $+7$
    $+(-12)$
    $\overline{-5}$

5. -9
    $-(+7)$
    $\overline{-16}$

6. $+13$
    $-(-13)$
    $\overline{+26}$

7. -13
    $-(-11)$
    $\overline{-2}$

8. -8
    $-(+9)$
    $\overline{-17}$

9. $+5$
    $\times 0$
    $\overline{0}$

10. -4
    $\times(-8)$
    $\overline{+32}$

11. $+5$
    $\times(-9)$
    $\overline{-45}$

12. -9
    $\times(+6)$
    $\overline{-54}$

13. $(-3)(+2)(-1)(+2)(+1)(-2) =$ __−24__

14. $(-4)(+1)(-2)(+1)(0)(-2) =$ __0__

15. $+12 \div (-4) =$ __−3__

16. $-9 \div (-3) =$ __+3__

17. $+8 \div (-2) =$ __−4__

18. $-7 \div (-1) =$ __+7__

Lesson 160 ~ Quiz   Name _____

                              Date _____

Scientific Notation   Score _____

A. Write these large numbers, using scientific notation.

1. 460,000,000,000 = _____4.6×10^{11}_____

2. 73,000,000,000,000 = _____7.3×10^{13}_____

3. 7,100,000,000,000,000,000,000 = _____7.1×10^{21}_____

4. 762,000,000,000,000,000,000 = _____7.62×10^{20}_____

B. Expand these numbers expressed in scientific notation.

5. 3.3×10^6 = _____3,300,000_____

6. 1.7×10^5 = _____170,000_____

7. 2.09×10^7 = _____20,900,000_____

8. 3.28×10^8 = _____328,000,000_____

C. Use negative exponents to express these decimals.

9. 0.0001 = _____10^{-4}_____

10. 0.0000001 = _____10^{-7}_____

11. 0.000000001 = _____10^{-9}_____

12. 0.00000000001 = _____10^{-11}_____

D. Write these small decimals, using scientific notation.

13. 0.00056 = _____5.6×10^{-4}_____

14. 0.000023 = _____2.3×10^{-5}_____

15. 0.000000072 = _____7.2×10^{-8}_____

16. 0.00000000068 = _____6.8×10^{-10}_____

Lesson 163 ~ Quiz   Name _____

The Duodecimal System   Date _____

                              Score _____

A. Change these base twelve numbers to base ten.

1. 63_{twelve} = ____75____   2. 71_{twelve} = ____85____

3. 182_{twelve} = ____242____   4. 635_{twelve} = ____905____

5. ETE_{twelve} = ____1,715____   6. TTE_{twelve} = ____1,571____

B. Change these base ten numbers to base twelve.

7. 74 = ____62_{twelve}____   8. 254 = ____192_{twelve}____

9. 574 = ____$3ET_{twelve}$____   10. 1,699 = ____$E97_{twelve}$____

Lesson 167 ~ Quiz   Name _____

                         Date _____

The Binary System   Score _____

A. Change these binary numbers to base ten.

1. 11_{two} = ___3___   2. 101_{two} = ___5___

3. $1{,}100_{two}$ = ___12___   4. 111_{two} = ___7___

5. $1{,}000_{two}$ = ___8___   6. 10_{two} = ___2___

B. Write these base ten numbers as binary numbers.

7. 4 = ___100_{two}___   8. 6 = ___110_{two}___

9. 9 = ___$1{,}001_{two}$___   10. 11 = ___$1{,}011_{two}$___

Chapter Tests

Answer Key

Grade 8   *Applying Mathematics*

99. Chapter 7 Test

Name _____   Date _____   Score _____

A. Write the formula for each fact.

1. Perimeter of a square $p = 4s$

2. Circumference of a circle when the diameter is known $c = \pi d$

3. Area of a triangle $a = \tfrac{1}{2} bh$

4. Area of a trapezoid $a = \tfrac{1}{2} h(b_1 + b_2)$

5. Area of a circle $a = \pi r^2$

B. Do these exercises.

6. Use symbols to write "line AB is parallel to line CD." $\overleftrightarrow{AB} \parallel \overleftrightarrow{CD}$

7. Use words to write the following expression: $\overleftrightarrow{EF} \perp \overleftrightarrow{GH}$
   line EF is perpendicular to line GH

8. An 85° angle is an example of a(n) ___ angle.   acute

9. If a triangle has angles of 25°, 25°, and 130°,
   it is called (acute, obtuse, right).   obtuse

10. If two angles of a triangle have 28° and 46°, the third angle has ___.   106°

11. If a triangle has sides of 8 inches, 8 inches, and 12 inches,
    it is called (equilateral, isosceles, scalene).   isosceles

12. The ___ of a circle is the distance from edge to edge
    through the center.   diameter

32   Chapter 7 Test

13. Use your protractor to measure this angle.   143°

14. Write the number of degrees in each angle of the triangle below.
Check your work by finding the sum of the degrees in the three angles.

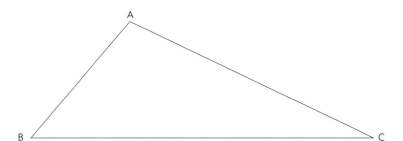

∠A = 106°

∠B = 49°

∠C = 25°

Total: 180°

C. Find the perimeter or circumference of each figure.

15. Square
 $s = 18$ cm

 $p =$ ___72 cm___

16. Circle
 $d = 14$ in.
 $\pi = 3\frac{1}{7}$
 $c =$ ___44 in.___

D. Find the area of each figure.

17. Rectangle
 $l = 28$ in.
 $w = 17$ in.
 $a =$ ___476 sq. in.___

18. Square
 $s = 14.5$ m
 $a =$ ___210.25 m²___

19. Parallelogram
 $b = 14.5$ cm
 $h = 16.2$ cm
 $a =$ ___234.9 cm²___

20. Triangle
 $b = 6\frac{1}{2}$ ft.
 $h = 3\frac{3}{4}$ ft.
 $a =$ ___$12\frac{3}{16}$ sq. ft.___

21. Trapezoid
 $h = 9$ in.
 $b_1 = 6$ in.
 $b_2 = 1$ in.
 $a =$ ___$31\frac{1}{2}$ sq. in.___

22. Circle
 $r = 18.5$ cm
 $\pi = 3.14$
 $a =$ ___1,074.665 cm²___

Grade 8   Applying Mathematics   Chapter 7 Test   33

E. Find the area of each compound figure. Use $3\frac{1}{7}$ for pi.

23. $a =$ __246$\frac{3}{4}$ sq. ft.__   24. $a =$ __358$\frac{3}{4}$ sq. ft.__

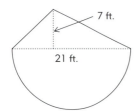
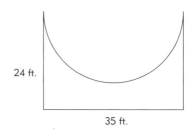

F. Solve these multistep reading problems.

25. In the grocery store, Mother noticed that 5-pound bags of sugar sold for $1.75 and 10-pound bags sold for $3.70. What was the difference between the two sizes in price per pound?

    __2 cents__

26. Daryl spent 45 minutes on his homework. Reviewing Bible memory took 5 minutes, finishing a spelling lesson took 10 minutes, and 20 math problems took the remainder of the time. What was the average amount of time he spent on each math problem?

    __$1\frac{1}{2}$ minutes__

27. Mr. Day pays $84 for a carton of 5 circuit breakers, which he sells for 20% more than his cost. How much does a customer pay for one breaker after 5% sales tax is added?

    __$21.17__

34   Chapter 7 Test

28. The Horsts' living room floor measures $11\frac{1}{2}$ feet by $19\frac{1}{2}$ feet. A stairway along one wall takes out an area 32 inches wide and 9 feet long. How many square yards of carpet will it take to cover the floor?

$$22\frac{1}{4} \text{ square yards}$$

29. The Wenger boys can stack 125 bales per layer in a mow that measures 18 feet by 25 feet. How many square feet is that per bale?

$$3\frac{3}{5} \text{ square feet}$$

30. There were still 72 gallons in the 500-gallon diesel fuel tank when the Elsons refilled it at $0.879 per gallon. The supplier allows a 2% discount if the bill is paid within 15 days. What is the amount of the discounted bill?

$$\$368.69$$

Grade 8   *Applying Mathematics*   Chapter 8 Test

116. Chapter 8 Test

Name _____   Date _____   Score _____

A. Write the formula for each fact.

1. Surface area of a rectangular solid $a_s = 2lw + 2wh + 2lh$

2. Surface area of a cylinder $a_s = 2\pi r^2 + 2\pi rh$

3. Volume of a rectangular solid $v = lwh$

4. Volume of a cylinder $v = \pi r^2 h$

5. Volume of a square pyramid $v = \frac{1}{3}lwh$

6. Volume of a sphere $v = \frac{4}{3}\pi r^3$

B. Find the surface area or volume of each figure. Use 3.14 for pi unless otherwise indicated.

7. Cube
   $e = 9$ in.
   $a_s =$ ___486 sq. in.___

8. Rectangular solid
   $l = 8$ cm
   $w = 5$ cm
   $h = 6$ cm
   $a_s =$ ___236 cm²___

9. Cylinder
   $r = 3$ in.
   $h = 4$ in.
   $a_s =$ ___131.88 sq. in.___

10. Square pyramid
    $b = 5$ cm
    $\ell = 6$ cm
    $a_s =$ ___85 cm²___

11. Sphere
    $r = 7$ in.
    $\pi = 3\frac{1}{7}$
    $a_s =$ ___616 sq. in.___

12. Cube
    $e = 14$ cm
    $v =$ ___2,744 cm³___

13. Rectangular solid:
    $l = 13$ cm
    $w = 11$ cm
    $h = 25$ cm
    $v =$ ___3,575 cm³___

14. Cylinder
    $r = 5$ ft.
    $h = 7$ ft.
    $\pi = 3\frac{1}{7}$
    $v =$ ___550 cu. ft.___

15. Cone
    $r = 12$ m
    $h = 18$ m
    $v =$ ___2,712.96 m³___

16. Sphere   (Use 4.18.)
    $r = 6$ in.
    $v =$ ___902.88 cu. in.___

36   Chapter 8 Test

C. Do these exercises.

17. Use estimation to find the square root of 1,225, which is a perfect square.   _____35_____

18. Extract the square root of 6,241, which is a perfect square.   _____79_____

Partial Table of Square Roots

77	8.775	85	9.220	93	9.644
78	8.832	86	9.274	94	9.695
79	8.888	87	9.327	95	9.747
80	8.944	88	9.381	96	9.798
81	9.000	89	9.434	97	9.849
82	9.055	90	9.487	98	9.899
83	9.110	91	9.539	99	9.950
84	9.165	92	9.592	100	10.000

19. Using the table above, find the hypotenuse of a right triangle with one leg 4 feet long and the other leg 9 feet long. (Answer to the nearest tenth.)   _____9.8 ft._____

20. Using the table above, find the length of leg a of a right triangle if the hypotenuse is 12 centimeters long and leg b is 8 centimeters long. (Answer to the nearest tenth.)   _____8.9 cm_____

D. For these sets of parallel reading problems, use your solution to the first problem as a help in solving the second problem.

21. On their way home from church one evening, the Erb family spent $\frac{1}{2}$ of the time singing. If they sang for 5 minutes, how long did it take them to get to church?
   _____10 minutes_____

22. On their way to church last Sunday, the Erb family spent $\frac{3}{4}$ of the time singing. If they sang for 9 minutes, how long did it take them to get to church?
   _____12 minutes_____

Grade 8   *Applying Mathematics*   Chapter 8 Test   37

23. Jeffrey worked 12 hours for his uncle and received $36.00. How much would he be paid for working 3 hours?

    $9.00

24. Jeffrey worked $9\frac{3}{4}$ hours for his uncle and received $66.30. How much would he be paid for working $1\frac{1}{2}$ hours?

    $10.20

25. A train travels 120 miles in 2 hours. What is its average speed?

    60 miles per hour

26. A train travels 51.15 miles in 0.55 hour. What is its average speed?

    93 miles per hour

27. David picked 30 quarts of strawberries in 3 hours. How many quarts was that per hour?

    10 quarts

28. David picked 7 quarts of strawberries in $\frac{3}{4}$ hour. How many quarts was that per hour?

    $9\frac{1}{3}$ quarts

E. Solve these reading problems.

29. Father is planning to pump out a water reservoir and clean it. If the reservoir measures 125 feet by 215 feet and the water is 35 feet deep, what is the volume to be pumped?

    940,625 cubic feet

30. A cylindrical feed bin is 50 feet high and has a radius of 7 feet. What is the capacity of the feed bin? (Use $3\frac{1}{7}$ for pi.)

    7,700 cubic feet

Grade 8   Applying Mathematics   Chapter 9 Test   39

131. Chapter 9 Test

Name _____   Date _____   Score _____

A. Use this information to complete the deposit ticket below. Use the current date.

1. Currency $25.00
   Coins 6.39
   Checks: 58-717 . . . . . 54.85
             62-111 . . . . . 48.96
   Cash received 50.00

DEPOSIT TICKET			
Donald Newman 23-334/234	CURRENCY	25	00
64 River Drive	COIN	6	39
Bryan, OH 43506	CHECKS		
	58-717	54	85
Account # 36-5740213	62-111	48	96
	SUBTOTAL	135	20
DATE [current date] 20 ___	LESS CASH RECEIVED	50	00
Bryan National Bank	TOTAL DEPOSIT	85	20

B. Use this information to complete the check below. Sign your own name.

	Check number	Date	Payee	Amount	Memo
2.	189	(Today's date)	Brian Minck	$28.87	Supplies

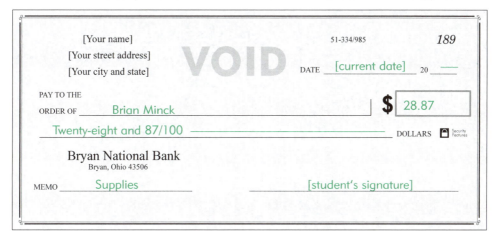

40   Chapter 9 Test

C. Use this information to complete the check register below. Record the balance after each transaction.

Balance carried forward: $488.43

3. Check number 523 on 2/25 to Mary's Fabrics for $6.78

4. Check number 524 on 2/26 to Aaron Zimmerman for $25.38

5. Deposit on 2/27 from sales: $121.38

NUMBER	DATE	DESCRIPTION	✓	FEE	CHECK AMOUNT	DEPOSIT AMOUNT	BALANCE
		Balance carried forward					$488 | 43
523	2/25	Mary's Fabrics			6 | 78		481 | 65
524	2/26	Aaron Zimmerman			25 | 38		456 | 27
D	2/27	Deposit—Sales				121 | 38	577 | 65

D. Use this information to complete the reconciliation form below. Write *yes* or *no* to tell whether the balances agree.

Ending bank balance	Deposits outstanding	Checks outstanding	Ending register balance
$295.69	$125.48	$141.52	$279.65

6.

Ending bank balance	$295.69
Add outstanding deposits	125.48
Total	421.17
Subtract outstanding checks	141.52
Adjusted balance	279.65

Do the balances agree?

yes

E. Calculate the simple interest.

7. p = $3,000
   r = 6%
   t = 3 yr.
   i = ___$540___

8. p = $425
   r = 8%
   t = 2 yr.
   i = ___$68___

F. Use the 360-day year to calculate simple interest.

9. p = $600
   r = 8%
   t = 150 days
   i = ___$20___

10. p = $900
    r = 8%
    t = 120 days
    i = ___$24___

Grade 8 *Applying Mathematics* Chapter 9 Test 41

G. Use the compound interest formula to find the total principal plus interest.

11. $p = \$2,500$
    $r = 8\%$ compounded annually
    $t = 2$ yr.
    $a =$ __$2,916.00__

12. $p = \$3,000$
    $r = 6\%$ compounded semiannually
    $t = 1$ yr.
    $a =$ __$3,182.70__

H. Find the principal plus interest by using the compound interest table. (Decimals from the table are given.)

13. $p = \$2,000$
    $r = 5\%$ compounded semiannually
    $t = 2$ yr.
    Decimal: 1.1038
    $a =$ __$2,207.60__

14. $p = \$3,100$
    $r = 8\%$ compounded quarterly
    $t = 2\frac{1}{2}$ yr.
    Decimal: 1.2190
    $a =$ __$3,778.90__

I. Find the sales tax and the total amount due on this sales receipt.

15.
	35.57
	14.68 T
	24.89 T
	17.95 T
Subtotal	93.09
4% Tax	2.30
Total	95.39

(taxable: $57.52)

J. Calculate the missing percents on this income statement.

Gary's Auto Body Shop
Income Statement
For the Year Ending
December 31, 20—

	Amount	Percent of Sales
Total sales	$43,000	100%
Cost of goods	12,000	16. 28%
Gross profit	31,000	17. 72%
Overhead	10,000	18. 23%
Net profit	21,000	19. 49%

Chapter 9 Test

K. Solve these problems mentally.

20. 237 + 86 =   323

21. 93 − 37 =   56

22. 8 × 62 =   496

23. 300 × 45 =   13,500

24. 128 ÷ 16 =   8

25. 3,600 ÷ 50 =   72

L. Solve these reading problems.

26. A property has an assessed value of $75,000. Find the property tax if the tax rate is 18 mills.

$1,350

27. Calculate the profit if income is $275 and expenses are $89.

$186

28. Calculate the gross profit if income is $325.43 and the cost of goods is $215.34.

$110.09

29. Calculate the net profit if income is $691.93, cost of goods is $399.78, and overhead is $178.71.

$113.44

30. Calculate the net loss if income is $727.45, cost of goods is $615.39, and overhead is $215.38.

$103.32

Grade 8 Applying Mathematics   Chapter 10 Test 43

143. Chapter 10 Test

Name _____ Date _____ Score _____

A. Write the answers.

1. Is $6n + 3mn$ a monomial or a polynomial?   polynomial

2. Is $6c + 3a - 4d$ a binomial or a trinomial?   trinomial

3. Write the numerical coefficient of $9xy$.   9

6. Simplify $4 + 21 - 7$.   18

7. Simplify $38 - [2(6 + 8)]$.   10

8. Evaluate $4x - 7y$ if $x = 9$ and $y = 4$.   8

9. Evaluate a^3b^2 if $a = 2$ and $b = 3$.   72

10. Which operation will solve the equation $8z = 96$?   division

11. Simplify $bc + 4a^2b + a^2b + 2c^3 + 3bc$. $5a^2b + 4bc + 2c^3$

12. Write $(2)(2)(n)(n)(n)(n)$, using exponents. 2^2n^4

13. Simplify $7a^3b^2 - 3a^3b^2$. $4a^3b^2$

14. Write the product of $6^4 \cdot 6^8$, using an exponent. 6^{12}

15. Write the product of $5^2 \cdot 5^2$ in standard form.   625

16. Write the product of $n^5 \cdot n^3$, using an exponent. n^8

17. Write the answer to $y^5 \div y^2$, using an exponent. y^3

18. Find this square root: $\sqrt{n^{10}}$ n^5

Chapter 10 Test

B. Use the correct operations to solve these equations. Show each step in your work.

19. $\frac{n}{3} = 5$ _n = 15_

     $\frac{n}{3} \cdot 3 = 5 \cdot 3$

     $n = 15$

20. $2x + 5 = 27$ _x = 11_

     $2x + 5 - 5 = 27 - 5$

     $2x = 22$

     $\frac{2x}{2} = \frac{22}{2}$

     $x = 11$

21. $n - 6 = 7 + 5$ _n = 18_

     $n - 6 = 12$

     $n - 6 + 6 = 12 + 6$

     $n = 18$

22. $2s - 3 = 7$ _s = 5_

     $2s - 3 + 3 = 7 + 3$

     $2s = 10$

     $\frac{2s}{2} = \frac{10}{2}$

     $s = 5$

C. Simplify these polynomials.

23. $8m + 3n - 2n - 9$ _$8m + n - 9$_

24. $6b + 7c + 4 - 4b + 4$ _$2b + 7c + 8$_

D. Write equations to solve these reading problems.

25. Heidi's grandmother is 2 years more than 5 times as old as Heidi. The sum of their ages is 80. How old is Heidi?

     Equation: _$h + 5h + 2 = 80$_

     Answer: _$h = 13$ years_

26. The sum of three consecutive numbers is 57. What is the smallest number?

     Equation: _$n + n + 1 + n + 2 = 57$_

     Answer: _$n = 18$_

27. Five more than 4 times Elsie's age is 73. How old is Elsie?

     Equation: _$4e + 5 = 73$_

     Answer: _$e = 17$ years_

Grade 8 Applying Mathematics   Chapter 10 Test 45

28. Susan was organizing the books on the living room bookshelf. She found that 3 times the number of children's books plus 1 was the number of other books. The total number of books on the shelf was 65. How many children's books were there?

    Equation: _____ $3c + 1 + c = 65$ _____

    Answer: _____ $c = 16$ books _____

29. The longest recorded rainless period in the world occurred at Arica, Chile. Two more than 7 times the length of the rainless period equals 100 years. For how many years did it not rain at Arica, Chile?

    Equation: _____ $7r + 2 = 100$ _____

    Answer: _____ $r = 14$ years _____

30. Although some hailstones reportedly weighing several pounds have fallen in the past, the average hailstone is much smaller than that. Ten average hailstones placed side by side, plus $1\frac{1}{2}$ inches, is equal to 4 inches. What is the diameter of an average hailstone in inches?

    Equation: _____ $10h + 1\frac{1}{2} = 4$ _____

    Answer: _____ $h = \frac{1}{4}$ inch _____

Grade 8 Applying Mathematics   Chapter 11 Test 47

154. Chapter 11 Test

Name _____   Date _____   Score _____

A. Write the answers.

1. Use a signed number to show a loss of $98.   −$98
2. Use a signed number to show a drop in temperature of 16°.   −16°
3. Write the absolute value of −44.   44
4. Compare −1 and −2 by using the < or > sign.   −1 > −2
5. Simplify $2a - 3b - 4a$. $-2a - 3b$

B. Solve these problems containing signed numbers.

6.   −8   7.   +5   8.   +6
   + (−4)   + (+2)   − (−13)
    −12   +7   +19

9.   +12   10.   +11   11.   +12
   − (−10)   × (+8)   × (−6)
    +22   +88   −72

12. $+1(-4)(-5)(+4) =$ +80

13. $-1(+3)(-3)(-1)(+3) =$ −27

14. $+14 \div (-2) =$ −7

15. $\frac{+45}{-15} =$ −3

16. $\frac{-70}{-14} =$ +5

C. Evaluate these expressions.

17. $s^3 t^2 =$ −9 $s = -1; t = 3$

18. $p^2 q^2 + p =$ 141 $p = -3; q = -4$

48   Chapter 11 Test

D. Complete this table.

Formula: $v = e^3$	Volumes of cubes			
e (in.)	4	8	12	16
v (sq. in.)	**19.** 64	**20.** 512	**21.** 1,728	**22.** 4,096

E. Prepare a graph from this table.

Formula: $a_s = 6e^2$	Surface areas of cubes					
e (in.)	0	2	4	6	8	10
a_s (sq. in.)	0	24	96	216	384	600

23–28.

Surface areas of cubes: $a_s = 6e^2$

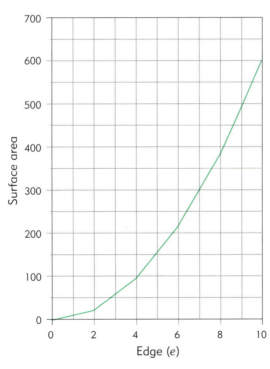

Grade 8   Applying Mathematics   Chapter 11 Test   49

F. Write equations to solve these reading problems.

29. Asa and Uzziah were two kings of Judah who had long reigns. Uzziah reigned 11 years more than Asa. Together they ruled a total of 7 less than 100 years. How long was Asa's reign?

    Equation: _____ $a + a + 11 = 100 - 7$ _____

    Answer: _____ $a = 41$ years _____

30. Baasha and Jehoahaz together reigned 41 years. Baasha reigned 10 years less than twice as long as Jehoahaz. How long did Jehoahaz reign?

    Equation: _____ $j + 2j - 10 = 41$ _____

    Answer: _____ $j = 17$ years _____

31. Last summer Mother froze twice as much corn as peas. Together, there were 186 quarts of these vegetables. How many quarts of peas did she freeze?

    Equation: _____ $p + 2p = 186$ _____

    Answer: _____ $p = 62$ quarts _____

32. Mother froze 6 quarts less than 3 times as many strawberries as blueberries. If she froze 114 quarts of these berries in all, how many quarts of blueberries did she freeze?

    Equation: _____ $b + 3b - 6 = 114$ _____

    Answer: _____ $b = 30$ quarts _____

33. Father's age is equal to 1 less than 3 times Beth's age. If Father is 41, how old is Beth?

    Equation: _____ $3b - 1 = 41$ _____

    Answer: _____ $b = 14$ years _____

34. Marian has a triangular flower bed with a perimeter of 27 feet. Side B is 2 feet longer than side A, and side C is 2 feet shorter than side A. How long is side A?

    Equation: _____ $a + a + 2 + a - 2 = 27$ _____

    Answer: _____ $a = 9$ feet _____

Grade 8   Applying Mathematics   Chapter 12 Test   51

170. Final Test

Name _____   Date _____   Score _____

A. Do these calculations.

1. 5,673
   + 3,876
   ─────
    9,549

2.   876
    × 76
    ─────
    66,576

3. $67\overline{)7,956}$ 118 R 50

4. $\frac{3}{8}$
   + $\frac{2}{5}$
   ───
   $\frac{31}{40}$

5. $\frac{7}{8}$
   − $\frac{5}{12}$
   ───
   $\frac{11}{24}$

6. $\frac{3}{4} \times 1\frac{1}{5}$ = $\frac{9}{10}$

7. $2\frac{3}{4} \div 2\frac{1}{5}$ = $1\frac{1}{4}$

8. 6 is $\frac{2}{3}$ of ___9___

9.   7.9
   − 2.9359
   ──────
    4.9641

10. 42.838
    × 0.08
    ──────
    3.42704

11.   −3
    + (−6)
    ─────
     −9

12.   +6
    − (−15)
    ─────
    +21

13. 2(+1)(−3)(+1)(+2)(−4) = ___+48___

14. 30 ÷ (−5) = ___−6___

B. Do these exercises relating to measures.

15. 1 fl. oz. = ___2___ tbsp.

16. 8 ft. 5 in. = $8\frac{5}{12}$ ft.

17. 3,500 cm² = ___0.35___ m²

18. 1 ft. = ___0.3___ m

19. 8 ft. 7 in.
   + 9 ft. 9 in.
   ──────────
   18 ft. 4 in.

20. 4 yd. 2 ft.
    ×   9
    ─────────
       42 yd.

21. 3.6 kg − 195 g = ___3,405___ g

22. 45 m = ___147.6___ ft.

52   Chapter 12 Test

C. Write the answers.

23. Write "thirty-two quadrillion, ten billion," using digits.   32,000,010,000,000,000

24. Round 372,992,111 to the nearest ten million.   370,000,000

25. Write the missing number: 20 × 8 + ___ × 8 = 24 × 8.   4

26. Write the metric prefix that means 0.001.   milli-

27. Reduce $\frac{50}{65}$ to lowest terms. $\frac{10}{13}$

28. What is the reciprocal of $4\frac{2}{3}$? $\frac{3}{14}$

29. Find 43.9 ÷ 0.19, to the nearest thousandth.   231.053

30. Find the distance represented by $2\frac{1}{2}$ inches if the scale is 1 inch = 16 miles.   40 mi.

31. In the problem below, 24 is the (base, rate, percentage).
35% of 24 = 8.4   base

32. 130% of 75 = ___   97.5

33. $3.28 increased by 35% = ___, to the nearest cent. $4.43

34. Express 628% as a decimal   6.28

35. Find $\frac{3}{4}$% of 640.   4.8

Grade 8   *Applying Mathematics*   Chapter 12 Test   53

36. Day 1 sales: $225; day 2 sales: $275;
rate of increase: ___ to the nearest whole percent   22%

37. 24 is 16% of ___   150

38. Find the median of this set.
96, 93, 95, 96, 91, 89, 100, 96, 94, 93 $94\frac{1}{2}$

39. Find the mode of the set in number 38 above.   96

40. Write the formula for finding the volume of a sphere. $v = \frac{4}{3}\pi r^3$

41. If two angles of a triangle have 18° and 38°,
what is the size of the third angle?   124°

42. A triangle with sides of seven inches, nine inches,
and seven inches is an example of a(n)
(scalene, equilateral, or isosceles) triangle.   isosceles

43. Find the circumference of this circle.
Diameter: 75 cm (pi = 3.14)   235.5 cm

44. Find the area of this circle.
Radius: 28 in. (pi = $3\frac{1}{7}$)   2,464 sq. in.

45. Find the area of this triangle.
Base: 8 in.; height: 4 in.   16 sq. in.

46. Find the volume of this cone.
Radius: 6 cm; height: 9 cm   339.12 cm³

47. Find the square root of 196 mentally.   14

54   Chapter 12 Test

48. Find the compound interest on this deposit.
p = $800; r = 8% compounded annually; t = 2 yr. $133.12

49. Use the compound interest formula to find the principal plus interest. p = $900; r = 6% compounded semiannually; t = 1 yr. $954.81

50. Find the property tax.
Assessment: $140,000; tax rate: 19 mills $2,660

51. Calculate the sales tax.
Sale: $225; tax rate: 6% $13.50

52. Simplify $12 - [\frac{12}{3} - (7 - 3)]$.   12

53. Simplify $xy + 2x^2y - 2xy + 5x^2y$. $7x^2y - xy$

54. Simplify $a^3b \cdot ab^2$, using exponents. a^4b^3

55. Write the absolute value of $|-12|$.   12

56. Evaluate $\frac{3y^2z^3}{yz^2}$ if $y = 3$ and $z = -2$. -18

57. Write 8,000,000,000,000,000,000, using scientific notation. 8×10^{18}

58. Write 0.0000005 using scientific notation. 5×10^{-7}

59. Write 111_{two} as a number in base ten.   7

Grade 8   Applying Mathematics   Chapter 12 Test   55

D. Use proportions to solve these reading problems.

60. If an 8-foot tree casts a 12-foot shadow, how tall is a tree casting a 48-foot shadow?

    $\dfrac{\text{height}}{\text{shadow}} \quad \dfrac{8}{12} = \dfrac{n}{48} \quad \dfrac{\text{height}}{\text{shadow}}$

    $n = \underline{\quad 32 \text{ feet} \quad}$

61. Sharon can get 18 slices from each loaf of homemade bread if the average slice is $\frac{1}{2}$ inch thick. How many slices can she get from each loaf if their average thickness is $\frac{9}{16}$ inch?

    $\dfrac{\text{thickness (1)}}{\text{thickness (2)}} \quad \dfrac{\frac{1}{2}}{\frac{9}{16}} = \dfrac{n}{18} \quad \dfrac{\text{slices (2)}}{\text{slices (1)}}$

    $n = \underline{\quad 16 \text{ slices} \quad}$

E. Use the data on this table to complete the double-line graph below.

Mean Temperatures at Miami, Florida, and Phoenix, Arizona

	J	F	M	A	M	J	J	A	S	O	N	D
Miami	67°	68°	72°	75°	79°	81°	83°	83°	82°	78°	73°	69°
Phoenix	52°	56°	61°	68°	77°	87°	92°	90°	85°	83°	61°	53°

Source: *1994 World Almanac*

62–68.

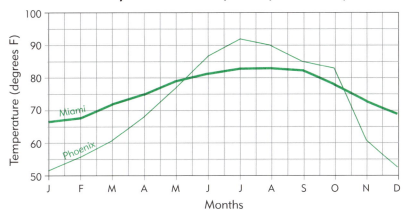

Source: *1994 World Almanac*

56 Chapter 12 Test

F. Draw a sketch for each problem, and use it to find the solution.

69. A child's playpen is a 4-foot square with sides 20 inches high. A plastic edging extends downward 4 inches from the top of the playpen, and another extends upward 6 inches from its base. In between is a soft netting. How many square feet of netting are on the 4 sides?

$13\frac{1}{3}$ square feet

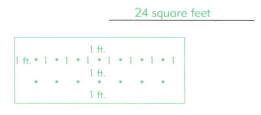

70. Andrea is planting petunias in a rectangular flower bed. She plants 14 of them in 2 rows. The plants are 1 foot apart in rows that are 1 foot apart. The rows are also 1 foot in from the edges of the flower bed. What is the area of the flower bed?

24 square feet

```
          1 ft.
1 ft. • 1 • 1 • 1 • 1 • 1 • 1 • 1
          1 ft.
      •   •   •   •   •   •   •
          1 ft.
```

G. Write equations to solve these reading problems.

71. The sum of three consecutive numbers is 69. What is the smallest number?

Equation: $n + n + 1 + n + 2 = 69$

Answer: $n = 22$

72. Leonard is 2 years older than Susan. Ten more than their combined ages is equal to their father's age, which is 40. How old is Susan?

Equation: $s + s + 2 + 10 = 40$

Answer: $s = 14$ years

INDEX

absolute value   483
account balance   401
acre   66, 84
acute angle   286
acute triangle   291
addition
    associative law   21
    axiom   458
    commutative law   21
    compound measure   87
    decimals   153
    fractions   124
    mental   24, 437
    monomials   455
    signed numbers   486
algebra   445–497
    equations   458–466
    evaluating   452
    exponents   470–476
    expressions   446, 447
    literal numbers   446
    order of operations   449
    signed numbers   482–497, 507
    terms   446
angle   282, 286
    acute   286
    complementary   286
    obtuse   286
    reflex   286
    right   282, 286
    sides   282
    straight   286
    supplementary   286
    vertex   282
antecedent   169, 191
Arabic number system   12, 15, 19

arc   308
area   66, 83
    circle   323
    compound figures   326
    metric   83
    parallelogram   315
    rectangle   312
    square   312
    trapezoid   319
    triangle   315
arithmetic mean   240
arithmetic notation   12
associative law   21
average   240
balancing accounts   401
bank number   395
bank statement   405
bar graph   257
base (of exponent)   470
base in numeration   527
    twelve   530, 537
    two   533, 537
base, percentage   204, 222
bath   102
Bible measure   102, 103
binary system   533, 537
binomial   447
bisecting
    angles   288, 300
    line segments   300
blueprint   191
bushel   103
cab   103

cancellation 127
capacity 63, 80
capitalism 426
caret 159
casting out nines 33, 44
Celsius 96
Centigram 77
centiliter 80
centimeter 73
century 69
check register 401, 405
checking
    casting out nines 33, 44
    division 44
    multiplication 33
checking account 394–406
chord 308
choosing correct operation 52
circle 283, 308
    area 323
    circumference 308
circle graph 265
circumference 308
coefficient 446
commission 225
common fraction 120
commutative law 21, 33
comparing
    common fractions 121
    decimal fractions 153
compass 297, 300
complementary angle 286
complex fraction 120
composite number 114
compound figures 326
consequent 169
cost of goods 426

common denominator 121
commutative law 21
complex fractions 134
compound interest 414, 417
compound interest table 418
compound measure
    adding, subtracting 87
    multiplying, dividing 90
cone 340
    volume 360
congruent figures 292
consequent 169, 191
constructing
    angles 287
    equilateral triangles 301
    hexagons 301
    graph from formula 503
    perpendicular lines 301
    table from formula 500
    triangles 296, 301
cost of goods 426
counting numbers 482
cube 340
    surface area 340
    volume 357
cubit 102
cylinder 340
    surface area 347
    volume 360
decade 69
decagon 283
decigram 77
deciliter 80
decimals 152, 200
    adding and subtracting 153
    dividing 159
    multiplying 156, 166
    nonterminating 163
    repeating 163
    terminating 162

decimal system   15
decimeter   73
decrease by percent   210, 213
dekagram   77
dekaliter   80
dekameter   73
deposit ticket   394
destination   490
diameter   308
direct proportion   173
directly proportionate   503
distance, rate, time   99
distributive law   39
divide-and-divide method   47
divisibility rules   47
division   43
   axiom   461
   checking   44
   compound measure   90
   decimals   159
   exponents   476
   fractions   131, 134
   mental   47–50, 139, 438
   signed numbers   494
double-and-divide-method   40
double-and-double method   50
dry measure   63
duodecimal system   530, 537
electric meter   518
English measure   60–69
English/metric conversion   93, 96
enlargement   191
ephah   103
equation   458
   solving   458, 461
   writing   465

equiangular triangle   291
equilateral triangle   291
estimation
   products   33
evaluating expressions   452
expanding fractions   120
expense   426
exponent   114
   in division   476
   in multiplication   473
   with literal numbers   470
exponentially proportionate   504
expressions, algebraic   446
extracting square root   373, 376
extremes   173
factors   114, 117, 446
factoring   114, 117, 118
Fahrenheit   96
fathom   60
finger   102
formulas   582, 583
fractional part   137
fractions   120, 200
   adding and subtracting   124
   comparing   121
   decimal equivalents   162, 163
   dividing   131, 134
   multiplying   127, 166
   percent equivalents   200, 201
   reciprocal   131
frequency distribution table   247
furlong   103
geometry   281
   plane   282–326
   solid   340–366
   terms   282–283, 340
gerah   102

gram 77

graph
    bar 257
    circle 265
    from formula 503
    histogram 247
    line 261
    picture 252
    rectangle 269

greatest common factor 117

gross profit 426

handbreadth 102

hectare 84

hectogram 77

hectoliter 80

hectometer 73

heptagon 283

hexagon 283

hin 102

Hindu-Arabic system 12

histogram 247

homer 102, 103

hypotenuse 379

improper fraction 120

income 426

increase by percent 210, 213

integers 482

interest 408–418
    compound 414, 417
    part-year 412
    simple 408

International System of Units 73

intersecting lines 283

inverse proportion 178

irrational number 308

irregular polygon 283

isosceles triangle 291

kilogram 77

kiloliter 80

kilometer 73

kilowatt-hour 518

lateral surface 341

league 60

leap year 69

like fractions 121

like terms 455

line 282
    intersecting 283
    parallel 283
    perpendicular 283
    segment 282

linear measure 60, 73

line graph 261

line segment 282

liquid measure 63

liter 80

literal coefficient 446

literal number 446
    with exponent 470

log 102

long division 43

long ton 60

lowest common multiple 118

maps 187

mathematical law
    associative 21
    commutative 21, 33
    distributive 39

means 173

measures
    adding, subtracting 87
    Bible 102, 103
    capacity 63, 80
    dry 63
    English 60–66
    linear 60, 73
    liquid 63
    metric 73–84
    metric/English conversion 93, 96
    multiplying, dividing 90
    square 66
    temperature conversion 96
    time 69
    weight 60, 77

median 243

mental math 437–438
    addition 24, 437
    dividing by proper fractions 139
    division 47–50, 438
    multiplication 36, 39, 437
    multiplying by fractions 127
    percent 228
    subtraction 30, 437

meter 73

metric/English conversion 93, 96

metric measure 73–84

metric ton 77

millennium 69

milligram 77

milliliter 80

millimeter 73

mixed number 120

mode 244

monomial 447

multiplication 32
    associative law 21
    axiom 461
    checking 33
    commutative law 21, 33
    compound measure 90
    decimals 156
    distributive law 39
    exponents 473
    fractions 127, 128
    mental 36, 437–438
    signed numbers 493

natural numbers 482

negative numbers 482

net profit 426

nonagon 283

nonterminating decimal 163

numbers
    Arabic 12
    counting 482
    integers 482
    natural 482
    negative 482
    place value 15
    positive 482
    real 482
    Roman 18
    rounding 16
    signed 482

numerical coefficient 446

obtuse angle 286

obtuse triangle 291

octagon 283

omer 103

order of operations 449

origin 490

578 *Index*

outstanding check/deposit 405
overhead 426
parallel lines 283
parallelogram 284
    area 315
pentagon 283
percent 200
    finding base 222
    finding percentage 204
    finding rate 216, 219
    fraction equivalents 201
    increase and decrease 210, 213, 219
percentage 204, 225
perfect square 369
perimeter 305
perpendicular lines 283
pi 308
picture graph 252
place value 15
plane 283
plane geometry 282
point 282
polygon 283
    regular/irregular 283
polynomial 447, 507
positive numbers 482
prime factor 114
prime number 114
principal, rate, time 408–414
profit 426, 430
proper fraction 120
property tax 422
proportion 173–191
    direct 173
    inverse 178
protractor 286, 287

pyramid 340
    surface area 351
    volume 363
Pythagorean rule 379, 382
quadrilateral 283
quadruple-and-quadruple method 50
radical sign 369
radicand 369
radius 308
ranking data 243
range of data 247
rate 204, 216, 219
greater than 100% or less than 1% 207
rate of commission 225
ratio 169, 200
ray 282
reading problems
    equations 465
    finances 434
    fractions in 141
    multistep 330
    missing information 145
    necessary information 106
    operation 52
    parallel problems 386
    proportions 183
    sketches 232
real numbers 482
reciprocal 131
reconciling account 405
rectangle 284
    area 312
    perimeter 305
rectangle graph 269
rectangular solid 340
    surface area 344
    volume 357
reducing fractions 120

reduction (drawing) 191
reed 102
reflex angle 286
regular polygon 283
repeating decimal 163
rhombus 284
right angle 282, 286
right triangle 291
rod 60
Roman numerals 18
rounding
   decimals 162
   picture graph symbols 253
   whole numbers 16
sales 225
sales tax 422
savings account 408
scale drawing 187, 191
scalene triangle 291
scientific notation 523
sector 266, 308
semicircle 308
shekel 102
short cut
   addition 24
   divide-and-divide 47
   double-and-divide 40
   double-and-double 50
   quadruple-and-quadruple 50
short division 44
signed numbers 482–497, 507
   adding 486
   dividing 494
   evaluating expressions 497
   multiplying 493
   polynomials 507
   subtracting 490
similar figures 292

simple interest 408
simplifying expressions 449
sketches 232
solid geometry 340
span 102
sphere 340
   surface area 354
   volume 366
square 284
   area 312
   perfect 369
   perimeter 305
square root 369
   extracting 373, 376
statistics 240, 247
   mean 240
   median 243
   mode 244
straight angle 286
substitution 452
subtraction 27
   axiom 458
   compound measures 87
   decimals 153
   fractions 124
   mental 30, 437
   monomials 455
   signed numbers 490
supplementary angles 286
surface area 340
   cube 340
   cylinder 347
   pyramid 351
   rectangular solid 344
   sphere 354
tables for formulas 500
talent 102
tax 422
temperature 96

terminating decimal 162
terms 120
time 69
time zones 69
trapezoid 284
    area 319
triangle 283
    area 315
    acute 291
    construction 296, 301
    equiangular 291
    equilateral 291
    isosceles 291
    obtuse 291
    right 291
    scalene 291
trinomial 447

unit fraction 120
unlike fractions 124
unlike terms 455
vertex 282
vinculum 18, 163
volume
    cone 360
    cube 357
    cylinder 360
    pyramid 363
    rectangular solid 357
    sphere 366
weight 60, 77
whole number 120
writing checks 397

SYMBOLS

Symbol	Name	Example	Meaning
>	greater than	7 > 3	"7 is greater than 3"
<	less than	2 < 5	"2 is less than 5"
·	multiplication	6 · 8	"6 times 8"
$\sqrt{}$	radical	$\sqrt{9}$	"the square root of 9"
∧	caret	2.05∧	insertion of decimal point
⟷	line	$\overleftrightarrow{AB}$	"line AB"
→	ray	$\overrightarrow{EF}$	"ray EF"
∠	angle	∠G	"angle G"
△	triangle	△JKL	"triangle JKL"
⌐	right angle	⌐	indication that an angle is a right angle
⊥	perpendicular	$\overleftrightarrow{ST} \perp \overleftrightarrow{UV}$	"line ST is perpendicular to line UV"
∥	parallel	$\overleftrightarrow{WX} \parallel \overleftrightarrow{YZ}$	"line WX is parallel to line YZ"
~	similar	△DEF ~ △GHI	"triangle DEF is similar to triangle GHI"
≅	congruent	△UVW ≅ △XYZ	"triangle UVW is congruent to triangle XYZ"
π	pi		relation of circumference to diameter of a circle; value near 3.14 or $3\frac{1}{7}$
—	vinculum	$\overline{FG}$	"line segment FG"
			$\overline{XXVI}$	thousands in Roman numerals
			$1.0\overline{298}$	non-terminating decimal repetition
			$\dfrac{a}{b+c}$	grouping and/or division in algebra
+	positive	+8	"positive 8"
−	negative	−8	"negative 8"
| |	absolute value	|−8|	"the absolute value of negative 8"

FORMULAS

Distance, rate, time

distance	$d = rt$
rate	$r = \frac{d}{t}$
time	$t = \frac{d}{r}$

Percent

percentage	$P = BR$
rate	$R = \frac{P}{B}$
base	$B = \frac{P}{R}$

Commission

commission	$c = sr$
rate	$r = \frac{c}{s}$
sales	$s = \frac{c}{r}$

Interest

interest	$i = prt$
rate	$r = \frac{i}{p}$
principal	$p = \frac{i}{r}$
compound interest	$a = p(1 + r)^n$

Temperature

Fahrenheit to Celsius	$C = \frac{5}{9}(F - 32)$
Celsius to Fahrenheit	$F = \frac{9}{5}C + 32$

FORMULAS

Geometric

PLANE GEOMETRY

	perimeter	*area*	
square	$p = 4s$	$a = s^2$	
rectangle	$p = 2(l + w)$	$a = lw$	
parallelogram		$a = bh$	
triangle	$p = a + b + c$	$a = \frac{1}{2}bh$	
right triangle			*hypotenuse* $c^2 = a^2 + b^2$
trapezoid		$a = \frac{1}{2}h(b_1 + b_2)$	
circle	$c = \pi d$	$a = \pi r^2$	*diameter* $d = \frac{c}{\pi}$
	$c = 2\pi r$		

SOLID GEOMETRY

	surface area	*volume*	
cube	$a_s = 6e^2$	$v = e^3$	
rectangular solid	$a_s = 2lw + 2wh + 2lh$	$v = lwh$	
cylinder	$a_s = 2\pi r^2 + 2\pi rh$	$v = \pi r^2 h$ or $v = Bh$	
cone		$v = \frac{1}{3}\pi r^2 h$ or $v = \frac{1}{3}Bh$	
square pyramid	$a_s = 4(\frac{1}{2}b\ell) + b^2$	$v = \frac{1}{3}lwh$	
sphere	$a_s = 4\pi r^2$	$v = \frac{4}{3}\pi r^3 = 4.18 \times r^3$	(for decimals)
		$v = \frac{4}{3}\pi r^3 = \frac{88}{21} \times r^3$	(for fractions and multiples of 7)

TABLE OF SQUARE ROOTS

Number	Square root	Number	Square root	Number	Square root	Number	Square root
1	1.000	51	7.141	101	10.050	151	12.288
2	1.414	52	7.211	102	10.100	152	12.329
3	1.732	53	7.280	103	10.149	153	12.369
4	2.000	54	7.348	104	10.198	154	12.410
5	2.236	55	7.416	105	10.247	155	12.450
6	2.449	56	7.483	106	10.296	156	12.490
7	2.646	57	7.550	107	10.344	157	12.530
8	2.828	58	7.616	108	10.392	158	12.570
9	3.000	59	7.681	109	10.440	159	12.610
10	3.162	60	7.746	110	10.488	160	12.649
11	3.317	61	7.810	111	10.536	161	12.689
12	3.464	62	7.874	112	10.583	162	12.728
13	3.606	63	7.937	113	10.630	163	12.767
14	3.742	64	8.000	114	10.677	164	12.806
15	3.873	65	8.062	115	10.724	165	12.845
16	4.000	66	8.124	116	10.770	166	12.884
17	4.123	67	8.185	117	10.817	167	12.923
18	4.243	68	8.246	118	10.863	168	12.961
19	4.359	69	8.307	119	10.909	169	13.000
20	4.472	70	8.367	120	10.954	170	13.038
21	4.583	71	8.426	121	11.000	171	13.077
22	4.690	72	8.485	122	11.045	172	13.115
23	4.796	73	8.544	123	11.091	173	13.153
24	4.899	74	8.602	124	11.136	174	13.191
25	5.000	75	8.660	125	11.180	175	13.229
26	5.099	76	8.718	126	11.225	176	13.266
27	5.196	77	8.775	127	11.269	177	13.304
28	5.292	78	8.832	128	11.314	178	13.342
29	5.385	79	8.888	129	11.358	179	13.379
30	5.477	80	8.944	130	11.402	180	13.416
31	5.568	81	9.000	131	11.446	181	13.454
32	5.657	82	9.055	132	11.489	182	13.491
33	5.745	83	9.110	133	11.533	183	13.528
34	5.831	84	9.165	134	11.576	184	13.565
35	5.916	85	9.220	135	11.619	185	13.601
36	6.000	86	9.274	136	11.662	186	13.638
37	6.083	87	9.327	137	11.705	187	13.675
38	6.164	88	9.381	138	11.747	188	13.711
39	6.245	89	9.434	139	11.790	189	13.748
40	6.325	90	9.487	140	11.832	190	13.784
41	6.403	91	9.539	141	11.874	191	13.820
42	6.481	92	9.592	142	11.916	192	13.856
43	6.557	93	9.644	143	11.958	193	13.892
44	6.633	94	9.695	144	12.000	194	13.928
45	6.708	95	9.747	145	12.042	195	13.964
46	6.782	96	9.798	146	12.083	196	14.000
47	6.856	97	9.849	147	12.124	197	14.036
48	6.928	98	9.899	148	12.166	198	14.071
49	7.000	99	9.950	149	12.207	199	14.107
50	7.071	100	10.000	150	12.247	200	14.142

Tables of Measure

Bible Measure

	Approximate English Equivalent	*Approximate Metric Equivalent*

Length

1 finger	$\frac{3}{4}$ inch	1.9 centimeters
1 handbreadth = 4 fingers	3 inches	7.6 centimeters
1 span = 3 handbreadths	9 inches	23 centimeters
1 cubit = 2 spans	18 in. or $1\frac{1}{2}$ feet	46 centimeters or 0.46 meters
1 fathom = 4 cubits	6 feet	1.8 meters
1 reed = 7 cubits	$10\frac{1}{2}$ feet	3.25 meters
1 furlong	606 feet or $\frac{1}{9}$ mile	184.7 meters or 0.18 kilometers

Weight

1 gerah	$\frac{1}{50}$ ounce	0.6 gram
1 bekah = 10 gerahs	$\frac{1}{5}$ ounce	$5\frac{2}{3}$ grams
1 shekel = 2 bekahs	$\frac{2}{5}$ ounce	11.3 grams
1 pound (maneh) = 50 shekels	20 ounces	566 grams
1 talent = 60 manehs	75 pounds	34 kilograms

Liquid Measure

1 log	almost 1 pint	$\frac{1}{3}$ liter
1 hin = 12 logs	$5\frac{1}{3}$ quarts	5 liters
1 bath = 6 hins	8 gallons	30.3 liters
1 firkin	9 gallons	34 liters
1 homer = 10 baths	80 gallons	303 liters

Dry Measure

1 cab	$2\frac{3}{4}$ pints	1.5 liters
1 omer = almost 2 cabs	5 pints	2.8 liters
1 seah = 6 cabs or $3\frac{1}{3}$ omers	1 peck	9.3 liters
1 ephah = 3 seahs or 10 omers	$3\frac{1}{4}$ pecks	28.2 liters
1 homer = 10 ephahs	8 bushels	282 liters

Tables of Measure

Metric Measure

Length

basic unit: **meter**

1 millimeter = 0.001 meter
1 centimeter = 0.01 meter
1 decimeter = 0.1 meter
1 dekameter = 10 meters
1 hectometer = 100 meters
1 kilometer = 1,000 meters

Area

1 hectare = 10,000 square meters

Capacity

basic unit: **liter**

1 milliliter = 0.001 liter
1 centiliter = 0.01 liter
1 deciliter = 0.1 liter
1 dekaliter = 10 liters
1 hectoliter = 100 liters
1 kiloliter = 1,000 liters

Weight

basic unit: **gram**

1 milligram = 0.001 gram
1 centigram = 0.01 gram
1 decigram = 0.1 gram
1 dekagram = 10 grams
1 hectogram = 100 grams
1 kilogram = 1,000 grams
1 metric ton = 1,000 kilograms

Temperature, Celsius scale

0° = freezing point
100° = boiling point

Metric-to-English Conversion

Linear Measure

1 centimeter = 0.39 inch
1 meter = 39.4 inches
1 meter = 3.28 feet
1 kilometer = 0.62 mile

Weight

1 gram = 0.035 ounce
1 kilogram = 2.2 pounds

Capacity

1 liter = 1.06 quarts (liquid)

Area

1 hectare = 2.5 acres
1 square kilometer = 0.39 square mile

Temperature

$\frac{9}{5}$ degrees Celsius + 32 = degrees F

English-to-Metric Conversion

Linear Measure

1 inch = 2.54 centimeters
1 foot = 0.3 meter
1 mile = 1.61 kilometers

Weight

1 ounce = 28.3 grams
1 pound = 0.45 kilogram

Capacity

1 quart (liquid) = 0.95 liter
1 tablespoon = 15 milliliters

Area

1 acre = 0.4 hectare
1 square mile = 2.59 square kilometers

Temperature

$\frac{5}{9}$ (degrees Fahrenheit - 32) = degrees C